Light and Video Microscopy

Light and Video Microscopy

Randy Wayne

AMSTERDAM • BOSTON • HEIDELBERG • LONDON • NEW YORK • OXFORD
PARIS • SAN DIEGO • SAN FRANCISCO • SINGAPORE • SYDNEY • TOKYO
Academic Press is an Imprint of Elsevier

ACADEMIC
PRESS

Academic Press is an imprint of Elsevier
30 Corporate Drive, Suite 400, Burlington, MA 01803, USA
525 B Street, Suite 1900, San Diego, California 92101-4495, USA
84 Theobald's Road, London WC1X 8RR, UK

Library of Congress Cataloging-in-Publication Data

Application Submitted

British Library Cataloguing-in-Publication Data
A catalogue record for this book is available from the British Library.

ISBN: 978-0-12-374234-6

For information on all **Academic Press** publications
visit our Web site at www.elsevierdirect.com

Typeset by Charon Tec Ltd., A Macmillan Company. (www.macmillansolutions.com)

Printed in The United States of America
09 10 9 8 7 6 5 4 3 2 1

Working together to grow
libraries in developing countries

www.elsevier.com | www.bookaid.org | www.sabre.org

ELSEVIER   BOOK AID   Sabre Foundation
            International

This book is dedicated to my brother Scott.

Contents

Visit our companion website for additional book content, including answers to the final exam in the book, all of the images from the book, and additional color images HYPERLINK "http://www.elsevierdirect.com/companions/9780123742346" www.elsevierdirect.com/companions/9780123742346

I am very lucky. I am sitting in the rare book room of the library waiting for Robert Hooke's (1665) *Micrographia*, Matthias Schleiden's (1849) *Principles of Scientific Botany*, and Hermann Schacht's (1853) *The Microscope*. I am thankful for the microscopists and librarians at Cornell University, both living and dead, who have nurtured a continuous link between the past and the present. By doing so, they have built a strong foundation for the future.

Robert Hooke (1665) begins the Micrographia by stating that "… the science of nature has already too long made only a work of the brain and the fancy: It is now high time that it should return to the plainness and soundness of observations on material and obvious things." Today, too many casual microscope users do not think about the relationship between the image and reality and are content to push a button, capture an image, enhance the image with Adobe Photoshop, and submit it for publication. However, the sentence that followed the one just quoted indicates that the microscope was not to be used in place of the brain, but in addition to the brain. Hooke (1665) wrote, "It is said of great empires, that the best way to preserve them from decay, is to bring them back to the first principles, and arts, on which they did begin." To understand how a microscope forms an image of a specimen still requires the brain, and today I am privileged to be able to present the work of so many people who have struggled and are struggling to understand the relationship between the image and reality, and to develop instruments that, when used thoughtfully, can make a picture that is worth a thousand words.

Matthias Schleiden (1849), the botanist who inspired Carl Zeiss to build microscopes, wrote about the importance of the mind of the observer:

> It is supposed that nothing more is requisite for microscopical investigation than a good instrument and an object, and that it is only necessary to keep the eye over the eye-piece, in order to be au fait. Link expresses this opinion in the preface to his phytotomical plates: 'I have generally left altogether the observation to my artist, Herr Schmidt, and the unprejudiced mind of this observer, who is totally unacquainted with any of the theories of botany, guarantees the correctness of the drawings.' The result of such absurdity is, that Link's phytotomical plates are perfectly useless; and, in spite of his celebrated name, we are compelled to warn every beginner from using them…. Link might just as well have asked a child about the apparent distance of the moon, expecting a correct opinion on account of the child's unprejudiced views. Just as we only gradually learn to see with the naked eye in our infancy, and often experience unavoidable illusions, such as that connected with the rising moon, so we must first gradually learn to see through the medium of the microscope….. We can only succeed gradually in bringing a clear conception before our mind….

Hermann Schacht (1853) emphasized that we should "see with intelligence" when he wrote,

> But the possession of a microscope, and the perfection of such an instrument, are not sufficient. It is necessary to have an intimate acquaintance, not only with the management of the microscope, but also with the objects to be examined; above all things it is necessary to see with intelligence, and to learn to see with judgment. Seeing, as Schleiden very justly observes, is a difficult art; seeing with the microscope is yet more difficult….Long and thorough practice with the microscope secures the observer from deceptions which arise, not from any fault in the instrument, but from a want of acquaintance with the microscope, and from a forgetfulness of the wide difference between common vision and vision through a microscope. Deceptions also arise from a neglect to distinguish between the natural appearance of the object under observation, and that which it assumes under the microscope.

Throughout the many editions of his book, *The Microscope*, Simon Henry Gage (1941) reminded his readers of the importance of the microscopist as well as the microscope (Kingsbury, 1944): "To most minds, and certainly to those having any grade of originality, there is a great satisfaction in understanding principles; and it is only when the principles are firmly grasped that there is complete mastery of instruments, and full certainty and facility in using them …. for the highest creative work from which arises real progress both in theory and in practice, a knowledge of principles is indispensable." He went on to say that an "image, whether it is made with or without the aid of the microscope, must always depend upon the character and training of the seeing and appreciating brain behind the eye."

This book is a written version of the microscopy course I teach at Cornell University. I introduce my students to the principles of light and microscopy through lecture-demonstrations and laboratories where they can put themselves in the shoes of the masters and be virtual witnesses to their original observations. In this way, they learn the strengths and limitations of the work, how first principles were uncovered, and, in some respects, feel the magic of discovery. I urge my students to learn through personal experience and to be skeptical of everything I say. I urge the reader to use this book as a guide to gain personal experience with the microscope. Please read it with a skeptical and critical mind and forgive my limitations.

Biologists often are disempowered when it comes to buying a microscope, and the more scared they are, the more likely it is that they will buy an expensive microscope, in essence, believing that having a prestigious brand name will make up for their lack of knowledge. So buying an expensive microscope when a less expensive one may be equally good or better may be more a sign of ignorance than a sign of wisdom and greatness. I wrote this book, describing microscopy from the very beginning, not only to teach people how to use a microscope and understand the relationship between the specimen and the image, but to empower people to buy a microscope based on its virtues, not on its name. You can see whether or not a microscope manufacturer is looking for a knowledgeable customer by searching the web sites to see if the manufacturer offers information necessary to make a wise choice or whether the manufacturer primarily is selling prestige. Of course, sometimes the prestigious microscope is the right one for your needs.

If you are ready to buy a microscope after reading this book, arrange for all the manufacturers to bring their microscopes to your laboratory and then observe your samples on each microscope. See for yourself: Which microscopes have the features you want? Which microscope gives you the best image? What is the cost/benefit relationship? I thank M. V. Parthasarathy for teaching me this way of buying a microscope.

Epistemology is the study of how we know what we know—that is, how reality is perceived, measured, and understood. Ontology is the study of the nature of what we know that we consider to be real. This book is about how a light microscope can be used to help you delve into the invisible world and obtain information about the microscopic world that is grounded in reality. The second book in this series, entitled, *Plant Cell Biology*, is about what we have learned about the nature of life from microscopical studies of the cell.

The interpretation of microscopic images depends on our understanding of the nature of light and its interactions with the specimen. Consequently, an understanding of the nature of light is the foundation of our knowledge of microscopic images. Appendix II provides my best guess about the nature of light from studying its interactions with matter with a microscope.

I thank David Bierhorst, Peter Webster, and especially Peter Hepler for introducing me to my life-long love of microscopy. The essence of my course comes from the microscopy course that Peter Hepler taught at the University of Massachusetts. Peter also stressed the importance of character in doing science. Right now, I am looking through the notes from that course. I was very lucky to have had Peter as a teacher. I also thank Dominick Paolillo, M. V. Parthasarathy, and George Conneman for making it possible for me to teach a microscopy course at Cornell and for being supportive every step of the way. I also thank the students and teaching assistants who shared in the mutual and never-ending journey to understand light, microscopy, and microscopic specimens. I have used the pictures that my student's have taken in class to illustrate this book. Unfortunately, I no longer know who took which picture, so I can only give my thanks without giving them the credit they deserve. Lastly, I thank my family: mom and dad, Scott and Michelle, for making it possible for me to write this book.

As Hermann Schacht wrote in 1853, "Like my predecessors, I shall have overlooked many things, and perhaps have entered into many superfluous particulars: but, as far as regards matters of importance, there will be found in this work everything which, after mature consideration, I have thought necessary."

Randy Wayne

Chapter 1

The Relation between the Object and the Image

And God said, "Let there be light," and there was light. God saw that the light was good, and he separated the light from the darkness.

Gen. 1:3-4

We get much of our information about the real world through our eyes, and we depend on the constancy of the interaction of light and matter to determine the physical and chemical characteristics of an object. Due to the constancy of the interaction of light with matter, we can determine the size, shape, color, transparency, chemical composition, and texture of objects with our eyes. After we understand the nature of the interaction of light with matter, we can use light as a tool to probe the properties of matter under the microscope. We can use a dark-field microscope or a phase-contrast microscope to see invisible (e.g., transparent) cells. We can use a polarizing microscope to determine the orientation of molecules in a cell and even determine the entropy and enthalpy of the polymerization reaction of the microtubules in the mitotic spindle. We can use an interference microscope to ascertain the mass of the cell's nucleus. We can use a fluorescence microscope to localize proteins in the cytoplasm or genes on a chromosome. We can also use a fluorescence microscope to determine the membrane potential of the endoplasmic reticulum or the free Ca^{2+} concentration and pH of the cytoplasm. We can use a laser microscope or a centrifuge microscope to measure the forces involved in cellular motility or to determine the elasticity and viscosity of the cytoplasm.

We can do all these things with a light microscope because the light microscope is a device that permits us to study the interaction of light with matter at a resolution much greater than that of the unaided eye. The light microscope is one of the most elegant tools available, and I wrote this book so that you can make the most of the potential of the light microscope and even extend its uses. To this end, the goals of this book are to:

- Describe the relationship between an object and its image.
- Describe how light interacts with matter to yield information about the structure, composition, and local environment of biological and other specimens.

- Describe how optical systems work. This will permit us to interpret the images obtained at high resolution and magnification.
- Give you the necessary procedures and tricks so that you can gain practical experience with the light microscope and become an excellent microscopist.

LUMINOUS AND NONLUMINOUS OBJECTS

All objects, which are perceived by our sense of sight, can be divided into two classes. One class of objects, known as luminous bodies, includes the sun, the stars, torches, oil lamps, candles, and light bulbs. These objects are visible to our eyes. The second class of objects is nonluminous. However they can be made visible to our eyes when they are in the presence of a luminous body. Thus the sun makes the moon, Earth, and other planets visible to us, and a light bulb makes all the objects in a room or on a microscope slide visible to us. The nonluminous bodies become visible by reemitting the light they absorb from the luminous bodies. A luminous or nonluminous body is visible to us only if there are sufficient differences in brightness or color between it and its surroundings. The difference in brightness or color between points in the image formed of an object on our retina is known as contrast.

OBJECT AND IMAGE

Each object is composed of many infinitesimally small points composed of atoms or molecules. Ultimately, the image of each object is a point-by-point representation of that object upon our retina. Each point in the image should be a faithful representation of the brightness and color of the conjugate point in the object. Two points on different planes are conjugate if they represent identical spatial locations on the two planes. The object we see may itself be an intermediate image of a real object. The intermediate image of a real object observed with a microscope, telescope, or by looking at a photograph, movie, or television screen should also be a faithful point-by-point representation of the brightness and color of each conjugate point of the real object. While we only see brightness and color, the mind interprets the relative brightness and colors of the points of light on the retina and makes a judgment as to the size, shape, location, and position of the real object.

What we see, however, is not a perfect representation of the physical world. First, our eyes are not perfect, and our vision is limited by physical factors (Inoué, 1986; Helmholtz, 2005). For example, we cannot see clearly things that are too far or too close, too dark or too bright, or things that emit radiation outside the visible range of wavelengths. Second, our vision is affected by psychological factors, and we can be easily fooled by our sense of sight (Russ, 2004). Goethe (1840) stressed the psychological component of color vision after noticing that when an opaque object is irradiated with colored light, the shadow appears to be the complementary color of the illuminating light even though no light exists in the shadow of the object.

Another famous example of the psychological component of vision is the "Moon Illusion". For example, the moon rising on the horizon looks bigger than the moon on the meridian, yet we can easily see that they are the same size by holding a quarter at arms length and observing that in both cases the quarter just obscures the moon (Molyneux, 1687; Wallis, 1687; Berkeley, 1709; Schleiden, 1849; Kaufman and Rock, 1962). When walking through a museum, it appears as if the eyes in the portraits seem to follow the viewer, yet the eyes do not move (Wollaston, 1824; Brewster, 1835). When a friend walks toward you, he or she appears to get taller, but does he or she actually get taller?

In order to demonstrate the effect of perspective on the appearance of size, hold one meter stick and look at another meter stick, parallel to the first and one meter further from your eyes. How long does ten centimeters on the distant stick appear to be when measured with the nearer stick? If we were to run two pieces of string from our eye to the two points 10 centimeters apart on the further meter stick, we would see that the string would touch the exact two points on the nearer stick that we used to measure how long 10 centimeters of the further stick appeared. It is as if light from the points on the two meter sticks traveled to our eyes along the straight lines defined by the strings. The relationship between distance and apparent size is known as *perspective*, and is used in painting as a way of capturing the world as we see it on a piece of canvas (da Vinci, 1970; Gill, 1974). Alternatively, *anamorphosis* is a technique devised by Leonardo da Vinci to hide images so that we can view them only if we know the laws of perspective (Leeman, 1977). Look at the following optical illusions and ask yourself, is seeing really believing? On the other hand, is believing seeing (Figure 1-1)?

Optical illusions are a fun way to remind ourselves that there can be a tenuous relationship between what we see and what we think we see. To further test the relationship between seeing and believing, look at the following books on optical illusions: Luckiesh, 1965; Joyce, 1995; Fineman, 1981; Seckel, 2000, 2001, 2002, 2004a, 2004b. Do you believe that all the people in da Vinci's *Last Supper* were men? Is that what you see? What do you think vision would be like if a blind person were suddenly able to see (Zajonc, 1993)?

THEORIES OF VISION

In order to appreciate the relationship between an object and its image, the ancient Greeks developed several theories of vision, which can be reduced into two classes (Priestley, 1772; Lindberg, 1976; Ronchi, 1991; Park, 1997):

- Theories that state that vision results from the emission of visual rays from the eye to the object being viewed (extramission theory).
- Theories that state that vision results from light that is emitted from the object and enters the eye (intromission theory).

The extramission theory was based, in part, on a comparison of the sense of vision with the sense of touch. It provided an explanation for the facts that we can see images when we sleep in the dark, we see light when we rub our eyes, and we can see only the surface of objects. The intromission theory was based on the idea that the image was formed from a thin skin of atoms that flew off the object and into the eye. Evidence supporting the intromission theory comes from the facts that we cannot see in the dark, we cannot see objects that are too close to the eye, and we can see the stars, and in doing so our eyes do not collapse from sending out an infinite number of visual rays such a vast distance (Sabra, 1989).

Historically, most theories of vision were synthetic theories that combined the two theses, suggesting that light emitted from the object combines with the visual rays in order for vision to occur (Plato, 1965). Many writers, from Euclid to Leonardo da Vinci, wavered back and forth between the two extreme theories. In 1088, Al-Haytham, a supporter of the intromission theory, suggested that images may be formed by eyes, in a manner similar to the way that they are formed by pinholes (Sabra, 1989). The similarity between the eye and a pinhole camera also was expressed by Giambattista della Porta, Leonardo da Vinci (1970), and Francesco Maurolico (1611). However they never were able to reasonably explain the logical consequence that, if an eye formed images just like a pinhole camera, then the world should appear upside down (Arago, 1857).

By 1604, Johannes Kepler developed, what is in essence, our current theory of vision. Kepler inserted an eyeball, whose back had been scraped away to expose the retina, in the pinhole of a *camera obscura*. Upon doing

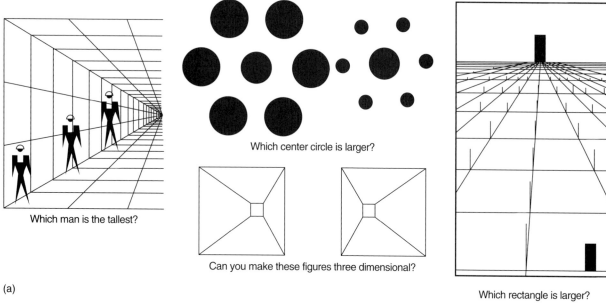

Which man is the tallest?

Which center circle is larger?

Can you make these figures three dimensional?

Which rectangle is larger?

(a)

"ALL IS VANITY"

(b)

FIGURE 1-1   (a) Optical illusions. Is seeing believing? (b) "All is vanity" by Charles Allan Gilbert (1892). When we look at this ambiguous optical illusion, our mind forms two alternative interpretations, each of which is a part of the single reality printed on the page. Instead of seeing what is actually on the page, our mind produces two independent images, each of which makes sense to us and each of which has meaning. When we look at a specimen through a microscope, we must make sure that we are seeing what is there and find meaning in what is there, as opposed to seeing only that which is already meaningful to us.

this, he discovered that the eye contains a series of hard and soft elements that act together as a convex lens, which projects an inverted image of the object on the concave retina. The image formed on the retina is an inverted point-by-point replica that represents the brightness and color of the object. Kepler dismissed the problem of the "upside up world" encountered by Porta, da Vinci, and Maurolico,

by suggesting that the brain subsequently deals with the inverted image. The importance of the brain in vision was expanded by George Berkeley (1709).

Before I discuss the physical relationship between an object and an image, I will take a step backward and discuss the larger philosophical problem of recognizing which is the object and which is the image. Plato illustrates

FIGURE 1-2   The troglodytes in a cave.

this point in the *Republic* (Jowett, 1908; also see Cornford, 1945) where he tells the following parable known as *The Allegory of the Cave* (Figure 1-2). Plato writes,

And now I will describe in a figure the enlightenment or unenlightenment of our nature: Imagine human beings living in an underground den which is open towards the light; they have been there from childhood, having their necks and legs chained, and can only see into the den. At a distance there is a fire, and between the fire and the prisoners a raised way, and a low wall is built along the way, like the screen over which marionette players show their puppets. Behind the wall appear moving figures, who hold in their hands various works of art, and among them images of men and animals, wood and stone, and some of the passers-by are talking and others silent .... They are ourselves ... and they see only the shadows of the images which the fire throws on the wall of the den; to these they give names, and if we add an echo which returns from the wall, the voices of the passengers will seem to proceed from the shadows. Suppose now that you suddenly turn them round and make them look with pain and grief to themselves at the real images; will they believe them to be real? Will not their eyes be dazzled, and will they not try to get away from the light to something which they are able to behold without blinking? And suppose further, that they are dragged up a steep and rugged ascent into the presence of the sun himself, will not their sight be darkened with the excess of light? Some time will pass before they get the habit of perceiving at all; and at first they will be able to perceive only shadows and reflections in the water; then they will recognize the moon and the stars, and will at length behold the sun in his own proper place as he is. Last of all they will conclude: This is he who gives us the year and the seasons, and is the author of all that we see. How will they rejoice in passing from darkness to light! How worthless to them will seem the honours and glories of the den! But now imagine further, that they descend into their old habitations; in that underground dwelling they will not see as well as their fellows, and will not be able to compete with them in the measurement of the shadows on the wall; there will be many jokes about the man who went on a visit to the sun and lost his eyes, and if they find anybody trying to set free and enlighten one of their number, they will put him to death, if they can catch him. Now the cave or den is the world of sight, the fire is the sun, the way upwards is the way to knowledge, and in the world of knowledge the idea of good is last seen and with difficulty, but when seen is inferred to be the author of good and right–parent of the lord of

light in this world, and of truth and understanding in the other. He who attains to the beatific vision is always going upwards....

Although this parable can be discussed at many levels, I will use it just to emphasize that we see images of the world, and not the world itself. Plato went on to suggest that the relationship between the image and its reality could be understood through study, particularly the progressive and habitual study of mathematics. In *Novum Organum*, Francis Bacon (in Commins and Linscott, 1947) described four classes of idols that plague one's mind in the scientific search for knowledge. One of these he called "the idols of the cave." He wrote,

The Idols of the Cave are the idols of the individual man. For everyone (besides the errors common to human nature in general) has a cave or den of his own, which refracts and discolors the light of nature; owing either to his own proper and peculiar nature or to his education and conversation with others; or to the reading of books, and the authority of those whom he esteems and admires; or to the differences of impressions, accordingly as they take place in a mind preoccupied and predisposed or in a mind indifferent and settled; or the like. So that the spirit of man (according as it is meted out to different individuals) is in fact a thing variable and full of perturbation, and governed as it were by chance. Whence it was well observed by Heraclitus that men look for science in their own lesser worlds, and not in the greater or common world.

Charles Babbage (1830) wrote, in *Reflections on the Decline of Science*, about the importance of understanding the "irregularity of refraction" and the "imperfections of instruments" used to observe nature. In his book, entitled, *The Image*, Daniel Boorstin (1961) contends that many of the advances in optical technologies have contributed to a large degree in separating the real world from our image of it. Indeed, the physical reality of our body and our own image of it does not have a one-to-one correspondence. In *A Leg to Stand On*, Oliver Sacks (1984) describes the neurological relationship between our body and our own image of our body.

Thus it is incumbent on us to understand that when we look at something, we are not directly sensing the object, but an image of the object projected on our retinas, and processed by our brains. The image, then, depends not only on the intrinsic properties of the object, but on the properties of the light that illuminates it, as well as the physical, physiological, and psychological basis of vision. Thus before we even prepare our specimen for viewing in the microscope, we must prepare our mind. While looking through the microscope, I would like you to keep the following general questions in mind:

1. How do we receive information about the external world?

2. What is the nature and validity of the information?

3. What is the relationship of the perceiving organism to the world perceived?

4. What is the nature and validity of the information obtained by using an instrument to extend the senses; and what is the relationship of the information obtained by the perceiving organism with the aid of an instrument to the world perceived?

LIGHT TRAVELS IN STRAIGHT LINES

It has been known for a long time that light travels in straight lines. Mo Tzu (470–391 BC) inferred that the light rays from luminous sources travel in straight lines because:

- A shadow cast by an object is sharp, and it faithfully reproduces the shape of the object.
- A shadow never moves by itself, but only if the light source or the object moves.
- The size of the shadow depends on the distance between the object and the screen upon which it is projected.
- The number of shadows depends on the number of light sources: if there are two light sources, there are two shadows (Needham, 1962).

The ancient Greeks also came to the conclusion that light travels in straight lines. Aristotle (384–322 BC, Physics Book 5, in Barnes, 1984) concluded that light travels in straight lines as part of his philosophical outlook that nature works in the briefest possible manner. Evidence, however, for the rectilinear propagation of light came in part from observing shadows. Euclid observed that there is a geometric relationship between the height of an object illuminated by the sun and the length of the shadow cast (Figure 1-3). Theon of Alexandria (335–395) amplified Euclid's conclusion that light travels in straight lines by showing that the size of a shadow depended on whether an object was illuminated by parallel rays, converging rays, or diverging rays (Lindberg and Cantor, 1985).

Mirrors and lenses have been used for thousands of years as looking glasses and for starting fires. Aristophanes (423 BC) describes their use in *The Clouds*. Euclid, Diocles, and Ptolemy used the assumption that a light ray (or visual ray) travels in a straight line in order to build a theory of geometrical optics that was powerful enough to predict the position of images formed by mirrors and refracting surfaces (Smith, 1996). According to geometrical optics, an image is formed where all the rays emanating from a single point on the object combine to make a single point of the image. The brighter the point in the object, the greater the number of rays it emits. Bright points emit many rays and darker points emit fewer rays. The image is formed on the surface where the rays from each point meet the other rays emitted from the same point. The success that the geometrical theory of optics had in predicting the position of images provided support that the assumption that light travels in straight lines, upon which this theory is based, must be true.

Building on the atomistic theories of Leucippus, Democritus, Epicurus, and Lucretius—and contrary to the

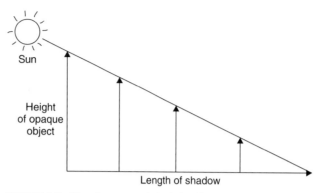

FIGURE 1-3   There is a geometrical relationship between the height of an object illuminated by the sun and the length of the shadow cast. Height$_{object 1}$/Length of shadow$_{object 1}$ = Height$_{object 2}$/Length of shadow$_{object 2}$ = constant.

continuous theories championed by Aristotle, Simplicus, and Descartes—Isaac Newton proposed that light traveled along straight lines as corpuscles.

Interestingly, the fact that light travels in straight lines allows us to "see what we want to see." The mathematician, William Rowan Hamilton (1833) began his paper on the principle of least action in the following way:

> The law of seeing in straight lines was known from the infancy of optics, being in a manner forced upon men's notice by the most familiar and constant experience. It could not fail to be observed that when a man looked at any object, he had it in his power to interrupt his vision of the object, and hide it at pleasure from his view, by interposing his hand between his eyes and it; and that then, by withdrawing his hand, he could see the object as before: and thus the notion of straight lines or rays of communication, between a visible object and a seeing eye, must very easily and early have arisen.

IMAGES FORMED IN A *CAMERA OBSCURA*: GEOMETRIC CONSIDERATIONS

Mo Tzu provided further evidence that rays emitted by each point of a visible object travel in a straight line by observing the formation of images (Needham, 1962; Hammond, 1981; Knowles, 1994). He noticed that although the light emitted by an object is capable of forming an image in our eyes, it is not able to form an image on a piece of paper or screen. However, Mo Tzu found that the object could form an image on a screen if he eliminated most of the rays issuing from each point by placing a pinhole between the object and the screen (Figure 1-4). The image that appears, however, is inverted. Mo Tzu (in Needham, 1962) wrote,

> An illuminated person shines as if he was shooting forth rays. The bottom part of the man becomes the top part of the image and the top part of the man becomes the bottom part of the image. The foot of the man sends out, as it were light rays, some of which are hidden below (i.e. strike below the pinhole) but others of which form an image at the top. The head of the man sends out, as it were light rays, some of which are hidden above (i.e. strike above the pinhole) but others of which form its image at the bottom. At a position farther or nearer from the source

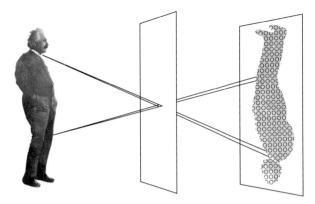

FIGURE 1-4 A pinhole forms an inverted image because light travels in straight lines. The pinhole blocks out the majority of rays that radiate from a single point on the object. The rays that do pass through the pinhole form the image. The smaller the pinhole, the smaller the circle of confusion that makes up each "point" of the image.

> of light, reflecting body, or image there is a point (the pinhole) which collects the rays of light, so that the image is formed only from what is permitted to come through the collecting-place.

The fact that the image can be reconstructed by drawing a straight line from every point of the outline of the object, through the pinhole, and to the screen, confirms that light does travel in straight lines According to John Tyndall (1887), "This could not be the case if the straight lines and the light rays were not coincident." Shen Kua (1086) extended Mo Tzu's work by showing the analogy between pinhole images and reflected images. However, Shen Kua's work could not go too far since it lacked a geometric foundation (Needham, 1962).

The Greeks also had discovered that images could be formed by a pinhole. Aristotle noticed that the light of the sun during an eclipse coming through a small hole made between leaves casts an inverted image of the eclipse on the ground (Aristotle; Problems XV:11 in Barnes, 1984).

The description of image formation based on geometric optics by Euclid and Ptolemy was extended by scholars in the Arab World. Al-Kindi (ninth century) in *De aspectibus* showed that light entering a dark room through windows travels in straight lines. Likewise the light of a candle is transmitted through a pinhole in straight lines (Lindberg and Cantor, 1985). Al-Kindi's work was extended by Al-Haytham, or Alhazen as he is often known (in Lindberg, 1968), who wrote in his *Perspectiva*,

> The evidence that lights and colors are not intermingled in air or in transparent bodies is that when a number of candles are in one place, [although] in various and distinct positions, and all are opposite an aperture that passes through to a dark place and in the dark place opposite the aperture is a wall or an opaque body, the lights of those candles appear on the [opaque] body or the wall distinctly according to the number of candles; and each of them appears opposite one candle along a [straight] line passing through the aperture. If one candle is covered, only the light opposite [that] one candle is extinguished; and if the cover is removed, the light returns.... Therefore, lights are not intermingled in air, but each of them is extended along straight lines.

The quality of the image formed by a pinhole depends on the size of the pinhole (Figure 1-4). When the pinhole is too small, not enough light rays can pass through it and the image is dark. However, if the pinhole is too large, too many light rays pass through and the image is blurry. Seeing this, Al-Haytham and his commentator Al-Farisi (fourteenth century) realized that the image formed by the pinhole was actually a composite of numerous overlapping images of the pinhole, each one originating from an individual luminous point on the object (Omar, 1977; Lindberg, 1983; Sabra, 1989).

Each and every point on a luminous object forms a cone of light that passes through the pinhole. The pinhole marks the tip of the cone and the light at the base of the cone forms the image. The fact that light originating from a point on an object forms a circle of light on the image leads to some blurring of the image known as the "circle of confusion" (Time-Life, 1970). The image will be distinct (or resolved) if the bases of the cones that originate from the two extreme points of the object do not overlap. Likewise the image will be clearer when the bases of cones originating from adjacent points on the object do not overlap. Given this hypothesis, the sharpness of the image would increase as the size of the aperture decreases. However, the brightness of the images also decreases as the size of the aperture decreases. Using geometry, Al-Haytham found the optimal diameter of an aperture when viewing an object of a given diameter (y_o) and distance (s_o) from the aperture. Al-Haytham showed, that when the object is circular, and the object, aperture, and plane of the screen are parallel, two light patches originating from two points on the object will touch when the ratio of the diameter of the aperture (a_o) to that of the object (y_o) is equal to the ratio of the distance between the image and the aperture (s_i), and the distance between the image and the object ($s_i + s_o$). That is:

$$a_o/y_o = s_i/(s_i + s_o)$$

The position of the optimal image plane (s_i) and the optimal size of the aperture (a_o) are given by the following analysis (Figure 1-5).

Since $\tan \theta = (\frac{1}{2}a_o)/s_i = (\frac{1}{2}y_o)/(s_i + s_o)$, then $a_o/y_o = s_i/(s_i + s_o)$ and $y_o/a_o = 1 + s_o/s_i$. For large distances between the object and the pinhole, $y_o/a_o \approx s_o/s_i$, and for a given s_o, the greater the aperture size, the greater is the distance from the aperture to a clear image.

Leonardo da Vinci (1970) also concluded that light travels through a pinhole in straight lines to form an image. He wrote, "All bodies together, and each by itself, give off to the surrounding air an infinite number of images which are all-pervading and each complete, each conveying the nature, colour and form of the body which produces it." da Vinci proved this hypothesis by observing that when one makes "a small round hole, all the illuminated objects will project their images through that hole and be visible inside

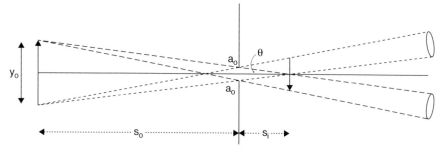

FIGURE 1-5 The position of the optimal image plane (s_i) and the optimal size of the aperture (a_o) for an object of height (y_o) placed at the object plane (s_o).

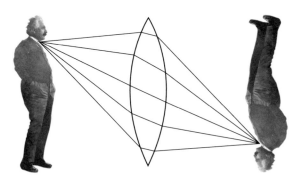

FIGURE 1-6 A converging lens can collect more of the rays that emanate from a point on an object than a pinhole can, thus producing a brighter image.

the dwelling on the opposite wall which may be made white; and there in fact, they will be upside down, and if you make similar openings in several places on the same wall you will have the same result from each. Hence the images of the illuminated objects are all everywhere on this wall and all in each minutest part of it." da Vinci (1970) also realized that the images formed by the pinhole were analogous to the images formed by the eye. He wrote,

> An experiment, showing how objects transmit their images or pictures, intersecting within the eye in the crystalline humour, is seen when by some small round hole penetrate the images of illuminated objects into a very dark chamber. Then, receive these images on a white paper placed within this dark room and rather near to the whole and you will see all the objects on the paper in their proper forms and colours, but much smaller; and they will be upside down by reason of that very intersection. These images being transmitted from a place illuminated by the sun will seem actually painted on this paper which must be extremely thin and looked at from behind.

Light rays that emanate from a point in an object separate from each other and form a cone. The pinhole sets a limit on the size of the cone that is used to form an image of any given point. When the aperture is large, the cone of light emanating from each point is large. Under this condition, light from every point on the object illuminates every part of the screen and there is no image. As the aperture decreases, however, the cone of light from each point illuminates a limited region of the screen, and an image is

formed. The screen must be far enough behind the pinhole so that the cones of light emanating from two nearby points do not overlap. The greater the distance between the screen and the pinhole, the larger the image will be, but it will also become dimmer. This dimness problem can be overcome by putting a converging lens over the pinhole (Wright, 1907; Figure 1-6).

Girolamo Cardano suggested in his book, *De subtilitate*, written in 1550, that a biconvex lens placed in front of the aperture would increase the brightness of the image (Gernsheim, 1982). In 1568, Daniel Barbaro, in his book on perspective, also mentioned that a biconvex lens increases the brightness of the image. The lens focuses all the rays emanating from each point of an object that it can capture and focuses them to form the corresponding conjugate point on the image. Thus a lens is able to capture a larger cone of light emitted from each point than an aperture can. In contrast to an image formed by a pinhole, an image formed by a lens is restricted to only one plane, known as the image plane. In front of or behind the image plane, the rays are converging to a spot or diverging from a spot, respectively. Consequently, the "out-of-focus" image of a bright spot is dim, and in the "out-of-focus" image there is no clear relationship between the brightness of the image and the brightness of the object. The distance of the image plane from the lens, as well as the magnification of the image depends on the focal length of the lens. For an object at a set distance in front of the lens, the image distance and magnification increases with an increase in the focal length of the lens (Figure 1-7).

With lenses of the same focal length, the brightness of the image increases as the diameter of the lens increases. This is because the larger a lens, the more rays it can collect from each point on the object. The sharpness of the image produced by a lens is related to the number of rays emanating from each point that is collected by that lens.

The *camera obscura* was popularized by Giambattista della Porta in his book *Natural Magic* (1589), and by the seventeenth century, portable versions of the *camera obscura* were fabricated and/or used by Johann Kepler (who coined the term *camera obscura*, which literally means dark room) for drawing the land he was surveying and for observing the sun. Kepler also suggested that

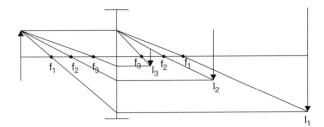

FIGURE 1-7 As the focal length of a lens increases ($f_1 > f_2 > f_3$), the image plane moves farther from the lens and the image becomes more magnified.

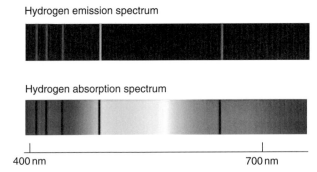

Hydrogen emission spectrum

Hydrogen absorption spectrum

400 nm   700 nm

FIGURE 1-8 A diffraction grating resolves the light emitted from an incandescent gas into bright lines. When a sample of the same gas is placed between a white light source and a diffraction grating, black lines appear at the same places as the emission lines occurred, indicating that gases absorb the same wavelengths as they emit.

the *camera obscura* could be improved by adding a second biconvex lens to correct the inverted image. Moreover, he suggested that the focal length of the lens could be reduced by combining a concave lens with the convex lens. Johann Zahn, Athanasius Kircher, and others used *camera obscuras* in order to facilitate drawing scenes far away from the studio, and Johann Hevelius connected a *camera obscura* to a microscope to facilitate drawing enlarged images of microscopic specimens (Hammond, 1981).

Some Renaissance painters, including Vermeer, used the camera obscura as a drawing aid. Indeed, it is thought that "A View of Delft" was painted with the aid of the *camera obscura* since the edges of the painting are out of focus. In 1681, Robert Hooke suggested that the screen of the *camera obscura* should be concave, since the image formed by either a pinhole or a simple lens does not form a flat field at sharp focus, but has a curved field of sharp focus. When a *camera obscura* was open to the public, the crowded dark room was used both as a venue to present shows of natural magic and as a convenient place to pick the pockets of the unsuspecting audience.

WHERE DOES LIGHT COME FROM?

Light comes from matter, the atoms of which are in an excited state, which has more energy than the most stable or ground state (Clayton, 1970). An atom becomes excited when one of its electrons makes a transition from an orbital close to the nucleus to an orbital further from the nucleus (Bohr, 1913; Kramers and Holst, 1923). Atoms can become excited by various forms of energy, including heat, pressure, an electric discharge, and by light itself (Wedgewood, 1792; Nichols and Wilber, 1921a, 1921b). Heating limestone ($CaCO_3$) for example gives off a bright light. Thomas Drummond (1826) took advantage of this property to design a spotlight that was used in theatrical productions in the nineteenth century. This is how we got the expression, "being in the limelight."

Although the ancient Chinese invented fireworks, the stunning colors were not added until the discovery and characterization in the nineteenth century of the optical properties of the elements. Various elements burned in a flame emit a spectacular spectrum of rich colors and each element

gives off a characteristic color. For example, the chlorides of copper, barium, sodium, calcium, and strontium give off blue, green, yellow, orange, and red light, respectively. This indicates that there is a relationship between the atomic structure of the elements and the color of light emitted. Interestingly, the structure of atoms has been determined to a large degree by analyzing the characteristic colors that are emitted from them (Brode, 1943; Serway et al., 2005).

In 1802, William Wollaston and, in 1816, Joseph von Fraunhöfer independently identified dark lines in the spectrum of the sun. Fraunhöfer identified the major lines with uppercase letters (A, B, C, D, E, F …) and the minor lines with lowercase letters. John Herschel (1827) noticed that a given salt gave off a characteristic colored light when heated and suggested that chemicals might be identified by their spectra. Fraunhöfer suggested that the colored lines given off by heated elements might be related to the dark lines observed in solar spectra, and subsequently he developed diffraction gratings to resolve and quantify the positions of the spectral lines (Figure 1-8). Independently, William Henry Fox Talbot (1834c) discovered that lithium and strontium gave off colored light when they were heated, and since the color of the light was characteristic of the element, Talbot also suggested that optical analysis would be an excellent method for identifying minute amounts of an element. Following this suggestion, Robert Wilhelm Bunsen and Gustav Kirchhoff used the gas burner Bunsen invented to determine the spectrum of light given off by each element (Kirchhoff and Bunsen, 1860; Gamow, 1988).

Fraunhöfer's A (759.370 nm) and B (686.719 nm) lines turned out to be due to oxygen absorption, the C (656.281) line was due to hydrogen absorption, the D_1 (589.592 nm) and D_2 (588.995 nm) lines were due to sodium absorption, the D_3 (587.5618 nm) line was due to hydrogen absorption, the E (546.073 nm) line was due to mercury absorption, the E_2 (527.039 nm) line was due to iron absorption, and the F (486.134 nm) line was due to hydrogen absorption. These lines are used as standards by lens makers to characterize

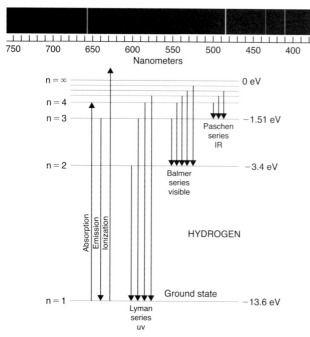

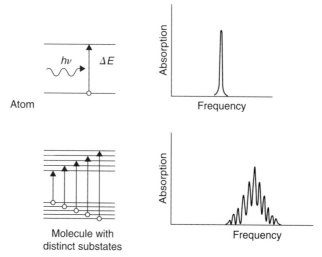

FIGURE 1-10   The absorption (and emission) spectra broaden and the peaks become less resolved, as a chemical gets more and more complex. This occurs because a complex molecule can utilize absorbed energy in more ways that a simple molecule by vibrating, rotating, and distributing the energy to other parts of the molecule. Likewise, the various vibrational, rotational, or conformational states of a molecule give rise to more complex spectra. The absorption and emission spectra of molecules are used to determine their chemical structure.

FIGURE 1-9   The bright spectral lines represent light emitted by electrons jumping from a higher energy level to a lower energy level. The dark absorption lines (shown in Figure 1-8) represent light absorbed by electrons jumping from a lower energy level to a higher energy level. The energy levels are designated by principal quantum numbers (n) and by binding energies in electron volts (1 eV = 1.6×10^{-19} J). Transitions in the ultraviolet range give rise to the Lyman series, transitions in the visible range give rise to the Balmer series, and transitions in the infrared range give rise to the Paschen series.

corrections for chromatic aberration in objective lenses used in microscopes (see Chapter 4).

When the emitted light from incandescent atoms or diatomic molecules is passed through a diffraction grating or a prism, the light is split into a series of discrete bands known as a line spectrum (Schellen, 1885; Schellen et al., 1872; Pauling and Goudsmit, 1930; Herzberg, 1944). The spectral lines represent the energy levels of the atom (Figure 1-9). When an excited electron returns to the ground state, the energy that originally was used to excite the atom is released in the form of radiant energy or light. The wavelength of the emitted light can be determined using Planck's Law:

$$\lambda = hc/\Delta E$$

where λ is the wavelength (in m), h is Planck's Constant (6.626×10^{-34} Js), c is the speed of light (3×10^{8} m s^{-1}), and ΔE is the transitional energy difference between electrons in the excited and the ground states (in J). Niels Bohr (1913) introduced the total quantum number (n) to describe the distance between the electron and its nucleus.

When gaseous atoms are combined into complex gaseous molecules, there is an increase in the number of spectral lines because of the formation of molecular orbitals, which exist in many vibrational and rotational states. Consequently, a gaseous molecule gives a banded spectrum

instead of a line spectrum (Figure 1-10). The spectra of liquids or solids are broadened further because a range of transition energies become possible as a consequence of the interactions between molecules. The various lines and bands become overlapping and the spectrum appears as a continuous spectrum. In the visible region, the spectrum appears as a continuous band of light, with colors that change smoothly from blue to red. The intensity of the various colors in a continuous spectrum depends on the temperature (Planck, 1949) and the relative velocity of the light source and the observer (Doppler, 1842). The sun and other stars can be considered to be black body radiators. Isaac Newton (1730) used a prism to resolve the sun's whitish-yellow glow, which is known as black body radiation, into its component parts.

Each point of an object emits light when an electron in an atom or molecule at that point undergoes a transition from a high energy level to a lower energy level. If any energy source besides light were used to excite the electron, then the object is known as a luminous source. If light itself is used to excite the electron, the object is known as a nonluminous source. The light emitted by the excited electron travels along rays emanating from that point. If the rays converge, an image is formed. In order to gain as much information as possible about the molecules that make up each point of the object, we have to understand the interaction of light with the atoms and molecules that make up that point; how the environment surrounding a molecule (e.g., pH, pressure, electrical potential, and viscosity) affects the emission of light from that molecule, how neighboring molecules influence each other, and finally how the light travels from the object in order to form an image.

HOW CAN THE AMOUNT OF LIGHT BE MEASURED?

The measurement of light, which is known as photometry, calorimetry, and radiometry, involves the absorption of light by a detector and the subsequent conversion of the radiant energy to another form of energy (Thompson, 1794; Talbot, 1834a; Johnston, 2001). A thermal detector converts light energy into thermal energy. A thermal detector is a type of thermometer whose detecting surface has been blackened so that it absorbs light from all regions of the spectrum.

A thermocouple is a thermal detector that consists of a junction of two metals coated with a black surface. When light strikes the blackened junction, a voltage is generated, a process discovered by Thomas Seebeck (1821). Often several (20–120) thermocouples are arranged in series to increase the response of the system. This arrangement is called a thermopile.

The bolometer is a thermal detector in which the detector is a thin strip of blackened platinum foil whose resistance increases with temperature. The bolometer was developed by Samuel Pierpont Langley (1881), the founder of the National Zoological Park in Washington, DC. Modern bolometers use thermistors made out of ceramic mixtures of manganese, nickel, cobalt, copper, and uranium oxides whose resistance decreases with temperature.

The amount of light can be measured by using chemical reactions whose rate is proportional to the amount of light that strikes the chemical substrates. This technique, known as chemical actinometry, can be done by using photographic paper, and then relating the amount of incident light to the darkening of the silver bromide impregnated paper.

I will discuss the use of electrical detectors, including photodiodes, photomultiplier tubes, video cameras, and charge coupled devices in Chapter 13.

Chapter 2

The Geometric Relationship between Object and Image

REFLECTION BY A PLANE MIRROR

In the last chapter I presented evidence that light travels in straight lines, and I used this assumption to describe image formation in a *camera obscura*. This hypothesis is limited, however, to light traveling through a homogeneous medium, and it is not true when light strikes an opaque body. After striking an opaque body, the light bounces back in a process known as reflection. Consider a flat surface that is capable of reflecting light. A line perpendicular to this surface is called the normal, from the Latin name of the carpenter's square used to draw perpendiculars. Experience shows that a ray of light that moves along the normal and then strikes the reflective surface head on will double back on its tracks. In general, if a ray of light strikes the reflective surface at an angle relative to the normal it will move away from the reflective surface at the other side of the normal at an angle equal to the angle the incident light beam made with the normal. The light beam moving toward the reflective surface is called the incident ray and its angle relative to the normal is called the angle of incidence. The light beam moving away from the reflective surface is called the reflected ray and its angle relative to the normal is called the angle of reflection. For all light

rays striking the surface at any angle, the angle of incidence equals the angle of reflection (Figure 2-1). That is, $<i = <r$, where $<i$ is the angle of incidence (in degrees) and $<r$ is the angle of reflection (in degrees). Although Euclid (third century BC) first described the law of reflection in his *Elements*, perhaps the most famous version of the law is found in literature. Dante Alighieri (1265–1321) described the law of reflection in *The Divine Comedy* (Longfellow's translation):

> As when from off the water, or a mirror,
> The sunbeam leaps unto the opposite side,
> Ascending upwards in the self-same measure
> That it descends, and deviates as far
> From falling of a stone in line direct,
> (As demonstrate experiment and art)....

To determine the position, orientation, and size of an image formed by a plane mirror, we can draw rays from at least two different points on the object to the mirror. Once the rays strike the mirror, we assume that they are reflected in such a way that the angle of reflection equals the angle of incidence. Practically, we can find an image point by drawing two characteristic rays from a point on the object using the following rules (Figures 2-2 and 2-3):

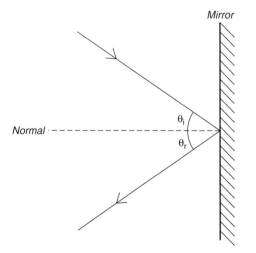

FIGURE 2-1 The law of reflection: The angle of reflection (θ_r) equals the angle of incidence (θ_i).

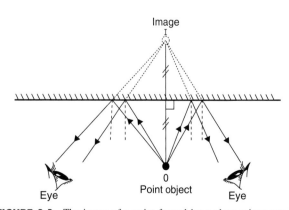

FIGURE 2-2 The image of a point formed by a plane mirror can be determined by using the law of reflection. Draw several rays that obey the law of reflection. The rays diverge when they enter the eye. The brain imagines that the diverging rays originated from a single point behind the mirror. The place where the rays appear to originate is known as the virtual image.

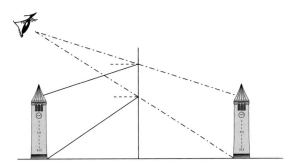

FIGURE 2-3 A virtual image produced by the eye and the brain of a person looking at a reflection of an object in a plane mirror.

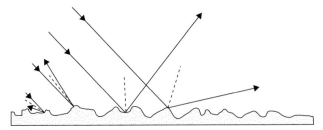

FIGURE 2-4 Diffuse reflection from a rough surface. The angle of reflection still equals the angle of incidence, but there are many angles of reflection. You can tell if a reflective surface is rough or smooth by observing if the reflection is diffuse or specular.

1. Draw a line from each point in the object perpendicular to the mirror. Since <i = 0, then <r = 0. Extend the reflected ray behind the mirror.

2. Draw another line from each point in the object to any point on the mirror. Draw the normal to the mirror at this point and then draw the reflected rays using the rule <i = <r. Extend the reflected rays behind the mirror, to the other reflected extension ray originating from the same point in the object. The point of intersection of the extension rays originating from the same object point is the position of the image of that object point.

If the reflected rays converged in front of the mirror, which they do not do when they strike a plane mirror, a real image would have been formed. A real image is an image that can be projected on a ground glass screen or a piece of paper, or captured by a camera; the light intensities of the points that make up a real image can be measured with a light meter. However, since the reflected rays diverge from the mirror, we extend the rays back from where they appear to be diverging. This is where the image appears to be, and thus is called a virtual image. A virtual image appears in a given place, but if we put a ground glass screen, a piece of paper, or a photographic film in that spot, nothing would appear.

Our eyes cannot distinguish whether light has been reflected or not. Many times, while watching a movie, we see an actor or actress and as the camera moves away we see that we have been fooled, and we saw only the reflection of that person in a mirror. When we look at ourselves in a mirror we see an image of ourselves behind the mirror—as if the image was actually behind the mirror and the light rays traveled in a straight line from it to our eyes. Because the image does not exist where we see it, the image is called a virtual image. That is, it has the virtues of an image without the image actually being there. Actually the image is reversed and the right of the object is on the left of the image and the left of the object is the right of the image. Perhaps this is the reason we usually do not like photographs of ourselves. We usually see a mirror image of ourselves where right and left are reversed. An image in

FIGURE 2-5 When light strikes a partially silvered mirror, some of the light is reflected and some of the light is transmitted. In this way, a partially silvered mirror functions as a beam splitter.

a photograph is in the correct orientation and thus seems strange to us. In older microscopes, plane mirrors were used to transmit sunlight to the specimen and for a drawing attachment known as a *camera lucida*.

I have been discussing front surface mirrors where the reflecting surface is deposited on the front of the glass. At home we use mirrors where the reflecting surface is deposited on the back of the glass. Therefore, there are two reflecting surfaces, the glass and the silvered surface. In this case, two images are formed, one from the reflection of each surface. Consequently, the image is a little blurred and much more complimentary. These are examples of specular reflection. By contrast, diffuse reflection is the reflection from a surface with many imperfections where parallel rays are broken up upon reflection. This occurs because one ray may strike an area where the angle of incidence is 0° whereas a parallel ray may strike an area where the angle of incidence is 10° (Figure 2-4).

Often two images need to be formed in a microscope: one portion of the image-forming rays goes to the eyepieces and the other portion goes to the camera. Partially silvered mirrors can be used to split the image-forming rays into two portions. Since the mirror is not fully silvered, part of the light passes directly through the mirror while the other part follows the normal law of reflection (Figure 2-5). If the mirror were 20 percent silvered, the image formed by the rays that go straight through would be four times brighter than the reflected image. Partially silvered mirrors are used in microscopes with epi-illumination and in some interference microscopes (Bradbury, 1988).

REFLECTION BY A CURVED MIRROR

Not all mirrors are planar and now I will describe images formed by concave, spherical mirrors. The center of curvature of a mirror is defined as the center of the imaginary sphere of which the curved mirror would be a part. The distance between the center of curvature of the spherical mirror and the mirror is equal to the radius of the sphere. The line connecting the midpoint of the mirror with the center of curvature is called the principal axis of the mirror. Consider a beam of light that strikes the mirror parallel to the principal axis. When a ray of light in this beam moves down the principal axis and strikes the mirror, it is reflected back on itself. When a ray of light in this beam strikes the mirror slightly above or below the principal axis, the ray makes a small angle with the normal and consequently the reflected ray is bent slightly toward the principal axis. If the incident ray strikes the mirror farther away from the principal axis, the reflected ray is bent toward the principal axis with a greater angle. In all cases $<i = <r$, and the reflected rays from every part of the mirror converge toward the principal axis at a point called the focus, which is midway between the mirror and the center of curvature. The focal length is equal to one-half the radius of curvature (Figure 2-6).

Figures 2-7 and 2-8 show examples of image formation by spherical mirrors when the object is placed behind or in front of the focus, respectively.

Using the law of reflection, we can determine the position, orientation, and size of the image formed by a concave spherical mirror. This can be done easily by drawing two or three characteristic rays using the following rules:

1. A ray traveling parallel to the principal axis passes through the focus after striking the mirror.

2. A ray that travels through the focus on the way to the mirror or appears to come from the focus travels parallel to the principal axis after striking the mirror.

3. A ray that strikes the point that is the intersection of the spherical mirror and the principal axis is reflected so that angle $<i = <r$ with respect to the principal axis.

4. A ray that passes through the center of curvature is reflected back through the center of curvature.

5. A real image of an object point is formed at the point where the rays converge. If the rays do not converge at a point, trace back the reflected rays to a point from where the extensions of each reflected ray seem to diverge and that is where a virtual image will appear.

A concave spherical mirror typically was mounted on the reverse side of the plane mirror on older microscopes without sub-stage condensers. When the concave mirror was rotated into the light path it focused the rays from the sun or a lamp onto the specimen to provide bright, even illumination (Carpenter, 1883; Hogg, 1898; Clark, 1925; Barer, 1956). Concave mirrors are still used to increase the intensity of the microscope illumination system; however, nowadays they are placed behind the light source in order to capture the backward traveling rays that would have been lost. For this purpose, the light source is placed at the center of curvature of the spherical mirror (Figure 2-9).

When the light source is placed at the center of curvature, the reflected rays converge on the light source

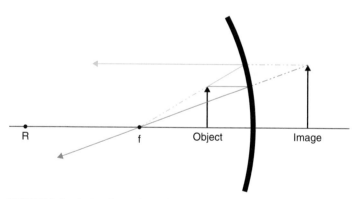

FIGURE 2-7 A virtual erect image is formed by the eye and brain of a person looking at the reflection of an object placed between the focus and a concave mirror. The virtual image of a point appears at the location from which the rays of that point appear to have originated.

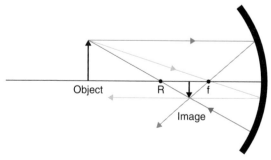

FIGURE 2-8 A real inverted image is produced by a concave mirror when the object is placed in front of the focal point. The further the object is from the focal point, the smaller the image, and the closer the image is to the focal plane.

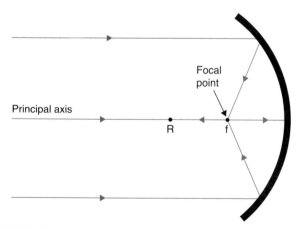

FIGURE 2-6 A beam of light, propagating parallel to the principal axis of a concave mirror, is brought to a focus after it reflects off the mirror. The focal point is equal to one-half the radius of curvature.

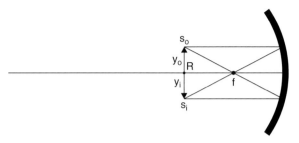

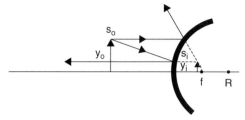

FIGURE 2-10 A virtual erect image is formed by a person looking at the reflection of an object placed anywhere in front of a convex mirror. The virtual image of a point appears at the location from which the rays of that point appear to have originated.

FIGURE 2-9 When an object is placed at the radius of curvature of a concave mirror, a real inverted image that is the same size as the object is produced at the radius of curvature by a concave mirror. When the object is the filament of a lamp, a concave mirror returns the rays going in the wrong direction so that the lamp will appear twice as bright.

itself and form an inverted image. If the light source were moved closer and closer to the focus of a concave mirror, the reflected rays would converge farther and farther away from the center of curvature. If the light source were placed at the focus, the reflected rays would form a beam parallel to the principal axis. This is the configuration used in searchlights.

As an alternative to drawing characteristic rays, we can determine where the reflected rays originating from a luminous or nonluminous object will converge to form an image with the aid of the following formula, which is known as the Gaussian lens equation:

$$1/s_o + 1/s_i = 1/f$$

where s_o is the distance from the object to the mirror (in m), s_i is the distance between the image and the mirror (in m), and f is the focal length of the mirror (in m). The transverse magnification (m_T), which is defined as y_i/y_o, is given by the following formula:

$$m_T = y_i/y_o = -s_i/s_o$$

where y_i and y_o are linear dimensions (in m) of the image and object, respectively. When using these formulae for concave and convex mirrors, the following sign conventions must be observed: s_o, s_i and f are positive when they are on the left of V, where V is the intersection of the mirror and the principal axis, and y_i and y_o are positive when they are above the principal axis. When the mirror is concave, the center of curvature is to the left of V and R is negative. When the mirror is convex, the center of curvature is to the right of V and R is positive. The analytical formulae used in geometric optics can be found in Menzel (1960) and Woan (2000).

When s_i is positive, the image formed by a concave mirror is real and when s_i is negative, the image formed by a concave mirror is virtual. The image is erect when m_T is positive and inverted when m_T is negative. The degree of magnification or minification is given by the absolute value of m_T. Let's have a little practice in using the preceding formulae:

- When an object is placed at infinity ($s_o = \infty$), $1/s_o$ equals zero, and thus $1/s_i = 1/f$ and $s_i = f$. In other words,

when an object is placed at an infinite distance away from the mirror, the image is formed at the focal point and the magnification ($-s_i/\infty$) is equal to zero.

- When an object is placed at the focus ($s_o = f$), $1/s_o = 1/f$. Then $1/s_i$ must equal zero and s_i is equal to infinity. In other words, when an object is placed at the focus, the image is formed at infinity, and the magnification ($-\infty/s_o$) is infinite.

- When an object is placed at the radius of curvature ($s_o = 2f$), then $1/s_o = 1/(2f)$. Then $1/s_i = 1/(2f)$, just as $\frac{1}{2} - \frac{1}{4} = \frac{1}{4}$. Thus $s_i = 2f$, and the image is the same distance from the mirror as the object is. The magnification ($-2f/2f$) is one, and the image is inverted.

- In any case where $s_o > f$, the image will be real and inverted.

What happens when the object is placed between the focus and the mirror? In this case the reflected rays diverge. These diverging rays appear to originate from behind the mirror. Thus a virtual image is formed. The virtual image will be erect. We can determine the nature of the image analytically:

- When an object is placed at a distance $\frac{1}{2}f$, then

$$2/f + 1/s_i = 1/f$$
$$1/s_i = 1/f - 2/f$$
$$1/s_i = -1/f$$
$$s_i = -f$$

Since s_i is a negative number, the image is behind the mirror. Since $(-(-f))/(\frac{1}{2}f)$ equals $+2$, the image is erect, virtual, and twice the height as the object.

Concave mirrors are spherical mirrors, which, by convention, have a negative radius of curvature. By contrast, convex mirrors are spherical mirrors with a positive radius of curvature (Figure 2-10). When a beam of light parallel to the principal axis strikes a convex mirror, the rays are reflected away from the principal axis, and therefore diverge. If we follow these rays backward, they appear to originate from a point behind the mirror. This point is the focus of the convex mirror. Since it is behind the mirror, it is known as a virtual focus and f is negative.

Using the law of reflection, we can determine the position, orientation, and size of the image formed by a convex spherical mirror. This can be done easily by drawing two or three characteristic rays using the following rules:

1. A ray traveling parallel to the principal axis is reflected from the mirror as if it originated from the focus.

2. A ray that travels toward the focus on the way to the mirror is reflected back parallel to the principal axis after striking the mirror.

3. A ray that strikes the point that is the intersection of the spherical mirror and the principal axis is reflected so that angle $<i = <r$ with respect to the principal axis.

4. A ray that strikes the mirror as it was heading toward the center of curvature is reflected back along the same path.

5. A real image of an object point is never formed. If we trace back the reflected rays to a point from where the extensions of each reflected ray seem to diverge, we will find the virtual image of the object point that originated the rays.

The focus of a convex mirror is negative and since an object must be placed in front of a convex mirror, where s_o is positive, to form an image, then it follows from the Gaussian lens equation, that s_i will always be negative. This means that the image formed by a convex mirror will always be virtual. Since $(-s_i/s_o)$ will always be positive, the virtual image formed by a convex mirror will always be erect. Moreover, since s_o is positive and $(1/s_o + 1/s_i)$ must be negative, then the absolute value of s_i must be smaller than the absolute value of s_o, and in all cases, the image formed by a convex mirror will be minified.

The Gaussian lens equation can also be used to determine the characteristics of an image formed by a plane mirror analytically. When light rays, parallel to the normal strike the mirror, they are reflected back along the normal and remain parallel. That is, they never converge and the focal length of a plane mirror is equal to infinity. Therefore $1/f$ is equal to zero and the Gaussian lens equation for a plane mirror becomes:

$$1/s_o + 1/s_i = 0$$

Since, for a plane mirror, s_i must equal $-s_o$, the image formed by a plane mirror will always be virtual, erect, and equal in size to the object. Table 2-1 summarizes the nature of the images formed by spherical reflecting surfaces for an object at a given location.

As long as we consider only the rays that emanate from a given point of an object and strike close to the midpoint of the mirror, we will find that these rays converge at a point. However, when the incident rays hit the mirror far from the midpoint, they will not be bent sharply enough and will not converge at the same point as the rays that strike close to the midpoint of the mirror. Thus even though all rays obey the law of reflection where $<i = <r$, with a spherical mirror, a zone of confusion instead of a point results. The inflation of a point into a sphere by a spherical mirror results in spherical aberration, from the Latin word *aberrans*, which means wandering. Even though spherical mirrors give rise to images with spherical aberration, they often are used because they are easy to make and are thus inexpensive. Francesca Maurolico (1611) and René Descartes (1637) found that spherical aberration could be eliminated by replacing a spherical mirror with a parabolic mirror. In contrast to a sphere, where the radius of curvature is constant, the radius of curvature at a point on a parabola increases as the distance between the point and the vertex increases. This relationship between radius of curvature and position on a parabolic ensures the elimination of spherical aberration. The rules used to

TABLE 2-1 Nature of Images Formed by Spherical Mirrors

Object					Image in a concave mirror
	Location	Type	Location	Orientation	Relative Size
$\infty > s_o > 2f$	Real	$f < s_i < 2f$	Inverted		Minified
$s_o = 2f$	Real	$s_i = 2f$	Inverted		Same size
$f < s_o < 2f$	Real	$\infty > s_i > 2f$	Inverted		Magnified
$s_o = f$			∞		
$s_o < f$	Virtual	$|s_i| > s_o$	Erect		Magnified

Object					Image in a convex mirror
	Location	Type	Location	Orientation	Relative Size
Anywhere	Virtual	$|s_i| < |f|$	Erect		Minified

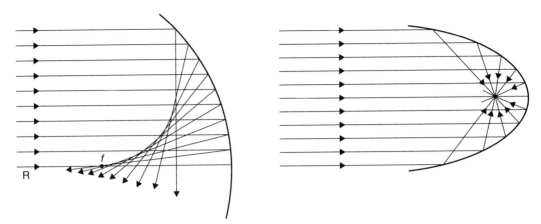

FIGURE 2-11 Because of the law of reflection, where the angle of reflection equals the angle of incidence, a spherical mirror does not focus parallel rays to a point, but instead produces a zone of confusion. This spherical aberration results because the rays that strike the distal regions of the mirror are bent too strongly to go through the focus. Spherical aberration can be prevented by gradually and continuously decreasing the radius of curvature of the distal regions of a concave mirror. Decreasing the radius of curvature gradually and continuously results in a parabolic mirror without any spherical aberration.

characterize images formed by a spherical mirror can also be used for characterizing the images formed by parabolic mirrors (Figure 2-11).

REFLECTION FROM VARIOUS SOURCES

Catoptrics is the branch of optics dealing with the formation of images by mirrors. The name comes from the Greek word *Katoptrikos*, which means "of or in a mirror." There are many chances to have fun studying image formation by mirrors. Plato in his *Timaeus*, Aristotle in his *Meteorologia*, and Euclid in his *Catoptrica* all describe various examples of reflection in the natural world. Hero of Alexander, who lived around 150 BC, wrote in his book, *Catoptrics*, about the enjoyment people have in using mirrors "to see ourselves inverted, standing on our heads, with three eyes and two noses, and features distorted as if in intense grief" just like in a fun house (Gamow, 1988). You can also have fun understanding the use of mirrors in image formation by studying kaleidoscopes (Brewster, 1818, 1858; Baker, 1985, 1987, 1990, 1993, 1999).

You can see your reflection in a plate glass window, in pots and pans, in either side of a spoon, and in pools of water. But don't get too caught up in studying reflections. Remember the story of Narcissus, the beautiful Greek boy, who never missed a chance to admire his own reflection? One day he saw his reflection in a cool mountain pool at the bottom of a precipice. Seeing how beautiful he was, he could not resist bending over and kissing his reflection. However, he lost his balance, fell over the precipice, and died. As a memoriam to the most beautiful human being that had ever lived on Earth, the gods turned Narcissus into a beautiful flower that, to this day, blossoms in the mountains in spring, and is still called *Narcissus*.

There is a close relationship between painting and geometrical optics (Hecht and Zajac, 1979; Summers, 2007). Jan Van Eyck painted the reflection, in a convex mirror, of *John Arnolfini and His Wife* in a painting by the same name. In *Venus and Cupid*, Diego Rodriguez de Silva y Veláquez painted Cupid holding a plane mirror so that Venus could look at the viewer. Edouard Manet painted a plane mirror that unintentionally did not follow the laws of geometrical optics in *The Bar at the Folies Bergères*, to give the viewer a more intimate feeling about the barmaid.

IMAGES FORMED BY REFRACTION AT A PLANE SURFACE

In ancient times, it was known already that the position of an image not only depended on the properties of opaque surfaces, but also depended on the nature of the transparent medium that intercedes between the object and the observer. In *Catoptrica*, Euclid explicitly stated as one of his six assumptions: "If something is placed into a vessel and a distance is so taken that it may no longer be seen, with the distance held constant if water is poured, the thing that has been placed will be seen again."

Claudius Ptolemy (150 AD) described a simple party trick, which would easily illustrate Euclid's sixth assumption (Figure 2-12). Ptolemy wrote in his *Theory of Vision* (Smith, 1996), that we could understand the

… breaking of rays… by means of a coin that is placed in a vessel called a baptistir. For, if the eye [A] remains fixed so that the visual ray [C] passing over the lip of the vessel passes above the coin, and if the water is then poured slowly into the vessel until the ray that passes over the edge of the vessel is refracted toward the interior to fall on the straight line extended from the eye to a point [C] higher than the true point [B] at which the coin lies. And it will be supposed not that the ray is refracted toward those lower objects but, rather, that the objects themselves are floating

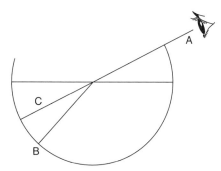

FIGURE 2-12 An observer (A) is looking over the rim of a dish so he or she can just not see a coin placed at B when the dish is full of air. As water is gradually added to the dish the rays coming from the coin are refracted at the water–air interface so that they will enter the eye. Consequently, the coin will become visible to the observer. Thinking that light travels in straight lines, the observer will think that the coin is at C.

and are raised up to meet the ray. For this reason, such objects will be seen along the continuation of the incident visual ray, as well as along the normal dropped from the visible object to the water's surface….

We see the coin suspended in the water and not at its true position at the bottom of a bowl because our visual system, which includes our eye and our brain, works on the assumption that light travels in straight lines. Consequently, we see the apparent position and not the true position of the coin. This brings up the question, what assumptions about light are made by our visual system when we look through a microscope?

Ptolemy's interest in the bending of light rays came from his deep interest in astrology. He knew that light had a big effect on plants for example, so it seemed reasonable to assume that the star light present at the time of one's birth would have a dramatic influence on a person's life (Ptolemy, 1936). Ptolemy knew, however, that since light bends as it travels through different media of different densities, he saw only the apparent positions of the stars, and not their true positions. Thus if he wanted to know the effect of star light on a person at the time of his or her birth, he must know the real position of the stars and not just the apparent positions he would observe after the rays of starlight were bent as they traveled through the Earth's atmosphere (Figure 2-13). Again, we "see" the star in the apparent position, instead of the real position because our visual system made up of the eyes and brain "believes" that light travels in a straight line, whether the intervening medium is homogeneous or not.

Another common example, according to Cleomedes (50 AD) where the assumption that light travels in straight lines gives us a misleading view of the world is when the mind "sees" a straight stick emerging from a water-air interface as bent. In order to understand the relationship between reality and the image, Ptolemy studied the relationship between the angle of incidence and the angle of transmission.

Ptolemy noticed that when light travels from one transparent medium to another it travels forward in a straight

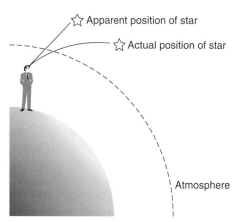

FIGURE 2-13 Rays from stars are refracted as they enter the Earth's atmosphere. Since we think that light travels in straight lines, we see the image of the star higher in the sky than it actually is.

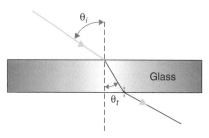

FIGURE 2-14 When light travels from air to glass it is bent or refracted toward the normal. By contrast, when light travels from glass to air, it is bent away from the normal. This behavior is codified by the Snell-Descartes law that states that the sine of the angle of incidence times the refractive index of the incident medium equals the sine of the angle of transmission times the refractive index of the transmission medium.

line, if and only if it enters the second medium perpendicular to the interface of the two media. However if the light ray impinges on the second medium at an angle greater than zero degrees relative to the normal, its direction of travel, although still forward, will change. This phenomenon is known as refraction and the rays are said to be refrangible.

When an incident light ray traveling through air strikes a denser medium (e.g., water or glass) at an oblique angle (θ_i) with respect to the normal, the ray is bent toward the normal in the denser medium (Figure 2-14). The angle that the light ray makes in the denser medium, relative to the normal, is known as the angle of transmission (θ_t). Ptolemy found that the angle of transmission is always smaller than the angle of incidence. He made a chart of the angles of incidence and transmission for an air-glass interface, but even though he knew trigonometry, he never figured out the relationship between the angle of incidence and the angle of transmission. Likewise, Vitello, Kircher, and Kepler also tried, but never discovered the relationship between the angle of incidence and the angle of transmission (Priestley, 1772).

The mathematical relationship between the angle of incidence and the angle of transmission was first worked out by Willebrord Snell in 1621 (Shedd, 1906). Snell, however, did not publish his work, and René Descartes, who independently worked out the relationship, first published the law of refraction in 1637. The Snell-Descartes Law states that when light passes from air to a denser medium, the ratio of the sine of the angle of incidence to the sine of the angle of transmission is constant. The Snell-Descartes Law can be expressed by the following equation:

$$\sin \theta_i / \sin \theta_t = n$$

where n is a constant, known as the refractive index. It is a characteristic of a given substance, and is correlated with its density. Descartes (1637) assumed that light, like sound, traveled faster in a more viscoelastic medium than in a lesser one. He wrote that light

> … was nothing else but a certain movement or an action, received in a very subtle material that fills the pores of other bodies; and you should consider that, as a ball loses much more of its agitation in falling against a soft body than, against one that is hard, and as it rolls less easily on a carpet than on a totally smooth table, so the action of this subtle material [light] can be much more impeded by the particles of air…than by those of water…. So that, the harder and firmer are the small particles of a transparent body, the more easily do they allow the light to pass; for this light does not have to drive any of them out of their places, as a ball must expel those of water, in order to find passage among them.

Isaac Newton read Descartes' work and after analyzing the refraction of light rays through media of differing densities with his newly developed laws of motion, Isaac Newton (1730) concluded that when light struck an interface between two media of different densities, the corpuscles of light were accelerated by the high density media such that the component of the velocity perpendicular to the interface, but not the component parallel to the interface, increased. Newton (Book II, Proposition X) assumed that the relative velocity of light could be determined by comparing the distance light traveled in the two media perpendicular to the interface at a given distance parallel to the interface from the point of incidence. Once the velocities were obtained, according to Newton, the attractive force in each medium could be determined by taking the square of the normal component of velocity in that medium. By assuming that the refractive index was the ratio of the force of attraction between the light corpuscles and the medium of transmission relative to the force of attraction between the light corpuscles and the medium of incidence and proportional to $v_{incident}^2 / v_{transmission}^2$, Newton could use his theory to obtain the known refractive indices of transparent media and to explain the cause of the refraction of light.

Newton's analysis led him to the conclusion that light travels faster in the denser medium than in the rarer medium, a conclusion that no one thought to test for approximately 150 years. Ultimately, Foucault showed that the speed of light is faster in rarer media than it is in denser media, a conclusion that was contrary to Newton's hypothesis. However, Foucault's data were consistent with the wave nature of light (see Chapter 3) and now we define the index of refraction according to the wave theory of light. That is, the index of refraction is now defined as the ratio of the velocity of light in a vacuum to the velocity of light in the medium in question. That is,

$$n_i = c/v_i$$

where n_i is the index of refraction of medium i (dimensionless), c is the speed of light (2.99792458×10^8 m/s which is almost equal to 3×10^8 m/s), and v_i is the velocity of light in medium i (in m/s). Table 2-2 lists the refractive indices of various media. As you can see from the following table, the refractive index of a substance is correlated with its density (in kg/m^3) and indeed, the refractive index depends on environmental variables like temperature and pressure that affect the density. The temperature coefficient of the refractive index is the amount the refractive index changes for each degree of temperature. The temperature coefficients of refractive indices are approximately 0.000001–0.00001 for solids and 0.0003–0.0009 for liquids (McCrone et al., 1984).

The law of refraction or the Snell-Descartes Law can be generalized to describe the bending of light by any two media by including both of their refractive indices:

$$n_i \sin \theta_i = n_t \sin \theta_t$$

where n_i and n_t are the refractive indices of the incident and transmitting medium, respectively.

What is the physical meaning of the index of refraction? The index of refraction is a dimensionless measure of the optical density of a material. The optical density is essentially the concentration of electrons that can absorb and reemit photons in the visible range. That is why the refractive index is correlated with the density of the substance. However, there is more to the optical density than the density of electrons since the optical density depends on the color (e.g., wavelength) of the light. That is, each

TABLE 2-2 Refractive Indices of Various Media (measured at 589.3 nm, which is the D line from a sodium vapor lamp)

Medium	n	Approximate density (kg/m^3)
Vacuum	1.00000	0
Air	1.00027	1.25
Water	1.3330	1000
Glass	1.515	2600
Diamond	2.42	3500

medium has a cross-section, given in units of area, that describes how much the atoms in it interfere with the forward motion of light of different colors. The greater the cross-section of the atoms for light of a given color, the greater the light is slowed down or bent by the atoms.

The absorption and subsequent reemission of photons in the visible light range takes approximately 10^{-15} s per interaction. Therefore light of a given wavelength traveling through a medium with a high index of refraction travels slower than light traveling through a medium with a lower index of refraction. The variation of refractive index with wavelength is known as dispersion. Glass makers go to great lengths varying the chemical composition of glass to produce transparent lenses with minimal dispersion (Hovestadt, 1902). The wavelength-dependence of the refractive index of glass and water is given in Table 2-3.

Dispersion by the glass that makes up a lens results in unwanted chromatic aberration. On the other hand, dispersion is desirable and welcome in prisms where it results in the separation of light by color (Figure 2-15).

Refraction causes an object that is immersed in a liquid to appear closer than it would if it were immersed in air (Clark, 1925; McCrone et al., 1984). To see the effect of refractive index on the apparent length, we can measure the actual height of a cover slip and the apparent height that it seems to have when light passes right through it. To compare the actual height with the apparent height of a cover slip, focus on a scratch on the top of a microscope slide and read the value of the fine focus adjustment knob (height a). Then place the cover slip over the scratch and take another reading of the fine focus adjustment knob (height b). Lastly, focus on a scratch on the top of the

cover slip and take a third reading of the fine focus adjustment knob (height c). The difference between (a) and (c) gives the actual height of the cover slip, and the difference between (b) and (c) gives the apparent height. This means that the fine focus adjustment knob, which is calibrated in micrometers/division for objects immersed in air, will not directly give the actual thickness of a transparent specimen if we focus on the top and bottom of it, but will give us only the apparent thickness due to the "contraction effect" of the refractive index.

This effect can be used to estimate the refractive index of a substance. I say estimate, because this technique is accurate only to within 5 to 10 percent of the refractive index. In Chapter 8, I will discuss a more accurate method to measure thickness using an interference microscope. The refractive index can be estimated from the following formula:

n = actual thickness/apparent thickness

When a light ray travels from a medium with a higher refractive index to a medium with a lower refractive index it is possible for the angle of refraction to be greater than 90 degrees. This means that the incident ray will never leave the first medium and enter the second medium (Figure 2-16). This is known as total internal reflection (Pluta, 1988). And since the rays undergoing reflection travel in the same medium as the incident rays, the Snell-Descartes Law reduces to the Law of Reflection, where $<i = <r$. The angle of incidence that will cause an angle of refraction of 90 degrees is called the critical angle. When θ_t is 90 degrees, the sine of θ_t equals one and the critical angle is given by the following formulae:

$$n_t/n_i = \sin \theta_i$$

or

$$\theta_i = \arcsin (n_t/n_i) = \sin^{-1}(n_t/n_i)$$

	486.1 nm (blue)	589.3 nm (yellow)	656.3 nm (red)
TABLE 2-3 Refractive Indices of Crown Glass, Flint Glass, and Water for Different Wavelengths			
Crown Glass	1.5240	1.5172	1.5145
Flint Glass	1.6391	1.6270	1.6221
Water	1.3372	1.3330	1.3312

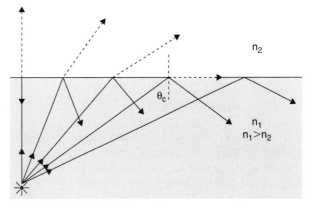

FIGURE 2-16 When a light ray travels from a medium with a higher refractive index to a medium with a lower refractive index, the angle of refraction can be greater than 90°, resulting in internal reflection. The critical angle θ_c is the incident angle that gives an angle of refraction of 90°.

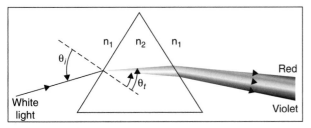

FIGURE 2-15 The refractive index of a medium is a function of the wavelength of light. This is the reason that a prism can disperse or resolve white light into its various color components.

TABLE 2-4 Critical Angles

Medium	Refractive index	Critical angle (degrees)
Water	1.3330	48.6066
Crown Glass	1.5172	41.2319
Flint Glass	1.6270	37.9249
Diamond	2.42	24.4075

Table 2-4 gives the critical angle for the air-medium interface for media with various refractive indices.

Diamonds are cut in such a way that the incoming light undergoes total internal reflection within the diamond. The increased optical path length allows the light to take maximal advantage of the dispersion of a diamond to break up the spectrum and give more "fire." Total internal reflection also redirects the light so that it is emitted in the direction of the observer. Prisms take advantage of total internal reflection to reorient light by ninety degrees. Fiber optic cables also take advantage of total internal reflection to transmit light down a cable from one place to another without any loss in intensity. The optical fibers within a bundle can be configured in parallel to transmit an image or arranged randomly to scramble and homogenize an image.

Total internal reflection is really a misnomer, since if we move another refracting medium within a few hundred nanometers of the air-medium interface, the light will jump from the first medium in which it was confined to the second refracting medium. The ability to jump across the forbidden zone is known as frustrated internal reflection and the waves that jump across are known as evanescent waves. The transfer of light trapped within a glass slide to an object of interest can be visualized in a total internal reflection microscope (TIRM; Temple, 1981). When combined with fluorescence microscopes (TIRFM), the evanescent wave can be used to visualize single molecules with high contrast (Axelrod, 1990; Steyer and Almers, 1997; Tokunaga and Yanagida, 1997; Gorman et al., 2007).

IMAGES FORMED BY REFRACTION AT A CURVED SURFACE

Glass that is curved into the shape of a lentil seed is known as a lens, the Latin word for lentil. Lenses typically are made of glass, a silicate of sodium and calcium, but optical glass may include oxides of lead, barium, antimony, and arsenic. According to Pliny (Nat Hist XXXVI: 190, in Needham, 1962), glass was discovered accidentally by Phoenician traders who needed something to prop up their cooking pots while they were camping on a sandy spot where the Belus River meets the sea. They used the bags of natron (sodium carbonate) they were carrying; the heat fused the sand (SiO_2) and the natron along with some lime (calcium carbonate) into small balls of glass. Glass manufacturing began in Mesopotamia some time around 2900 BC.

In ancient Greece, lenses were used for starting fires. Aristophanes (423 BC) wrote about the use of glass for starting fires in *The Clouds*:

Strepsiades. "I say, haven't you seen in druggists' shops
That stone, that splendidly transparent stone,
By which they kindle fire?"
Socrates "The burning glass?"
Strepsiades. "That's it: well then, I'd get me one of these,
And as the clerk was entering down my case,
I'd stand, like this, some distance towards the sun,
And burn out every line."

Not only have lenses been used to burn bills, but lenses have long been used to improve our ability to see the world. Using the laws of dioptrics, the study of refraction, inventors have been able to develop spectacles, telescopes, and microscopes. It is not clear who invented spectacles and when people began to wear them. Perhaps Roger Bacon made a pair in the thirteenth century. It is inscribed on a tomb, that Salvinus Armatus, who died in 1317, was the inventor of spectacles. In any case, by the mid sixteenth century, Francesco Maurolico (1611) already understood and wrote about how concave and convex lenses can be used to correct nearsightedness and farsightedness, respectively, and the time was ripe for the invention of telescopes and microscopes. It is thought that the children of spectacle makers playing with the lenses made by their fathers, including James Metius, John Lippersheim, and Zacharias Joannides (=Jansen), may accidentally have looked through two lenses at the same time and discovered that objects appeared large and clear. Perhaps such playing led to the invention of the telescope that Galileo (1653) used to increase our field of clear vision to Jupiter and Saturn.

Soon after the invention of the telescope, the microscope, which Robert Hooke (1665) used to extend our vision into the minute world of nature, was invented by Zacharias Jansen and his son, Hans. The priority of discovery is not certain: Francis Fontana claims to have invented the microscope in 1618, three years before the Jansens (Priestley, 1772).

Thus the extent of our vision has been increased orders of magnitude, thanks to a little grain of sand, the main component of glass lenses. William Blake (1757–1827) wrote:

To see a world in a grain of sand
And a heaven in a wild flower
Hold infinity in the palm of your hand
And eternity in an hour.

When a light ray passes from air through a piece of glass with parallel edges and returns to the air, the refraction at the far edge reverses the refraction at the near edge and the ray emerges parallel to the incident ray, although slightly displaced (Figure 2-17). The amount of displacement

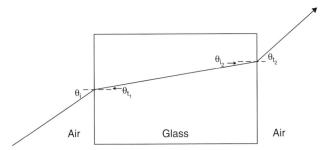

FIGURE 2-17 Light traveling from air through a piece of glass with parallel sides and back through air is slightly displaced compared with where the light would have been had it passed through only air. The degree of displacement depends on the thickness of the glass and its refractive index. The light that leaves the glass is parallel to the light that enters the glass because the refraction at the far side of the glass reverses the refraction at the near side.

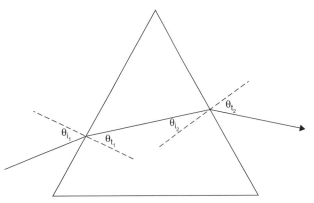

FIGURE 2-18 When the two surfaces of the glass are not parallel, but form a prism, the refraction that takes place on the far side does not reverse the effect of the refraction that takes place on the near side. The second refraction amplifies the first refraction and the incident light is bent toward the base of the prism.

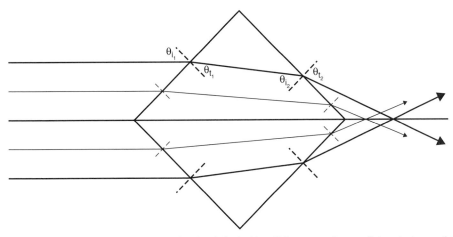

FIGURE 2-19 Two prisms, with their bases cemented together, bend the incident light propagating parallel to the bases of the prisms toward the bases. The prisms do not have the correct shape to focus parallel light to a point since the rays that strike the two corresponding prisms farther and farther from the principal axis will converge at greater and greater distances from the double prism.

depends on two things: the refractive index of the glass and the distance the beam travels in the glass. However, when the edges are not parallel, the refraction at the far edge will not reverse the effect of the refraction at the near edge. In this case, the light ray will not emerge parallel to the incident light ray, but will be bent in a manner that depends on the shape of the edges.

Consider a ray of light passing through a prism oriented with its apex upward (Figure 2-18). If the ray of light hits the normal at an angle from below, it crosses into the glass above the normal but makes a smaller angle with respect to the normal since the glass has a higher refractive index than the air. When the ray of light reaches the glass–air interface at the far side of the prism, it makes an angle with a new normal. As it emerges into the air it bends away from the normal since the refractive index of air is less than the refractive index of glass. The result is the ray of light is bent twice in the same direction.

What would happen to the incident light rays when they strike two prisms whose bases are cemented together (Figure 2-19)? Suppose that a parallel beam of light impinges on both prisms with an orientation parallel to the bases. The light that strikes the upper prism will be bent downward toward its base and the light that strikes the lower prism will be bent upward toward its base. The two halves of the beam of light will converge and cross on the other side.

The beam emerging from this double prism will not come to a focus since the rays that strike the two corresponding prisms farther and farther from the principal axis will converge at greater and greater distances from the double prism. However, imagine that the front and back surfaces of the prisms were smoothed out to form a "lentil-shaped" lens (Figure 2-20). Now suppose that a parallel beam of light impinges on the near edge of the glass. The light ray that goes through the thickest, center portion of the glass will enter parallel to the normal at that point and thus will go straight through the glass. Light rays that impinge on the glass just above this point will make a small angle with the normal and thus will be bent toward the axis. Light rays that impinge on the glass even higher up will make an even larger angle with the normal and thus will be bent even more toward the normal. This behavior continues as the parallel light rays impinge farther and farther from the axis. That is, as the parallel rays strike farther and farther from the axis, the rays are bent more and more toward the axis. The same is true for the light rays that strike the glass below the thickest point.

As the light rays reach the other side of the glass they will be bent away from the normal since they will be traveling from a medium with a higher refractive index to a medium with a lower refractive index. Thus, the light rays that travel through the thickest part of the glass will travel straight through since they make a zero degree angle with the normal. The imaginary line coincident with this ray is known as the principal axis. The rays that pass through the thinner part of the glass arrive at the glass–air interface at some angle to the normal. Thus, they will be refracted toward the principal axis when they emerge from the lens.

The further from the principal axis the rays emerge, the more they will be bent by the lens. Consequently, all the rays converge at one point known as the focus.

The surface of a lens can be convex, flat, or concave. Lenses can be biconvex, plano-convex, or concavo-convex (also called a meniscus lens). All these lenses are thickest at the center and thinnest at the edges, and thus, they typically act as converging lenses. Alternatively, lenses can be biconcave, plano-concave, or convexo-concave. All these lenses are thinnest at the center and thickest at the edges, and consequently they typically act as diverging lenses by causing the rays to diverge from the principal axis. Converging lenses and diverging lenses can act as diverging lenses and converging lenses, respectively; but only if their refractive index is smaller than the refractive index of the medium in which they are used (Figure 2-21).

The ability of a lens to bend or refract light rays is characterized by its focal length; the shorter the focal length, the greater the ability of the lens to bend light. The focal length is related to the radius of curvature of the lens, the refractive index of the lens (n_l), and the refractive index of the medium (n_m). The focal length of a lens is given by the lens maker's equation:

$$1/f = ((n_l/n_m) - 1)(1/R_1 - 1/R_2)$$

where R_1 is the radius of curvature of the first surface and R_2 is the radius of curvature of the second surface. For a biconvex lens, R_1 is right of V, the intersection of the lens with the principal axis, so R_1 is positive and R_2 is left of V so it is negative. For a biconcave lens, R_1 is negative and R_2 is positive. When one surface of the lens is planar, $R = \infty$ and $1/R = 0$. For a biconvex lens made of glass (n = 1.515) surrounded by air (n = 1), $1/f = 1.03/R$. That is, the focal length is approximately equal to the radius of curvature. The focal length of a plano-convex lens is approximately equal to half the radius of curvature.

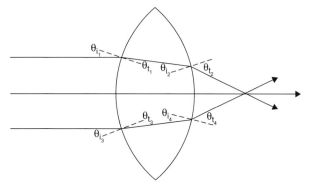

FIGURE 2-20 A lentil-shaped surface has the correct geometry to focus parallel rays to a point.

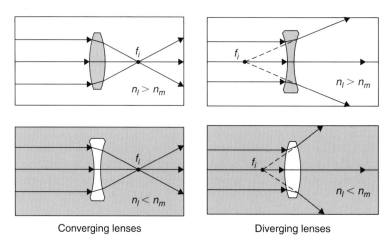

Converging lenses   Diverging lenses

FIGURE 2-21 A lentil-shaped biconvex piece of glass focuses parallel rays when the refractive index of the lens n_l is greater than the refractive index of the medium n_m. When the refractive index of the medium is greater than the refractive index of the lens, the lens must be biconcave to focus parallel rays. Whether a lens is converging or diverging is not a function of the lens alone but of the lens *and* its environment.

Every lens has a unique distance, called the object focal length (f_o). If an object is placed at this distance from a converging lens, the image will appear an infinite distance from the other side of the lens. In other words, if a point source of light is placed on the principal axis at f_o, a beam of light parallel to the principal axis will emerge from the other side of the lens.

Every lens has another unique distance called the image focal length (f_i). If an object is placed an infinite distance in front of a converging lens, the image will appear at the image focal length on the other side of the lens. In other words, if a bundle of light impinges on the lens parallel to the principal axis, it will converge on the principal axis, at a distance f_i from the lens. The focal planes are the two planes, which are parallel to the lens, perpendicular to the principal axis, and include a focal point.

Parallel rays emerging from a diverging lens appear to come from a source placed at the object focal point. Parallel rays impinging on a diverging lens appear to focus at the image focal point. Light diverges from a real object and converges toward a real image. By contrast, the light converges to a virtual object and diverges from a virtual image.

In order to determine where an image formed by a lens will appear, we can use the method of ray tracing and draw two or three characteristic rays. Remember:

- A ray that strikes a converging lens parallel to the principal axis goes through the focus (f_i).
- A ray that strikes a diverging lens parallel to the principal axis appears to have come from the focus (f_i).
- A ray that strikes a converging lens after it passes through the focus (f_o) emerges parallel to the principal axis.
- A ray that strikes a diverging lens on its way to the focus (f_o) emerges parallel to the principal axis.
- A ray that passes through the center of a converging or diverging lens (V) passes through undeviated.

Table 2-5 characterizes the type, location, orientation, and relative size of images formed by converging and diverging lenses. Note the similarity between the images formed by concave mirrors and converging lenses and the images formed by convex mirrors and diverging lenses.

Just as we could determine the characteristics of images formed by mirrors analytically, we can use the Gaussian lens equation and the magnification formula to determine the characteristics of images formed by lenses analytically.

$$1/f = 1/s_i + 1/s_o$$
$$m_T = y_i/y_o = -s_i/s_o$$

We must, however, know the sign conventions for lenses: s_o and f_o are positive when they are to the left of V (the intersection of the lens and the principal axis); s_i and f_i are positive when they are to the right of V; y_i and y_o are positive when they are above the principal axis; x_o is positive when it is to the left of f_o; x_i is positive when it is to the right of f_i; and R is positive when the center of curvature is to the right of V and negative when the center of curvature is to the left of V, above the principal axis.

When s_i is positive, the image formed by a spherical lens is real and when s_i is negative, the image formed by a spherical lens is virtual. The image is erect when m_T is positive and inverted when m_T is negative. The degree of magnification or minification is given by the absolute value of m_T. Notice the similarities between mirrors and lenses in the sign conventions.

Not only can we use the Gaussian lens equation to predict and describe the images formed by lenses, but if we know the relationship between the object and the image, we can use the Gaussian lens equation to determine the focal lengths of lenses. Next, I will derive the Gaussian lens equation from geometrical optics. Consider the following optical situation (Figure 2-22):

TABLE 2-5 Nature of Images Formed by Spherical Lenses

Object		Image formed by a converging lens		
Location	Type	Location	Orientation	Relative Size
$\infty > s_o > 2f$	Real	$f < s_i < 2f$	Inverted	Minified
$s_o = 2f$	Real	$s_i = 2f$	Inverted	Same size
$f < s_o < 2f$	Real	$\infty > s_i > 2f$	Inverted	Magnified
$s_o = f$		∞		
$s_o < f$	Virtual	$|s_i| > s_o$	Erect	Magnified
Object		Image formed by a diverging lens		
Location	Type	Location	Orientation	Relative Size
Anywhere	Virtual	$|s_i| < |f|$	Erect	Minified

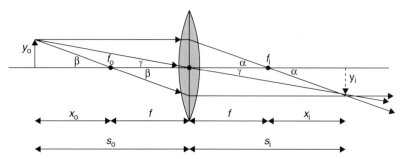

FIGURE 2-22 Image formation by a converging lens with focal length f. The object with height y_o and the image with height y_i are distances s_o and s_i from the lens, respectively. $x_o = s_o - f$ and $x_i = s_i - f$.

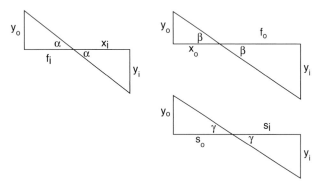

FIGURE 2-23 Three pairs of similar triangles made by the rays shown in Figure 2-22.

Look at the pairs of similar triangles (Figure 2-23). Remember that the tangent is the ratio of the length of the opposite side to the length of the adjacent side. Since tan $\alpha = y_o/f_i = y_i/x_i$, and tan $\beta = y_o/x_o = y_i/f_o$, then

$$y_i/y_o = x_i/f_i = f_o/x_o$$

From Figure 2-22, we see that:

$$s_o = f_o + x_o$$
$$s_i = f_i + x_i$$

Rearranging we get:

$$x_o = s_o - f_o$$
$$x_i = s_i - f_i$$

Since $x_i/f_i = f_o/x_o$, then

$$(s_i - f_i)/f_i = f_o/(s_o - f_o)$$

For a biconvex or biconcave lens $f_o = f_i = f$, therefore

$$(s_i - f)/f = f/(s_o - f)$$
$$f_2 = (s_o - f)(s_i - f)$$
$$f_2 = s_o s_i - s_o f - s_i f + f_2$$
$$f_2 - f_2 = s_o s_i - s_o f - s_i f = 0$$
$$s_o s_i = s_o f + s_i f = f(s_o + s_i)$$

Multiply both sides by $(1/f)$ $(1/s_o s_i)$

$$(s_o s_i)(1/f)(1/s_o s_i) = f(s_o + s_i)(1/f)(1/s_o s_i)$$

Cancel like terms.

$$1/f = (s_o + s_i)/s_o s_i$$
$$1/f = s_o/s_o s_i + s_i/s_o s_i$$

Cancel like terms again and we get:

$$1/f = 1/s_i + 1/s_o$$

which is the Gaussian lens equation.

Since tan $\gamma = y_o/s_o = y_i/s_i$, then $y_i/y_o = s_i/s_o$. The transverse magnification (y_i/y_o) is given by the following equation:

$$m_T = -s_i/s_o$$

The minus sign comes from including the vectorial nature of the distances and applying the sign conventions for lenses. A positive magnification means the image is erect, a negative magnification means the image is inverted.

The Gaussian lens equation is an approximation that applies only to "thin lenses." The equations used to determine the characteristics of an image made by a real or "thick lens" can be found in Hecht and Zajac (1974).

Most lenses used in microscopy are compound lenses; that is, they are composed of more than one refracting element (Figure 2-24). Microscope lenses include the eyepiece or ocular, the objective lens, the sub-stage condenser lens, and the collecting lens. We can use the ray tracing method to predict the type, location, orientation, and size of the image formed by compound lenses. When using the ray tracing method for compound lens, we follow the same rules as for a single, thin lens. When the two lenses are separated by a distance greater than the sum of their focal lengths we can assume that the real image formed by the first lens serves as a real object for the second lens.

In Figure 2-25, the compound lens system forms a real, erect, magnified image. Here I would like to introduce four new terms that characterize an optical system composed of more than one lens. The distance from the object focus to the first surface of the first optical element is called the front focal length. The front focal plane occurs at this distance, perpendicular to the principal axis. The back (or rear) focal length is the distance between the last optical

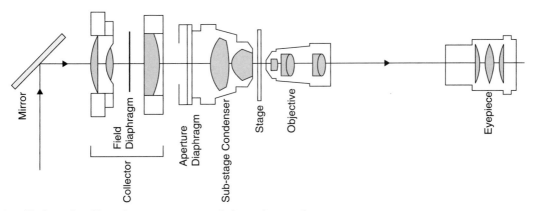

FIGURE 2-24 The lenses found in a microscope are composed of more than one element.

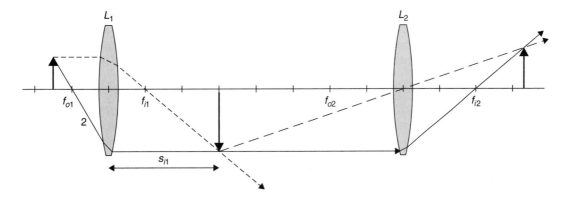

FIGURE 2-25 Two converging lenses that are separated by a distance greater than the sum of their focal lengths form a real erect image.

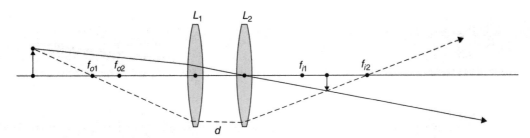

FIGURE 2-26 Image formation by two converging lenses separated by a distance smaller than either of their focal lengths.

surface and the second focal point (F_{i2}). The plane perpendicular to the principal axis that includes F_{i2}, is known as the back (or rear) focal plane.

Now let us consider the case of two thin lenses that are separated by a distance smaller than either of their focal lengths (Figure 2-26). How do we know where the image will be in this case? First consider ray 2; it goes through the focus (f_{o1}) of lens 1 and thus emerges parallel to the principal axis. Since it enters lens 2 parallel to the principal axis, it will pass through the focus of lens 2 (f_{i2}).

Now consider ray 1. It travels through the center of lens 2 and thus does not deviate—it is as if lens 2 were not there. However we do not know the angle that ray 1 makes as it goes through lens 1 or lens 2. So, imagine that lens 2 is not there and construct two characteristic rays: one that

strikes the lens after it passes through the focus (f_{o1}), and one that passes through the center of lens 1 (Figure 2-27). These two characteristic rays converge at P'_2.

Now we can easily construct ray 1 by tracing it backward from P'_2 through O_2 through L_1 to S_2. We can also draw ray 1 on the original figure where we easily drew ray 2 and see where ray 1 and ray 2 converge. This is where the image is. It is real, inverted, and minified.

Just like we can determine the characteristics of images formed by mirrors and single lenses analytically, we can determine the position of images formed by compound lenses analytically with the aid of the following equation:

$$s_i = \frac{f_2 d - [f_1 f_2 s_o / (s_o - f_1)]}{d - f_2 - [f_1 s_o / (s_o - f_1)]}$$

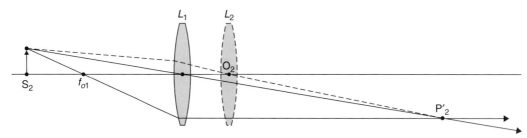

FIGURE 2-27 Finding the image produced by the first lens of the pair shown in Figure 2-26, if the second lens were not there and deducing the ray that would go through the center of the second lens if it were there.

where s_o is the usual object distance (in m), s_i is the usual image distance (in m), f_1 is the focal length of lens 1 (in m), f_2 is the focal length of lens 2 (in m), and d is the distance between the two lens (in m).

The total transverse magnification (m_T) of the optical system is given by the product of the magnification due to lens 1 and the magnification due to lens 2.

$$m_{Total} = (m_{T1})(m_{T2})$$

Thus, the first lens produces an intermediate image of magnification m_{T1}, which is magnified by the second lens by m_{T2}. The total transverse magnification can be determined analytically by the following equation:

$$m_{Total} = \frac{f_1 s_i}{d(s_o - f_1) - s_o f_1}$$

The image is erect when m_{Total} is positive and inverted when m_{Total} is negative. The front and back focal lengths of the compound lens is given by the following formulae:

$$front\ focal\ length = \frac{f_1(d - f_2)}{d - (f_1 + f_2)}$$

$$back\ focal\ length = \frac{f_2(d - f_1)}{d - (f_1 + f_2)}$$

When d = 0, the front and back focal lengths are equal, and the focal length (f) of the optical system is given by the following formula:

$$1/f = 1/f_1 + 1/f_2$$

where f_1 is the focal length of lens 1, and f_2 is the focal length of lens 2.

The dioptric power of a lens system (in m^{-1} or diopters) is defined as the reciprocal of the focal length (in m) and is given by the following formula:

$$D = 1/f$$

Therefore the total dioptric power of the optical system (D_{Total}) is given by the following formula:

$$D_{Total} = D_1 + D_2$$

where D_1 is the dioptric power of lens 1, and D_2 is the dioptric power of lens 2. The greater the dioptric power of a lens, the shorter its focal length and the more it bends light. The strength of spectacles often is given in diopters.

A compound lens can also be described by its f number, or f/#, where f/# is the ratio of the focal length to the diameter of the compound lens. The strengths of camera lenses often are given in f numbers. A compound lens, 50 mm in diameter, with a focal length of 200 mm, has an f/# of f/4. The f-stops on a camera or on a photographic enlarger are selected specifically so that every time you close down the lens by one stop, you decrease the light by one-half.

FERMAT'S PRINCIPLE

By now we know how to use the Snell-Descartes Law to predict the behavior of light rays through thin, thick, and compound lenses. Now we will ask why light follows the Snell-Descartes Law. According to Richard Feynman (Feynman et al., 1963), the development of science proceeds in the following manner: First we make observations; then we gather numbers and quantify the observations. Third, we find a law that summarizes all observations, and then comes the real glory of science: we find a new way of thinking that makes the law self-evident. The new way of thinking that made the Snell-Descartes Law evident came from Pierre de Fermat in 1657. It is known as the principle of least time, or Fermat's Principle. Fermat's Principle states that of all the possible paths that the light may take to get from one point to another, light takes the path that requires the shortest time.

First I will demonstrate that Fermat's Principle is true in the case of reflection in a plane mirror. Consider the following situation (Figure 2-28).

Which is the way to get from point A to point B in the shortest time if we say that the light must strike the mirror (MM′)? Remember, the speed of light (c) in a vacuum is a constant, so the distance light travels is related to the duration of travel by the following equation:

$$duration = distance/c$$

The light could go to the mirror as quickly as possible and strike at point D, but then it has a long way to go to get to point B. Alternatively, the light can strike the mirror at point E and then continue to point B. The time it takes to take this path is less than the time it takes to take

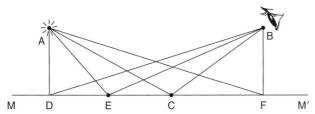

FIGURE 2-28 Why does the reflected light go from A to B by striking point C on the mirror instead of rushing to the mirror and striking it at D or rushing from the mirror after it strikes it at F?

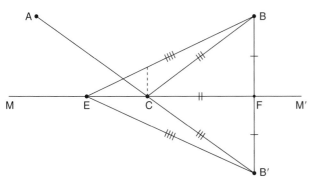

FIGURE 2-29 B and B′ are equidistant to the mirror. BC = B′C and EB = EB′.

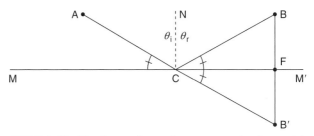

FIGURE 2-30 The shortest distance between two points is a straight line. Since AB′ is a straight line and AB′ = AB, the shortest distance between A and B is when <ACN = <BCN, or the angle of incidence equals the angle of reflection.

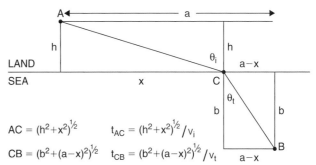

$$AC = (h^2+x^2)^{\frac{1}{2}} \qquad t_{AC} = (h^2+x^2)^{\frac{1}{2}}/v_i$$

$$CB = (b^2+(a-x)^2)^{\frac{1}{2}} \qquad t_{CB} = (b^2+(a-x)^2)^{\frac{1}{2}}/v_t$$

FIGURE 2-31 Using Fermat's Principle as the basis of the Snell-Descartes law, a = x + (a–x).

the first path; however, this still isn't the shortest path. Remembering that light travels at a constant velocity in a given medium, we can find the point where the light will strike the mirror in order to get to point B in the shortest possible time by using the following trick (Figure 2-29).

Consider an artificial point B′, which is on the other side of the mirror, and is the same distance below MM′ as point B is above it. Draw the line EB′. Because BFE is a right angle and since BF = FB′ and EF = EF, then EB is equal to EB′ (using the Pythagorean Theorem). Therefore the distance AE + EB is equal to AE + EB′. This distance is proportional to the time it will take the light to travel. Since the smallest distance between two points is a straight line, the distance AC + CB′ is the shortest line. How do we find where point C is? Point C is the point where the light will strike the mirror if it heads toward the artificial point B′.

Now we will use geometry to find point C (Figure 2-30). Since BF = B′F, CF = CF, and CB = CB′, FCB and FCB′ are similar triangles. Therefore <FCB = <FCB′. Also <FCB′ is equal to <ACM since they are vertical angles. Thus <FCB = <ACM. Since both <MCA and <ACN, and <FCB and <BCN are pairs of complementary angles, <ACN = 90° – <ACM and <BCN = 90°– <FCB. Since <FCB = <ACM, it follows that <ACN is equal to <BCN. Thus, the light ray that will take the shortest time to get from point A to point B by striking the mirror will make an angle where the angle of incidence equals the angle of reflection. Therefore, according to Fermat, the reason the angle of incidence equals the angle of reflection is because light takes the path that requires the shortest time.

Hero of Alexandria proposed in his book, *Catoptrics*, that light takes the path that is the shortest distance between

two points when it is reflected from a mirror. This explanation would also be valid for the example with the mirror given earlier; however, Hero's explanation could not be applied to refraction since the shortest distance between two points is a straight line and light bends upon refraction. This was the impetus for Fermat to come up with a principle that could be generalized for both reflection and refraction. Let us use Fermat's Principle to derive the Snell-Descartes Law (Figure 2-31). We must assume that the speed of light in water is slower than the speed of light in air by a factor, n.

According to Fermat's Principle we must get from point A to point B in the shortest possible time. According to Hero of Alexandria we must get from point A to point B by the shortest possible distance. The Snell-Descartes Law tells us that Hero cannot be right. Let's use an analogy to illustrate that Fermat's Principle will lead to the Snell-Descartes Law. Assume that your boyfriend or girlfriend fell out of a boat and he or she is screaming for help at point B. You are standing at point A. What do you do? You have to run and swim to the poor victim. Do you follow a straight line? (Yes, unless you use a little intelligence and apply Fermat's Principle first). If you think about it, you realize that you can run faster than you can swim, so it would be advantageous to travel a greater distance on land than in the sea. Where is point C, where you should enter the water?

The time that it will take to go from point A to point B through point C will be equal to:

$$t = t_{AC} + t_{CB} = \frac{(h^2 + x^2)^{1/2}}{v_i} + \frac{(b^2 + (a-x)^2)^{1/2}}{v_t}$$

where v_i is the speed of travel on land (in m/s), and v_t is the speed of travel in the sea (in m/s).

If ACB is the quickest path, then any other path will be longer. So if we graph the time required to take each path and plot these values against various points on the land/sea interface, then point C will appear as the minimum. Near point C the curve is almost flat. In calculus this is called a stationary value. In order to minimize t with respect to variations in x we must set dt/dx = 0. The minimum (and maximum) time is where dt/dx = 0. (A full derivation requires taking the second derivative, which will determine whether C is a minimum *or* a maximum. If the second derivative is positive, C is a minimum.)

$$dt/dx = \frac{d}{dx} \frac{(h^2 + x^2)^{1/2}}{v_i} + \frac{(b^2 + (a - x)^2)^{1/2}}{v_t} = 0$$

To solve this equation, use the chain rule. First differentiate the first term:

$$\frac{d}{dx} \frac{(h^2 + x^2)^{1/2}}{v_i} = (1/2) \frac{(h^2 + x^2)^{-1/2}}{v_i} (2x)$$
$$= \frac{(h^2 + x^2)^{-1/2}}{v_i} (x)$$

Use the chain rule for the second term:

$$\frac{d}{dx} \frac{(b^2 + (a - x)^2)^{1/2}}{v_t} = \frac{d}{dx} \frac{(b^2 + a^2 - 2ax + x^2)^{1/2}}{v_t}$$

$$\frac{d}{dx} \frac{(b^2 + a^2 - 2ax + x^2)^{1/2}}{v_t} =$$
$$\frac{(1/2)(b^2 + (a - x^2)^{-1/2}}{v_t} (2x - 2a)$$

Since $2x - 2a = -2a + 2x = 2(x - a)$, then

$$\frac{d}{dx} \frac{(b^2 + a^2 - 2ax + x^2)^{1/2}}{v_t} =$$
$$(1/2) \frac{(b^2 + a^2 - 2ax + x^2)^{-1/2}}{v_t} (2)(x - a)$$
$$= (x - a) \frac{(b^2 + (a - x)^2)^{-1/2}}{v_t}$$

Thus

$$dt/dx = (x) \frac{(h^2 + x^2)^{-1/2}}{v_i}$$
$$+ (x - a) \frac{(b^2 + (a - x)^2)^{-1/2}}{v_t} = 0$$

$$dt/dx = \frac{x}{v_i (h^2 + x^2)^{1/2}}$$
$$+ \frac{(x - a)}{v_i (b^2 + (a - x)^2)^{1/2}} = 0$$

and

$$\frac{x}{v_i (h^2 + x^2)^{1/2}} = \frac{-(x - a)}{v_t (b^2 + (a - x)^2)^{1/2}}$$
$$\frac{x}{v_i (h^2 + x^2)^{1/2}} = \frac{(a - x)}{v_t (b^2 + (a - x)^2)^{1/2}}$$

Since $\sin \theta_i = \frac{x}{(h^2 + x^2)^{1/2}}$ and

$$\sin \theta_t = \frac{(a - x)}{(b^2 + (a - x)^2)^{1/2}}$$

then

$$\frac{\sin \theta_i}{v_i} = \frac{\sin \theta_t}{v_t}$$

Multiply both sides by c.

$$c \frac{\sin \theta_i}{v_i} = c \frac{\sin \theta_t}{v_t}$$

Since $n_i = \frac{c}{v_i}$ and $n_t = \frac{c}{v_t}$

then

$$n_i \sin \theta_i = n_t \sin \theta_t$$

which is the Snell-Descartes Law.

I have just derived the Snell-Descartes Law using the assumption that, when light travels from point A to point B, it takes the path that gives the minimum transit time. It is clear that the whole beautiful structure of geometric optics can be reduced to a single principle: Fermat's Principle of Least Time.

OPTICAL PATH LENGTH

The optical path length (OPL) through a homogeneous medium is defined as the product of the thickness (s) of the medium and the refractive index (n) of that medium:

$$OPL = ns$$

If the medium is not homogeneous, but composed of many layers, each having a different thickness and refractive index (Figure 2-32), then the time it takes light to pass from the beginning of the first layer through the mth layer is given by the following formula:

$$t = (1/c) \sum_{j=1}^{m} n_j s_j = (1/c)\, OPL$$

and after rearranging,

$$OPL = \sum_{j=1}^{m} n_j s_j$$

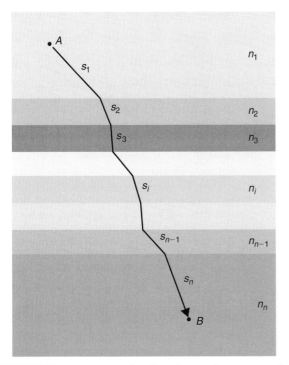

FIGURE 2-32 A ray propagating through a specimen composed of layers with various refractive indices and thicknesses has an optical path length. The optical path length differs from the length itself. The optical path length is obtained by finding the product of the refractive index and thickness of each layer and then summing the products for all the layers.

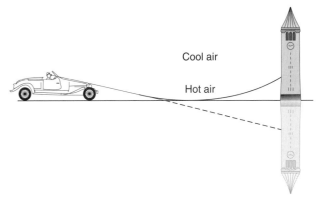

FIGURE 2-33 On a hot day, when we look down at the road ahead of us, we see the image of a tree or clouds on the road because light obeys Fermat's Principle and travels to our eyes in an arc. However, we think that light travels through a medium with a continuously varying refractive index in straight lines.

The optical path length (OPL, in m) is defined as the sum of the products of the refractive index of a given medium and the distance traveled in that medium. In a perfect lens, the optical path lengths of each and every ray emanating from a given point on the object and going to the conjugate point on the image are identical. That is, in a perfect lens, the optical path difference (OPD) between all rays vanishes. We can also restate Fermat's Principle by saying that light, in going from point A to point B, traverses the route having the shortest optical path length.

We can observe Fermat's Principle in the world around us. For example, when the sun begins to set, it looks like it is above the horizon. However, it is actually already below it. This is because the earth's atmosphere is rare at the top and dense at the bottom and the light travels faster through the rarer medium than through the denser medium. Thus the light can get to us more quickly if it does not travel in a straight line, but travels a short distance through the denser atmosphere and a longer distance through the rarer atmosphere. Since our visual system is hardwired to believe that light travels in straight lines, the rays from the setting sun appear to come from a position higher in the sky than the sun actually is.

Another everyday example of Fermat's Principle is the mirage we see when we are driving on hot roads (Figure 2-33). From a distance we see water on the road

but when we get there it is as dry as a desert. What we are really seeing is the skylight reflected on the road. How does the skylight reflected from the road end up in our eyes? The air is very hot and rarer just above the road and cooler and denser higher up. The light comes to our downward-looking eyes in the least amount of time by traveling the longest distance in the rarer air and the shortest distance in the denser air. Since our visual system is hardwired to believe that light travels in straight lines, the image of the sky appears on the road ahead of us.

For other examples of natural phenomena that can be explained by Fermat's Principle, see Minnaert (1954) and Williamson and Cummins (1983). And of course we see Fermat's principle every time we look through a camera lens, spectacles, a microscope, a telescope, or any instrument that has a lens.

LENS ABERRATIONS

In order to get a perfect image of a specimen, all the rays that diverge from each point of the object must converge at the conjugate point in the image. However a lens may have aberrations that cause some of the rays to wander (Gage and Gage, 1914). The rays will wander if the focal length varies for the different rays that come from each point on the object. Just as is the case for mirrors, spherical aberration occurs when using spherical lenses. Spherical aberration occurs because the rays from any given object point that hit the lens far from the principal axis are refracted too strongly (Figure 2-34). This results in a zone of confusion around each point in the image plane and a point is inflated into a sphere. Spherical aberration can be reduced by replacing a biconvex lens with two plano-convex lenses, or by using an aspherical lens. Lenses that have been corrected for spherical aberration are known as aspheric, aplanactic, achromatic, fluorite, and apochromatic, in order of increasing correction.

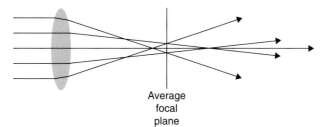

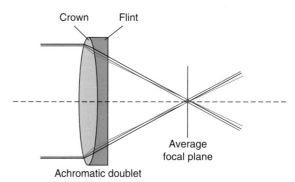

FIGURE 2-34 Spherical aberration occurs because the rays from any given object point that hit a lens with spherical surfaces far from the principal axis are refracted too strongly. This results in a circle of confusion. Spherical aberration can be reduced by grinding the lens so that it has aspherical surfaces.

FIGURE 2-36 Chromatic aberration can be reduced by combining a diverging lens made of flint glass with a converging lens made of crown glass. Because the flint glass has a greater dispersion than the crown glass, the chromatic aberration produced by the crown glass is reduced more than the magnification produced by the crown glass is reduced.

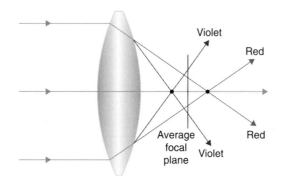

FIGURE 2-35 Chromatic aberration occurs because the refractive index of glass is color-dependent. This results in the violet-blue rays being more strongly refracted by glass than the orange-red rays.

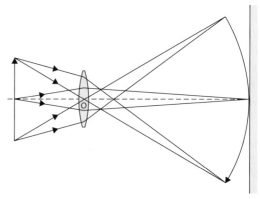

FIGURE 2-37 The image plane of a converging lens is not flat. Additional lens elements must be added to the converging lens to decrease the focal length of the image close to the axis.

Rays of every wavelength are focused to the same point by mirrors. However, since the refractive index of a transparent medium depends on wavelength, the lenses show chromatic aberration. That is, rays of different colors coming from the same point on the object disperse and do not focus at the same place in the object plane. Consequently, instead of a single image, multiple monochromatic images with varying degrees of magnification are produced by a lens with chromatic aberration (Figure 2-35). Newton believed that all transparent materials had an equal ability to disperse white light into colored light and therefore chromatic aberrations could not be corrected. However, Newton did not have sufficient observational data and John Dollond (1758) showed that by combining two materials with different dispersive powers, for example crown glass and flint glass, color-corrected lenses in fact could be made (Figure 2-36).

Lenses corrected for chromatic aberration are labeled achromatic, fluorite, and apochromatic, in order of increasing correction. These compound lenses are made by putting together a plano-concave lens made out of flint glass with a biconvex lens made out of crown glass, such that each lens cancels the chromatic aberration of the other one while still focusing the rays. Perhaps two types of plastic with complementary dispersion properties could be put together to make color-corrected lenses. Semiconductor technology has been used to lightly coat lenses with silicon

to correct for chromatic aberration. This technique is based on the principles of diffraction and not on typical geometric optics (Veldkamp and McHugh, 1992). This correction works because refraction causes blue light to be bent stronger than red, whereas diffraction causes red light to be bent stronger than blue.

In order to get a perfect image, all the light rays emanating from each point of the object, must arrive at the image plane by following equal optical path lengths. As we observed with the *camera obscura*, the only way this is possible is by curving the image plane (Figure 2-37). However, unlike our retina, film and silicon imaging chips are flat. Therefore if we want the image at the center of the field and the edge of the field to be in focus at the same time, we must use a lens that has been corrected for "curvature of field." Lenses that are corrected to have a flat field have the prefix F- or Plan-. A Plan-apochromat is the most highly corrected lens.

An image is a point-by-point representation of an object. We have learned how to determine the position, orientation, and magnification of an image formed by reflection in a mirror, or by refraction through a lens. We can

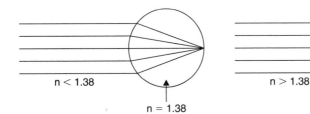

FIGURE 2-38 The lonely fungi obey the laws of geometric optics. In air, light is focused on the far side of the cylindrical sporangiophore, and the fungus bends toward the light source. In a high refractive index medium, light is dispersed over the far side of the sporangiophore and is brightest on the side closest to the light source. In this case, the fungus bends away from the light source.

determine the properties of an image graphically, using characteristic rays, or analytically, using the Gaussian lens equation. We have determined that light travels from the object to the image as if it were taking the path that takes the least time. As long as all the rays emanating from each point in an object have the same optical path length when they meet on a flat image plane, the image will be perfect. However, we see that lenses can have spherical and chromatic aberrations and they can cause curvature of field, which results in an image that is not a perfect point-by-point replica of the object. In Chapter 3, I will show that even if the lens system were perfect, the very nature of light would result in an imperfect image. In Chapter 4, I will show you how to select lenses that minimize these aberrations while using a microscope.

GEOMETRIC OPTICS AND BIOLOGY

The principles of geometric optics can be applied to many biological systems. Birds that catch fish in water and fish that catch insects in air must have an instinctive understanding, far better than our own, of the laws of refraction.

One of the best understood examples of a biological organism using geometric optics to complete its life cycle is the sporangiophore of the fungus *Phycomyces*. This cylindrical cell that makes up the sporangiophore typically bends toward sunlight (Castle, 1930, 1932, 1938, 1961, 1966; Dennison, 1959; Dennison and Vogelmann, 1989; Shropshire, 1959, 1963; Zankel et al., 1967). In 1918, Blaauw suggested that this cylindrical cell might act like a converging lens that focuses the light to the back of the cell (Figure 2-38) and that light stimulates growth on the so-called "dark side" resulting in the cell bending toward the light.

Buder (1918, 1920) and Shropshire (1962) varied the refractive index of the medium in which the cells grew and found that when the refractive index of the medium was less than the refractive index of the cell (1.38), parallel rays caused the cell to bend toward the light. However, when the refractive index of the medium was greater than that of the cell, parallel rays caused the cell to bend away from the light. This occurs because in the former case, the cell acts like a converging lens and the light is focused on dark side of the cell. In the latter case, the cell acts like a diverging lens and the light is focused more on the light side of the cell than on the dark side of the cell.

Plants also take advantage of geometric optics since the epidermal cells of many plants form lenses that focus the sun's light onto the chloroplasts (Vogelmann, 1993; Vogelmann et al., 1996) to enhance photosynthesis.

GEOMETRIC OPTICS OF THE HUMAN EYE

The two balls in our head, known as eyes are the interface though which we receive visual information about the rest of the world (Young, 1807; Huxley, 1943; Polyak, 1957; Gregory, 1973; Inoué, 1986; Ronchi, 1991; Park, 1997; Helmholtz, 2005). Information-bearing light enters our eyes through the convex surface of the cornea, a transparent structure with a refractive index of 1.377. The cornea, which is composed of cells and extracellular fibrous protein, acts like a converging lens (Figure 2-39). The rays refracted by the cornea pass through the anterior (between the cornea and the iris) and posterior (between the iris and the crystalline lens) chambers filled with a dilute salt solution known as the aqueous humor (n = 1.337) and are further refracted by the biconvex crystalline lens, which has an index of refraction of 1.42–1.47. The rays refracted by the crystalline lens pass through a jelly-like substance called the vitreous humor (n = 1.336) and come to a focus on the photosensitive layer on our retina that contains color-sensitive cones used in bright light and light-sensitive rods used in dim light. The image on the retina is inverted. Neurons transmit signals related to the inverted image from the retina to the visual cortex of the brain. The brain then interprets the image and makes an effigy of the object that we see with the mind's eye. In creating this effigy, the brain is able to make inverted images on the retina upright, but is not able to lower a coin covered with water, unbend a stick passing through a water–air interface, or place the setting sun beneath the horizon.

The blue, green, gray, amber, hazel, or brown part of the eye situated between the cornea and the crystalline lens is known as the iris. The iris is a variable aperture whose

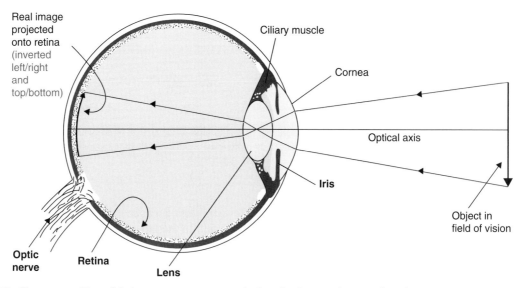

FIGURE 2-39 The cornea and lens of the human eye act as a converging lens that throws an image on the retina.

diameter varies between 2 mm and 8 mm, in bright light and dim light, respectively. The hole in the center of the iris through which light passes is called the pupil. The best images are produced when the pupil diameter is 3 to 4 mm because the central regions of the eye suffer from the fewest aberrations compared with the peripheral regions.

The length of the eye from the cornea to the retina is approximately 23 mm. Together, the optical elements of the eye act as a converging lens with a focal length of about 2 cm to project an image on the retina. The large difference in refractive index between the air and the cornea means that most (80%) of the refraction by the eye takes place at the cornea. The crystalline lens accounts for the other 20 percent.

When we look at distant objects, the ciliary muscles attached to the crystalline lens are relaxed, and if we have normal vision, an "in-focus" image of the distant object appears on the retina. When looking at objects up close, the ciliary muscles contract, causing the crystalline lens to become more rounded. This reduces the focal length of the crystalline lens to about 1.8 cm in a process known as accommodation so that "in-focus" images of the near objects appear on our retina (Peacock, 1855; Wood and Oldham, 1954; Robinson, 2006). In people over 40, a loss in the ability to accommodate, or presbyopia, occurs because the crystalline lens becomes too rigid and/or the ciliary muscles become too weak. Presbyopia can be corrected with the convex lenses placed in reading glasses or bifocals.

If our corneas are too convex (focal length = 1.96 cm), images of distance objects are not in focus on the retina, but closer to the crystalline lens, and we are nearsighted or myopic. Myopia can also result when the length of the

eyeball is too great. Nearsightedness can be corrected by wearing spectacles with diverging lenses or through laser surgery that reduces the convexity of the cornea.

If our corneas are too flat (focal length = 2.04 cm), images of near objects are not in focus on the retina, but farther away, and we are farsighted or hyperopic. Hyperopia can also result when the length of the eyeball is too short. Farsightedness can be corrected by wearing spectacles with converging lenses or through laser surgery that increases the convexity of the cornea.

If either the cornea or the lens is not spherical, but ellipsoidal, horizontal objects and vertical objects are brought to a focus at different image planes. This is known as astigmatism, and can be corrected through the use of cylindrical lenses or laser surgery.

Together the optical elements of the eye make up a compound converging lens that forms an image on the retina. When objects are placed at infinity, the eye, like any other converging lens, forms images at the focal plane, where the retina is located. When the objects are brought closer and closer to the converging lens of the eye, the image becomes more and more magnified; but it also forms further and further from the lens since the retina is a fixed distance from the lens, and the eye changes its focal length through accommodation and forms an image on the retina. However, the eye cannot accommodate without limit, and consequently there is a minimum distance that an object can be observed by the eye and still be in focus on the retina. This distance is known as the distance of distinct vision, the comfortable viewing distance, or the near point of the eye. It is approximately 25 cm in front of the eye.

When we look at an object from a great distance away, the rays emanating from the borders of the object make

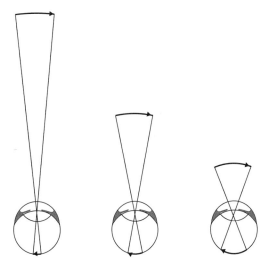

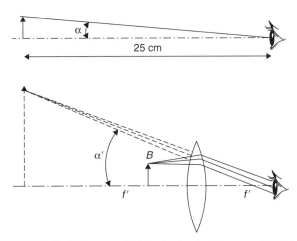

FIGURE 2-40 The closer an object is to our eye, the larger is the visual angle made by the object and the eye, and the larger the image is on the retina.

FIGURE 2-41 When a small object is placed at the near point of the naked eye, we still cannot see it clearly, because the visual angle is too small and the image falls on a single cone. A lens placed between the object and the eye produces an enlarged image on the retina. The size of the image on the retina is the same as that that would be produced by a magnified version of the object placed at the near point of our eye.

a miniscule angle at the optical center of the eye and the image of the object formed on the retina is minute (Figure 2-40). When the object is moved closer to the eyes, the rays emanating from the borders of the object make a larger visual angle and the image on the retina becomes slightly larger. When the object is placed 25 cm from the naked eye, the rays emanating from the object make an even larger visual angle at the optical center of the eye and consequently the image on the retina is even greater. However, if the object is microscopic, even if we bring it to the distance of distinct vision, we will not be able to see it because the visual angle will be too small to form a sizeable image on the retina. The visual angle must be at least 1 minute of arc (1/60 of a degree) for us to be able to form an image of the object on our retina.

When the visual angle is too small, the microscopic object can be observed through a microscope so that the rays will make a large visual angle at the optical center of the eye (Figure 2-41). With a microscope with 100× magnification, the image of the microscopic object on the retina will be as large as if the object were 100 times larger than its actual size and placed at the distance of distinct vision. The magnified apparent image that occurs at the distance of distinct vision is known as the virtual image.

The microscope is thus a tool that can be used to project a larger and more magnified image of an object on the

retina than would be projected in the absence of a microscope. According to Simon Henry Gage (1917, 1941):

> In considering the real greatness of the microscope and the truly splendid service it has rendered, the fact has not been lost sight of that the microscope is, after all, only an aid to the eye of the observer, only a means of getting a larger image on the retina than would be possible without it, but the appreciation of this retinal image, whether it is made with or without the aid of the microscope, must always depend upon the character and training of the seeing and appreciating brain behind the eye. The microscope simply aids the eye in furnishing raw material, so to speak, for the brain to work upon.

According to Sigmund Freud (1989), "With every tool, man is perfecting his own organs, whether motor or sensory, or is removing the limits to their functioning … by means of the microscope he overcomes the limits of visibility set by the structure of his retina."

WEB RESOURCES

I think you will enjoy the following web sites, which provide information, animations, and Java applets about geometric optics.

- http://www.educypedia.be/education/physicsjavalabolenses.htm
- http://www.educypedia.be/education/physicsjavacolor.htm
- http://www.educypedia.be/education/physicsjavalabooptics.htm
- http://hyperphysics.phy-astr.gsu.edu/hbase/ligcon.html

The Dependence of Image Formation on the Nature of Light

In the last chapter I discussed geometric optics and the corpuscular theory of light. In this chapter I will discuss physical optics and the wave theory of light. Geometric optics is sufficient for understanding image formation when the objects are much larger than the wavelength of light. However, when the characteristic length of the object approaches the wavelength of light, we need the principles of physical optics, developed in this chapter, to understand image formation.

CHRISTIAAN HUYGENS AND THE INVENTION OF THE WAVE THEORY OF LIGHT

Up until now, we have assumed that light travels as infinitesimally small corpuscles along infinitesimally thin rays. This hypothesis has been very productive; having allowed us to predict the position, orientation, and magnification of images formed by mirrors and lenses. In the words of Christiaan Huygens (1690):

> As happens in all the sciences in which Geometry is applied to matter, the demonstrations concerning Optics are founded on truths drawn from experience. Such are that the rays of light are propagated in straight lines; that the angles of reflexion and of incidence are equal; and that in refraction the ray is bent according to the law of sines, now so well known, and is no less certain than the preceding laws.

However, Huygens recognized a problem with the assumption that light was composed of material particles (Figure 3-1). He went on to say:

> … I do not find that any one has yet given a probable explanation of the first and most notable phenomena of light, namely why is it not propagated except in straight lines, and how visible rays, coming from an infinitude of diverse places, cross one another without hindering one another in any way.

Huygens realized that light must be immaterial since light rays do not appear to collide with each other. He concluded that light consists of the motion of an ethereal medium. Here is how he came to this conclusion: Fire produces light, and likewise, light, collected by a concave mirror, is capable of producing fire. Fire is capable of melting, dissolving, and burning matter, and it does so by disuniting the particles of matter and sending them in motion. According to the mechanical philosophy of nature championed by Descartes, anything that causes motion must itself be in motion, and therefore, light must be motion. Since two beams of light crossing each other do not hinder the motion of each other, the components of light that are set in motion must be immaterial and imponderable. The motion of the ether causes an impression on our retina, which results in vision much like vibratory motion of the air causes an impression on our eardrum, which results in hearing. Huygens considered that a theory of light and vision might have similarities to the newly proposed theory of sound and hearing (Airy, 1871; Millikan and Gale, 1906; Millikan and Mills, 1908; Millikan et al., 1920, 1937, 1944; Miller, 1916, 1935, 1937; Poynting and Thomson, 1922; Rayleigh, 1945; Helmholtz, 1954; Lindsay, 1960, 1966, 1973; Kock, 1965; Jeans, 1968).

Since hearing is important for communication and the ability to enjoy the aural world around us, the studies of sound and acoustics have been important to astute observers and inquisitive people since ancient times. It is a commonplace that there is a relationship between sound and vibration. Pythagoras (sixth century BC) noticed that the pitch of the sound emitted by a vibrating string was related to the length of the string—low pitches came from

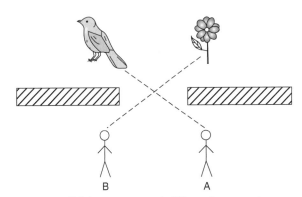

FIGURE 3-1 If light were composed of Newtonian corpuscles, corpuscles propagating from the bird to observer A should make it more difficult for observer B to see the flower, since the corpuscles from the flower will cause the corpuscles coming from the bird to scatter.

plucking long strings and high pitches came from plucking short strings. The long thick strings that produced low notes vibrated with low frequency and the short thin strings that produced high notes vibrated with high frequency. It seemed reasonable to the ancients that the vibrations produced by the strings were transmitted through air to the ear the way vibrations produced by stones thrown into water were visibly transmitted across water. We can realize how well developed the theory of acoustics must have been when visiting an ancient theater or amphitheatre and experiencing its fine acoustic qualities.

Marin Mersenne in 1625 and Galileo Galilei in 1638 independently published theories of sound, both of which stated that sound was caused by the vibration of strings and other bodies, and that the pitch of a note depended on the number of vibrations per unit time (i.e., the frequency). Galileo came to this conclusion after he rubbed the rim of a goblet filled with water and produced a sound—a sound that could be visualized by looking at the ripples produced in the water. He noticed that the number of ripples appearing on the surface of the water was related to the pitch of the sound. Sauver and John Wallis, the famous brain anatomist, independently noticed that when a string was plucked in various places along the length of the string, certain regions, known as nodes, did not move, while the rest of the string did. A single plucked string could produce a simple vibration, known as the fundamental vibration. It could also produce a complex vibration composed of a series of harmonic vibrations or overtones. The harmonics, which had higher pitches than the fundamental vibration, were characterized as having an integral numbers of nodes. A mathematical description of the movement of strings and their creation of various pitches was given in the eighteenth century by Daniel Bernoulli, Jean d'Alembert, and Leonhard Euler. Euler described the complex sound produced by a vibrating string as the superposition of the fundamental and harmonic vibrations.

The propagation of sound was also studied by the ancients, including the architect Vitruvius, who in the first century BC compared the propagation of sound with the propagation of water waves. In 1660, Robert Hooke discovered that sound can not travel in a vacuum when he found that the intensity of the sound of a bell, placed in a glass jar, decreased as he evacuated the air from the bell jar with a vacuum pump. Gassendi, Mersenne, and others assumed that the speed of light was infinite, and determined the speed of sound in air by shooting guns and measuring the time between when they saw the flash and when they heard the explosion.

The speed of sound through air, which depends on the temperature and wind speed, is approximately 340 m/s. In general, the speed of sound through a medium depends on the elasticity and density of that medium. By comparing the time it takes for sound to travel along long pipes with the time it takes sound to travel in air, in 1808, J. B. Biot and others showed that the velocity of sound is faster in liquids and solids than it is in air. Daniel Colladon and Charles Strum determined the elasticity (compressibility) of water in 1826 by measuring the speed of sound produced by a bell under the water of Lake Geneva.

The fact that the ear is involved in the reception of sound is well known and in 1650, Athanasius Kircher designed a parabolic horn as an aid to hearing. In 1843 Georg Ohm proposed that the ear can analyze complex sounds as many fundamental pitches in the manner that a complex wave can be expanded mathematically using Fourier's theorem. By 1865 Hermann von Helmholtz gave a complete theory of how the ear works and in 1961, Georg von Békésy won the Nobel Prize for showing how the ear works.

Making the comparison between the mysterious properties of light and the better understood properties of sound, Huygens (1690) went on to say:

> We know that by means of the air, which is an invisible and impalpable body, Sound spreads around the spot where it has been produced, by a movement which is passed on successively from one part of the air to another; and that the spreading of this movement, taking place equally rapidly on all sides, ought to form spherical surfaces ever enlarging and which strikes our ears. Now there is no doubt at all that light also comes from the luminous body to our eyes by some movement impressed on the matter which is between the two; since, as we have already seen, it cannot be by the transport of a body which passes from one to another. If, in addition, light takes time for its passage—which we are going to examine—it will follow that this movement, impressed on the intervening matter, is successive; and consequently it spreads, as sound does, by spherical surfaces and waves: for I call them waves from their resemblance to those which are seen to be formed in water when a stone is thrown into it, and which present a successive spreading as circles….

Since waves travel at a finite speed, it seemed before 1676 that wave theory was applicable to sound, but not to light since it appeared that the speed of light was infinite (Descartes, 1637). However, in 1676, Ole Römer, an astronomer who was studying the eclipses of Io, one of the moons of Jupiter, proposed that the speed of light was finite in order to make sense of the variation he observed in the duration of time between eclipses. Since the relative distance between Io and the Earth changed throughout the year as the planets orbited the sun, Römer realized that the discrepancies in his observations could be accounted for if light did not travel infinitely fast; but it took light several minutes to travel from Io to the Earth, the exact time depending on their relative distance. Although he never published any calculations on the speed of light, from his numbers, Huygens could calculate that the speed of light was approximately 137,879 miles per second—more than 600,000 times greater than the speed of sound.

With this new data in hand, Huygens concluded that light indeed travels as a wave (Figure 3-2). However, light waves and sound waves differ in many respects. For example, sound waves must be longer than light waves since sound waves result from the agitation of an entire body and light waves result from the independent agitation of each point on the body. Huygens wrote, "… the movement of the light must originate as from each point of the luminous object, else we should not be able to perceive all the different parts of that object." Moreover, "… the particles which engender the light ought to be much more prompt and more rapid than is that of the bodies which cause sound, since we do not see that the tremors of a body which is giving out a sound are capable of giving rise to light, even as the movement of the hand in the air is not capable of producing sound."

In order to transmit light from every point of an object, the ether had to be composed of extremely small particles. Moreover, Huygens knew that the speed of sound depended on the compressibility (elasticity) and penetrability (density) of the medium through which it moved. The velocity of a sound wave or any mechanical wave is equal to the square root of the ratio of the elasticity to the density. He thus assumed that the speed of light would also depend on the elasticity and density of the ether and consequently, the ether would have to be very elastic and not very dense. He imagined what the properties of the ether would be like (Figure 3-3):

> When one takes a number of spheres of equal size, made of some very hard substance, and arranges them in a straight line, so that they touch one another, one finds, on striking with a similar sphere against the first of these spheres, that the motion passes as in an instant to the last of them, which separates itself from the row, without one's being able to perceive that the others have been stirred. And even that one which was used to strike remains motionless with them. Whence one sees that the movement passes with an extreme velocity which is the greater, the greater the hardness of the substance of the spheres.

He went on to say, "Now in applying this kind of movement to that which produces light there is nothing to hinder us from estimating the particles of the ether to be a substance as nearly approaching to perfect hardness and possessing a springiness as prompt as we choose." The ether had some unusual properties. Not only must it be extremely elastic, but it also must be very thin, since it could penetrate through glass. This conclusion came from an experiment done by Galileo's assistant, Evangelista Torricelli, who showed that light could penetrate through a tube from which the air had been exhausted. It is ironic that, in order to circumvent the problems he found associated with the corpuscular nature of light, Huygens had to propose that the medium through which light traveled was corpuscular.

Huygens concluded that each minute region of a luminous object generates its own spherical waves, and these waves, which continually move away from each point, can be described, in two dimensions, as concentric circles around each originating point. The elementary waves

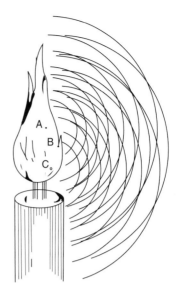

FIGURE 3-2 According to Christiaan Huygens, light radiates from luminous sources as waves. The waves must have small wavelengths since we can resolve points A, B, and C in the candle.

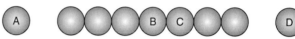

FIGURE 3-3 According to Huygens, in order for light to travel so fast, the aether must be composed of highly elastic diaphanous particles. The particles must be smaller than the minimum distance between two clearly visible points. The motion from sphere A is passed to sphere D without any perceptible change in the intervening spheres B and C.

emanating from each particle of a luminous body travel through the ether, causing a compression and rarefaction of each ethereal particle. The excitation of each particle causes each of them to emit a spherical wave that excites the neighboring ethereal particles. Consequently, each ethereal particle can be considered as a source of secondary wavelets whose concentric circles coincide with those of the primary wave. The wavelets that coincide with each other reinforce each other, producing an optically effective wave. The optically effective wave can be demonstrated by drawing a circle that envelopes the infinite number of circles centering about each point on the original front (Mach, 1926). This envelope, which overlies the parts of the wavelets that coincide with each other, represents the optically effective front, which then serves as the starting line for another infinite number of points that act as the source of an infinite number of spherical waves (Figure 3-4). Thus the propagation of light is a continuous cycle of two transformations—one involving the fission of the primary wave into numerous secondary wavelets, and the other involving the fusion of the secondary wavelets into a new primary wave.

Huygens' wave theory of light described the manner in which light spread out from a point source and provided a geometrical reason why the intensity of light decreased

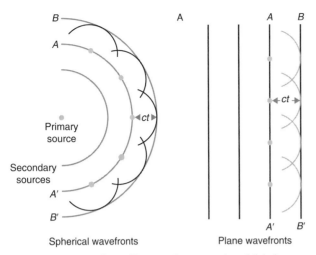

FIGURE 3-4 According to Huygens, the propagation of light is a continuous cycle of two transformations—one involving the fission of the primary wave into numerous secondary wavelets, and the other involving the fusion of the secondary wavelets into a new primary wave.

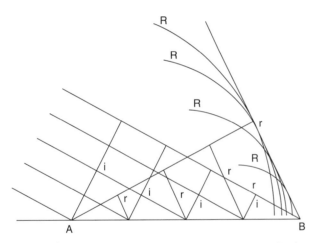

FIGURE 3-5 The wave theory can explain the law of reflection. According to Huygens, a wave front incident on a mirror produces secondary wavelets. By the time the last part of the incident wave strikes the mirror at B and begins producing secondary wavelets, the wavelets initiated by the portion of the wave that first struck the mirror already at A have produced many secondary wavelets. The secondary wavelets (R) formed from each consecutive region of the mirror reinforce each other (r), leading to a wave front that propagates away from the mirror so that the angle of reflection equals the angle of incidence. The incident wavelets are represented with i's and the reflected wavelets are represented with r's.

with the square of the distance. Moreover, since each particle in the ether could simultaneously transmit waves coming from different directions, Huygens' wave theory of light, as opposed to Newton's corpuscular theory, could explain why light rays cross each other without hindering one another. "Whence also it comes about that a number of spectators may view different objects at the same time through the same opening, and that two persons can at the same time see one another's eyes."

Newton's corpuscular theory of light provided a clear explanation of why light traveled in straight lines, but it was not so clear how light, if it were wave-like, could propagate in straight lines. Huygens realized that spherical waves can be approximated by plane ways as they travel far from the source, so he could provide some explanation for why light traveled in straight lines. Huygens imagined that when a spherical wave approached a boundary with an opening, the secondary wavelets that passed through the aperture would align and reinforce each other so that the majority of the wave traveled straight through the opening and the few waves that bent around the edges of the boundary would be "too feeble to produce light there." As we will see later in this chapter, Huygens explanation is accurate only in cases where the characteristic length of the opening is much larger than the wavelength of light.

In order to explain reflection using Huygens' Principle, imagine a wave front that impinges on a mirror where the incident front makes an angle relative to the normal. The first part of the wave to hit the mirror begins to produce secondary wavelets, and then the next part of the wave to hit the mirror produces secondary wavelets. By the time the last part of the incident wave strikes the mirror and begins producing secondary wavelets, the wavelets initiated by the portion of the wave that first struck the mirror already have produced many secondary wavelets. The secondary wavelets formed from each consecutive region of the mirror reinforce each other, leading to a wave front that propagates away from the mirror at the same angle as the incident wave approached (Figure 3-5).

Huygens used the wave theory of light to explain refraction by assuming that "the particles of transparent bodies have a recoil a little less prompt than that of the ethereal particles ... it will again follow that the progression of the waves of light will be slower in the interior of such bodies than it is in the outside ethereal matter." Thus, the secondary wavelets formed in the refracting medium in a given period of time will be smaller than the secondary wavelets produced in air (Figure 3-6). The first part of the incident wave to strike the refracting surface produces secondary wavelets, and then the next part of the wave to hit the refracting surface produces secondary wavelets. By the time the last part of the incident wave strikes the refracting surface and begins producing secondary wavelets, the wavelets initiated by the portion of the wave that first struck the refracting surface already have produced many secondary wavelets. The secondary wavelets formed from each consecutive region of the refracting surface reinforce each other, leading to a wave front that propagates through the refracting medium at an angle that follows the Snell-Descartes law as well as Fermat's Principle of least time.

Huygens' wave theory could also explain the action of lenses on light. Imagine a plane wave emanating from a distant source striking a biconvex converging lens. Since the speed of light is lesser in the glass than in the air, the peripheral region of the wave travels faster than the central portion and the waves emerging from the lens are concave instead of planar. The converging lens transforms the waves from no curvature to positive curvature. The concave

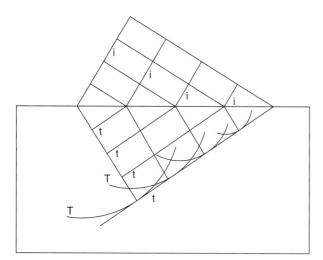

FIGURE 3-6 The wave theory can explain the law of refraction. Huygens believed that the waves traveled slower in media with higher refractive indices. This resulted in the wavelets forming closer together in the medium with a higher refractive index and the consequent bending of the wave toward the normal. The secondary wavelets (T) formed from each consecutive region in the transmitting medium reinforce each other (t), leading to a wave front that propagates into the transmitting medium at an angle consistent with the Snell-Descartes Law.

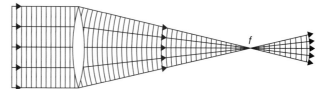

FIGURE 3-7 Because light travels more slowly through glass than through air, a converging lens converts a plane wave to a spherical wave. To visualize how a converging lens transforms spherical waves into plane waves, imagine the source being placed at f, and then reverse the direction of all the arrows to the left of f.

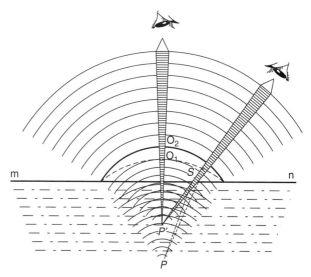

FIGURE 3-8 The wavelength of a light wave increases as the light passes from water to air. This causes the waves to have a greater wavelength and curvature when in the air. Since we do not realize that the wavelength and curvature change, we imagine that the object that produced the image on our retina is along a straight line, perpendicular to the wave front that enters our eyes. This is the reason we see objects in water as being at P' instead of at P and why objects appear closer to us.

waves leaving the lens come to a point at the focus (Figure 3-7). Now imagine the reverse situation with convex waves emanating from the focus and striking the biconvex lens. Again the peripheral region of the wave travels faster than the central portion of the wave. Thus the waves that emerge from the lens are planar. The converging lens transforms the waves from negative curvature to no curvature.

In general, converging lenses (as well as concave mirrors), which have a positive radius of curvature, increase the curvature of the incident waves. By contrast, diverging lenses (as well as convex mirrors), which have a negative radius of curvature, and which transform plane waves to convex waves and concave waves to plane waves, decrease the curvature of the incident waves (Millikan et al., 1944). We can relate geometrical optics to wave optics by drawing lines that run perpendicular to the wave fronts. Each line normal to the wave front represents a ray that radiates out from the source and is composed of corpuscles. The infinitesimal thinness of the rays is an approximation and an abstraction as is the infinite lateral extension of plane waves. Waves that have a finite wave width are intermediate between the two pictures of light and could represent the radial extensions or widths of the corpuscular rays.

The wave theory of light can also explain why objects viewed through inhomogeneous media appear to be in a different place than they actually are (Figure 3-8). Imagine a coin placed on the bottom of a dish of water. According to the wave theory of light, the speed of light is slower in water than it is in air, and consequently the spherical waves emanating from each point in the coin are compressed. That is, the wavelength in water is shortened relative to the wavelength in air. When the waves hit the water–air interface, they decompress. That is, the wavelength becomes

longer and the wave hits a point o_2 instead of the point o_1, which it would have hit if the waves traveled the same speed in air as they do in water. Because of the air, the waves now have greater curvature than they would have had, had the wavelength not changed at the surface. Apparently our visual system is hardwired to believe that light travels as plane waves at the speed of light in air (which is negligibly different than the speed of light in a vacuum). For this reason, we see the coin floating at P' instead of at its true location P. In his treatise, Huygens showed how the wave theory can explain the apparent position of the setting sun and other visual illusions.

THOMAS YOUNG AND THE DEVELOPMENT OF THE WAVE THEORY OF LIGHT

The wave theory was as good as the corpuscular theory in describing reflection and refraction, but it did not provide a satisfying description of shadows. Isaac Newton (1730) thought that if light were a wave, then it should be able to bend behind obstacles just as sound waves and water waves are able to bend behind obstacles. Newton did not realize that although all waves have similar behaviors, the specific behaviors of waves depend on the relative dimensions of the wave (e.g., their wavelength) compared with the characteristic lengths of the structures they encounter.

Casual inspection shows that water waves that would bend behind a stick will not bend behind an ocean liner. Science, however, advances when casual inspection is supplemented with attention to detail. Newton did not apply his usually powerful observational and analytical powers to understand the importance of relative lengths when it came to waves. Indeed, when we look at the edges of an ocean liner, we see that the water waves do bend around it. Likewise when we look at the edges of an opaque object, as Franciscus Maria Grimaldi (1665) did, we see that light waves bend or diffract around the object.

Grimaldi noticed that the shadow formed by a small opaque body (FE) placed in a cone of sunlight that had entered a room though a small aperture was wider (MN) than it would be (GH) if light propagated in straight lines (Figure 3-9). Grimaldi found that the anomalous shadow was composed of three parallel fringes. Grimaldi assumed that these fringes were caused by the bending of light away from the body. He called this effect, which differed from reflection and refraction, the diffraction of light from the Latin *dis*, which means "apart" and *frangere*, which means "to break" (Meyer, 1949). Grimaldi noticed that the fringes disappeared when he increased the diameter of the aperture that admitted the light, and when the summer Italian sun was bright enough, he could distinguish that the fringes between M and N were brightly colored (Priestley, 1772; Mach, 1926). Grimaldi also observed that light striking a small aperture cut in an opaque plate illuminated an area that was greater than would be expected if light traveled in straight lines.

Robert Hooke (1705) and Isaac Newton (1730) repeated and extended Grimaldi's observations on diffraction by studying the influence of a human hair on an incident beam of light (Figure 3-10). Newton noticed that the hair cast a shadow on a white piece of paper that was wider than would have been expected given the assumption of the rectilinear propagation of light. Newton concluded that the shadow was broadened because the hair repelled the corpuscles of light with a force that fell off with distance.

Newton also observed the shadow cast on a piece of white paper, by a knife-edge illuminated with parallel rays. Newton noticed that the image of the knife-edge was not sharp, but consisted of a series of light and dark fringes. Then he placed a second knife parallel to the first so as to form a slit. As he decreased the width of the slit, the fringes projected on the white paper, moved further and further from the bright image of the slit. By comparing the position of the fringes formed at various distances from the

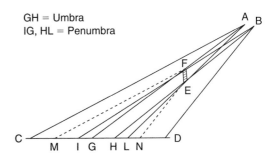

GH = Umbra
IG, HL = Penumbra

FIGURE 3-9 Grimaldi saw that the shadow formed by a small opaque body was larger than it should be if light traveled only in straight lines. He noticed that the additional shadow was composed of colored fringes.

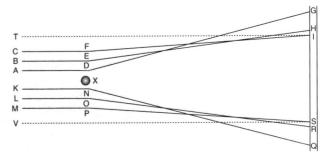

FIGURE 3-10 Isaac Newton noticed that the shadow of a hair (x) was larger than would be expected if light traveled in straight lines. He concluded that the broadened shadow occurred because the hair exerted a repulsive force on the corpuscles that fell off with distance. Newton did not see any light fringes inside the geometrical shadow.

slit, Newton concluded that the corpuscles repelled by the knife-edges followed a hyperbolic path. Using monochromatic light, Newton noticed that the fringes made in red light were the largest, those made in violet light were the smallest, and those made in green light were intermediate between the two.

After performing experiments with results that could not be explained easily with his corpuscular theory, Newton (1730) ended the experimental portion of his *Opticks* with the following words:

> When I made the foregoing Observations, I design'd to repeat most of them with more care and exactness, and to make some new ones for determining the manner how the Rays of Light are bent in their passage by Bodies, for making the Fringes of Colours with the dark lines between them. But I was then interrupted, and cannot now think of taking these things into farther Consideration. And since I have not finish'd this part of my Design, I shall conclude with proposing only some Queries, in order to a farther search to be made by others.

In the queries at the end of the book, Newton (1730) wondered:

> Are not the Rays of Light in passing by the edges and sides of Bodies, bent several times backwards and forwards, with a motion like that of an Eel? and Do not several sorts of Rays make Vibrations of several bignesses, which according to their

bignesses excite Sensations of several Colours, much after the manner that the Vibrations of the Air, according to their several bignesses excite Sensations of several Sounds? And particularly do not the most refrangible rays excite the shortest Vibrations for making a Sensation of deep violet, the least refrangible the largest for making a Sensation of deep red ...?

He went on to ask, "And considering the lastingness of the Motions excited in the bottom of the Eye by Light, are they not of a vibrating nature?"

Newton went on to conclude that light was not a wave, but was composed of corpuscles that traveled through an ether that could be made to vibrate. The vibrations were equivalent to periodic changes in the density of the ether; and these variations put the light corpuscles into "easy fits of reflection or transmission." Newton could not believe that light was a wave because he felt that if light in fact did travel as a wave, it should not only bend away from an opaque object, but it should also bend into the geometrical shadow.

As a consequence of the great achievements of Isaac Newton and the hagiographic attitude and less than critical thoughts of the followers of this great man, the corpuscular theory of light predominated, and the wave theory of light lay fallow for almost 100 years. The wave theory was revived by Thomas Young (1794, 1800, 1801), a botanist, a translator of the Rosetta stone, and a physician who was trying his hand at teaching Natural Philosophy at the Royal Institution (Peacock, 1855). While preparing his lectures, Young reviewed the similarities between sound and light, and reexamined the objections that Newton had made to the wave theory of light. Young, who studied the master, not the followers, concluded that the wave theory in fact could describe what happens to light when it undergoes diffraction as well as reflection and refraction. Here is how Young (1804a) came to this conclusion:

> I made a small hole in a window-shutter, and covered it with a piece of thick paper, which I perforated with a fine needle. For

greater convenience of observation, I placed a small looking glass without the window-shutter, in such a position as to reflect the sun's light, in a direction nearly horizontal, upon the opposite wall, and to cause the cone of diverging light to pass over a table, on which were several little screens of card-paper. I brought into the sunbeam a slip of card, about one-thirteenth of an inch in breadth, and observed its shadow, either on the wall, or on other cards held at different distances. Besides the fringes of colours on each side of the shadow, the shadow itself was divided by similar parallel fringes, of smaller dimensions, differing in number, according to the distance at which the shadow was observed, but leaving the middle of the shadow always white.

That is, Young observed something that Newton had missed. Young noticed that the light in fact did bend into the geometrical shadow of the slip of card (Figure 3-11). Young went on to describe the origin of the white fringe in the middle of the geometrical shadow:

> Now these fringes were the joint effects of the portions of light passing on each side of the slip of card, and inflected, or rather diffracted, into the shadow. For, a little screen being placed a few inches from the card, so as to receive either edge of the shadow on its margin, all the fringes which had before been observed in the shadow on the wall immediately disappeared, although the light inflected on the other side was allowed to retain its course, and although this light must have undergone any modification that the proximity of the outer edge of the card might be capable of occasioning. When the interposed screen was more remote from the narrow card, it was necessary to plunge it more deeply into the shadow, in order to extinguish the parallel lines; for here the light, diffracted from the edge of the object, had entered further into the shadow, in its way towards the fringes. Nor was it for want of a sufficient intensity of light, that one of the two portions was incapable of producing the fringes alone; for, when they were both uninterrupted, the lines appeared, even if the intensity was reduced to one-tenth or one-twentieth.

Young shared his ideas and results with Francois Arago, who told him that Augustin Fresnel had also been doing similar experiments on diffraction (Arago, 1857). Subsequently a fruitful collaboration by mail ensued

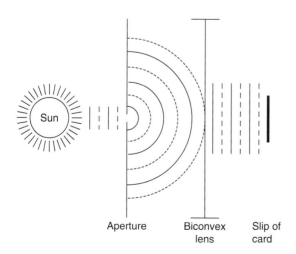

Sun   Aperture   Biconvex   Slip of
                                lens   card

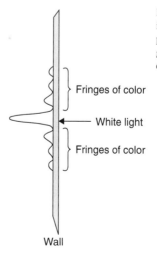

FIGURE 3-11 Thomas Young illuminated a slip of card with parallel light and observed light fringes in the geometrical shadow of the card.

Fringes of color

White light

Fringes of color

Wall

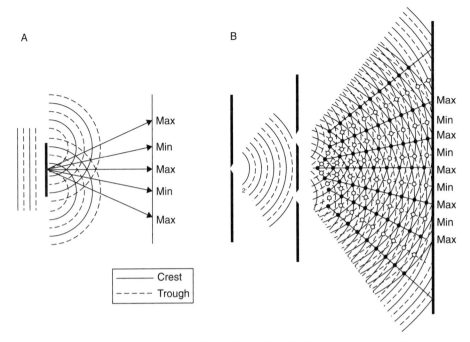

FIGURE 3-12 (A) According to Huygens' Principle, the two edges of the card used by Young act as sources of secondary wavelets. The bright spots appear where the wavelets reinforce each other. (B) Two slits in a card also act a sources of secondary wavelets forming alternating light and dark fringes.

between Arago, Fresnel, and Young. Although Young was a great experimentalist and theoretician, Fresnel was also a great mathematician, and between 1815 and 1819, he constructed mathematical formulae that could accurately give the positions of the bright and dark fringes observed by Young in his experiments on diffraction. Fresnel's formulae were based on Young's theory of interference (Buchwald, 1983; see later).

While Fresnel was working on his formulae, the *Académe des Sciences*, chaired by Arago, announced that the Grand Prix of 1819 would be awarded for the best work on diffraction. Fresnel (1819) and one other contender submitted their essays in hopes of winning the Grand Prix. The judges included Arago, who had come to accept the wave theory of light, as well as Siméon Poisson, Jean-Baptiste Biot, and Pierre-Simon LaPlace, who were advocates of Newton's corpuscular theory of light. After calculating the solutions to Fresnel's integrals, Poisson was unable to accept Fresnel's theory because if Fresnel's ideas about diffraction were true, then there should be a bright spot in the center of the shadow cast by a circular mirror, and to everyone's knowledge, such a bright spot did not exist. Poisson wrote (Baierlein, 1992), "Let parallel light impinge on an opaque disk, the surrounding being perfectly transparent. The disk casts a shadow- of course- but the very centre of the shadow will be bright. Succinctly, there is no darkness anywhere along the central perpendicular behind an opaque disk (except immediately behind the disk)."

Arago subjected Poisson's prediction to a test, and found that indeed there was a bright spot in the center of the shadow. Arago wrote in a report about the Grand Prix (Baierlein, 1992), "One of your [*Académe des Sciences*] commissioners, M Poisson had deduced from the integrals reported by [Fresnel] the singular result that the centre of the shadow of an opaque circular screen must, when the rays penetrate there at incidences which are only a little more oblique, be just as illuminated as if the screen did not exist. The consequence has been submitted to the test of direct experiment, and observation has perfectly confirmed the calculation." Fresnel won the Grand Prix of 1819. Although this was a victory for the wave theory of light, it did not result in an immediate and wide acceptance of the reality of the wave nature of light.

In order to understand how the fringes are formed by diffraction, we can use Huygens' Principle and posit that the edges of the card used by Young to diffract the light act as sources of secondary wavelets (Figure 3-12). The bright spots appear where the wavelets coincide with each other and reinforce each other, producing an optically effective wave. Each minute portion of a spherical wavelet can be considered to represent the crest of a sine wave (Figure 3-13). By combining Young's qualitative treatment of waves with Jean d'Alembert's mathematical treatment, Fresnel was able to simplify the analysis of diffraction.

D'Alembert (1747) derived and solved a wave equation that described the motion of a vibrating string. According to d'Alembert, the standing wave produced by a vibrating

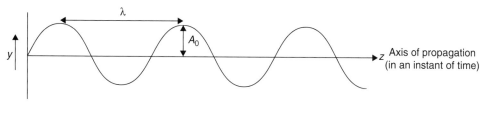

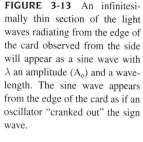

FIGURE 3-13 An infinitesimally thin section of the light waves radiating from the edge of the card observed from the side will appear as a sine wave with λ an amplitude (A_o) and a wavelength. The sine wave appears from the edge of the card as if an oscillator "cranked out" the sign wave.

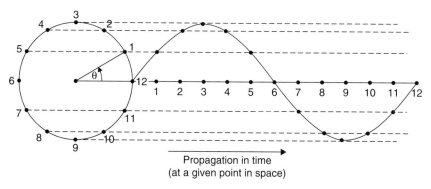

string resulted from two traveling waves propagating in opposite directions. He described the propagation of the two waves with functions whose arguments included the wavelength (λ, in m), the frequency (ν, in s^{-1}) of the wave, and its speed (c, in m/s). The wavelength is the distance between two successive peaks in a wave. The frequency is related to the wavelength by the following formula, known as the dispersion relation:

$$\lambda\nu = c$$

The arbitrary functions that satisfy d'Alembert's wave equation must be periodic, and typically sinusoidal functions, including sine and cosine, are used to describe the propagation of sound and light waves (Crawford, 1968; Elmore and Heald, 1969; French, 1971; Hirose and Lonngren, 1991; Georgi, 1993) through space (x, in m) and time (t, in s). Thus the time variation of the amplitude (Ψ) of a light wave with wavelength λ and frequency ν at a constant position x, or the spatial variation of the amplitude of a light wave with wavelength λ and frequency ν at constant time t can be described by the following equation (Figure 3-14):

$$\Psi(x, t) = \Psi_0 \sin 2\pi (x/\lambda \pm \nu t)$$

The spatiotemporal varying height of the sine wave, whether a light wave or a water wave, is known as its amplitude and Ψ_0 is the maximal amplitude. The brightness or intensity of a light wave is related to the square of its amplitude. We perceive differences in the wavelength of light waves as differences in color. The minus sign is used to describe a wave traveling toward the right and the plus sign is used to describe a wave traveling toward the left.

Although traditionally light waves are treated as if they have length but not width, Hendrik Lorentz (1924) insisted that light must have extension. It is possible to model a

plane wave with nonvanishing width by using the angular frequency (ω, in radians/sec) instead of the frequency and the angular wave number (k, in radians) instead of the wavelength. The angular frequency is related to the frequency by the following equation:

$$\omega = 2\pi\nu$$

and the angular wave number is related by wavelength by the following equation:

$$k = 2\pi/\lambda$$

The dispersion relation, given in terms of angular parameters is:

$$\omega/k = c$$

The traveling wave is then described by:

$$\Psi(x, t) = \Psi_0 \sin(kx \pm \omega t)$$

where ω can be interpreted as the angular frequency, in which the quantity $\lambda/2\pi$ rotates around a circle, such that the product of ω and $\lambda/2\pi$ equal the speed of light. If such a wave were to translate at velocity c along the axis of propagation, $\lambda/2\pi$ would represent the radius of the wave (see appendix).

Waves are not only described by their intrinsic qualities, including their velocity, wavelength (angular wave number), frequency (angular frequency), and amplitude, but also by relative or "social" qualities, including phase (Ψ, dimensionless). The phase of a wave is its position relative to other waves. When two waves interact, their phase has a dramatic effect on the outcome.

Young (in Arago, 1857) wrote:

It was in May of 1801, that I discovered, by reflecting on the beautiful experiments of Newton, a law which appears to me to account for a greater variety of interesting phenomena than any other optical

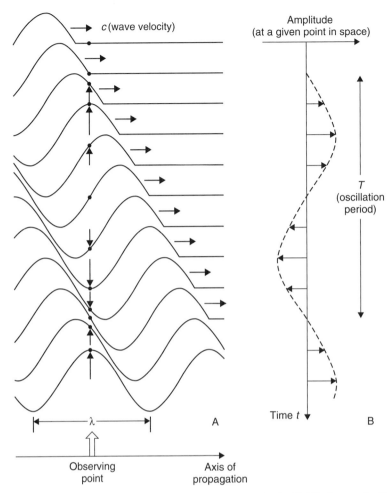

FIGURE 3-14 (A) At a single instant of time, we see a wave as a spatial variation in amplitude. (B) Whether we visualize the wavelength of a wave or the frequency of a wave depends on the mode of observation. At a single point in space, we see a wave as a time variation in amplitude.

principle that has yet been made known. I shall endeavour to explain this law by a comparison:– Suppose a number of equal waves of water to move upon the surface of a stagnant lake, with a certain constant velocity, and to enter a narrow channel leading out of the lake;– suppose, then, another similar cause to have excited another equal series of waves, which arrive at the same channel with the same velocity, and at the same time with the first. Neither series of waves will destroy the other, but their effects will be combined; if they enter the channel in such a manner that the elevations of the one series coincide with those of the other, they must together produce a series of greater joint elevations; but if the elevations of one series are so situated as to correspond to the depressions of the other, they must exactly fill up those depressions, and the surface of the water must remain smooth; at least, I can discover no alternative, either from theory or from experiment. Now, I maintain that similar effects take place whenever two portions of light are thus mixed; and this I call the general law of interference of light.

The eye cannot perceive the absolute phase of a light wave; but it can distinguish the difference in phase between two waves because the intensity that results from the combination of two or more waves depends on their relative phase (Figure 3-15). When two waves are in phase, their combined intensity is bright since the intensity depends on the square of the sum of their amplitudes. This is known as constructive interference. When two waves are one-half wavelength out-of-phase, the two waves cancel each other and darkness is created. This is known as destructive interference. Any intermediate difference in phase results in intermediate degrees of brightness. Young (1802) wrote: "Wherever two portions of the same light arrive at the eye by different routes, either exactly or very nearly in the same direction, the light becomes most intense when the difference in the routes is any multiple of a certain length, and least intense in the intermediate state of the interfering portions; and this length is different for light of different colors."

We use the principle of superposition to determine the intensity of the resultant of two or more interfering waves (Jenkins and White, 1937, 1950, 1957, 1976; Slayter, 1970). According to the principle of superposition, the time- or position-varying amplitude of the resultant of two or more waves is equal to the sum of the time- or position-varying amplitudes of the individual waves. The intensity of the resultant is obtained by squaring the summed

A

B

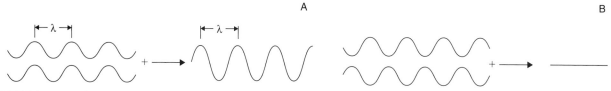

FIGURE 3-15 (A) Constructive interference occurs between two waves that are in phase. The amplitude of the resultant is equal to the sum of the individual amplitudes. Destructive interference occurs between two waves that are $\lambda/2$ out-of-phase. (B) The amplitude of the resultant vanishes. Since the intensity of light is related to the square of the amplitude of the resultant, constructive interference produces a bright fringe and destructive interference produces a dark fringe.

amplitudes. The intensity of the resultant of two or more waves can be determined graphically by adding the amplitudes and then squaring the sum of the amplitudes. The intensity of the resultant wave can also be determined analytically. Consider two waves with equal amplitudes, both with angular wave number k, traveling to the right; let wave two be out-of-phase with wave one by the phase factor φ (in radians). Since 360° is equivalent to 2π radians, one radian is equivalent to 57.3°.

$$\Psi(x, t) = \Psi_{o1} \sin(kx - \omega t) + \Psi_{o1} \sin(kx - \omega t + \varphi)$$

Since $\sin A + \sin B = 2 \cos[(A - B)/2] \sin[(A + B)/2]$, this equation can be simplified to yield:

$$\Psi(x, t) = 2\Psi_{o1} \cos(\varphi/2) \sin(kx - \omega t + \varphi/2)$$

When the two waves are in-phase and $\varphi = 0$, cos (0) = 1 and

$$\Psi(x, t) = 2\Psi_{o1} \sin(kx - \omega t)$$

The resultant has twice the amplitude and four ($=2^2$) times the intensity of either wave individually. The resultant also has the same phase as the individual waves that make up the resultant. This is the situation that leads to a bright fringe. When the two waves are completely out-of-phase, where $\varphi = \pi$, cos ($\pi/2$) = 0 and

$$\Psi(x, t) = 0$$

The resultant has zero amplitude and intensity ($=0^2$). This is the situation that leads to a dark fringe. The phase of the resultant is not always the same as the phase of the component waves. The amplitude of the resultant of two similar waves that are $\pi/4$ ($=45°$) out-of-phase with each other is:

$$\Psi(x, t) = 1.8476\Psi_{o1} \sin(kx - \omega t + 0.38)$$

The intensity (3.414) is given by the square of the amplitude and the resultant is out-of-phase with both component waves. This is the situation that leads to the shoulder of a fringe. The amplitude of the resultant of two similar waves that are $\pi/2$ ($=90°$) out-of-phase with each other is:

$$\Psi(x, t) = 1.4142\Psi_{o1} \sin(kx - \omega t + 0.707)$$

The intensity (1.999) is given by the square of the amplitude and is almost indistinguishable from the sum of the intensities of the component waves since the interference is neither constructive nor destructive. The resultant is out-of-phase with the two component waves.

Let us continue to consider Young's diffraction experiment in terms of Huygens' Principle and the principle of interference (Figure 3-16). Let the waves emanate from a source and pass through a slit. Place a converging lens so that the slit is at its focus and plane waves leave the lens and strike the strip made out of cardboard. Each edge of the card acts as a source of secondary wavelets. The waves from each secondary source radiate out and interfere with each other. Where the crests of the waves radiating from the two sides of the card come in contact, they will constructively interfere. Where the troughs of each set of waves come in contact, they will constructively interfere. Where a crest from one set meets a trough from the other set, they will destructively interfere. We can see that there is a region, equidistant from both edges of the slip of card, where the crests from one secondary wave interact only with the crests from the other secondary wave. Thus in this region, light interferes only constructively. These rays will give rise to a bright spot on the screen known as the zeroth-order band.

We can also see that just to the left or right of the middle, there are regions where the crests of one secondary wavelet always meet the troughs of the other secondary wavelet and thus they always destructively interfere. This gives rise to one dark area on the left of the zeroth-order band, and another on the right of the zeroth-order band.

Again as we move further from the middle of the slip of card, we see that on the left, the crests of the wave from the near edge meet the crests of the wave from the far edge. The waves from the far edge are one full wavelength behind the waves from the near edge. The waves constructively interfere and give rise to a bright band. Likewise, on the right, we see that the crests of the wave from the near edge meet the crests of the wave from the far edge that are also retarded by one full wavelength. These waves also constructively interfere and give rise to a bright band. These are called the first-order bands.

In a similar manner, the second-order, third-order (and so on) bright fringes are formed when the crests of each secondary wavelet meet, but one wave is retarded by m = 2, 3 (and so on) full wavelengths relative to the other one. The bright fringes (maxima) alternate with dark fringes (minima). The maxima appear where the optical path lengths (OPL) of the component waves differ by an integral number (m) of wavelengths. The minima appear where the

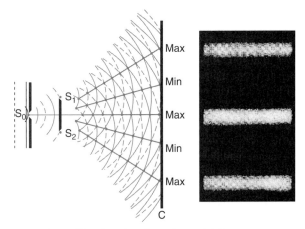

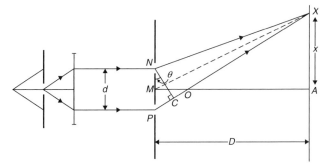

FIGURE 3-17 The rays NX and PX are perpendicular to the wave fronts emanating from N and P, respectively. If x is the first-order maximum, then PC is equal to 1λ.

FIGURE 3-16 The slip of card can be modeled as two sources of Huygens' wavelets. Bright fringes are formed where the waves from the two sources constructively interfere and dark fringes are formed where the waves from the two sources destructively interfere.

optical path lengths of the component waves differ by $m \pm \frac{1}{2}$ wavelengths.

The zeroth-order band is the brightest, the first-order bands are somewhat dimmer, the second-order bands are dimmer still, and so on . We can explain this by saying that each secondary wavelet is a point source of light whose energy falls off with distance from the source of secondary wavelets. Therefore there is the most energy in the light that interacts close to the source and there is less energy in the light that interacts further from the source.

The distance between the fringes depends on the size of the object, the distance between the object and the viewing screen, and the wavelength of light. If we know any three of these parameters, we can deduce the fourth. How can we determine the distance between the fringes? Let's consider an object of width d that is a distance D from the viewing screen (Figure 3-17).

Waves arriving at point A from points N and P have identical optical path lengths and are thus in-phase. At any other point X on the screen, separated from A by a distance x, waves from N and P differ in optical path length. The value of this optical path difference (OPD) is:

$$\text{Optical path difference} = PC$$

The maxima occur where the optical path difference equals an integral number of wavelengths; that is, where $PC = m\lambda$, where m is an integer. Likewise, the minima occur where the optical path difference equals $m \pm \frac{1}{2}$ wavelengths. Intermediate intensities occur where the optical path difference equals $(m + n)$ wavelengths where $0 < n < \frac{1}{2}$.

It is true the $PC = d \sin \theta$, since, for a right triangle, the sine of an angle equals the ratio of the length of the

opposite side to the length of the hypotenuse. The first-order maximum occurs when $PC = (1)\lambda$ and must satisfy the following condition:

$$d \sin \theta = m\lambda$$

Note that $<NPC$ is the same as $<MPO$, and since $<NCP$ and $<OMP$ are right angles, and all triangles have π radians $(=180°)$, then

$$<\theta = <MOP$$

$<MOP = <AOX$ because they are vertical angles. Thus,

$$<AOX = <\theta$$

$$\tan (AOX) = \tan \theta \approx x/D$$

As long as θ is small (that is $D \gg MO$ and $D > d$), then $D \approx OA$. Using the small angle approximation:

$$\sin \theta \approx \tan \theta \approx x/D.$$

Therefore the maxima occur where $d (x/D) = m\lambda$, or

$$x = (D/d) \, m\lambda$$

and minima occur where

$$x = (D/d) (m + \frac{1}{2})\lambda$$

Thus, the following three statements are true:

1. The smaller the object, the greater the distance between the first-order band and the zeroth-order band. Thus, a short distance in the object plane results in a long distance in the diffraction plane, and a long distance in the object plane results in a short distance in the diffraction plane. This is known as the concept of reciprocal space.

2. The shorter the wavelength, the smaller the distance between the first-order band and the zeroth-order band. Thus diffraction can be used to separate the wavelengths of light, and in fact, diffraction gratings are used to separate light into various colors (see Chapter 8). Moreover,

if the size of the object, the length between the object and the screen, and the distance between the diffraction bands are known, the wavelength of the light illuminating the object can be determined. Thomas Young (1804a) determined the wavelengths of visible and UV waves with this method.

3. The greater the distance from the object to the screen, the greater the distance between the first-order band and the zeroth-order band. Thus, the greater the distance, the easier it is to distinguish a greater number of diffraction bands.

Interference occurs because the intensity of a resultant wave depends on the square of the sum of the amplitudes of the component waves and not on the sum of the squares of the amplitudes. When we consider two waves, the intensity of one wave alone is proportional to Ψ_1^2 and the intensity of the other wave alone is proportional to Ψ_2^2, but the intensity of the two interfering waves is proportional to $(\Psi_1 + \Psi_2)^2$. Thus when $\Psi_1 = -\Psi_2$, the intensity is 0. When $\Psi_1 = \Psi_2$, the intensity is $4\Psi_1$, and when Ψ_1 and Ψ_2 have any other relationship, the intensity is intermediate between 0 and 4Ψ. These relationships hold for any number of waves where the intensity is proportional to the square of the sum of the amplitudes of all the waves that strike a given point.

$$I = (\Psi_1 + \Psi_2 + \Psi_3 + \Psi_4 + \cdots + \Psi_n)^2$$

Young also investigated the diffraction of light through two slits. Again, we can determine the positions of the bright and dark fringes graphically or analytically. If we analyze the two-slit diffraction graphically using Huygens' Principle, we consider each of the two slits to be sources of secondary wavelets, and the bright fringes appear where the crests of the waves emanating from both slits reinforce each other, and the dark fringes appear where the crests of the waves from one slit coincide with the troughs of the waves from the other slit.

When we analyze two-slit diffraction analytically, we use the same formulae as we used to describe the diffraction of light around a card, except that d represents the distance between the slits instead of the width of the card.

We can look at the distribution of light passing around the card or through two slits in term of the radial width ($\lambda/2\pi$) or radial diameter (λ/π) of a light wave (see Appendix II). The amplitudes of the light that passes through slit one ($\Psi(r_1, t)$) and slit two ($\Psi(r_2, t)$) are given by the following equations:

$$\Psi(r_1, t) = \Psi_0 \sin(kr_1 - \omega t)$$
$$\Psi(r_2, t) = \Psi_0 \sin(kr_2 - \omega t)$$

where r_1 is the distance from slit 1 to the screen and r_2 is the distance from slit 2 to the screen. The phase difference (φ) between the two waves is $kr_1 - kr_2$ ($=k(r_1 - r_2)$),

which is equal to $(r_1 - r_2)(2\pi/\lambda)$. Consequently the sum of the amplitudes can be written:

$$\Psi(r_{1+2}, t) = \Psi_0[\sin(kr_1 - \omega t) + \sin(kr_1 - \omega t + \varphi)]$$

Since $\sin A + \sin B = 2\cos[(A - B)/2]\sin[(A + B)/2]$,

$$\Psi(r_{1+2}, t) = 2\Psi_0 \cos(\varphi/2)\sin(kr_1 - \omega t + \varphi/2)$$

The resultant is a propagating wave. The properties of propagation are determined by the sine term and the amplitude of the resultant is strongly influenced by the cosine term (Hirose and Lonngren, 1991). Thus the amplitude of the resultant wave depends on the phase difference between the component waves according to the following equation:

$$\Psi(r_{1+2}, t) = 2\Psi_0 \cos((r_1 - r_2)(2\pi/\lambda)/2)$$

and the intensity ($I = \Psi(r_{1+2}, t)^2$) of the resultant wave is given by

$$\begin{aligned}I &= 4\Psi_0^2 \cos^2((r_1 - r_2)(2\pi/\lambda)/2) \\ &= 4\Psi_0^2 \cos^2((OPD)(\pi/\lambda))\end{aligned}$$

where $OPD = r_1 - r_2$. We could say that interference effects are a function of the ratio of the OPD to the diameter of the wave. Since the $OPD = m\lambda$ and $m\lambda = d\sin\theta$, the interference effects for a given angle (θ) are a function of the distance between slits and the diameter of the wave. Diffraction usually is given as a function of the length of light waves that have no radial dimension. However, in this unorthodox introduction of wave width, we can almost visualize a single wave wrapping around the card or simultaneously going through the two slits. The smaller the wave width, the more the light wave approximates a light ray. The smaller the wave width and wavelength, the more a light wave approximates a corpuscle of light.

The diffraction pattern of an object is related to the shape of the object. Compare the diffraction patterns of a slit (whose length is much greater than its width), a square (which is a slit whose length and width are equal), and a circle (which is a slit rotated π radians ($=180°$) (Figure 3-18).

The diffraction pattern of a circular pinhole is called an Airy disc, named after Sir George Airy. His observation that the size of the image of a star depended on the diameter of the lens used as an object glass in a telescope led him to study the diffraction pattern of circular apertures (Airy, 1866; Rayleigh, 1872). Airy found that maxima resulted when the optical path difference from two diametrically opposed points on the circle to the viewing screen was a nonintegral number of wavelengths. The reason that the maxima occur at nonintegral numbers is because the width of a circular aperture varies from 0 to d and waves originating from the thin regions interfere with waves originating from the wide areas. The values of $m = [d\sin\theta/\lambda]$, which correspond to minima or maxima of an Airy Disc,

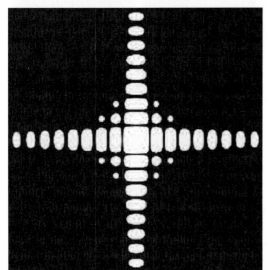

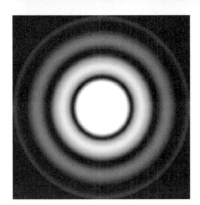

FIGURE 3-18 Fraunhöfer diffraction patterns of a vertical slit, a square, and a circle.

TABLE 3-1 Coefficients of Wavelength that Give Rise to Minima and Maxima		
Order	Minima (λ)	Maxima (λ)
0	0	0
1	1.220	1.635
2	2.233	2.679
3	3.238	3.699
4	4.241	4.710

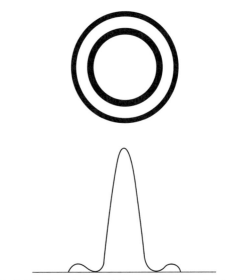

FIGURE 3-19 The point spread function of an Airy disc. The maxima occur at non-integral numbers.

are given in Table 3-1. The function that describes the distribution of intensities in an Airy disc often is called the point spread function (Figure 3-19).

These coefficients can be obtained experimentally, or they can be obtained analytically using one of the Bessel functions. A Bessel function is a convenient and useful function that describes damped periodic functions. Friedrich Bessel used this function to understand the disturbance of the elliptical motion of one planet by another planet (Smith, 1959). We will use the Bessel function to describe the distribution of light intensity (I) in the plane perpendicular to the optical axis when light hits a circular aperture. The intensity at position x depends on the wavelength of light, the radius of the aperture, the distance of the point from the axis, and the light intensity at the aperture. The light intensity at any point x away from the center of the diffraction pattern is given by the following equation (Menzel, 1955, 1960):

$$I_x = I_0[2J_1(x)/x]^2$$

where

$$x = (\pi/\lambda)d \sin \theta$$

and $J_1(x)$ is the Bessel function of the first kind. The Bessel function, in general, is defined as:

$$J_n(x) = \sum_{k=0}^{\infty} \frac{(-1)^k (x/2)^{n+2k}}{k! \, (k + n)!}$$

When n = 1, the Bessel function of the first kind is written as:

$$J_1(x) = \sum_{k=0}^{\infty} \frac{(-1)^k \, (x/2)^{1+2k}}{k! \, (k+1)!}$$

When the Bessel function of the first kind is expanded, it looks like this:

$$J_1(x) = \frac{(x/2)}{1!} - \frac{(x/2)^3}{1!2!} + \frac{(x/2)^5}{2!3!}$$
$$- \frac{(x/2)^7}{3!4!} + \frac{(x/2)^9}{4!5!} - \ldots$$

When we solve the factorials, it looks like this:

$$J_1(x) = (x/2) - \frac{(x/2)^3}{2} + \frac{(x/2)^5}{12} - \frac{(x/2)^7}{144} + \frac{(x/2)^9}{2880} - \ldots$$

This equation gives an infinite series. When x is large and positive, we can use the following asymptotic series to simplify the answer:

$$J_1(x) = (2/\pi x)^{1/2} [P_1(x) \cos(x - (3\pi/4)) - Q_1(x) \sin(x - (3\pi/4))]$$

where

$$P_1(x) = 1 + \frac{(1^2)(3)(5)}{2!(8x)^2} - \frac{(1^2)(3^2)(5^2)(7)(9)}{4!(8x)^4}$$
$$+ \frac{(1^2)(3^2)(5^2)(7^2)(9^2)(11)(13)}{6!(8x)^6}$$

and

$$Q_1(x) = \frac{(1)(3)}{1!(8x)} - \frac{(1^2)(3^2)(5)(7)}{3!(8x)^3}$$
$$+ \frac{(1^2)(3^2)(5^2)(7^2)(9)(11)}{5!(8x)^5}$$

Thus, the maxima and minima can be obtained for apertures of various diameters (d) illuminated with a given wavelength (λ) using this equation:

$$I_x = I_0 [2J_1(x)/x]^2$$

where $x = (\pi/\lambda) \, d \sin \theta$, and is the ratio at a given angle θ of the diameter of the circular disc to the radial wave diameter.

The diffraction pattern of a circle consists of a central disk, which contains 84 percent of the intensity and the first-order maximum, which contains 7 percent of the intensity. The higher-order maxima contain 3 percent, 1.5 percent, 1 percent, and so on, of the light intensity transmitted by the aperture. For practical purposes, we can consider the diffraction pattern of a circle to consist of a central disk that is surrounded by a minimum. The angular position (θ) of the first minimum is obtained from the following equation:

$$\sin \theta = 1.220 \, \lambda/d$$

where d is the diameter of the aperture. At small angles, sin θ is equal to tan θ. And since tan θ = x/D, where x is the distance on the screen between the center and edge of the central spot and D is the distance from the aperture to the screen, then sin θ = x/D and

$$x = 1.220 \, \lambda \, D/d$$

Thus circular apertures can be characterized by the following rules:

1. The greater the distance between the aperture and the screen, the greater the diameter of the central spot.
2. The smaller the diameter of the aperture, the greater the diameter of the central spot.
3. The shorter the wavelength, the smaller the diameter of the central spot.

A biological specimen can be considered a series of points, dots, and holes, and the image of a biological object can be considered a series of overlapping Airy discs of various sizes. Consider the nucleus to be a large pinhole ($d = 10^{-5}$m), a mitochondrion to be a medium pinhole ($d = 10^{-6}$m), and a vesicle to be a small pinhole ($d = 10^{-7}$m) illuminated with 500×10^{-9}m light (radial wave diameter = 1.59×10^{-7}m). Table 3-2 gives the angular position of the first minimum and first maximum produced by microscopic objects of various sizes (Figure 3-20).

We can see from Table 3-2 that the smaller the aperture, the less the rays travel in straight lines. The larger the aperture, relative to the radial wave diameter, the more the diffraction pattern looks like the image of the aperture. When I discussed the *camera obscura* or the pinhole camera in Chapter 1, I showed that the smaller the aperture, the more distinct the image. However, there is a limit as to how small the aperture can get before the image of each point on the object is "inflated" by diffraction (Rayleigh, 1891). As a rule of thumb, the optimal pinhole diameter (in μm)

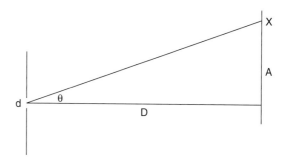

FIGURE 3-20 The angle described by the first order maximum depends on the diameter of an organelle.

TABLE 3-2 Angles Subtending the First Minimum and Maximum Produced by Microscopic Objects of Various Sizes

d	First minimum (degrees)	First maximum (degrees)
10^{-7} m	–	–
10^{-6} m	37.589	54.84
10^{-5} m	3.497	4.689
10^{-4} m	0.349	0.468
10^{-3} m	0.0349	0.047
10^{-2} m	0.0035	0.0048

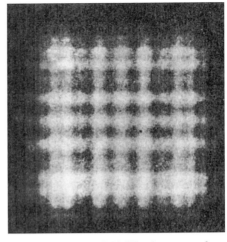

FIGURE 3-21   Fresnel or near-field diffraction pattern of a square.

for 0.5 μm light is (1.1) $\sqrt{s_i}$, where s_i is the distance from the pinhole to the object.

I have been discussing diffraction patterns of objects illuminated with plane waves, and the diffraction patterns are viewed an infinite distance from the object. This is known as far-field or Fraunhöfer diffraction. If we were to reduce the wavelength of light to zero, the Fraunhöfer diffraction pattern would turn into an image that exactly reproduced the object. Fresnel diffraction, also known as near-field diffraction, occurs when an object is illuminated with plane waves and the diffraction pattern is observed a short distance away from the object (Fresnel, 1819). The Fresnel diffraction pattern is clearly recognizable as an imperfect and somewhat fuzzy image of the object (Figure 3-21).

Young realized that, as long as the wavelength of light was not infinitesimally short, the image of a specimen would be affected by diffraction. That is, slits will appear as fringes, squares will appear as crosses, and circles will appear as Airy discs. Young (1804a) wrote:

> The observations on the effects of diffraction and interference, may perhaps sometimes be applied to a practical purpose, in making us cautious in our conclusions respecting the appearances of minute bodies viewed in a microscope. The shadow of a fibre, however opaque, placed in a pencil of light admitted through a small aperture, is always somewhat less dark in the middle of its breadth than in the parts on each side. A similar effect may also take place, in some degree, with respect to the image on the retina, and impress the sense with an idea of a transparency which has no real existence: and, if a small portion of light be really transmitted through the substance, this may again be destroyed by its interference with the diffracted light, and produce an appearance of partial opacity, instead of uniform semitransparency. Thus, a central dark spot, and a light spot surrounded by a darker circle, may respectively be produced in the images of a semitransparent and an opaque corpuscle; and impress us with an idea of a complication of structure which does not exist.

If light is a wave, then what is waving and what is the nature of the medium through which it propagates? In the early part of the nineteenth century, Young, Arago, and Fresnel considered light, like sound, to be a mechanical longitudinal compression wave. Longitudinal waves, which vibrate parallel to the direction of propagation, can propagate in gases, liquids, and solids. However, after considering the polarization of light (see Chapter 7), Young and Fresnel realized that light, unlike sound, must be, at least in part, a transverse mechanical wave. Transverse waves, which vibrate perpendicular to the direction of propagation, can propagate only through a highly elastic solid medium or through a liquid medium subjected to gravitational forces. This seemed to indicate that the luminous ether had to be elastic enough to propagate light from distant stars without impeding the movement of the planets around the sun or causing the planets to snap back in the direction they came from! In addition, the ether had to be thin enough to pass though the glass walls of a vacuum jar.

Since the square of the velocity of a wave propagating through a mechanical medium is equal to the ratio of the elasticity to the density of the medium, in order for the speed of light to be 299792458 m/s, the elasticity had to be enormously high and the density had to be infinitesimally low. David Brewster told John Tyndall (1873) that he objected to the wave theory of light because "he could not think the Creator guilty of so clumsy a contrivance as the filling of space with ether in order to produce light."

Throughout the nineteenth century, much effort was expended trying to resolve the requirements of this enigmatic luminous ether though the merciless addition of putative physical properties and the use of inexorable mathematics (Airy, 1866; MacCullagh, 1880; Lodge, 1907, 1909, 1925; Lorentz, 1927; Whittaker, 1951, 1953; Schaffner, 1972; Swenson, 1972; Hunt, 1991). Then, at the turn of the twentieth century, Einstein (1905a) proclaimed by fiat that the ether was unnecessary for the propagation

Chapter | 3   The Dependence of Image Formation on the Nature of Light

of light, if we accept, in exchange, the idea that space and time are only relative concepts.

Although we can clearly see that Thomas Young gave us great insight into the nature of light and its practical applications in microscopy and vision, in his lifetime, this Englishman was a *persona non grata*, since he did not accept the corpuscular nature of light proffered a century before by Newton. Newton was held in high regard in England as can be seen by the following epitaph written for Newton in 1727 by Alexander Pope:

Nature and Nature's laws lay hid in night:
God said, "Let Newton be!" and all was light.

Thomas Young was viciously attacked anonymously in the Edinburgh Review for being "Anti-Newtonian" (Anonymous, 1803, 1804; Young, 1804; Peacock, 1855; Wood and Oldham, 1954; Klein, 1970). The anonymous reviewer (1803, 1804b), most likely Lord Brougham (Tyndall, 1873), wrote about Young and his wave theory of light:

A mere theory is in truth destitute of all pretensions to merit of every kind, except that of a warm and misguided imagination. It demonstrates neither patience of investigation, nor rich resources of skill, nor vigorous habits of attention, nor powers of abstracting and comparing, nor extensive acquaintance with nature. It is the unmanly and unfruitful pleasure of a boyish and prurient imagination, or the gratification of a corrupted and depraved appetite.

The anonymous reviewer went on to say:

We take our leave of this paper with recommending it to the Doctor to do that which he himself says would be very easy; namely, to invent various experiments upon the subject. As, however, the season is not favourable for optical observation, we recommend him to employ his winter months in reading the "Optics", and some of the plainer parts of the "Principia", and then to begin his experiments by repeating those which are to be found in the former of these works.

Young decided that the Royal Institution was no place for him and resigned from his post. However, his work has held up over time, and it is an essential aspect of understanding image formation in the microscope. In 1850 Léon Foucault (1850, 1862) measured the speed of light in air and in water and found that it was slower in water than it was in air, a finding that was consistent with the wave theory of light, but inconsistent with the corpuscular theory of light championed by Newton (1730). This result, along with the theoretical work done by James Clerk Maxwell and the Maxwellians, led to a widespread acceptance of the wave theory of light (Lloyd, 1873; Herschel, 1876; Lommel, 1888). Since the corpuscular theory was able to explain interference phenomena only by inventing *ad hoc* hypotheses, the corpuscular theory was no longer thought of as a contending theory, but as a "mob of hypotheses" (Tyndall, 1873).

JAMES CLERK MAXWELL AND THE WAVE THEORY OF LIGHT

Michael Faraday (1845) was looking for the relationship between electricity, magnetism, and light, and eventually found that he could influence the plane of polarization of a light beam when he placed the glass through which the beam traveled in a magnetic field (Tyndall, 1873; Thompson, 1901; Williams, 1987; Day, 1999). Other nineteenth century scientists, including Wilhelm Weber, also were interested in unifying electricity, magnetism, and optics. While Maxwell was working on the equations of electricity (E) and magnetism (B), he serendipitously found the relationship between electricity, magnetism, and light (Maxwell, 1891; Niven, 2003). These equations, known as Maxwell's Equations, combined an extended version of Ampere's Law ($\nabla \times B = \mu_o J + \mu_o \varepsilon B_o \partial E / \partial t$), which relates the magnitude of the magnetic field to the electrical currents, displacement currents, and the magnetic permeability of the medium (μ); Faraday's Law ($\nabla \times E = -\partial B / \partial t$), which relates the magnitude of the electric field to a time-varying magnetic field; Gauss's Law of Electricity ($\nabla \cdot E = (\rho / \varepsilon_o)$), which relates the electric field to the charge density and electric permittivity (ε); and Gauss's Law of Magnetism ($\nabla \cdot B = 0$), which states that there are no magnetic monopoles (Heaviside, 1893; Hertz, 1893; Thomson, 1895; Poynting, 1920; Lorentz, 1923).

By combining these laws, Maxwell and the Maxwellians (Campbell and Garnett, 1884; Lodge, 1889; Heaviside, 1892, 1922; Hertz, 1893; Maxwell, 1931; Cohen, 1952; Everitt, 1975; Hunt, 1991; Yavetz, 1995; Darrigol, 2000; Niven, 2003) showed that the electric (E) and magnetic (B) fields, like sound, could be described by a wave equation:

$$\partial^2 E_y / \partial x^2 + \partial^2 E_y / \partial y^2 + \partial^2 E_y / \partial z^2 = \mu_o \varepsilon_o \partial^2 E_y / \partial t^2$$

$$\partial^2 B_z / \partial x^2 + \partial^2 B_z / \partial y^2 + \partial^2 B_z / \partial z^2 = \mu_o \varepsilon_o \partial^2 B_z / \partial t^2$$

Combining the four laws of electricity and magnetism resulted in equations for the electric and magnetic field that had the same form as a wave equation. By solving the wave equation, Maxwell found that the speed of electromagnetic waves in a vacuum is equal to $\sqrt{(1/(\varepsilon_o \mu_o))}$, where ε_o is the electric permittivity of a vacuum (8.85×10^{-12} s^2 C^{2}m^{-3}kg^{-1}) and μ_o is the magnetic permeability of a vacuum ($4\pi \times 10^{-7}$ m kg C^{-2}), and moreover, the speed turned out to be 2.9986×10^8 m/s, which was the same value that Hippolyte Fizeau (1849a,b) and Fizeau and Bréguet (1850) found for the speed of light (Frercks 2000)! That is, when Maxwell solved the wave equation to find the speed of propagation of an electromagnetic wave, he found that electromagnetic waves travel at the speed of light. This meant to Maxwell, that light itself must be an electromagnetic wave (Figure 3-22).

Fizeau measured the speed of light by passing light to a distant mirror (8.6 km) through a rotating wheel with an

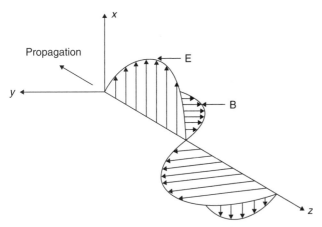

FIGURE 3-22 A light wave at a given instant of time according to Maxwell. The electric (E) and magnetic (B) fields are transverse, in-phase with each other, and orthogonal to each other.

TABLE 3-3 A comparison between $\sqrt{K_e}$ and n		
Substance	$\sqrt{K_e}$	n
Air	1.000294	1.000293
Hydrogen	1.000131	1.000132
Helium	1.000034	1.000035
Carbon dioxide	1.00049	1.00045
Water	8.96	1.333
Fused silica	1.94	1.458

opening cut into it. A pulse of light leaving the wheel traveled to the mirror where it was reflected back to the rotating toothed wheel. Fizeau adjusted the speed of rotation of the wheel (N turns/second) so that the returning light pulse either went through the hole or did not. From knowing the distance the light traveled (D), the number of teeth in the wheel (n), and the speed of rotation, Fizeau could calculate the sped of light from the following formula:

$$c = NnD$$

Fizeau found the speed of light to be 3.153×10^8 m/s, a speed consistent with that found from astronomical measurements done by Römer and Bradley (Michelson, 1907, 1962; Jaffe, 1960; Livingston, 1973).

Maxwell (1865) was so excited when he saw this "coincidence" he said: "This velocity is so nearly that of light, that it seems we have strong reason to conclude that light itself (including radiant heat, and other radiations, if any) is an electromagnetic disturbance in the form of waves propagated through the electromagnetic field according to electromagnetic laws."

One important consequence of Maxwell's electromagnetic wave theory of light for microscopists is that we can use it to predict and understand the refractive index of various media (Maxwell, 1891). Since, according to Maxwell's wave equation, the velocity of light in a vacuum (c) is given by:

$$c = (\varepsilon_0 \mu_o)^{-1/2}$$

and the speed of light in a dielectric medium is given by:

$$v = (\varepsilon \mu)^{-1/2}$$

where ε is the electric permittivity of the dielectric (in $s^2 C^2 m^{-3} kg^{-1}$) and μ is its magnetic permeability (in m kg C^{-2}). And since the refractive index (n) of a medium is defined as:

$$n = c/v$$

then

$$n = (\varepsilon_0 \mu_o)^{-1/2}/(\varepsilon \mu)^{-1/2} = (\varepsilon \mu)^{1/2}/(\varepsilon_0 \mu_0)^{1/2}$$
$$= [(\varepsilon/\varepsilon_0)(\mu/\mu_0)]^{1/2}$$

The dimensionless ratio of ε to ε_o is known as the relative permittivity (K_e), which is a measure of the electric characteristics of a substance; and the dimensionless ratio of μ to μ_o is known as the relative permeability (K_m), which is a measure of the magnetic characteristics of a substance. Therefore

$$n = (K_e K_m)^{1/2}$$

Since most substances are only weakly magnetic (except ferromagnetic materials), $K_m \approx 1$, which means

$$n \approx \sqrt{K_e}$$

This is known as Maxwell's Relation, and it tells us that the refractive index is a measure of the electrical properties of a material. That is, it is a measure of concentration, distribution, and movement of the electrons in the material. The concentration of electrons is roughly related to the density of the substance. Table 3-3 compares $\sqrt{K_e}$ with n.

The discrepancies between the square root of the relative permittivity and the refractive index in Table 3-3 results from the fact that the relative permittivity is measured at 60 Hz, whereas the refractive index is measured at 5×10^{14} Hz. When K_e of glass is measured with D.C. fields, it is 12.61; when the field varies at a frequency of 1 Hz, the index of refraction decreases to 2.739, at 25 Hz, $K_e = 2.51$, and at 1.2×10^4 Hz, $K_e = 2.404$. When the frequency of an electric field is increased to approximately 10^{10} Hz, K_e of glass falls to 2.04. Thus K_e decreases as the frequency approaches that of visible light ($\sim 10^{14}$ Hz; Bose, 1927) and in fact $\sqrt{K_e}$ approaches n. In most cases K_e has been determined by ASTM (American Society for Testing Materials) test methods at room temperature under standard conditions; however, like the values of the refractive index, the values of $\sqrt{K_e}$ vary with temperature.

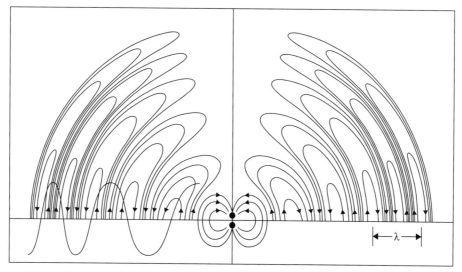

FIGURE 3-23 According to Heinrich Hertz, incoming electric waves will cause an electron to oscillate. The energy of the oscillating electron can be dissipated by the atoms or molecules that house the electron or the oscillating electron can act as a secondary source of electromagnetic waves by reradiating the energy. Electromagnetic waves can also be radiated when the electrons oscillate as a result of other energy inputs (e.g., heat, friction, etc.).

Following Maxwell's work, the wave theory of light became the only accepted theory of light. However, light was no longer treated as it was in the dynamical wave theory, as mechanical waves traveling through a very complicated and paradoxical ether as Huygens, Young, Fresnel, Arago, McCullagh, Neumann, and Kirchhoff visualized them, but as electromagnetic waves (The Classical Wave Theory), which may or may not have a mechanical action. Maxwell still invoked the ether as a mechanical medium necessary to transmit the electromagnetic waves (Whittaker, 1951, 1953), but Hendrik Lorentz removed the mechanical properties of the ether, leaving only its electrical permittivity and magnetic permeability.

We can now look at diffraction in terms of the electromagnetic theory of light. The edges of an obstruction or a slit are composed of electrons in atoms or molecules that can interact with the incident light. The electrons in the material behave as oscillators whose phase and amplitude of vibration depend on the electric field surrounding it. The incoming electromagnetic wave drives the electrons into oscillation and then the electrons reemit radiation at the same frequency (Figure 3-23). The direction of the reemitted, scattered, or diffracted light is a function of the spatial frequency of the object with which the light interacts.

We now know that gamma rays, X-rays, ultraviolet rays, visible rays, infrared rays (heat), microwaves, and radio waves are all electromagnetic waves that differ only in their wavelengths and frequencies (Figure 3-24). Each type of electromagnetic wave interacts with charged particles within atoms and molecules that are capable of interacting with the frequency of the incident electromagnetic wave. Heinrich Hertz (1893) demonstrated that radio waves could be reflected and refracted just like light waves.

Moreover, he showed that the electric field of the electromagnetic waves had transverse polarization as predicted by Maxwell's equations. Most textbook of optics present optical phenomena in terms of the electromagnetic theory of light (Stokes, 1884, 1885; Preston, 1895; Kelvin, 1904; Schuster, 1904, 1909; Wood, 1905, 1914, 1961; Schuster and Nicholson, 1924; Mach, 1926; Hardy and Perrin, 1932: Drude, 1939; Robertson, 1941; Jenkins and White, 1957; Strong, 1958; Bitter and Medicus, 1973; Hecht and Zajac, 1974; Born and Wolf, 1980), and many textbooks on electricity and magnetism are helpful in understanding optics (Thompson, 1904; Thomson, 1909; Lorentz, 1923, 1952; Haas, 1925; Jeans, 1927; Abraham and Becker, 1932; Planck, 1932; Frank, 1940; Stratton, 1941; Skilling, 1942; Harnwell, 1949; Panofsky and Phillips, 1955; Corson and Lorrain, 1962; Jackson, 1962; Sommerfeld, 1964; Jefimenko, 1966; Lindsay, 1969; Pauli, 1973; Shadowitz, 1975; Purcell, 1985; Griffiths, 1989; Heald and Marion, 1989; Pramanik, 2006). Textbooks concerning thermodynamics can also be consulted to understand optics since Macedonio Melloni and John William Draper (1878) independently demonstrated that thermal radiation follows the same laws as visible radiation. Currently almost the whole spectrum of electromagnetic radiation is used in microscopes to study the microscopic structure of matter (see Chapter 12).

ERNST ABBE AND THE RELATIONSHIP OF DIFFRACTION TO IMAGE FORMATION

We have reached what I think is a really exciting point where we can now apply the basic principles of geometric

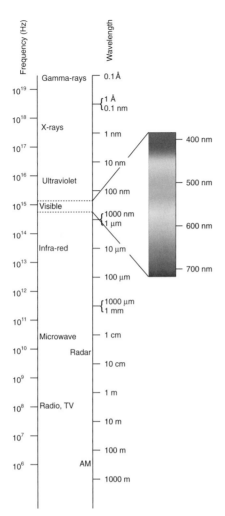

FIGURE 3-24 The electromagnetic spectrum.

and physical optics to understand how a microscope works and how it renders an image of a biological specimen. Let's assume that all the lenses in our microscope are perfect and free from aberrations. However, even if our lenses were perfect in reality, diffraction would still limit our ability to render a perfectly faithful image of our specimen. This is a result of the fact that the wavelength—and if you will, the wave width—of light is finite and not infinitesimally small.

Diffraction is an extremely important concept in microscopy. Interestingly, we still use Kirchhoff's mechanical view of diffraction (Braddick, 1965; Baker and Copson, 1987) to predict the position of spots since the methods consistent with Maxwell's equations require difficult mathematics (Sommerfeld, 2004) or unreasonable physical assumptions (Bethe, 1944).

Normally, in the absence of a lens, an object illuminated with plane waves forms a Fraunhöfer diffraction pattern at infinity. However, when an object is placed in the object space in front of a lens so that the lens forms an image in the image space behind the lens, a diffraction pattern that is equivalent to a Fraunhöfer diffraction pattern of the object is produced at the back focal plane of the lens. The pattern

of light spots found in the diffraction plane can be described by a mathematical function called a Fourier transform. For this reason, the back focal plane of the lens is commonly called the diffraction plane or the Fourier plane.

Let's assume that the object is an amplitude grating with an alternating pattern of sharp dark bands and sharp transparent bands. The object can be described by a square wave with an amplitude that varies between $+1$ and -1 (Figure 3-25). The function that characterizes this square wave is:

$$f(x) = \begin{array}{l} +1 \text{ when } 0 < x < \pi \\ -1 \text{ when } \pi < x < 2\pi \end{array}$$

According to Joseph Fourier (1822), who was interested in modeling the movement of heat through a material of any shape, any shape can be described by a complex wave, and that a complex wave is the sum of an infinite series of sine waves. Consequently, we can approximate the square wave with a sine wave that has the same period or angular wave number as the square wave and an amplitude that varies between $+1$ and -1 (Figure 3-26).

We can increase the fidelity of the representation of the square wave if we add to the fundamental sine wave, a sine wave with a higher spatial angular wave number and a smaller amplitude (Figure 3-27). We can increase the fidelity of the representation even more if we add to the two sine waves a third sine wave with an even greater spatial wave number and a smaller amplitude (Figure 3-28). We can increase the fidelity of the representation by adding more and more sine waves with higher and higher spatial angular wave numbers or frequencies and smaller and smaller amplitudes (Figure 3-29). A completely faithful representation requires the addition of an infinite number of sine waves. However, by summing a series of four sine waves, we already have obtained a "reasonable" likeness of the square wave. According to Fourier's Theorem, the square wave can be described by the following equation:

$$f(x) = (4/\pi)[(1/1)\sin(1kx) + (1/3)\sin(3kx) + (1/5)\sin(5kx) + (1/7)\sin(7kx) + \ldots]$$

where k is the fundamental spatial angular wave number (in m^{-1}), and x is the position on the horizontal axis. We use the term spatial angular wave number for k, since k represents an inverse distance, however in most texts, the spatial angular wave number is commonly called the spatial frequency.

Let's analyze the Fourier Transform bit by bit. Notice that the coefficient of each term follows the sequence 1/1, 1/3, 1/5, 1/7, 1/9, 1/11.... The coefficients in front of the sine function determine the amplitude of the wave. The coefficients of simple Fourier Transforms are either odd or even. If the original function is symmetrical above and below the x-axis, the coefficients are all even (and the function is composed of cosine waves instead of sine waves). However, when the original function is not symmetrical,

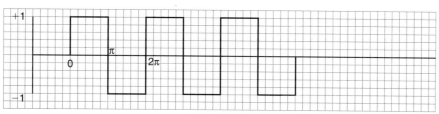

FIGURE 3-25 A square wave.

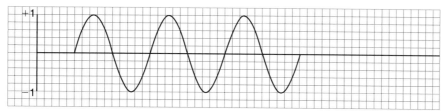

FIGURE 3-26 A sine wave that approximates the fundamental spatial angular wave number of the square wave.

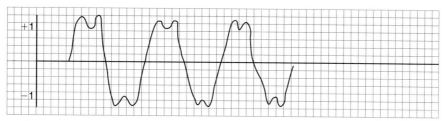

FIGURE 3-27 The sum of two sine waves with different spatial angular wave numbers better approximates the square wave.

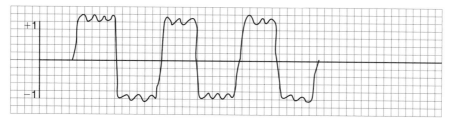

FIGURE 3-28 The sum of three sine waves with different spatial angular wave numbers approximates the square wave even better.

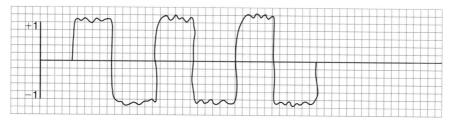

FIGURE 3-29 The sum of four sine waves with different spatial angular wave numbers approximates the square wave with high fidelity.

all the coefficients are odd (like our square wave, earlier) and the function is composed of sine waves (like our square wave, earlier). Complicated functions (like the wave that would describe the arrangements of organelles in a cell) have Fourier functions that are composed of sine waves and cosine waves. In theory, every function, whether it is periodic or nonperiodic, can be represented by the sum of an infinite number of sinusoidal waves as long as the function is everywhere finite and integratable.

Notice also that in the first term we take the sine of the fundamental spatial angular wave number (k) of the sine wave, which represents the fundamental spatial characteristics of the object. In the next terms, we take the sine of 3k, 5k, 7k, and so on. So the math really says: We can approximate a square wave by adding sine waves together. The rules are that we find the most fundamental spatial angular wave number that best describes the square wave. Then we add sine waves with greater and greater spatial angular wave numbers. The spatial angular wave numbers are preceded by a coefficient so that every time the spatial angular wave number is multiplied by that coefficient, the amplitude is divided by that same coefficient. The fundamental spatial angular wave number is related to the reciprocal of the distance of an object that determines the position of the first-order diffraction spot.

The spatial angular wave numbers of a square wave are relatively simple and straightforward, however the spatial angular wave numbers of more complex objects like the cityscape shown in Figure 3-30, or the structure of a cell can also be described by Fourier transforms, albeit more complicated ones.

To understand image formation, we must understand the relationship between the diffraction pattern and the image. The terms of the Fourier transform, with greater and greater spatial angular wave numbers, are equivalent to the diffraction spots of higher and higher orders found farther and farther from the principal axis. The small spatial angular wave number terms of the Fourier transform are related to the lower-order diffraction spots and the large spatial angular wave number terms of the Fourier transform are related to the higher-order diffraction spots. To make a perfect image we must add an infinite number of Fourier terms, which is equivalent to capturing an infinite number of orders of the diffracted light.

Ernst Abbe, while working for Carl Zeiss (Auerbach, 1904; Schütz, 1966; Smith, 1987), noticed in 1873 that when an image is in focus, the diffraction pattern appears in the back focal plane of the objective lens. When a specimen is illuminated, each point on the specimen can be considered to act as a point source of light and give rise to Huygens' spherical wavelets. The light waves travel in all possible directions from each point. The light, which travels in a given direction, from each and every individual point source, forms a collimated beam. Therefore light emanating from the specimen can be considered to consist of a series of collimated beams traveling in each and every direction.

In a perfect converging lens, by definition, a collimated beam of light converges to a point at the back focal plane of the lens. In a perfect lens, the light traveling parallel to the optical axis focuses at the back focal point of the lens. The other collimateds beams, which impinge on the lens, from other directions converge at other points in the back focal plane of the lens (Figures 3-31 and 3-32).

According to Abbe, points s_0, s_1, and s_2 at the back focal plane act as point sources that give rise to Huygens' spherical wavelets. The waves emanating from these points interfere in an image plane to form an image of the original amplitude grating. Abbe realized that the image arises directly from the diffraction spots on the back focal plane and only indirectly from the points on the object. It is as if the image is formed through two sequential optical processes. In the first process, the object diffracts the illuminating light into the objective lens, and in the second process, the objective lens moves the diffraction pattern from infinity to the back focal plane of the lens. The light

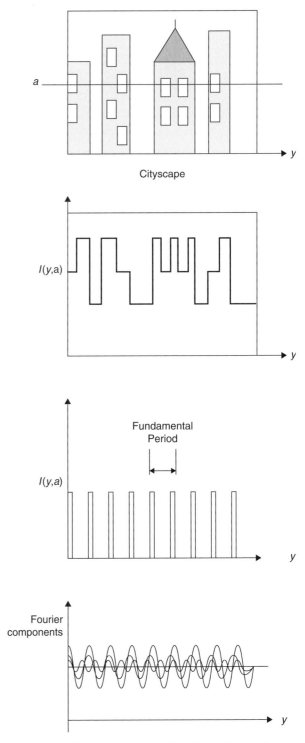

FIGURE 3-30 No matter how complicated an object is, it can be resolved into its Fourier components.

emanating from the diffraction pattern in the back focal plane of the objective lens interferes to form an image on the image plane. The mathematical process that resolves the object into components of various spatial frequencies in the back focal plane of the objective is called a forward Fourier transform. The mathematical process that recombines

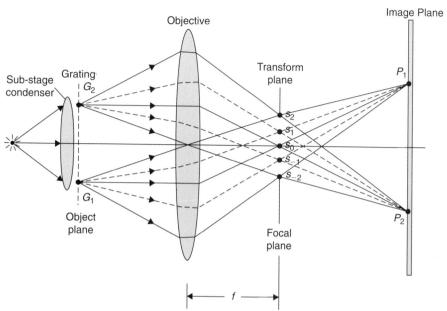

FIGURE 3-31   The object (G) diffracts the illuminating light. The objective lens collects the diffracted light and produces a diffraction pattern at its back focal plane. The spherical waves that emanate from the spots (s) at the back focal plane of the objective interfere with each other to produce an image at the image plane (P). This diagram emphasizes the rays normal to the wave fronts.

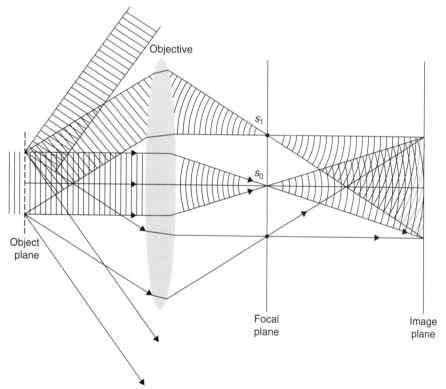

FIGURE 3-32   The object diffracts the illuminating light. The objective lens collects the diffracted light and produces a diffraction pattern at its back focal plane. The spherical waves that emanate from the spots (s) at the back focal plane of the objective interfere with each other to produce an image at the image plane. This diagram emphasizes the wave fronts.

the components of various spatial frequencies into an image is called an inverse Fourier transform.

Physically, the illuminating light is diffracted by the specimen and depending on the radius of the objective lens and its distance from the specimen, fewer or more orders of diffracted light are captured by the objective lens. The objective lens then forms the diffraction pattern at its back focal plane. Since the diffraction pattern lacks the diffracted light of the orders that were not captured by the objective lens, the diffraction pattern does not represent the diffraction pattern of the object, but of a somewhat incomplete or blurred version of the object. The dark and light regions of the image are formed by the interference of waves emanating from the diffraction spots. The brightness of a given spot depends on the amplitude and phase of the interfering waves. The bright spots of the image are formed where the wavelets originating from the back focal plane constructively interfere and the dark spots occur where the wavelets destructively interfere.

Let's consider the formation of an image of a biological specimen with a periodic structure such as a diatom. When the diatom is illuminated, light radiates from each point in the diatom in all possible directions. In Figure 3-33, we will make use of the characteristic rays with which we are familiar from Chapter 2. AB and A'B' are rays, perpendicular to their wave fronts, that strike the lens parallel to the principal axis. AC is a ray, perpendicular to its wave front, which travels through the front focal point. All other rays parallel to AC will pass through the lens and converge at point D in the back focal plane. All the light rays, perpendicular to their associated light wave fronts, which leave the object parallel to AB, converge at point f_i in the back focal plane of the objective. Subsequently light rays, with their associated wave fronts, diverge from these positions on the back focal plane to the image plane, where an image is produced at the position where two rays and their associated wave fronts (which originate at the same point in the object) converge. Bright spots are formed where the rays have optical path lengths that are equal to an integral numbers of wavelengths. Dark spots are formed where the rays vary by half-wavelengths.

In his initial experiments, Abbe noticed that when he increased the diameter of the objective lens, he increased the fidelity of the resulting image even though the apparent cone of incident light (zeroth-order light) filled only a small portion of the objective lens. Abbe figured that the "dark space" of the objective lens must contribute to the fidelity of the image. He proposed that specimens, whose size were close to the wavelength of light, diffracted light into the dark space of the objective lens. Thus, if the aperture were not large enough, some of these rays would not enter the objective lens, and the diffraction pattern at the back focal plane of the lens would not contain the higher-order diffraction spots. As a result of the incomplete capture of the diffracted light, the image was not a faithful reproduction of the object.

He then said that the image actually is related to a fictitious object whose complete diffraction pattern exactly matches the one collected by the objective lens. Through experimentation he realized that the light (diffracted at large angles) that is missed by the objective lens represents the higher spatial angular wave numbers of the specimen, and their removal results in an image that does not have as much high spatial angular wave number information as is present in the object. As it performs an inverse Fourier transform on the diffraction pattern, the objective lens also acts as a low pass filter that removes the higher spatial angular wave number terms and the higher-order diffracted light that they represent (Stoney, 1896).

Let us discuss in detail the experiments that serve at the basis for Abbe's theory of image formation. Abbe performed these experiments in collaboration with Carl Zeiss and Otto Schott in order to produce the best objective lens that could be made (von Rohr, 1920). Although excellent objective lenses were already being made by craftsmen working for small microscope companies, the high quality objective lenses were obtained through trial and error and without the benefit of physical theory.

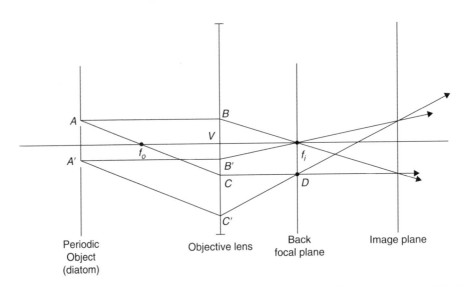

FIGURE 3-33 The object diffracts the illuminating light. The objective lens collects the diffracted light and produces a diffraction pattern at its back focal plane. The spherical waves that emanate from the spots at the back focal plane of the objective interfere with each other to produce an image at the image plane. This diagram emphasizes the characteristic rays.

In order to simplify the conditions used in image formation, Abbe used high contrast amplitude gratings for specimens and observed their diffraction patterns in the back focal plane of the objective lens and their images in the image plane. When he observed amplitude gratings consisting of widely spaced or narrowly spaced slits, Abbe noticed that there was a mathematical relationship between the amplitude grating and the diffraction pattern in that the spacing between the dots in the diffraction pattern was inversely proportional to the spacing between the slits in the object (Figure 3-34).

Abbe also found that he could change the nature of the image by altering the diffraction pattern. For example he could block out the first order diffraction spots that result from a widely spaced grating and the image would look as if it came from a narrowly spaced grating (Figure 3-35). Abbe also found that when he altered the diffraction pattern, in a process now known as spatial filtering, he obtained a modified image.

Albert Porter (1906) extended Abbe's experiment using a grid for an object and obtained a diffraction pattern like that shown in Figure 3-36 (Meyer, 1949; Bergmann et al.,

1999). Porter blocked out everything in the diffraction pattern except for the three central, horizontal spots, and saw an image of vertical slits (Figure 3-37). When he blocked out everything in the diffraction pattern but the three central, vertical spots, he saw an image of horizontal slits (Figure 3-38). When he blocked only the central spot and let the light from all the other spots to form an image, he saw a bright grid on a dark background. When he blocked out everything but the central spot, he saw a uniformly lit background without any image at all and when he allowed more and more orders of diffracted to pass and form the image, the image became a more and more faithful representation of the object.

We can really see that the image is related directly to the diffraction pattern in the back focal plane of the objective by making a mask that represents the diffraction pattern of the grid. Then, without putting the grid in the microscope, we insert the mask in the back focal plane of the objective that mimics the diffraction pattern of the grid. Even without the grid in the object plane, we see an image of the grid in the image plane. Many more examples of spatial filtering can be found in *Optical Transforms* by Taylor and Lipson (1964) and in *Atlas of Optical Transforms* by Harburn et al. (1975). Ernst Abbe and Albert Porter established the relationships between the object, the diffraction pattern, and the image by showing:

- Two different objects can be made to produce the same image by altering the diffraction pattern.
- Two identical objects can be made to produce different images by altering the diffraction pattern.
- The removal of the diffracted light destroys the image. We must collect at least the zeroth- and the first-order diffracted light in order to construct a bright-field image.

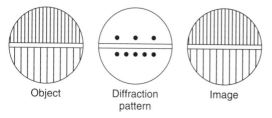

FIGURE 3-34 Ernst Abbe's experiment viewing the image and diffraction pattern of a grating.

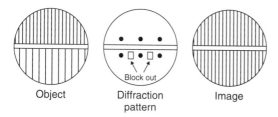

FIGURE 3-35 Abbe used a mask to block out the first-order diffraction spots produced by the coarse grating and obtained an image of the fine grating.

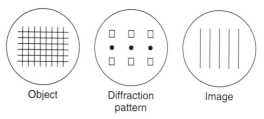

FIGURE 3-37 Porter produced an image of a vertical grating by masking certain diffraction spots produced by the grid.

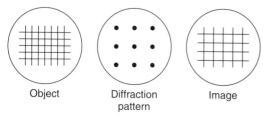

FIGURE 3-36 Albert Porter's extension of Abbe's experiment. Porter viewed the image and diffraction pattern of a grid.

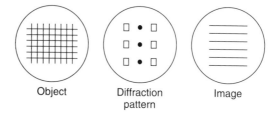

FIGURE 3-38 Porter produced an image of a horizontal grating by masking certain diffraction spots produced by the grid.

• The more orders of diffraction that an objective lens captures, the more faithful the image. The additional orders can be represented mathematically by higher-order harmonics of a Fourier Transform. Each higher order diffraction spot represents a higher-order spatial angular wave number.

• The image contrast is improved by the removal of some of the zeroth-order light. Removing all the zeroth-order light results in black objects appearing white, and white objects appearing black. We will remove the zeroth-order light completely when we do dark-field microscopy.

Abbe's theory "applied not merely to very small objects, to which he had first limited its application but that it must supply the ultimate explanation of the images of all objects seen by borrowed light; even of fence-poles, as he fiercely put it in a reply to one of his critics" (Spitta, 1907). Abbe's studies and conclusions are central when it comes to us incorporating Plato's teachings and Bacon's admonitions about the idols of the cave (see Chapter 1) when we do microscopy. Artifacts introduced by diffraction alter the relationship between the reality of the specimen and the image of the specimen.

RESOLVING POWER AND THE LIMIT OF RESOLUTION

The ability to see an isolated object is called detection. There is not any lower limit to the size of a bright object that can be detected against a black background as long as the light radiated from the object is bright enough for us to see or for a camera to capture. Resolving power, on the other hand, is the ability to distinguish closely spaced points as separate points. Our ability to resolve two bright dots on a black background with the naked eye depends on the distance the dots are from our eyes. When the closely spaced dots are far from our eyes, they appear as a single dot, but as we bring the dots closer and closer to our eyes, the two dots appear to separate and we say we can resolve the two dots as two separate dots. We can resolve the two dots when they are close to us because we increase the angle that subtends the rays that enter our eyes. The greater the angular aperture, the greater the number of diffraction orders emanating from a specimen we collect. Consequently, the image of the specimen on our retina becomes a more and more faithful representation of the object.

Under ideal conditions, where the objects are bright on a black background, our eyes have the power to resolve two objects that are about $70\,\mu m$ apart at a distance $25\,cm$ from our eyes. When objects are less than $70\,\mu m$ apart, our eyes are unable to collect more than the zeroth-order diffracted light and there is no structure in the image on our retinas. Seventy micrometers is the limit of resolution of the human eye. Squinting results in an increase in contrast, but a decrease in the resolving power of the eye (Figure 3-39).

The limit of resolution of a microscope is defined as the smallest distance apart two points in an object may be, with it still being possible to distinguish them as two separate points in the image. The resolving power of the light microscope is limited by diffraction. In this chapter, I will discuss three approaches used to characterize the resolving power of the light microscope—one by Ernst Abbe, one by John Strutt, also known as Lord Rayleigh, and one by Sparrow. In Chapter 12, I will describe the "near-field" approach to increasing the resolving power of the light microscope (Synge, 1928; Ash and Nicholls, 1972).

Abbe based his criterion for the limit of resolution of a light microscope on his use of high-contrast amplitude gratings as objects. He illuminated the gratings with axial, coherent light, and the objective lens captured the diffracted rays to form a Fraunhöfer diffraction pattern at

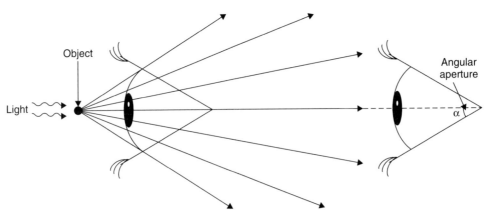

FIGURE 3-39 The ability of the human eye to create a faithful image of an object by collecting as many orders of diffracted light as possible depends on twice the angular aperture of the eye. The closer we move toward an object, the more diffraction orders we collect and the better we see the object.

the back focal plane of the objective lens and an image of the grating at the image plane. Abbe concluded that, in order to resolve the bars of the grating, the objective lens must be wide enough and/or close enough to the object to catch the first-order diffracted rays (Figure 3-40). If the spatial frequency of the specimen were too high for the objective lens to capture the first-order diffracted light, there would be no interference at the image plane, and a spherical wave originating at the back focal plane of the objective lens would illuminate the whole image plane.

The ability of the lens to catch the first order rays is characterized by the angular aperture of the objective lens (Figure 3-41). The angular aperture is defined by $\angle ACB$. If we define $\angle ACD$ as half the angular aperture and call it α, then we see that $\tan \alpha = r/s_o$, where r is the radius of the lens and s_o is the distance between the object and the lens. For infinity-corrected lenses, where the object is placed at the focal point, $\tan \alpha = r/f_o$. If α is small enough, we can use the small angle assumption: $\tan \alpha \approx \sin \alpha$.

A light microscope has the resolving power to form a distinct image of an amplitude grating with a distance d between the black bands when the angle between the first-order diffracted light and the zeroth-order diffracted light equals the angular aperture of the objective lens. The sine of the angle necessary to capture the first-order diffracted light is given by the ratio of the wavelength of light to the characteristic dimension of the object.

$$\sin \theta = \lambda/d$$

The limit of resolution can be determined by replacing θ with α, which yields:

$$d = \lambda/(\sin \alpha)$$

This equation serves only as an estimate of the limit of resolution, since it is based on the assumption of small angles. Abbe (1881) noticed that this equation is limited further by the assumption that the diffracted rays are not refracted as they pass from the cover glass to the air on the way to the objective lens. However, Abbe knew that when microscopists replaced the air between the specimen and the glass objective lens with immersion oil, they increased the resolving power of the microscope (Figure 3-42). The resolving power increases because the rays that were refracted by the air out of the collecting range of the objective lens could enter the objective lens in the presence of immersion oil with a refractive index of 1.515. Thus the resolving power of the lens depended on both the angular aperture and the refractive index of the medium between the specimen and the lens.

Abbe incorporated the effect of the medium between the specimen and the lens on the limit of resolution (d). Thus,

$$d = \lambda/(n \sin \alpha)$$

Abbe (1881) gave the name *numerical aperture* (NA) to $n \sin \alpha$. According to Abbe, the limit of resolution of the microscope when illuminating the specimen with axial coherent light is:

$$d = \lambda/NA$$

Given that the maximum angle that α can be is 90 degrees ($\sin \alpha = 1$), and given that the medium that would be most effective in refracting the diffracted rays into the lens will have the same refractive index as the lens (1.515), the maximal NA $=1.5$. Using blue light of 400 nm, the limit of resolution of the light microscope will be:

$$d = 400\,nm\,/\,1.515 = 264\,nm$$

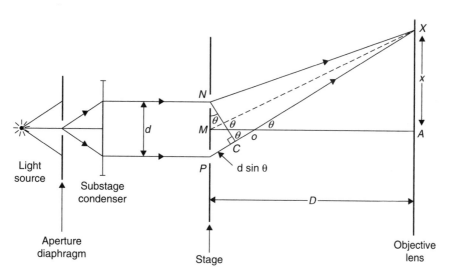

FIGURE 3-40   According to Ernst Abbe, a microscope cannot resolve objects smaller than a certain length (d), because the angle of the first-order diffracted light is too great for that light to be captured by the objective lens.

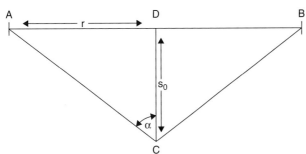

FIGURE 3-41 The angular aperture (α) of a lens depends on the radius (r) of the lens and the distance (s_o) between the object and the lens.

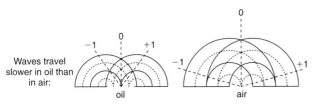

FIGURE 3-42 Because waves travel more slowly through oil than through air, the angle between the first-order diffracted wave and zeroth-order diffracted wave is smaller in oil than in air. This allows oil-immersion objectives to resolve finer details than dry objectives.

The resolving power of the light microscope can be increased by using light of shorter wavelength. The wavelength can not be shortened without limit, because as the wavelength decreases, the energy of each photon increases, as will be discussed in Chapter 12, and consequently, the short wavelength light can damage living specimens.

The resolving power of the light microscope can also be increased by illuminating the object with oblique coherent light (oblique illumination) by using a sub-stage condenser (Figure 3-43). While the zeroth-order light travels through one edge of the objective lens, the first-order light travels through the opposite edge of the lens, effectively doubling the angular aperture. In the case of oblique illumination, the limit of resolution is estimated by the following formula:

$$d = \lambda/(2NA)$$

Given that the maximum angle that α can be is 90 degrees ($\sin \alpha = 1$), and given that the medium that would be most effective in refracting the diffracted rays into the lens will have the same refractive index as the lens (1.515), the maximal NA =1.5. Using blue light of 400 nm, then the limit of resolution of the light microscope will be about 132 nm.

In Chapter 4, I will discuss Köhler illumination, where the specimen is illuminated with a cone of light whose rays illuminate the specimen at a variety of angles. When illuminating the specimen with Köhler illumination, the limit of resolution is given by the following inequalities:

$$0.5\lambda/NA < d < \lambda/NA$$

If lenses were to be made from materials with higher refractive indices, then we could use immersion media with higher refractive indices, and, consequently, we could reduce the limit of resolution of the light microscope. Lenses with a refractive index of 1.515 were chosen because the early microscopists were restricted to using polychromatic sunlight to get bright enough images at high magnification (especially for making photomicrographs). This required the development of achromatic or apochromatic lenses made out of crown ($n \approx 1.5$) and flint ($n \approx 1.6$) glass, which in combination, reduced the chromatic aberration; but had the unintended consequence of limiting the resolving power of the microscope. However, now that high-intensity, solid state, monochromatic light sources are available, we could increase the resolving power of the light microscope by using monochromatic light with aspherical lenses made of high refractive index glass or easy-to-mold high refractive index plastic. If we made lenses out of plastics (Roukes, 1974) or other polymers that have refractive indices of 2.5, the limit of resolution could be as low as 80 nm. This is made even more feasible by the development of solid state imaging chips that respond to very low light levels (see Chapter 13).

Lord Rayleigh and Hermann von Helmholtz (Wright, 1906) took another approach to understanding the limit of resolution of a microscope. They considered that each point of an object acted as an independent point source of light, much like the stars in the night sky. According to this view, each point source of light sends out spherical waves that illuminate the lens. Since the lens is an aperture, the image of each point of light would actually be a diffraction pattern of the lens (diameter = 2r). Given that the apertures of lenses are wide compared to the point sources in question, the diffraction fringes would be close to the image that would have been made if the wavelength of light were zero. However, even when the fringes are close, they still exist, and consequently, each object point is inflated at the image plane and surrounded by concentric fringes. That is, they form a diffraction pattern that looks like an Airy disc.

Rayleigh then considered how an image of two nearby self-luminous dots would appear. Rayleigh assumed that the light emanating from the two points were incoherent and thus did not interfere on the image plane (Figure 3-44). There may be some coherence between points that are so close to each other, however.

The fact that a single point of light produces an Airy disc when the light passes through the aperture of a lens is significant when we want to localize objects such as proteins or quantum dots that are below the limit of resolution of the light microscope. As a result of diffraction, any point of light will be enlarged in the image plane. For example, a 10 nm object will be inflated to about 218 nm by an objective lens with a numerical aperture of 1.4. An object smaller than the limit of resolution is delocalized as

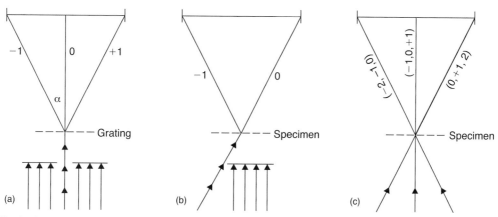

FIGURE 3-43 Illuminating a specimen with oblique illumination (b) effectively doubles the angular aperture of the objective lens compared with axial illumination (a). Illuminating a specimen with a solid cone of light (c), as is done when using Köhler illumination, produces a composite image composed of a mixture of high- and low-resolution images.

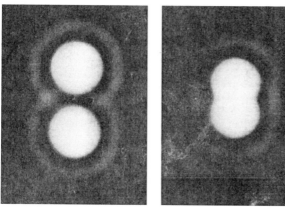

FIGURE 3-44 Two Airy discs that are clearly resolved in the image plane (left) and two Airy discs that overlap in the image plane (right).

a result of diffraction and takes up more area in the image plane than it should.

As the distance between the two self-luminous objects decreases, the Airy discs overlap more and more, until we are unable to distinguish whether we are looking at one object or two. Lord Rayleigh suggested a criterion, which can be used to determine whether or not we have the ability to tell whether we are looking at one object or two when their Airy discs overlap. Rayleigh suggested that there is sufficient contrast to distinguish two points in the image plane if, in intensity scans, the central maximum of one point lies over the first minimum of the other (Figure 3-45). Under this condition, the intensity of the valley between the two points is approximately 80 percent of the intensities of the peaks. Since the first minimum occurs a distance of 1.22λ away from the peak intensity, the limit of resolution is given by the following formula:

$$d = 1.220\,\lambda/(2NA)$$

When using 400 nm light and a lens with an NA of 1.4, the limit of resolution is 174 nm. The resolving power given

by Rayleigh's criterion is less than that given by the Abbe criterion because Rayleigh assumed that contrast must also be considered when you consider resolution. Rayleigh proscribed that there must be a 20 percent dip in intensity between two points in order to resolve them. Thus, whereas Abbe considered image formation to be diffraction limited, Rayleigh considered image formation to be diffraction- and contrast-limited.

C. Sparrow has shown that as long as the two peaks in the image plane are separated enough to produce a minute dip in intensity between the two points, the two points can be resolved using analog and/or digital contrast-enhancement techniques (see Chapters 13 and 14).

CONTRAST

In the absence of contrast, resolution is meaningless. If there is not enough contrast to distinguish two points, it does not matter how widely they are separated. Contrast is defined as the difference in intensity between the image point (I_i) and the background points (I_b). The percent contrast is given by the following formula:

$$\% \text{ contrast} = (I_b - I_i)/I_b \times 100\%.$$

Absorption contrast is generated when the object or the background differentially absorbs the incident light. In terms of wave theory, absorption results in a reduction in the amplitude of the waves. In absorption contrast, part of the energy of the illuminating wave is absorbed by the specimen and thus removed from the zeroth- and higher-order diffracted waves. When these waves reunite in the image plane, the image is darker than the surround. Since absorption often varies with wavelength, the specimen may decrease the amplitude of a limited range of wavelengths giving color contrast either naturally or through staining

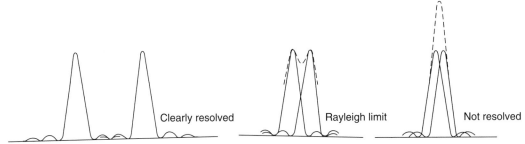

FIGURE 3-45 According to Lord Rayleigh, two points that produce Airy discs can just be resolved in the image plane when the central maximum of one point lies over the first minimum of the other. Under this condition, there is sufficient contrast to resolve the two points because the sum of the point's intensities midway between the peaks is about 80% of the intensity of each peak. When the points are too close, the Airy discs overlap so that the intensity midway between the two points is equal to or greater than the intensity of the individual points.

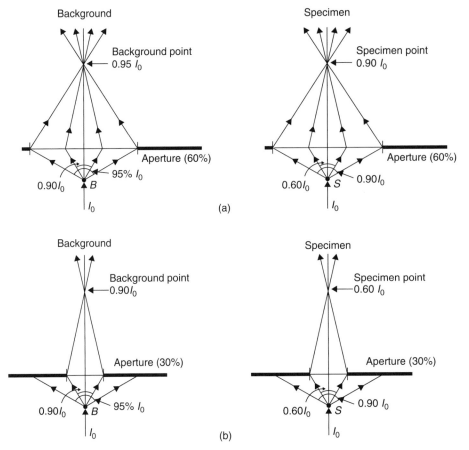

FIGURE 3-46 Reducing the opening of the aperture diaphragm from 60° (a) to 30° (b) increases the scattering contrast of the image.

(Conn, 1933). Consequently, the image is a different color than the surround. Since the absorption of visible light is determined by the structure of the absorbing molecule, absorption contrast provides information on the chemical structure of the specimen. This can be quantified when combined with microspectrophotometry.

Scattering contrast can arise from making use of the differential diffraction of light that occurs when objects with different spatial angular wave numbers are illuminated. An object with high spatial angular wave numbers diffracts

part of the incident intensity beyond the aperture of the objective lens, and thus, its image appears darker than the surround. Therefore contrast increases when the aperture of the objective lens decreases because fewer and fewer diffraction orders are captured. The orders that are not captured are not able to recombine to form the image. Thus a transparent object would have no contrast if every diffracted ray were captured by the objective lens. Scattering contrast can be increased by decreasing the effective NA of the objective. Increasing the contrast by decreasing the NA directly decreases the resolution of the light microscope. This is just one of the many tradeoffs that occur in microscopy. Finding the best compromise comes from experience.

In order to see how stopping down the aperture of the objective lens increases the contrast (Figure 3-46), consider a point in a homogeneous background that transmits 100 percent of the intensity of the illuminating system (I). Ninety and 95 percent of the light from this point is scattered within an angle of 30 and 60 degrees, respectively. Consider a point in the specimen that absorbs 5 percent of the light from the illuminating system and scatters 60 percent of the light within 30 degrees and 90 percent within 60 degrees. The percent contrast due to absorption would be 5 percent, according to the following formula:

$$\text{percent contrast} = (1.00I - 0.95\,I)/(1.00I)$$
$$\times 100\% = 5\%$$

If we use a lens with a 60 degree aperture, then the intensity of the surround is equal to $1.00I \times 0.95 = 0.95I$, the intensity of the specimen is equal to $0.95I \times 0.90 = 0.855I$, and the percent contrast is 10 percent, according to the following formula:

$$\text{percent contrast} = (0.95I - 0.855I)/0.95I$$
$$\times 100\% = 10\%$$

When we close down the aperture, the resolving power of the light microscope decreases and the limit of resolution becomes 282 nm according to the following formula:

$$\text{limit of resolution} = 1.22(400\,\text{nm})/(2\sin 60°)$$
$$= 282\,\text{nm}$$

If we reduce the aperture to a 30 degree angle, then the intensity of the surround equals $1.00I \times 0.90 = 0.90I$, the intensity of the specimen = $0.95I \times 0.60 = 0.57I$, and

the percent contrast is about 37 percent, according to the following formula:

$$\text{percent contrast} = (0.90I - 0.57I)/0.90I$$
$$\times 100\% = 36.67\%$$

The limit of resolution, according to Rayleigh will become $1.22(400\,\text{nm})/(2\sin 30°) = 488\,\text{nm}$.

We have learned that the wave nature of light affects our ability to resolve objects with the light microscope. We have found that we can increase contrast and thus detect transparent objects with the light microscope if we decrease the aperture of the lens. However, as we increase the contrast in this manner, *pari passu*, we decrease the resolving power, and are not able to resolve the fine details of the object. Resolution and contrast are complementary properties.

- Good resolution is meaningless in the absence of contrast.
- The conditions that tend to enhance contrast are exactly those that tend to destroy resolving power.

One of the main goals of light microscopy is to increase contrast while maintaining the diffracted-limited resolving power of the light microscope. One of the advantages of the light microscope over the electron microscope is that living organisms can be observed in the light microscope. Therefore it is our goal to increase contrast, while maintaining the diffraction-limited resolving power of the light microscope in order to visualize living specimens and the processes that occur within them.

WEB RESOURCE

http://www.kettering.edu/~drussell/Demos.html

Chapter 4

Bright-Field Microscopy

Our naked eye is unable to resolve two objects that are separated by less than 70 μm. Perhaps we are fortunate that, without a microscope, our eyes are unable to resolve small distances. John Locke (1690) wrote in *An Essay Concerning Human Understanding*,

> We are able, by our senses, to know and distinguish things.... if that most instructive of our senses, seeing, were in any man a thousand or a hundred thousand times more acute than it is by the best microscope, things several millions of times less than the smallest object of his sight now would then be visible to his naked eyes, and so he would come nearer to the discovery of the texture and motion of the minute parts of corporeal things; and in many of them, probably get ideas of their internal constitutions: but then he would be in a quite different world from other people: nothing would appear the same to him and others: the visible ideas of everything would be different. So that I doubt, whether he and the rest of men could discourse concerning the objects of sight, or have any communication about colours, their appearances being so wholly different. And perhaps such a quickness and tenderness of sight could not endure bright sunshine, or so much as open daylight; nor take in but a very small part of any object at once, and that too only at a very near distance. And if by the help of such microscopical eyes (if I may so call them) a man could penetrate further than ordinary into the secret composition and radical texture of bodies, he would not make any great advantage by the change, if such an acute sight would not serve to conduct him to the market and exchange; if he could not see things he was to avoid, at a convenient distance; nor distinguish things he had to do with by those sensible qualities others do. He that was sharp-sighted enough to see the configuration of the minute particles of the spring of a clock, and observe upon what peculiar structure and impulse its elastic motion depends, would no doubt discover something very admirable: but if eyes so framed could not view at once the hand, and the characters of the hour-plate, and thereby at a distance see what o'clock it was, their owner could not be much benefited by that acuteness; which, whilst it discovered the secret contrivance of the parts of the machine, made him lose its use.

Alexander Pope (1745) considered the same question in *An Essay on Man*:

> *Why has not Man a microscopic eye?*
> *For this plain reason, Man is not a Fly.*
> *Say what the use, were finer optics given,*
> *T'inspect a mite, not comprehend the heaven.*

The eye is a converging lens that produces a minified image of the object on the retina. The dimensions of the cells that make up the light-sensitive retina limit the ability of the human eye to resolve two dots that are closer than 70 μm from each other. The retina contains approximately 120 million rods and 8 million cones packed into a single layer. In the region of highest resolution, known as the fovea, the cones are packed so tightly that they are only 2 μm apart. Still, this distance between the cones limits the resolving power of our eyes. If a point of light originating from one object falls on only one cone, that object will appear as a point. If light from two objects that are close together fall on one cone, the two objects will appear as one. When light from two points fall on two separate cones separated by a third cone, the two points can be clearly resolved. The resolving power of the eye can be increased slightly by eye movements that vary the position of the cones.

The numerical value of the limit of resolution of the human eye was first discovered by Robert Hooke in 1673. Birch (1968) wrote:

> Mr. Hooke made an experiment with a ruler divided into such parts, as being placed at a certain distance from the eye, appeared to subtend a minute of a degree; and being earnestly and curiously viewed by all the persons present, it appeared, that not any one present, being placed at the assigned distance, was able to distinguish those parts, which appeared of the bigness of a minute, but that they appeared confused. This experiment he produced, in order to shew, what we cannot by the naked eye make any astronomical or other observations to a greater exactness than that of a minute, by reason, that whatever object appears under a lens angle, is not distinguishable by the naked eye; and therefore he alleged, that whatever curiosity was used to make the divisions of an instrument more nice, was of no use, unless the eye were assisted by other helps from optic glasses.

In order for two points to appear as separate points, light from those points must enter the eye forming an angle greater than one minute of arc. Thus the object must be brought very close to the eye. However, due to the limitation of our eye to focus at close distances, a specimen can be brought up only to the near point of the eye, which is about 25 cm from our eye (see Chapter 2, Figure 2-40). A microscope makes it possible to increase the visual angle, so that light, emanating from two near, but separate points, can enter the eye, forming an angle that subtends more than one minute of arc such that the light from the two separate points fall on separate cones (Figure 4-1; Gage, 1908). The eye of the observer plays an integral role as the optical interface between the microscope and the effigy of the specimen produced by the brain.

To make a very simple microscope, place an index card that contains a minute pinhole, smaller than your pupil, between your eye and an object closer than the near point of your eye. Then look at an object that would be blurry at this distance without the very simple pinhole microscope. An object that was blurry at this close distance will appears clear through the pinhole. Moreover, since it is close to your eye, it also appears magnified, compared to how it would look when it was placed at the distance where it would look sharp without the pinhole. The pinhole acts as a microscope by reducing the angle of the cone of light that comes from each point in the object without decreasing the visual angle of an object placed so close to your eye. Since, with the pinhole, the image on your retina is formed by cones of light with smaller zones of confusion, objects closer than the near point of your eye that would have formed blurry images on the retina now form sharp images.

Thus, microscopes are necessary to create large visual angles that allow us to resolve microscopic objects. We need a microscope if we want to see a mite, or any other microscopic aspect of the natural world. Indeed the microscope opened up a whole new world to seventeenth and eighteenth century microscopists (Hooke, 1665, 1678, 1705; Leeuwenhoek, 1673, 1798; Malpighi, 1675–1679, 1686;

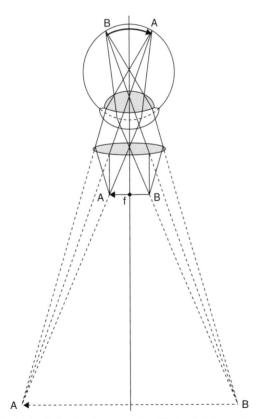

FIGURE 4-1   A simple microscope placed in front of the eye increases the visual angle, thus producing an enlarged image of a microscopic specimen on the retina. The specimen appears to be located at the near point of the relaxed eye and magnified.

Grew, 1672, 1682; Swammerdam, 1758). Augustus de Morgan (1872) wrote:

Great fleas have little fleas upon their backs to bite e'm,
And little fleas have lesser fleas, and so on ad infinitum.

According to Emily Dickinson (1924):

Faith is a fine invention
When gentlemen can see,
But microscopes are prudent
In an emergency.

The word microscope comes from the Greek words μικρος (small) and σκοπειν (to see). The word microscope was coined by Giovanni Faber on April 13, 1625. The bright-field microscope is, perhaps, one of the most elegant instruments ever invented, and the first microscopists used the technologically advanced increase in the resolving power of the human eye to reveal that the workmanship of the Creator can be seen at the most minute dimensions (Hooke, 1665; Grew, 1672, 1682; Leeuwenhoek, 1674–1716; Malpighi, 1675–1679; Swammerdam, 1758). The bright-field microscope also has been instrumental in revealing the cell as the basic unit of life (Dutrochet, 1824; Schwann, 1847; Schleiden, 1849), the structural basis for the transmission of inherited characteristics (Strasburger, 1875; Flemming, 1880), and the microscopic basis of infectious diseases (Pasteur, 1878, 1879; Koch, 1880; Dobell and O'Connor, 1921). Color plates 1 and 2 show examples of cork cells and chromatin observed with bright-field microscopy.

The bright-field microscope also has had a tremendous influence in physics and chemistry in that it made it possible for Robert Brown (1828, 1829) to discover the incessant movement of living and nonliving particles, now known as Brownian motion or Brownian movement (Deutsch, 1991; Ford, 1991, 1992a, 1992b, 1992c; Rennie, 1991; Bown, 1992; Wheatley, 1992; Martin, 1993). In 1905, Albert Einstein (1926) analyzed Brownian motion and concluded that the movement occurred as a result of the statistical distribution of forces exerted by the water molecules surrounding the particles. Jean Perrin (1909, 1923) confirmed Einstein's hypothesis by observing Brownian motion under the microscope and used his observations, along with Einstein's theory, to calculate Avogadro's number, the number of molecules in a mole. Ernst Mach and Wilhelm Ostwald, who were the last holdouts to accept the reality of atoms and molecules, became convinced in the reality of molecules from the work done on Brownian motion. These influential scientists were held back from accepting the evidence of the existence of molecules from other kinds of physicochemical data because of their positivist philosophy, which could be summed up by the phrase "seeing is believing."

COMPONENTS OF THE MICROSCOPE

A simple bright-field microscope, like that used by Leeuwenhoek (1674–1716), Jan Swammerdam, Robert Brown, and Charles Darwin consists of only one lens, which forms a magnified, erect, virtual image of the specimen (Ford, 1983, 1985). By contrast, the compound microscope uses two lens systems to form an image. The primary lens system (object glass or objective lens) captures the light that is diffracted by the object and then forms a real intermediate image that is further magnified by a second lens system, known as the eyepiece or ocular.

There are currently two standard optical configurations. Traditional objectives are designed to produce a real, magnified, inverted image 16 or 17 cm behind the objective lens when the specimen is placed in front of the front focal plane of the objective (Gage et al., 1891). Most microscope objectives manufactured today produce an image of an object at infinity when the object is placed at the front focal plane of the objective (Lambert and Sussman, 1965). When infinity-corrected objectives are used, an intermediary lens, known as the tube lens, is placed behind the objective lens. The tube lens focuses the parallel rays leaving the objective lens, and produces a real image at the back focal plane of the tube lens. This arrangement, by which parallel rays pass from the objective lens, minimizes the introduction of additional aberrations introduced by rays coming from every conceivable angle going through extra optical pieces placed between the objective lens and the ocular. Objectives that produce an intermediate image 160 mm or 170 mm behind the lens are marked with 160 or 170, respectively. Objectives that produce the intermediate image at infinity are marked with ∞.

Each objective lens is labeled with a wealth of information (Figure 4-2). The most prominent number signifies the transverse magnification (m_T) of the intermediate image. Remember from Chapter 2, that

$$m_T = y_i/y_o = -s_i/s_o$$

The objective lenses often are surrounded by a thin band whose color represents the magnification of the objective.

From the beginning, microscopists and inventors realized that microscopes do not magnify objects faithfully. In fact the microscope itself introduces fictions or convolutions into the image that are not part of the object itself. Over the years, lenses were developed that were as free as

possible from spherical aberrations (Kepler, 1604; Descartes, 1637; Molyneux, 1692, 1709; Gregory, 1715, 1735; Smith, 1738; Martin, 1742, 1761, 1774; Adams, 1746, 1747, 1771, 1787, 1798; Baker, 1742, 1743, 1769; McCormick, 1987) and chromatic (Dollond, 1758; Amici, 1818; Lister, 1830; Beck, 1865; Abbe, 1887; Cheshire, 1905; Disney et al., 1928; Clay and Court, 1932; von Rohr, 1936; Payne, 1954; Feffer, 1996)—aberrations that were once thought to be absent in the human eye (Paley, 1803).

Due to dispersion, which is the wavelength-dependence of the refractive index (see Chapter 2), a single lens will not form a single image, but a multitude of images, each of which is a different color and each of which is offset axially and laterally from the others. For example, when we use an aspheric objective that is not color-corrected, the microscopic image will go from bluish to reddish as we focus through the object. This chromatic aberration can be mitigated, but not eliminated by using an achromatic doublet, which was invented by Chester Moor Hall, John Dolland, and James Ramsden. The achromatic doublet is made by cementing a diverging lens made of high-dispersion flint glass to the converging lens made of low-dispersion crown glass. The flint glass mostly cancels the dispersion due to the crown glass, while only slightly increasing the focal length of the lens.

When we plot the focal length of an aspheric lens as a function of wavelength, we get a monotonic plot, where wavelengths in the blue range experience shorter focal lengths, wavelengths in the red range experience longer focal lengths, and wavelengths in the green range experience intermediate focal lengths (Figure 4-3). When we plot the focal length of an achromatic lens as a function of wavelength, we

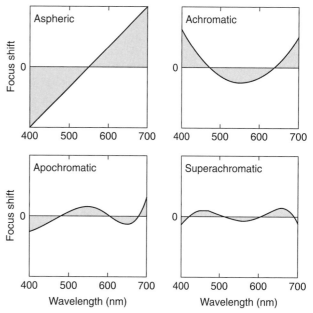

FIGURE 4-3 Chromatic aberration means that the focal length of an objective lens is wavelength dependent. There is a large variation in focal lengths with wavelength in aspheric objectives, a smaller variation in achromatic objectives, an even smaller variation in apochromatic objectives, and the smallest variation in superachromatic objectives.

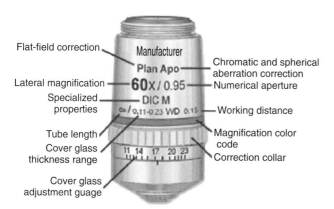

FIGURE 4-2 Each objective is labeled with an abundance of useful information.

get a parabolic plot, where the focal length for a wavelength in the blue region (spectral line F) is similar to the focal length for a wavelength in the red region (spectral line C).

Ernst Abbe developed an objective lens that introduced less chromatic aberration than the achromats and coined the term apochromat, which means "without color," to characterize his lens design (Haselmann, 1986; Smith, 1987). He defined apochromats as lenses that had the same focal length for three widely spaced wavelengths in the visible range. Apochromats can be defined as objective lenses whose differences in focal lengths do not exceed λ/4 throughout the spectral range from 486.1 nm (spectral line F), through 546.1 nm (spectral line E) to 6563 nm (spectral line C). However, this definition remains fluid as microscope lens makers are striving to make "super" apochromatic objective lenses that each have a single wavelength-independent focal length all the way from the ultraviolet to the infrared wavelengths.

More lens elements are required to effectively reduce chromatic aberration and this increases the cost of the objective dramatically. Consequently, apochromats are the most expensive lenses and achromats are the least expensive objective lenses available in good microscopes. Microscope makers make a variety of lenses that are more corrected than achromats, but less corrected than apochromats. These semi-apochromats and fluorites, which go by a variety of manufacturer-specific names, are intermediate in cost.

A single lens also introduces spherical aberrations because the rays that enter the peripheral region of the lens experience a shorter focal length than do the paraxial rays. Spherical aberration often is corrected at the same time as chromatic aberration so that spherical aberration in aspheric objective lenses is absent only for green wavelengths. Green light was chosen, not because it produces diffraction-limited images with the highest resolution, but because it is the color that people can see with the least amount of eye strain. Spherical aberration is nearly absent for all wavelengths for apochromatic objective lenses. Again, the spherical aberration is intermediate in achromatic and semi-apochromatic objective lenses. Objectives can be readily tested for chromatic and spherical aberration by using a homemade Abbe-test plate (Sanderson, 1992).

The resolving power of the objective lens is characterized by its numerical aperture (NA; Chapter 3). The numerical aperture of the lens describes the ability of the lens to resolve two distinct points. The numerical apertures of objectives typically vary from 0.04 to 1.40. Objectives with NAs as high as 1.6 were made over a century ago (Spitta, 1907), but were not popular as a result of their requirement for special immersion oils and cover glasses. However, objectives with NAs between 1.49 and 1.65, which are used for Total Internal Reflection Fluorescence Microscopy (TIRFM; Chapter12), are being reintroduced. The NA of an objective is given by the small print number that either follows or is printed under the magnification.

The brightness of the image, in part, is proportional to the square of the numerical aperture and inversely proportional to the square of the transverse magnification (Naegeli and Schwendener, 1892; Beck, 1924).

$$\text{Brightness} \propto (NA/m_T)^2$$

In order to produce a maximally bright image we may want to use an objective lens with the highest possible NA and the lowest possible magnification. However, the brightness of the image does not only depend on the geometry of the lens but also on the transparency of the optical glasses used to construct the lenses (Hovestadt, 1902). In the past there has been a tradeoff between the transparency of an objective lens and the number of corrections. For example, the fluorites, which are composed of lens elements made of calcium fluoride, were more transparent to the ultraviolet light used for fluorescence microscopy than were the apochromats made with glass lens elements. Manufacturers are striving to make highly corrected lenses that are transparent throughout the spectrum from ultraviolet to infrared.

Objective lenses must also be corrected so that the specimen is in sharp focus from the center to the edge of a flat image plane. Objectives that are designed to produce flat fields are labeled with F- or Plan-; for example, F-achromat or Plan-Apochromat. The Plan-objectives are more highly corrected for curvature of field than the F-objectives. It is the nature of optics that it takes one lens element to produce a magnified image of the object and many lens elements to eliminate or perform a deconvolution on the aberrations or convolutions introduced by the "imaging" lens element. Following are some examples of the lens combinations used to make highly corrected objective lenses (Figure 4-4).

Objectives are made with features that are useful for doing specific types of microscopy. For example, some objectives have an iris at the back focal plane, which can reduce the NA of the objective. These objectives may have the word Iris printed on them. This feature, which increases contrast at the expense of resolution, is useful when doing dark-field microscopy (see Chapter 6).

Some objectives are made out of completely homogeneous glass, known as strain-free glass. A lens made out of strain-free glass does not depolarize linearly polarized

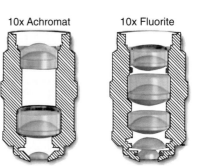

10x Achromat   10x Fluorite   10x Apochromat

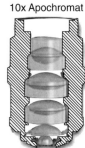

FIGURE 4-4 Chromatic aberration is reduced by building an objective lens with additional achromatic doublets and triplets.

light that passes through the lens. These objectives may have the word Pol printed on them. They are useful when doing polarization microscopy (see Chapter 7).

The working distance of an objective lens is defined as the distance between the front of the lens and the top of a cover glass (Gage, 1917). Typically, the working distances, which vary between 0.1 and 20 mm, are inversely proportional to the magnifications and the numerical apertures of objective lenses. However, some long working distance objectives are made that also have high magnifications and numerical apertures. These objectives, which may be marked with an LD, are especially useful for doing micromanipulation.

When doing micromanipulation, sometimes it is helpful to immerse the lens directly in the dilute aqueous solution that bathes the specimen. Water immersion objectives, which are marked with a W, are good for this use. Objectives also are made that can be used in solutions with higher refractive indices, like glycerol. Glycerol immersion objectives are marked with a Glyc. Lastly, immersion oils are used to increase the resolving power of the microscope. Objective lenses that are made to be used with oil are marked with the word Oil. Some objective lenses can be immersed in media with refractive indices from 1.333 to 1.515. Depending on the manufacturer, these objectives are marked with Imm or W/Glyc/Oil.

Objectives are designed so that the cover glass acts as the first lens in objectives that are corrected for spherical aberration. Most objectives used in transmitted light microscopy are marked with 0.17, which means it is corrected for use with a 0.17 mm (#1½) cover glass. Some objectives are made for use without cover slips and are marked with /0. Objectives, which are insensitive to cover glass thickness, are marked with /-. Some objectives can be used with a variety of cover glasses. These objectives have a correction collar and may be marked with korr.

The cover glass introduces an increase in the optical path length of the diffracted rays that pass through it (Figure 4-5). The magnitude of the increase depends on the angle or order of the diffracted rays. The more oblique the rays are; the greater the increase in optical path length. The thicker the cover glass, the greater the difference between rays emanating at different angles from the same point. The highly diffracted rays appear to come from a nearer object than the rays that are diffracted from a smaller angle. Consequently, the different diffraction order rays will be focused at different distances from the image plane. This results in a zone of confusion instead of a single point, and contributes to spherical aberration. The manufacturers of the objective lenses design the objectives to compensate for the increase in the optical path induced by a given cover glass thickness. The Abbe Test Plate can be used to determine the effect of cover glass thickness on spherical aberration.

One of the characteristics of objective lenses is their cost; and unfortunately cost will probably play the biggest part in your choice of objectives. As I discussed in the last

chapter, resolution and contrast are often competing qualities. However, as I will discuss in Chapter 14, we can use analog or digital image processing to enhance the contrast in electronically captured images. Thus, in our constant tug of war between contrast and resolution, we can opt for an objective that will provide the highest resolution and/or the greatest brightness at the expense of contrast and then enhance the contrast of the high resolution image using image processing techniques.

A real image of the specimen formed by the objective lens falls on the field diaphragm between the front focal plane of the ocular and the eye lens of the ocular itself. The ocular-eye combination forms a real image of the intermediate image on the retina, which appears as a magnified virtual image 25 cm in front of our eyes. Since the intermediate image is inverted with respect to the specimen, the virtual image also is inverted with respect to the specimen. Oculars typically add a magnification of 5x to 25x to that produced by the objective. Moreover, a turret that contains a series of low magnification lenses can be inserted into the microscope just under the oculars. These ancillary magnification lenses increases the magnification of the virtual image by one or two times. Most oculars used in binocular microscopes can be moved laterally to adjust for your interpupillary distance. One or both oculars will have a diopter adjustment ring, which can be turned to compensate for difference in magnification between your two eyes. When the interpupillary distance and the diopter adjustment are set correctly, it is actually relaxing to sit in front of a microscope all day.

When correcting aberrations in microscopes, designers take into consideration the objectives, tube lens, and oculars. Depending of the objective lens, special matching

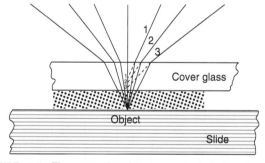

FIGURE 4-5 The cover glass introduces spherical aberration. The higher-order diffracted rays are refracted more than the lower-order diffracted rays, resulting in the point being imaged as a zone of confusion. This occurs because the lower order-diffracted rays appear to come from a position close to the object and the higher order-diffracted rays, when they are traced back through the cover glass, seem to come from a position between the real object position and the objective lens. The greater the cover glass thickness, the greater the amount of spherical aberration that will be introduced by the cover glass. A correction for the spherical aberration introduced by an objective lens also corrects the spherical aberration introduced by a cover glass of a certain thickness (e.g., 0.17 mm). Objectives with correction collars can correct for the spherical aberration introduced by a range of cover glass thicknesses.

oculars may have to be used with it in order to obtain the best image. There are several common types of oculars in use, and they fall into two categories: negative and positive (Figure 4-6). The Huygenian eyepiece, which was designed by Huygens for a telescope ocular, is composed of two plano-convex lenses. The upper lens is called the eye lens and the lower lens is called the field lens. The convex sides of both lenses face the specimen. Approximately midway between the two lenses there is a fixed circular aperture that defines the field of view and holds an ocular micrometer. This is where the intermediate image formed by the objective lens is found. Since the object focus is behind the field lens, the Huygenian eyepiece is an example of a negative ocular. Huygenian oculars are found on relatively routine microscopes with achromatic objectives.

The Ramsden eyepiece is an example of a positive ocular, whose object focal plane is in front of the field lens. The Ramsden eyepiece consists of two plano-convex lenses where the convex side of both lenses face the inside of the eyepiece. The circular aperture that defines the field of view and holds the ocular micrometer is below the field lens.

Compensating eyepieces, which can be either negative or positive, contain a number of lens elements. Compensating oculars are important for correcting the residual chromatic aberration inherent in the design of some objective lenses from the same manufacturer. As digital imaging techniques develop (see Chapters 13 and 14), fewer and fewer people are looking at microscopic images through the oculars and more and more people are looking at the images displayed on a monitor. Consequently, optical corrections to mitigate aberrations are no longer included in the oculars. The corrections are completed in the objective lenses or the objective lens-tube lens combination.

Oculars may be labeled with a field of view number, which represents the diameter (in millimeters) of the field that is visible in the microscope when using those oculars. The diameter of the field can be obtained by dividing the field of view number by the magnification of the objective lens and any other lenses between the objective and the ocular. This is helpful in estimating the actual size of objects. The field of view numbers vary from 6.3 to 26.5. The fields of view in a microscope equipped with an ocular with a field of view number of 20, and 10x, 20x, 40x, and 100x objectives is 2, 1, 0.5, and 0.2 mm, respectively.

The total transverse magnification of the compound microscope is given by the product of the magnification of the objective lens (obj), the ocular (oc), and any other intermediate pieces (int), including the optivar.

$$m_{total} = (m_{obj})(m_{int})(m_{oc})$$

In microscopy, there is a limit to the amount of magnification that is useful, and beyond which, the image quality does not improve. This is reached when two points in the image appear to send out rays that subtend one minute of arc, which is the resolution of the human eye.

What is the maximum useful magnification of a light microscope? Let us assume that the final image is formed 25 cm from the eye and the smallest detail visible in the specimen is given by the following equation: $d = 1.22 \lambda/(2NA)$, and $d = 0.161 \mu m$. Since the eye can just resolve two points, $70 \mu m$ apart, the magnification necessary for the eye to resolve two points (i.e., useful magnification) is $70 \mu m/0.161 \mu m = 435x$. It is uncomfortable to work at the limit of resolution of the eye, so we typically use a magnification two to four times greater than the calculated useful magnification, or up to 1740x. Higher magnifications can result in a decrease in image quality since imperfections of the optical system and vibration become more prominent at high magnifications. As a rule of thumb, the optimal magnification is between 500 (NA) and 1000 (NA). However, when working with good lenses and stable microscopes, it is possible to increase the magnification to 10,000x (Aist, 1995; Aist and Morris, 1999).

The specimen is illuminated by the sub-stage condenser (Wenham, 1850, 1854, 1856). There are a variety of sub-stage condensers that have different degrees of corrections (Figure 4-7). The Abbe condenser, which originally was designed to provide an abundance of axial rays, is neither

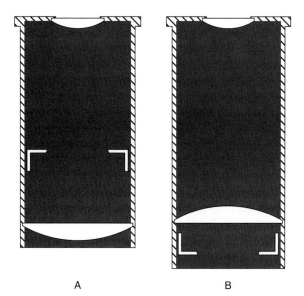

A   B

FIGURE 4-6 A negative ocular (A) has the field diaphragm between the eye lens and field lens. A positive ocular (B) has the field diaphragm in front of the field lens.

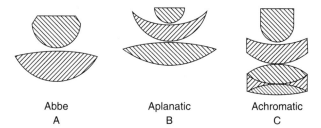

Abbe   Aplanatic   Achromatic
A   B   C

FIGURE 4-7 Lens elements in sub-stage condensers. An Abbe chromatic sub-stage condenser (A), an aplanatic sub-stage condenser (B), and an achromatic sub-stage condenser (C).

achromatic nor aplanatic (Martin, 1993). Other condensers designed to use oblique rays as well as axial rays are corrected for spherical aberration and/or chromatic aberration. The aplanatic condenser is corrected for spherical aberration and the achromatic condenser is corrected for both spherical and chromatic aberration.

A sub-stage condenser may contain a rotating ring with annuli, prisms, or slits at its front focal plane that allows us to do many forms of light microscopy, including dark-field and phase-contrast (see Chapter 6), differential interference contrast (see Chapter 9), and amplitude modulation contrast (see Chapter 10). Some condensers are made with long working distances, which are convenient to use when performing micromanipulations on inverted microscopes.

The objectives, oculars, and sub-stage condenser usually are supported by either an upright or inverted microscope stand. The typical microscope, which is on an upright stand, is ergonomically designed for most uses. However, inverted microscopes make manual manipulations, including micro-injection, microsurgery, and electrophysiological measurements much easier. The microscope stand often is overlooked as an important part of the microscope. It provides stability for photomicrography or video microscopy and flexibility to convert the stand for various types of optical microscopy. In the long run, it is worth getting the most stable, flexible, and, consequently, most expensive microscope stand. The stand contains the coarse and fine focus knobs that raise or lower the nosepiece that holds the objectives.

The stage is where the specimen is placed. Inexpensive glide stages let you position the specimen by sliding the stage plate over a greased plate placed below it. In order to exactly position a specimen, you need a stage that has controls to translate the specimen in the XY direction. Such a stage may be motorized and controlled by computers. Rotating stages are useful for polarized light microscopy (see Chapter 7) and other forms of microscopy that utilize polarized light or depend on the direction of shear (see Chapters 9 and 10).

Originally the specimen in a bright-field microscope was illuminated by sunlight and a heliostat, which rotated with the speed of the rotation of the earth so that the sunlight would remain stationary with respect to the microscope. Alcohol, kerosene, and gas lamps were used on bleak days or after the sun set at night. Eventually electric lights became standard (Beale, 1880; Bracegirdle, 1993). Currently a bright-field microscope is equipped with one or more electric light sources to illuminate the specimen (Davidson, 1990). The light source is usually a quartz halogen bulb, although mercury vapor lamps, xenon lamps, lasers, and light-emitting diodes may also be used in special cases.

The light source usually is placed between a parabolic mirror and a collector lens. The light source is placed at the center of curvature of the parabolic mirror so that the rays that go backward are focused back on the bulb. The collector lens is used to project an image of the filament onto the front focal plane of the condenser (Köhler illumination) or onto the specimen itself (critical or confocal illumination).

The diameter of the field illuminated by the light source is controlled by the field diaphragm, and the number and angle of the illuminating rays is controlled by the aperture diaphragm.

THE OPTICAL PATHS OF THE LIGHT MICROSCOPE

I will briefly discuss two types of microscope illumination: Köhler and critical. In practice, Köhler illumination is used in most microscopes, and a specialized form of critical illumination is used in confocal microscopes. Köhler illumination provides a uniformly illuminated, bright field of view, which is important when using an uneven light source, like a coiled tungsten filament. At the end of the nineteenth century, microscopists used sunlight or oil lamps to illuminate their specimens, and very slow film to photograph them. The exposures needed to expose the film were as long as five hours. Thus August Köhler (1893) was motivated to find a way to obtain the brightest image possible so that he could continue his work investigating the taxonomic position of the mollusk, *Syphonaria*, by taking good photomicrographs of the taxonomically important gills.

Köhler devised a method in which an image of the source is formed by a converging lens, known as the collector lens, at the front focal plane of the sub-stage condenser, while an image of the field diaphragm is formed in the plane of the specimen by the sub-stage condenser. The sub-stage condenser produces collimated light beams, each of which originates from a point on the source. Each point on the source forms a collimated beam of light that illuminates the entire field of view. The points on the center of the source form a collimated beam that is parallel to the optical axis. The points farther and farther away from the optical axis make collimated beams that strike the object at greater and greater angles. Thus, the specimen is illuminated with a cone of light composed of both parallel and oblique illumination (Evennett, 1993; Gundlach, 1993; Haselmann, 1993).

In critical illumination, an image of the light source is focused in the plane of the specimen. The illumination is intense, but it is uneven unless a ribbon filament is used. Critical illumination does not require a sub-stage condenser. In critical illumination, each point in the object acts as a point source of light. If the light radiating from two nearby points is truly incoherent, it will form two overlapping images of Airy discs, the intensity of which will be the sum of the two intensities. Since light from two nearby points will be somewhat coherent and will interfere, the intensity of each point will not be exactly the sum of the two intensities, but will, in part, be described by the square of the sum of the amplitudes of the light radiating from both points.

When the microscope is set up for Köhler illumination, the following optical conditions result. The collector lens focuses an image of the light source onto the front focal plane of the sub-stage condenser where the aperture diaphragm resides. The sub-stage condenser turns each point of light at its front focal plane into a beam of light whose angle is proportional to the lateral distance of the point from the principle axis. These beams of light are focused on the rear focal plane of the objective by the objective lens itself. The relative positions of the points of light on the back focal plane of the objective are identical to the relative positions they had when they originated at the front focal plane of the sub-stage condenser. The ocular makes a real image of this light disc at the eye point, also known as the exit pupil or Ramsden disc of the ocular. The eye point is where we place the front focal point of our eye. In Köhler illumination, light originating from each and every point of the light source illuminates our entire retina.

At the same time as the illuminating rays illuminate our retina, the sub-stage condenser lens focuses an image of the field diaphragm on the plane of the specimen. The objective lens forms an intermediate image on the field diaphragm of the ocular. Together, the ocular and the eye form an image of the specimen on the retina. In Köhler illumination, the light that falls on any point on the retina originated from every point of the filament.

Köhler illumination gives rise to two sets of optical paths and two sets of conjugate image planes—the illuminating rays, which are equivalent to the zeroth-order diffracted light, and the image-forming rays, which are equivalent to the sum of the first-order and higher-order diffracted light (Figure 4-8). When the microscope is adjusted for Köhler illumination, we get the following advantages:

1. The field is homogeneously bright even if the source is inhomogeneous (e.g. a coiled filament).

2. The working NA of the sub-stage condenser and the size of the illuminated field can be controlled independently. Thus, glare and the size of the field can be reduced without affecting resolution.

3. The specimen is illuminated, in part, by a converging set of plane wave fronts, each arising from separate points of the light source imaged at the front focal plane of the condenser. This gives rise to good lateral and axial resolution. Good axial resolution allows us to "optically section."

4. The front focal plane of the sub-stage condenser is conjugate with the back focal plane of the objective lens, a condition needed for optimal contrast enhancement.

To achieve Köhler illumination, the light source is placed a distance equal to twice the focal length of the parabolic mirror so that the rays that travel "backward" are focused back onto the filament (Figure 4-9). The collector lens produces a magnified, inverted, real image of the light source onto the front focal plane of the sub-stage condenser where the aperture diaphragm resides. That is, any given point of the filament is focused to a point at the aperture diaphragm.

Light emanating from a point in the plane of the aperture diaphragm emerges from the sub-stage condenser as a plane wave (Figure 4-10). All together, the points in the front focal plane of the sub-stage condenser give rise to a converging set of plane waves. The angle of each member of the set of plane waves is related to the distance of the point to the center of the aperture. In order to produce radially symmetrical cones of light from the sub-stage condenser, the filament in the bulb should also be radially symmetrical.

The plane waves emerging from the sub-stage condenser traverse the specimen and enter the objective lens. The objective lens converts the plane waves to spherical waves, which converge on the back focal plane of the objective lens. Thus each point at the back focal plane of the objective lens is conjugate with a corresponding point in the plane of the sub-stage condenser aperture diaphragm as well as a point on the filament. The sub-stage condenser and the objective together form a real inverted image of the filament at the back focal plane of the objective (Figure 4-11).

The back focal plane of the objective lens and the front focal plane of the sub-stage condenser can be visualized by inserting a Bertrand lens between the oculars and the objective, or by replacing an ocular with a centering telescope. The ocular forms a real minified image of the uniformly illuminated back focal plane of the objective lens at the eye point. The eye point is located just beyond the back focal point of the ocular, and it is where the front focal point of the eye is placed. The object (i.e., the filament in this case) and the image (of the filament) lie in conjugate planes. In Köhler illumination, the light source, the aperture diaphragm, the back focal plane of the objective lens, and the eye point of the ocular lie in conjugate planes called the aperture planes. In each of these planes, the light that does not interact with the specimen, that is the zeroth-order light, is focused.

Now I will trace the path of the waves that interact with the specimen. These are called the image-forming waves and they represent the superposition of all diffracted waves. When the microscope is set up for Köhler illumination, the condenser lens forms a minified, inverted real image of the field diaphragm on the specimen plane. Each point on the field diaphragm is illuminated by every point of the filament (Figure 4-12). The specimen is then focused by the objective lens, which produces a magnified, inverted image of the specimen and the field diaphragm in the optical tube, past the back focal plane of the objective. This plane is just behind the front focal plane of the ocular (Figure 4-13).

The lenses of the ocular and the eye together form an image on the retina as if the eye were seeing the virtual image of the specimen (Figure 4-14). These four conjugate planes are called the field planes. With Köhler illumination there are two sets of conjugate planes, the aperture planes and the field planes (Figure 4-8). The two sets of conjugate planes are reciprocally related to each other.

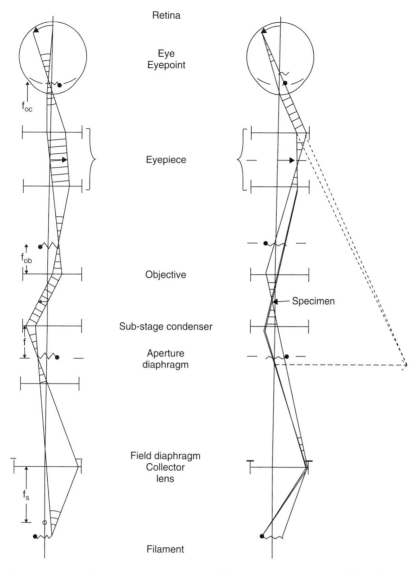

Retina

Eye
Eyepoint

f_{oc}

Eyepiece

f_{ob}

Objective

Specimen

Sub-stage condenser

f

Aperture
diaphragm

Field diaphragm
Collector
lens

f_s

Filament

FIGURE 4-8   Paths of the illuminating rays (A) and the image forming rays (B) in a microscope set up with Köhler illumination. The conjugate aperture planes are shown in (A) and the conjugate field planes are shown in (B).

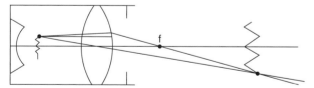

FIGURE 4-9   The lamp is placed at the center of curvature of a concave mirror to capture the backward-going light. The collecting lenses focus an image of the filament onto the aperture plane at the front focal plane of the sub-stage condenser.

To find illuminating rays between the light source and the specimen plane (Figure 4-15):

1. Draw two or three characteristic rays from each of three points on the filament. Find the image plane. For Köhler illumination, we move the filament and collector lens so that the image plane is on the aperture diaphragm.

2. Draw two to three characteristic rays from each of three points on the image of the filament on the aperture diaphragm. Since the aperture diaphragm is at the front focal plane of the sub-stage condenser, all rays from a single point come through the condenser as parallel pencils of light. Draw each pencil till it reaches the objective lens.

To find image-forming rays between the light source and the specimen plane:

3. Draw two or three characteristic rays from two or three points on the field diaphragm to the image plane, where the specimen is placed. For Köhler illumination, the sub-stage condenser is adjusted so that the image plane is identical to the specimen plane on the stage.

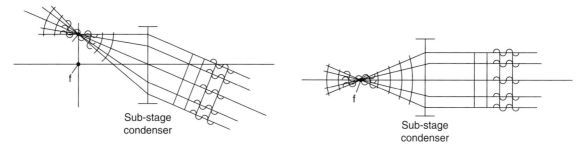

FIGURE 4-10   Light emanating from all the points in the plane of the aperture diaphragm give rise to a converging set of plane waves. The angle of each plane wave relative to the optical axis of the microscope is a function of the distance of the point giving rise to the plane wave from the optical axis.

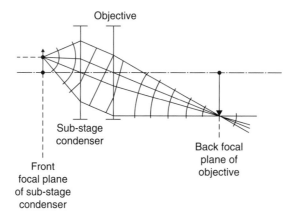

FIGURE 4-11   Together, the sub-stage condenser and the objective lenses produce an image of the filament at the back focal plane of the objective.

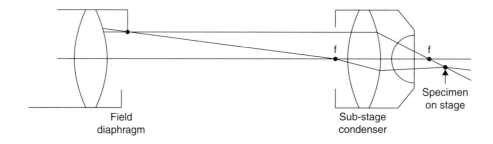

FIGURE 4-12   The sub-stage condenser focuses an image of the field diaphragm onto the focused specimen. Each and every point of the filament contributes to illuminating each and every point on the field diaphragm and the specimen.

As the aperture diaphragm is closed, points 1 and 3 and bundles 1 and 3 are eliminated. Thus, the angle of illumination and the NA of the condenser are decreased. As the field diaphragm is closed, the size of the viewable field is decreased.

USING THE BRIGHT-FIELD MICROSCOPE

When using a microscope, it is as important to prepare your mind and eyes as it is to prepare the specimen and the microscope (Brewster, 1837; Schleiden, 1849; Schacht, 1853; Gage, 1941). Chapters 1 through 3 set the foundation for preparing your mind. In order to prepare your eyes,

make sure that you are comfortable sitting at the microscope. Adjust your seat to a comfortable height. Adjust the interpupillary distance of the oculars for your eyes. Set the diopter adjustment to correct for any optical differences between your two eyes. Make sure that the room is dark, and your eyes are relaxed. Focusing with a relaxed eye will prevent eyestrain and prolong your eyes' ability to accommodate.

Place the specimen on the stage, and focus the specimen with the coarse and fine focus knobs using a low magnification objective. Then close down the field diaphragm, and adjust the height of the sub-stage condenser until the leaves of the diaphragm are sharply focused in the specimen plane. Center the sub-stage condenser if the field diaphragm is not in the center of the field. Open the field diaphragm so that the light just fills the field. This will minimize glare, which is light that is out of place, just as dirt is matter that is out of place. Then adjust the aperture diaphragm to give optimal contrast and resolution. The cone of light that enters the objective typically is controlled by the aperture diaphragm. Determine which

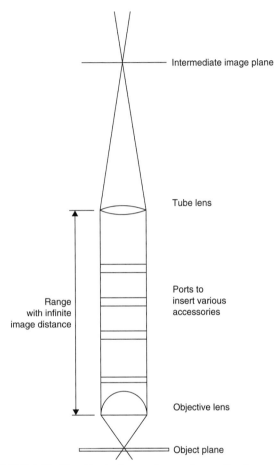

FIGURE 4-13 The objective lens produces an intermediate image of the specimen and the field diaphragm at the field plane of the ocular. If the objective is marked with a 160 or 170, the field plane is 160 or 170 mm behind the objective. If the objective is marked with an ∞, the objective lens produces an intermediate image at infinity and a tube lens is inserted so that the intermediate image is produced at the field plane of the ocular.

position gives optimal resolution and contrast. Repeat the process with higher magnification objectives. There is no need to raise the objectives before rotating them into the optical path, since all the objectives are parfocal and will give a nearly in-focus image at the position the lower magnification lens gave a focused image. Many excellent books describe the theory and practice of the microscope (Martin, 1966; Slayter, 1970; Zieler, 1972; Rochow and Rochow, 1978; Kallenbach, 1986; Spencer, 1982; Richardson, 1991; Oldfield, 1994; Murphy, 2001).

Abbe (1889, 1906, 1921), a physical (and social) experimentalist and theorist, recommended using small cones of light because "the resulting image produced by means of a broad illuminating beam is always a mixture of a multitude of partial images which are more or less different and dissimilar from the object itself." Moreover, Abbe did not see any reason for believing "that the mixture should come nearer to a strictly correct projection of the object ... by a narrow axial illuminating pencil" since the image of an object actually is formed by double diffraction. The image of an object illuminated with a cone of light will be formed from many different diffraction patterns, and the image will be "a mixture of a multitude of partial images."

On the other hand, many leading microscopists, including E. M. Nelson and the bacteriologist, Robert Koch, suggested that using a broad cone of light instead of axial illumination gives a more faithful image without ghosts. Here is what Nelson (1891) had to say:

> The sub-stage condenser is nearly as old as the compound Microscope itself. The first microscopical objects were opaque, and in very early times a lens was employed to condense light upon them. It was an easy step to place the lens below the stage when transparent objects were examined.
>
> On the Continent, where science held a much more important place, the real value of the Microscope was better understood, and it at once took an important place in the medical schools. But the increase of light due to the more perfect concentration of rays by achromatism enabled objects to be sufficiently illuminated by the concave mirror to meet their purposes. Therefore, we find that on the Continent the Microscope had no condenser.
>
> England followed the Continental lead, and now the "foolish philosophical toy" has entirely displaced in our medical schools the dog-Latin text-book with its ordo verborum. But the kind of

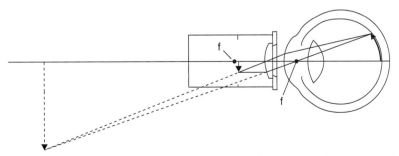

FIGURE 4-14 The intermediate image is formed between the focal plane and the eye lens of a Ramsden ocular. Together, the eye lens and the eye produce a real image of the specimen, any reticle in the ocular and the field diaphragms on the retina. Without the eye lens, the visual angle of the intermediate image would be tiny. With the eye lens, the visual angle is large and we imagine that we see the specimen enlarged 25 cm from our eye.

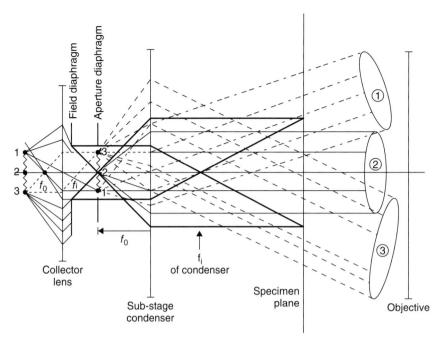

FIGURE 4-15 Adjusting the field diaphragm changes the size of the field and adjusting the aperture diaphragm changes the angle of illumination.

Microscope adopted was not that of the English dilettanti, but the condenserless Continental. It may be said that the Microscope for forty years—that is, from the time it was established in the schools in, say, 1810 to 1880, has been without a condenser.

In 1880 a change came from two separate causes—first, the rise of bacteriology; secondly, the introduction of a cheap chromatic condenser by Abbe in 1873.

Taken by itself, the introduction of the Abbe condenser had not much effect, but as Zeiss's Microscopes had for some time been displacing the older forms, and when the study of bacteriology arose, oil-immersion objectives of greater aperture than the old dry objectives (especially those of the histological series) were used, illumination by the mirror was soon discovered to be inefficient, so a condenser became a necessity. The cheap Abbe condenser was the exact thing to meet the case.

The real office of the sub-stage condenser being a cone-producer, the first question that arises is, What ought to be the angle of the cone?

This is really the most important question that can be raised with regard to microscopical manipulation. To this I reply that a 3/4 cone is the perfection of illumination for the Microscope of the present day. By this I mean that the cone from the condenser should be of such a size as to fill 3/4 of the back of the objective with light, thus N.A. 1.0 is a suitable illuminating cone for an objective of 1.4 N.A. (dark grounds are not at present under consideration). This opinion is in direct opposition to that of Prof. Abbe in his last paper on the subject in the December number of the R.M.S. Journal for 1889, where he says:'The resulting image produced by means of a broad illuminating beam is always a mixture of a multitude of partial images which are more or less different (and dissimilar to the object itself). There is not the least rational ground—nor any experimental proof—for the expectation that this mixture should come nearer to a strictly correct project to the object (be less dissimilar to the latter) than that image which is projected by means of a narrow axial illuminating pencil.'

This paper I consider to be the most dangerous paper ever published, and unless a warning is sounded it will inevitably lead to erroneous manipulation, which is inseparably connected with erroneous interpretation of structure.

If you intend to carry out his views and use narrow-angled cones, you do not need a condenser at all—more than this, a condenser is absolutely injurious, because it affords you the possibility of using a large cone, which, according to Prof. Abbe, yields an image dissimilar to the object. If there is the slightest foundation for Prof. Abbe's conclusion, then a condenser is to be avoided, and when a mirror is used with low powers care must be exercised to cut the cone well down by the diaphragm.

Let me at the place state that I wish it to be distinctly understood that I am not, in this paper, attacking Prof. Abbe's brilliant discovery that the image in the Microscope is caused by the reunion of rays which have been scattered by diffraction, neither do I question what I venture to think is his far more brilliant experiment, which exhibits the duplication of structure, when the spectra of the second order are admitted, while those of the first are stopped out. I regard these facts as fundamental truths of microscopy.

What is a microscopist to do when the experts disagree? Trust experience. According to Spitta (1907):

The situation then is exceedingly difficult to deal with; for, when the result of direct experiment, conducted with all the refinement and skill of a master hand like that of Mr. Nelson, coupled with a full scientific appreciation of the situation, seems to point absolutely and directly in the opposite direction to the teaching of a mathematical expert and philosopher such as the late Professor Abbe, undoubtedly was, one who has never been surpassed, if ever equaled, in acuteness of thought coupled with resourcefulness of investigation in all matters concerning the microscope—we repeat, when these opinions are positively at variance, the onlooker is compelled, from shear inability, to wait and consider. We are bound to confess, however, after several years of attention to this difficult and far-reaching problem, the weight of evidence in our opinion, taken for what it is worth, certainly rests in favor of Mr. Nelson's view, and we venture to suggest that, perhaps, the data upon which the learned Professor built his theoretical considerations may not have included a sufficient "weight" to the teachings of actual experiment; and hence that, although the theory deduced was undoubtedly correct, the data from which it was made were insufficiently extensive. For example, bacteriologists seem mostly agreed that the Bacillus tuberculosis is probably not an organism likely to have a capsule under ordinary conditions, and yet with a narrow cone, whether the specimen be stained or unstained, a very pronounced encircling capsule, as bright and clear as possible to the eye, appears in every case; yet, as the cone is steadily and slowly increased, so does this mysterious capsule disappear! ... The question then arises, how far can we reduce the aperture of the sub-stage diaphragm without the risk of introducing false images? To this we reply that, speaking in general, it must not be curtailed to a greater extent than a cutting off of the outer third of the back lens of any objective as seen by looking down the tube of the instrument, the ocular having been removed.

DEPTH OF FIELD

Geometrical optics tells us that there is an image plane where the specimen is in focus. Physical optics tells us that even if we used aberration-free lenses, each point in the object plane is inflated by diffraction to a spheroid at the image plane, and consequently, the resolution that can be obtained at the image plane is limited. Putting these two ideas together, we see that there is a distance in front of and behind the image plane where we will not be able to resolve whether or not the image is in focus. The linear dimension of this region is known as the depth of focus. The depth of field is a measure of how far away from either side of the object plane an object will be in focus at the true image plane. Thus, the numerical aperture will affect both the depth of focus and the depth of field (Slayter, 1970).

The depth of field (or axial resolution) is defined as the distance between the nearest and farthest object planes in which the objects are in acceptable focus (Delly, 1988). Here we derive the relationship between depth of field and the numerical aperture (NA) using Abbe's criterion and simple trigonometry (Staves et al., 1995). According to Abbe's criterion, the minimum distance (d) between two points in which those two points can be resolved is:

$$d = \lambda/(2NA)$$

We make the assumption that d is the zone of confusion surrounding a point and represents the size of an object that will be in acceptable focus at the nearest and farthest object plane. The depth of field (Y) is the distance between the plane of nearest and farthest acceptable focus (Figure 4-16). Let $x = d/2$ and $y = Y/2$. Using the Abbe criterion for resolution, an object that has a linear dimension of d will appear as a point as long as $2x \leq \lambda/(2NA)$. The largest that x can be is $x = \lambda/(4NA)$. Given the definition of tangent $(\tan \theta = x/y)$,

$$x = y \tan \theta$$

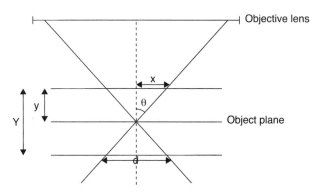

FIGURE 4-16   Depth of field.

Given the two definitions of x,

$$x = y \tan \theta = \lambda/(4NA).$$

After solving for y we get:

$$y = \lambda/((4NA)(\tan \theta))$$

Since depth of field (Y) is equal to 2y,

$$Y = \lambda/((2NA)(\tan \theta))$$

Remember that $\tan \theta = (\sin\theta)/(\cos\theta)$. Multiply the right side of the equation by one (n/n):

$$\tan \theta = (n \sin\theta)/(n \cos\theta)$$

Since $NA = (n \sin \theta)$, then

$$\tan \theta = NA/(n \cos\theta)$$

Remember that $\cos^2\theta + \sin^2\theta = 1$, and $\cos^2\theta = (1 - \sin^2\theta)$. Thus,

$$\cos \theta = \sqrt{(1 - \sin^2\theta)}$$

After substitution, we get

$$\tan \theta = NA/[n\sqrt{(1 - \sin^2\theta)}]$$

Since $n = \sqrt{n^2}$,

$$\tan \theta = NA/[\sqrt{(n^2(1 - \sin^2\theta))}]$$

Distribute the n^2 on the right side to get

$$\tan \theta = NA/[\sqrt{(n^2 - n^2\sin^2\theta)}]$$

Simplify, since $n^2\sin^2\theta = (n \sin \theta)^2 = NA^2$:

$$\tan \theta = NA/[\sqrt{(n^2 - NA^2)}]$$

Substitute into $Y = \lambda/((2NA)(\tan \theta))$:

$$Y = \lambda[\sqrt{(n^2 - NA^2)}]/[(2NA)(NA)]$$

Simplify:

$$Y = \lambda[\sqrt{(n^2 - NA^2)}]/(2NA^2)$$

This equation relates the depth of field to the numerical aperture of the objective lens. This equation is based on the validity of the Abbe criterion, the assumption that the "zone of confusion" is equal to d, and the use of illumination where the full NA of the lens is utilized. This equation states that the depth of field is proportional to the wavelength of light and decreases as the numerical aperture of the objective lens increases. Thus for a narrow depth of field, as is prerequisite for the observation of a localized plane, we need an objective lens with a fairly high numerical aperture.

When θ approaches 90 degrees and NA approaches n, the objective lens tends to form an image at a single plane. This is known as optical sectioning. The higher the NA of the objective, the smaller the depth of field and the more we are able to optically section. As the NA increases, the contrast and depth of field decrease, which makes it more difficult to see the specimen. When you first observe a specimen, it is good to close down the aperture diaphragm and get the maximal contrast and depth of field. As you get to know a specimen, you should aim for the greatest axial and transverse resolution.

As we will discuss later in the book, the depth of field can be decreased using illumination methods, such as two-photon confocal microscopy (see Chapter 12), and image processing methods (see Chapter 14).

OUT-OF-FOCUS CONTRAST

I have been discussing the observation of specimens that show amplitude contrast or sufficient scattering-contrast. However, highly transparent objects would be almost invisible in a perfectly focused, aberration-free microscope that captures most of the diffracted rays. However, the bright-field microscope can detect pure phase objects when you defocus the specimen (Figure 4-17). The contrast arises

because waves that originate from nearby points in the specimen are partially coherent and can interfere before and behind the image plane. The degree of interference and the intensity of the light at these other planes depend on the relative phase and amplitude of the interfering waves. Since the relative amplitude and phase of these waves depend on the nature of the points from which they originate, they contain some information about the object. So slightly defocusing allows us to see a pure phase object in a bright-field microscope. In the next chapter, I will discuss ways of viewing perfectly focused images of phase objects.

USES OF BRIGHT-FIELD MICROSCOPY

The bright field microscope can be used to characterize chemicals (Chamot, 1921; Schaeffer, 1953), minerals (Smith, 1956; Adams et al., 1984), natural and artificial fibers in textiles (Schwarz, 1934), food (Winton, 1916; Vaughan, 1979; Flint, 1994), microorganisms (Dobell, 1960), and cells and tissues in higher organisms (Lee, 1921; Chamberlain, 1924; Kingsbury and Johannsen, 1927; Conn, 1933; McClung, 1937; Johansen, 1940; Jensen, 1962; Berlyn and Miksche, 1976; Harris, 1999). It has been used in the study of The Shroud of Turin, and in the identification of art forgeries (McCrone, 1990; Weaver, 2003).

Bright-field microscopy has long been used to furnish strong evidence in criminal trials (Robinson, 1935). Typically hair and fibers are identified with a light microscope to see, for example, if the hair of the accused is at the crime scene or if the hair of the victim or a fiber from the victim's house carpet can be found on the accused (Smith and Glaister, 1931; Rowe and Starrs, 2001). Moreover, since plants have indigestible cell walls, the food that an autopsied homicide victim last ate can readily be identified (Bock et al., 1988). Light microscopes are becoming useful in the United States' counterterrorism program (Laughlin, 2003).

CARE AND CLEANING OF THE LIGHT MICROSCOPE

First of all, try to keep the microscope and the area around it clean, but let's face it, we live in a world of dirt and dust. When dirt and dust do fall on the microscope, localize the surface that contains the dirt or dust by rotating or raising or lowering the various components of the microscope and looking through the microscope. The surface that has the dirt on it is the one, which when moved, causes the dirt to move. Remove what dust you can with a dust blower. The dirt can be removed by breathing on the lens, or wetting the lens surface with distilled water of a lens cleaning solution used for cleaning camera lenses. Then gently wipe the lens using a spiral motion moving from the center

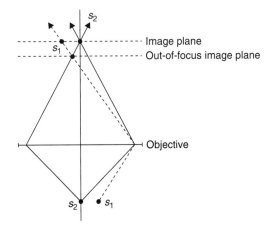

FIGURE 4-17 Out-of-focus contrast in an invisible object. The waves coming from the closely spaced points s_1 and s_2 in a transparent image interfere below the image plane to provide enough contrast to make the nearly-faithful, out-of-focus image visible.

of the lens toward the edge with "no-glue" cotton-tipped cleaning sticks or lint-free lens paper. Immersion oil can be removed the same way. Be careful if you use organic solvents to clean the objectives because the cement that holds a given lens together may dissolve in that solvent. Each microscope manufacturer has its own recommendations on which solvents to use. You can easily check how clean the front surface of the objective lens is by inspecting them by looking through the wrong end of the ocular.

WEB RESOURCES

Molecular Expressions. Exploring the world of optics and microscopy: http://micro.magnet.fsu.edu/

Nikon U: http://www.microscopyu.com/

Olympus Microscopy Resource Center: http://www.olympusmicro.com/

Carl Zeiss Microimaging: http://www.zeiss.com/micro

Leica Microsystems: http://www.leica-microsystems.com/

The Moody Medical Library's Collection of Microscopes can be viewed online at: http://ar.utmb.edu/areas/informresources/collections/blocker/microscopes.asp

Microscopy UK: http://www.microscopy-uk.org.uk/index.html

McCrone Research Institute: www.mcri.org

Southwest Environmental Health Sciences Center Microscopy & Imaging Resources on the web: http://swehsc.pharmacy.arizona.edu/exppath/micro/index.html

Chapter 5

Photomicrography

Photomicrography is the technique of taking pictures through the microscope. It is almost as old as photography itself. Some of the first photomicrographs were taken in 1839 by J. B. Reade, and by William Henry Fox Talbot, the inventor of the positive-negative photographic process (Arnold, 1977). Prior to the invention of photography, all images viewed with the microscope could be captured for posterity only with the aid of an artist's pencil or pen. The *camera lucida*, a drawing aid that was invented by William Wollaston in 1808, was not applied to the microscope until 1880, more than 40 years after a photographic camera was first used.

The goal of photomicrography is to form a complete and faithful point-by-point reproduction of an object on paper. (I will discuss digital images more fully in Chapters 13 and 14). Two things should be kept in mind when taking a photomicrograph. First and foremost, a photomicrograph is a research record that documents, through illustration, a particular phenomenon or structure. Second, photomicrographs are also art, and should look beautiful. Remember, though, that although it is important to make each picture as beautiful as possible, the most important aspect of the photomicrograph is the scientific content, and the artistic approach should enhance and not replace or distract from the scientific content. Photographic techniques are used for quantifying light intensities as well as for the qualitative application of recording an image. However, when recording an image for qualitative purposes, the quantitative aspects of photomicrography should be kept in mind to optimize the image quality because the image intensities should be related to the intensities of each point of the object.

I would like to emphasize that a photomicrograph can be only as good as the image produced by the microscope. So when taking photographs of a microscopic object, make sure that there is no dust on the microscope or slide, shut off the room lights, focus accurately, use Köhler illumination, close the field diaphragm to the appropriate size to prevent glare, and make the best possible choice (compromise) for the diameter of the aperture diaphragm.

SETTING UP THE MICROSCOPE FOR PHOTOMICROGRAPHY

When we look through the oculars of a microscope we see a virtual image approximately 25 cm from our eyes. The oculars produce a virtual image because the intermediate image formed by the objective falls between the focal plane of the ocular and the eye lens of the ocular itself. The virtual image cannot be captured on film; consequently, when doing photomicrography, we must change the optical setup that we use for viewing the image with our eyes, and we must project a real image on the film (Figure 5-1).

A clever person can turn an ocular into a photographic eyepiece that creates a real image beyond the back focal plane of the lens by pulling out the ocular a little from its normal position and fixing it in its new position with tape. In this position, the intermediate image formed by the objective lens falls in front of the front focal plane of the ocular. This leads to the formation of a real image. This is how a solar microscope works. A solar microscope is a microscope that projects an image of a specimen illuminated by sunlight (or an artificial light source) onto a screen so that many people can view the image (McCormick, 1987). Photo eyepieces, with magnifications from about 2x to 6x, are constructed with the correct

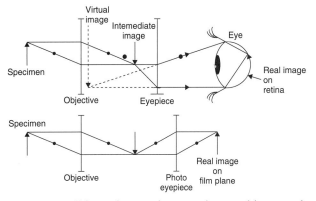

FIGURE 5-1 Using a photo eyepiece to produce a real image on the film (or imaging tube or chip) plane.

geometry to project a real image onto film or the imaging chip of a digital camera.

We can estimate the total magnification of the image on the negative by multiplying the magnification of the objective lens by the magnification of the projection lens and any other tube length factor. The magnification of the printed image is given by multiplying the magnification of the image on the negative by the magnification used to print the picture. It is more accurate and convenient to photograph a stage micrometer on the microscope stage using the same objectives used to observe the specimens of interest, and then print the negatives of the stage micrometer using the same magnification used to make the prints. We must also take a picture of a stage micrometer when using a digital camera and ensure that we are using the same optical and digital zoom values for photomicrography and calibration. The optical zoom increases the amount of detail that can be captured in the photograph, whereas the digital zoom only enlarges the details obtained using the optical zoom.

I discussed depth of field in Chapter 4, but for photomicrography, depth of focus is also important. The depth of field is the total axial distance from the object plane where we see the object in focus at the image plane. The depth of focus is the total axial distance from the image plane where we can place film or an imaging chip and still capture the image in focus. In the light microscope, the depth of focus, which is in the micrometer range, is approximately equal to the depth of field (Slayter, 1970).

A microscope that has been focused correctly for visual observation is not necessarily in focus for photographic recording. Always focus the camera! You will always get in-focus pictures if you focus the camera with a relaxed eye. I imagine I am in a peaceful field or lying on a beach in order to relax my eyes before I focus the camera. When we look at an image visually, our eyes should be completely relaxed; yet, when we focus up and down, we may squint or strain our eyes, and our eyes may accommodate to put out-of-focus objects in focus. Squinting reduces the numerical aperture of our eyes and thus increases the depth of focus. The ability of our eyes to accommodate is disadvantageous for photomicrography because an object may appear in focus to our accommodated eyes, but is out-of-focus on the film, because the film does not have the ability to accommodate. Digital cameras, on the other hand, have the ability to accommodate when used in auto focus mode.

When using a film camera, we must focus the camera telescope so that the virtual image we see is parfocal with the film plane. Always focus the reticule of the camera by moving the diopter adjustment toward you. Never focus out, then in, then out, and so on, because your eye will accommodate. Move the diopter adjustment toward you, when you pass the in-focus point, screw it back in, look toward infinity, and try again. Repeat until you can end by stopping at perfect focus. At the risk of being repetitive, relax your eyes!

For black-and-white photography, an achromatic objective lens used with a green interference filter is perfect. However, instead of using a green filter, which takes advantage of the achromat's correction for spherical aberration, you may want to use a complementary-colored filter to increase the contrast of a low-contrast colored object. In this case, it would be better to use a fluorite or apochromatic objective lens. Using orange filters instead of green filters will increase contrast of regions that are stained blue, while reducing resolution. Using a blue filter instead of a green filter may increase resolution and contrast in yellow/orange specimens, but the spherical aberration will reduce the image quality when using a blue filter with an achromatic lens. Everything is a compromise. Microscopy, like politics, is the art of compromise. When doing black-and-white photography, the type of lamp used is not important as long as it is bright enough. The brightness of the lamp can be varied by adjusting the voltage.

When doing color photography, it is best to use an apochromatic objective lens. The lamp is important when doing color photography. The color temperature of an illuminator is determined by comparing the spectrum of colors it radiates to the spectrum of colors that come from a black body heated to a given temperature (see Chapter 11). To achieve the correct color balance in the photograph, the color temperature of the film must match the color temperature of the illuminator. The color temperature of the film is a measure of the film's sensitivity to each color. The voltage of the lamp determines its temperature and its temperature determines the spectrum of the emitted light. Since changing the voltage changes the color temperature of the lamp, the voltage of the lamp should remain constant when you do color photography. The intensity can be reduced by inserting neutral density filters into the light path. These filters have an unvarying cross-section for visible light and thus uniformly decrease the intensity all across the visible spectrum.

Daylight film is the most versatile for photomicrography, but using daylight film requires using a lamp that has the same color temperature as the sun (5500–6000 K). This color temperature is obtained with xenon arc lamps (5500 K). However, the most typical lamps found in light microscopes are tungsten-halogen lamps. These have color temperatures of only 2800 to 3400 K. Type A films are made for tungsten-halogen lamps, which have a color temperature of 3400 K, and type B film is made for tungsten lamps, which have a color temperature of 3200 K. The voltage necessary to maintain a given color temperature increases as a bulb ages. The color temperature of a given bulb can be converted to the color temperature of a given film by using filters that alter the spectrum of the illuminating light (Delly, 1988).

In a digital camera, the white balance controls let you adjust the color temperature so that the color temperature of the digital camera is matched with the color temperature of the illuminating light. Digital cameras have settings that include automatic, daylight, and incandescent, which may be good matches to the color temperature of the microscope's illuminating system. Minor adjustments can also be made to the daylight and incandescent settings to slightly blue-shift or red-shift the color temperature of the digital camera. If you cannot balance the color temperature correctly, you can always input a preset white balance to the camera by allowing the illuminating light to pass through a clear portion of the slide.

Vibration is a concern to the photomicroscopist. Vibration can come from several sources, including the camera, humans, and building construction. Many microscopes were made especially for photomicrography. These microscopes had a heavy base and the camera either was built into the microscope or mounted lower than the center of gravity to minimize vibration problems. Mounting a camera above the microscope is inherently unstable, unless the camera is not very heavy. To minimize problems with vibration, a heavy camera can be supported mechanically.

It is possible to take pictures on a microscope with your own 35 mm camera on the microscope. However, if the camera has a focal plane shutter, it will vibrate during the exposure and thus blur the image. When using cameras with focal plane shutters, it is necessary to decrease the light intensity and increase the exposure time to "burn in" a good image while the shutter is open. Cameras with shutters that are composed of leaves can be used successfully for photomicrography since the vibration is equal in all directions and thus cancels out. Due to the lightness of digital cameras and their lack of mechanical shutters, they do not suffer from vibration problems.

Buildings may cause a lot of vibration. You can help solve this vibration problem by using a commercially available vibration-free table, or by putting the microscope on a heavy table with a lot of mass. An excellent vibration-free table can be made inexpensively by making a table with cinder block legs, and covering the cinder blocks with a layer of lab bench. Then put some bicycle inner tubes pumped up to a few pounds of pressure on top of that lab bench. Then mount a second lab bench over the inner tubes. I have used a bench like this. It is excellent and inexpensive if the raw materials are available to you. Some people prefer tennis balls to the inner tubes. If vibration still remains a problem, you must use the fastest exposures possible.

Vibration has plagued photomicroscopists from the beginning. Following is an excerpt from Robert Koch's (1880) book, *Investigations into the Etiology of Traumatic Infective Diseases*, which describes his problems with vibration:

> With respects to the illustrations accompanying this work I must here make a remark. In a former paper on the examination and photographing of bacteria I expressed the wish that observers would photograph pathogenic bacteria, in order that their representations of them might be as true to nature as possible. I thus felt bound to photograph the bacteria discovered in the animal tissues in traumatic infective diseases, and I have not spared trouble in the attempt. The smallest, and, in fact, the most interesting

bacteria, however, can only be made visible in animal tissues by staining them, and by thus gaining the advantage of colour. But in this case the photographer has to deal with the same difficulties as are experienced by photographing coloured objects, e.g., coloured tapestry. These have, as in well known, been overcome by the use of coloured collodion. This led me to use the same method for photographing stained bacteria, and I have, in fact, succeeded, by the use of eosin-collodion, and by shutting off portions of the spectrum by coloured glasses, in obtaining photographs of bacteria which had been stained with blue and red aniline dyes. Nevertheless, from the long exposure required and the unavoidable vibrations of the apparatus, the picture does not have sharpness of outline sufficient to enable it to be of used as a substitute for a drawing, or indeed even as evidence of what one sees. For the present therefore, I must abstain from publishing photographic representations; but I hope at a subsequent period, which improved methods allow a shorter exposure, to be able to remedy this defect.

SCIENTIFIC HISTORY OF PHOTOGRAPHY

The fact that light coming through a small hole in a cave, tent, room, or even the leaves on a tree casts an inverted image on the opposite wall has been known since antiquity (Aristotle, Problems XV:11 in Barnes, 1984). In medieval times, a chamber with a pinhole was used to view eclipses of the sun (Eder, 1945). The *camera obscura*, which literally means dark chamber, was later exploited by the Renaissance artists as a drawing aid (da Vinci, 1970). The *camera obscura* was made by putting a small pinhole on one side of a darkened room. Light emanating from an object or scene outside the pinhole formed an inverted image of the illuminated objects outside the hole. An image of the object was captured by tracing the image on a piece of paper, which was attached to the wall opposite the pinhole. Girolamo Cardano suggested in his book *De subtilitate*, written in 1550, that a biconvex lens placed in front of the aperture would increase the brightness of the image (Gernsheim, 1982). Daniel Barbaro (1568) suggested putting an aperture in front of the lens so that only the part of the lens with the least aberrations would be used.

The *camera obscura* was popularized by Giambattista della Porta in his book, *Natural Magic* (1589), and by the seventeenth century, portable versions of the *camera obscura* were fabricated and/or used by Johann Kepler for drawing the land he was surveying and observing the sun, and by Johann Zahn, Athanasius Kircher, and others in order to facilitate drawing scenes far away from the studio. By the eighteenth century, the insides of these cameras were painted black and a mirror was installed at a 45-degree angle so that the image could be viewed on a translucent screen right side up. An artist would aim the camera lens at the object and manually trace the image on a thin piece of paper, which was placed over the translucent glass.

The optical properties of *camera obscura* lenses were improved by William Hyde Wollaston, George B. Airy, and Joseph Petzval in the beginning of the nineteenth century

(Eder, 1945). The development of aberration-free lenses with large apertures allowed shorter exposure times due to the greater amount of light that was captured by these "fast" lenses. However, before the invention of a photographic plate, the ability to capture the elusive image of a picturesque scene on paper required the drawing skills of an artist. Automatic capture of the image had to wait for the invention of light-sensitive plates and film (Newhall, 1937, 1949).

Discoveries that led to the invention of light-sensitive plates proceeded independently of the development of the *camera obscura* (Vogel, 1889). The ability of light to change the color or hue of matter has been known since ancient times. The most obvious example is that sunlight causes skin to tan. It also causes fabric to bleach. Aristotle knew that light caused plants to turn green, and the yellowish secretion of snails (*Murex*) to turn purple. The purple dye became the famous purple of Tyre (Eder, 1945). It was only a matter of time before this process could be developed to a stage where images formed by the *camera obscura* could be permanently captured.

In 1727, Johann Heinrich Schulze discovered that silver salts were sensitive to light. He came across this discovery accidentally while trying to produce a phosphorescent stone. He found that sunlight caused a mixture of chalk, silver, and nitric acid to change from whitish to deep purple due to the reduction of silver ions to metallic silver. After discovering the scotophore (carrier of darkness) instead of a phosphor (carrier of light), Schulze wrote that "often we discover by accident what we could scarcely have found by intention or design" (in Eder, 1945). Schulze concluded that the sun's light and not its heat was the effective agent since the heat of a fire had no such effect. In a later experiment, Schulze covered one side of the bottle with a stencil cut out of opaque paper, and found that when he exposed the stencil side to sunlight and then carefully removed the stencil, a temporary image was formed by the mixture.

Karl Wilhelm Scheele (1780) separated sunlight into its component colors with a prism and found that blue rays were more effective than red rays in reducing the silver salts (by taking up phlogiston). This work was repeated by Jean Senebier in 1782. By 1801, Johann Wilhelm Ritter (1968), stimulated by Sir William Herschel's (1800a, 1800b) discovery of invisible heat rays beyond the red end of the spectrum, found invisible rays beyond the blue end of the spectrum by showing that the invisible rays of the spectrum were very effective in reducing silver salts. Thomas Young (1803) used paper impregnated with silver and his diffraction apparatus to measure the wavelength of the invisible UV light.

In an attempt to capture images formed in the *camera obscura*, Thomas Wedgwood worked with Humphry Davy to make photosensitive plates. They made these photosensitive plates by soaking paper or white leather in solutions of silver nitrate and letting them dry. They found that the images formed by the *camera obscura* were too faint to cause the silver nitrate to turn black. They did find, however, that these photosensitive plates were sensitive enough to capture images produced with a solar microscope, and suggested that this will "probably be a useful application of the method." Unfortunately, the paper was not very sensitive and thus it had to "be placed at but a small distance from the lens" (Wedgwood and Davy, 1802). They did find that their technique could be used to transfer images of paintings made on glass to the photosensitive paper or leather. But, since the areas through which the light passed caused the photosensitive plate to turn black, while the regions of the object through which no light passed remained white, the images were dark-light reversed. Unfortunately, the images were not permanent and had to be viewed by candlelight or else the whole image would eventually turn black.

Joseph Nicéphore Niépce was finally able to capture an image from the *camera obscura* on silver chloride-treated paper in 1816 using a technique he named heliography. However, these images could not be fixed either. By 1822 Niépce could capture a permanent image with the *camera obscura* by coating glass plates with bitumen of Judea, a form of asphalt dissolved in Dipple's oil. Dipple's oil is a complex substance obtained by distilling animal tissues, especially bones. Exposure of the plates to light in the *camera obscura* caused the bitumen of Judea to become insoluble in lavender oil. The areas not exposed to light remained soluble. Thus a reversed image appeared when the plate was washed with lavender oil. These plates captured permanent images, although they were not very light-sensitive and required several-hour-long exposures (Newhall, 1937, 1949; Mack and Martin, 1939; Eder, 1945; Arnold, 1977).

In 1829, Niépce joined forces with Louis Jacques Mandé Daguerre in order to develop more sensitive plates. Daguerre accidentally had found that a silver spoon left lying on an iodized silver plate left an image of itself on the plate. Following this discovery, Daguerre coated copper plates with highly polished silver and then exposed the plates to iodine vapor, which resulted in the formation of silver iodide. Daguerre discovered that he could form a latent image on these plates after only a few-minute exposure in the *camera obscura*. The latent image was formed by the light-induced formation of silver metal. A permanent image could be formed from the latent image by a development process that he could perform in the dark. Daguerre's development process consisted of treating the exposed plates to the vapors of heated mercury.

Daguerre accidentally found that mercury vapor could develop the latent images when he put undeveloped plates into a cupboard and found that when he took them out they had developed an image. They developed an image because some mercury had accidentally spilled in the cupboard giving off mercury vapor (James, 1952). The mercury vapor formed a whitish amalgam with the metallic silver. The unused silver iodide was washed away, and the resulting image appeared as a positive if the Daguerreotype was held such that the underlying silver did not reflect any light into the viewer's eyes. At the correct viewing angle, the white portion of the image traveled to the eye as a result of diffuse reflection, and the black (silver) portion of the image missed the eye due to specular reflection. The Daguerreotype process went through many improvements and was of great scientific and commercial value; unfortunately it allowed the formation of only one image.

This disadvantage was overcome by Fox Talbot (1839b, 1839c, 1844–1846). Talbot exposed a piece of silver halide-coated paper in the *camera obscura* to the light emanating from the object to form a negative image. After development, this piece of paper, which would be waxed or oiled to make it transparent, would be placed between the sunlight and an unexposed piece of silver halide-coated paper. In this way, a positive image would appear on the second piece of paper. This process could be repeated indefinitely to produce multiple copies of a single image. Talbot used the negative-positive process to capture images of landscapes, architecture, people, sculpture, and art taken with a *camera obscura*.

While developing this technique for use with a solar microscope, Talbot (1839b) developed paper that was very sensitive to light, and exposures as short as one-half second could be used. His technique involved first soaking the paper in a weak solution of common salt, wiping it dry, and then spreading a dilute solution of silver nitrate on one side of the paper. Upon drying, the paper was ready to use (Talbot, 1839c). He achieved a very light-sensitive paper when he included a potassium bromide rinse in the photographic paper-making process. In 1839, J. B. Reade (1854) surmised that silver-impregnated leather was more sensitive to light than silver-impregnated paper because of the chemicals used to tan the leather. Thus he applied an extract from galls (i.e., gallic acid), a solution used to tan leather, to paper to increase its sensitivity to light.

Taking steps to decrease the time needed to record the image, Talbot found that he could reduce the time by 100-fold if he exposed dry plates to light and subsequently developed them in gallic acid and silver nitrate. In 1851 Justus Liebig and Henri Victor Regnault independently found that pyrogallic acid was a faster developer than gallic acid, and by 1861 J. Mudd introduced pyrogallol as a developer. Major C. Russell introduced an alkaline form of this developer, which worked best with the very light-sensitive silver bromide.

The photographs made with the negative-positive process developed by Talbot could not be made permanent until a fixative was invented that would prevent the unexposed silver halide grains from turning black. Washes with NaCl worked partially; however in 1819, John Herschel discovered that silver halide could be effectively solubilized by hyposulfite of soda (sodium thiosulfate). In 1839, he started using sodium thiosulfate, which is known as fixer, to remove the silver salts that had not reacted with the starlight he was photographing (Herschel, 1840). Herschel introduced many terms into photography, including positive, negative, fixed, and photograph (Sutton, 1858).

By 1839, cameras, film, paper, developers, and fixers had reached a stage where photography could be used for portraits, capturing scenery, and taking pictures of wars. In 1839, J. B. Reade captured the first permanent image of a flea taken with a microscope (Wood, 1971a, 1971b), and over the next few years, Alfred Donné, Léon Foucault, Josef Berres, J. B. Dancer, and Richard Hogson captured permanent images taken with the microscope (Eder, 1945). Following the development of bright artificial light sources, high numerical aperture sub-stage condensers, new methods of illumination, and faster films, photography also could be used by all microscopists to capture microscopic images. E. B. Wilson (1895), Starr (1896), and Slater and Spitta (1898) presented some of the first photographically documented biology books. Walker (1905) illustrated mitosis in his textbook, with photomicrographs that could be viewed as stereo pairs.

Removing and replacing a lens cap manually to expose the film was no longer practical when high-speed films that reduced the exposure time from minutes to less than a second became available. Initially, mechanical leaf-shutters were placed in front of the camera lens, but eventually they became part of the lens system. In single lens reflex (SLR) cameras, in which lenses could be interchanged freely, the shutter was moved to the focal plane.

GENERAL NATURE OF THE PHOTOGRAPHIC PROCESS

The photographic emulsion of the film consists of small grains (about $0.05\,\mu$m in diameter) containing silver bromide crystals suspended in a gelatin matrix (James, 1952). The emulsion usually is supported by a plastic film. The silver bromide crystals are light sensitive. Upon exposure to light, the silver bromide crystals are altered to form a latent image. Chemical processing then turns the latent image into a visible image.

The steps in the latent image formation are not completely known (Neblette, 1952). It is possible that light causes the irreversible ejection of an electron from the bromide ion. In this photoelectric process, radiant energy is converted into the kinetic energy of the electron. The electron is then trapped by AgS contaminants in the AgBr crystal lattice. The electron then reduces the Ag^+ to Ag^0 (which is metallic silver). The precipitation of the free Ag^0 produces the latent image.

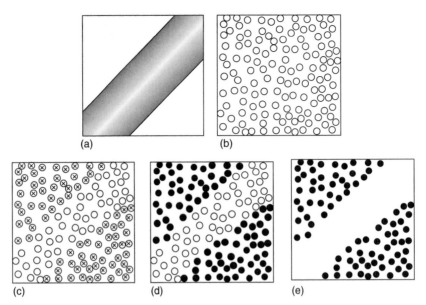

FIGURE 5-2 Stages in the photographic process. The object (a), the unexposed film (b), the exposed film with a latent image (c), the developed film (d), and the fixed film (e).

hν.

Step I:

$$Ag^+Br^- \rightarrow Ag^+ + Br^0 + e^- \text{(trapped by AgS)}$$

Step II:

$$Ag^+ + e^- \rightarrow Ag^0$$

During the development process, the film is washed in organic reducing agents related to hydroquinone (Figure 5-2). The silver salts, which are near the metallic silver that forms the latent image, are reduced by the developer. The quinones in the developer act as reducing agents at alkaline pH. A stop bath, composed of acetic acid, is used to terminate the development and to stop the cascading reactions by acidifying the solution. Fixer is then used to remove the unexposed, undeveloped silver salts while leaving the reduced silver on the film. The fixer is composed of sodium or ammonium thiosulfite (S_2O_2). The fixing process is very complicated due to the many oxidation states of thiosulfite. It is approximated by the following reaction:

$$AgBr + Na(S_2O_2)^- \rightarrow Ag(S_2O_2)^- + NaBr$$

The film is then washed with water to eliminate the fixer and unreduced silver. All the products are soluble in water and are removed in the wash. The film subsequently is dried and printed. To make color films, dyes are coupled to the silver halide grains.

Photographic density is a quantitative measure of the blackening of the photographic emulsion. Photographic density (D) is defined by the following equation:

$$D = \log(I_0/I)$$

where I_0 is the incident intensity, and I is the intensity transmitted through the film. The photographic density is another term for absorbance. Therefore, according to the Beer-Lambert Law, the photographic density is equal to εcd, where ε characterizes the ability of the silver grains to interact with light and is known as the extinction coefficient or molar cross-section (in m^2/mol), c is the concentration of silver grains (in mol/m^3), and d is the thickness of the film (in m) (Figure 5-3).

Films usually are characterized by a Hurter and Driffield curve, which often is referred to as an H-D curve or a characteristic curve. An H-D curve is a plot of the photographic density of a film as a function of the log of exposure. The exposure (E, in photons/m^2) is equal to the product of the incident light intensity (I, in photons m^{-2} s^{-1}) and the time (t in s):

$$E = It$$

This relationship, which shows the reciprocal relationship between intensity and time, is known as the Bunsen-Roscoe Law, and film should be tested only in the exposure range where this reciprocity law holds (James, 1980). When exposures are either extremely long or extremely short, reciprocity does not hold and in both cases the influence of light is lower than expected. At low light levels or very short times, the chance of a grain capturing light is so low that doubling the time or light intensity has no effect (this is one aspect of the quantum nature of light; a single photon can't reduce thousands of silver grains, a little bit each). At high light levels or long exposure times, the chance of a grain capturing light is so high, that doubling the time doesn't expose any more grains: the system is saturated.

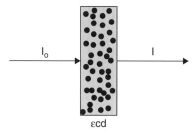

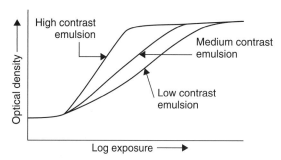

FIGURE 5-3 According to the Beer-Lambert Law, the photographic density is equal to the product of the extinction coefficient, the concentration, and the thickness.

FIGURE 5-5 Characteristic curves for film with the same speed, but different levels of contrast.

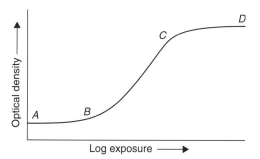

FIGURE 5-4 Characteristic curve for film.

Therefore, we must consider the H-D curves as being characteristic curves of the film in cases when reciprocity holds. The characteristic curves of various films are available from the manufacturers. The speed and contrast of a film for exposure by a given light source and upon development under a fixed set of conditions are specified by the characteristic curve. Figure 5-4 shows the general form of the H-D curve.

Some degree of blackening occurs at zero exposure; this is known as the fog density of the film (A). The fog density arises from the development of grains that have not interacted with light. The distance A-B is known as the toe of the curve. The film is relatively insensitive to light in this region—this is the region of underexposure. The optical density increases linearly with the log of the exposure between points B and C. This is the range where films should be used. If all the portions of the object produce an exposure within this range, then the photographic density level of each point in the image will be proportional to the brightness of each conjugate point in the object. This condition must be met when doing densitometry or when comparing two images quantitatively. The quantitative measurement of photographic density is known as densitometry and it is not uncommon to use a microdensitometer to quantitatively determine the amount of blackening of a film.

The speed of the film is a measure of the size of the toe of the curve. The higher the speed the smaller the toe, the lower the speed, the more the toe extends to the right. The speed is related to the size of the grains. For a given

emulsion type, the larger the grain, the faster the film will be. Larger grains produce more extensive blackening of the image compared to smaller grains for the same number of incident photons.

The film speed is given in ISO numbers. ISO stands for the International Organization for Standardization, which has superseded the American Standards Association (ASA), the Deutsches Institut für Normung (DIN), and the GOST, which was the system used by the former Soviet Union, all of which had scales for film speed. Fast film is necessary when photographing dim objects typically encountered when doing fluorescence microscopy. Fast film is also necessary when the organism is moving or there is a vibration problem. Because the film speed is proportional to the size of the grains and the resolution of the film is inversely proportional to the size of the grains, the speed and resolution are complementary properties of a film. There is always a tradeoff, but the tradeoff can be minimized by studying the characteristics of various films.

The contrast of the film is related to the slope of the linear portion of the curve. The slope of the linear portion of the characteristic curve also is known as the gamma of the film. The steeper the curve, the larger is the difference in photographic density for small differences in exposure and thus the greater the contrast. Figure 5-5 shows the H-D curves of three films that have the same speeds but different contrasts.

Notice that higher contrast can be obtained only by reducing the range of exposures in which the contrast is produced. When we use high-contrast films, we must be very careful in selecting exposures. The development process can also change the contrast of an emulsion. Figure 5-5 then can represent the characteristic curves of the same film developed under different conditions. A high-contrast film is usually important in microscopy since the microscope generally produces a low-contrast image. Image contrast can also be increased by using a high-contrast developer during processing. A gamma of 1.0 will give a one-to-one correspondence between the brightness of the object and the density of the image.

The characteristic curve of a digital camera can be varied using image adjustment controls. Digital cameras have an ISO option, which varies the sensitivity of the camera just as the ISO determined the sensitivity of the film. Digital cameras also have contrast and brightness controls that are useful for increasing the contrast of low-contrast specimens.

George Eastman who, like Ernst Abbe (1906, 1921), was also a practicing social philosopher, founded Kodak and pioneered the development of film (Ackerman, 1930; Brayer, 1996; Mattern, 2005). Film comes in many sizes, although 35 mm (24 $\times$ 36 mm) is the most commonly used for photomicrography since it is convenient, inexpensive, and there are many kinds available. However, it is possible to obtain higher resolution in a final enlarged print by using larger (e.g., 4 $\times$ 5 in., 5 $\times$ 7 in.) film formats.

THE RESOLUTION OF THE FILM

Resolution ultimately is limited by the granularity of the film. Since the smaller the grain size, the better the resolution, but the slower the film, we must make a trade off between resolution and speed. The resolution of film is given in lines per millimeter (lpm). It is determined by photographing a high-contrast object (white bars on a black background with a contrast of 1000:1). In film, a "line" consists of one black and one white line, and thus represents a pair of lines and is equivalent to a "line pair" used to measure the resolution of electronic imaging devices. The limit of resolution of a film for a low-contrast biological object will be about 40 to 60% of the value given for a high-contrast object. Kodak technical pan 2415 is a high resolution film (320 lpm) with excellent contrast. However, it is very slow (ISO 25) and thus not good for dim or moving objects or microscopes with a vibration problem.

How much film resolution in necessary? Suppose we view an object, illuminated obliquely with 550 nm light using a 100x (NA 1.4) objective. The limit of resolution will be about 0.2 μm. When the intermediate image is magnified 5x by a projection lens, the total transverse magnification becomes 500x. The apparent limit of resolution at the image plane is thus 0.2 $\times$ 500 = 100 μm or 0.1 mm. This is equivalent to 10 lpm. The resolutions of today's films and imaging chips are much greater than the resolution limit of the light microscope. As I will discuss later, it is important to use high-resolution film when you want to enlarge the negative for reproduction in a quality journal.

The Modulation Transfer Function (MTF) is a measure of how faithfully the image detail represents the object detail. In this case it is a measure of how the image detail of the film represents the real image projected on it. It is equivalent to the real component of the Optical Transfer Function (OTF). The real component of the optical transfer function characterizes the amplitude relations between the

(a)

(b)

FIGURE 5-6 (a) A sinusoidal test object used to determine the modulation transfer function and (b) the film image of the test object used to determine the modulation transfer function.

object and the image, and the imaginary component characterizes the phase relations between the two. Here, the modulation transfer function relates the contrast of the image in the film to the contrast of the object photographed. The MTF is given by:

$$MTF = \frac{(H'_{max} - H'_{min})/(H_{max} - H_{min})}{(H'_{max} + H'_{min})/(H_{max} + H_{min})}$$

where H is the intensity of light incident on the film and H′ is the intensity of light passing through the film. The subscripts min and max represent the minimum and maximum intensities, respectively. This equation usually is multiplied by 100% to give the MTF in percent response. Figure 5-6 shows a typical MTF square wave test object (a) and image (b).

The test object is made so that the contrast $((H_{max} - H_{min})/(H_{max} + H_{min}))$ is constant throughout the grating (Figure 5-7). The intensity modulation within the image is measured by scanning the film with a densitometer and the variations in density $((H'_{max} - H'_{min})/(H'_{max} + H'_{min}))$ are plotted as a function of the spatial angular wave number of the test object. The percent response is plotted as a function of spatial angular wave number. It is convenient to plot the percent response as a function of the spatial angular wave number using a log-log plot so that the function will intersect the x-axis instead of approaching asymptotically (Figure 5-8). Thus the resolution of the film can be described as the point on the modulation transfer curve where two lines can just be resolved visually (that is, according to Rayleigh's criterion, two objects can be resolved if there is a dip in intensity between them of 20%, or a 20% response). The resolution of a film often is given as a number in lines per mm that

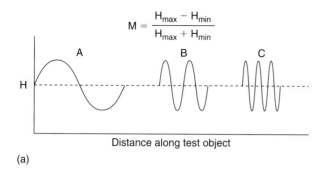

$$M = \frac{H_{max} - H_{min}}{H_{max} + H_{min}}$$

(a)

Distance along test object

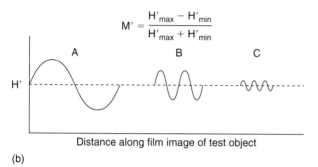

$$M' = \frac{H'_{max} - H'_{min}}{H'_{max} + H'_{min}}$$

Distance along film image of test object

(b)

FIGURE 5-7 Graphical description of densitometer measurements of the test object (a) and image (b) used to calculate the modulation transfer function.

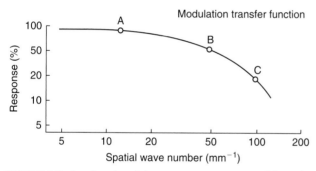

FIGURE 5-8 Log-log plot of the percent response vs. spatial angular wave number.

can be separated visually in the photographic image of a standard test object.

When a photograph is going to be reproduced for a journal, it is first digitized through a computer analyzing system and then printed. The printed photograph is composed of numerous dots, characterized by 256 or more gray values that go from pure black to pure white. The distance between the dots varies with the journal, but they range from 6 to 11 dots/mm. In other words, the limit of resolution of a photograph in a journal is about 6 to 11 lines/mm. As a consequence, any photograph with a limit of resolution equal to 6 to 11 lpm is of journal quality. If we use film that has a resolution of 100 lpm, we can enlarge the negative about 10 times. A larger format film with the same resolution can be magnified 10 times to give a larger high-quality image. The resolution and size of the film determines how magnified a journal-quality print can be.

EXPOSURE AND COMPOSITION

When using the exposure meter in a camera to determine the correct exposure, remember that the exposure selected by the camera is only an estimate. If the camera has an averaging meter, the meter determines the average exposure for a circle with a diameter of approximately 18 mm in the center of the film. The exposure is based on the assumption that the objects are gray and spread evenly throughout the field (Figure 5-9). When this is the case, the exposure will be good. However, when a bright-field background is dotted with fairly dense specimens, the camera will "think" that the specimen is bright, and the predicted exposure will be too short. If we do not want a thin low-contrast negative or a dark low-contrast print, we must increase the exposure time. If bright objects are scattered in a dark field, then the exposure meter will "think" we need more light than we really need, and the predicted exposure will be long. If we do not want a dark low-contrast negative or a bright low-contrast print, we must make the exposure faster than the predicted exposure.

A "spot" meter determines the correct exposure for the exact spot it is targeting. Digital cameras give the option of using a built-in average, spot, or center-weighted exposure meter. It is often better and more efficient to take a series of exposures in order to get the optimal results. Digital cameras make it easy to find the correct exposure for a given specimen by automatically taking a series of exposures with different shutter speeds, apertures, or both.

The composition or framing of a photograph to a large degree depends on what you put on the microscope slide. There are two tricks that you can do to optimally frame the specimen. One is to rotate the stage, and the other is to rotate the camera.

Michael Davidson (1990, 1998) has created beautiful "landscapes" by photographing crystals, using multiple exposures. He calls these photographs *microscapes*. Microscapes can be viewed online at http://www.microscopy. fsu.edu/microscapes/index.html.

Here is how you can make a microscape:

1. Expose the foreground. Use a low-power objective with crossed polars (see Chapter 7). The image will be completely black. Place a slide that has some recrystallized chemicals so that the birefringent image is only on the bottom one-third to one-half of the field. To make "wheat fields" in the foreground, use ascorbic acid crystals or recrystallized ascorbic acid. To make "plants at the edge of a lake," use DNA. To make a "sandy, rocky beach," use ascorbic acid that has been crystallized slowly from

Measuring area	Specimen and background	Exposure adjustment dial setting
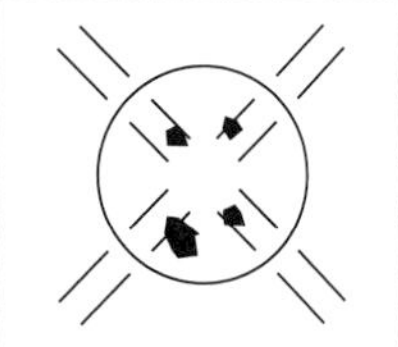	Dark specimens are sparsely distributed on bright background within the metering region	0.25X
	Dark specimens on a bright background take up about one fourth of the metering region	0.5X
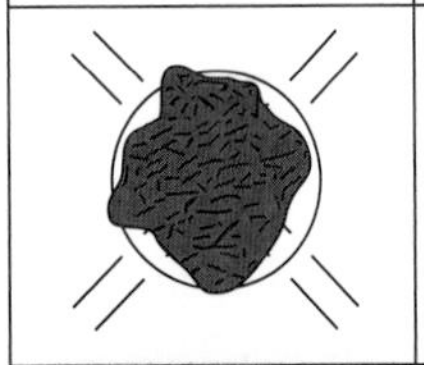	Specimen is evenly distributed within the metering region.	1X
	Bright specimens on a dark background take up about one half of the metering region	2X
	Bright specimens on a dark background take up about one fourth of the metering region	4X
	Bright specimens are sparsely distributed on a dark background within the metering region	Use the ISO sensitivity dial

FIGURE 5-9   How to correct the exposure for specimens that are not gray and uniform.

an aqueous ethanol solution and then smear it across the microscope slide as it dries. To make "sea oats," use melt-recrystallized ascorbic acid. To make a canyon, use sulfur crystals.

2. Cut a mask out of black poster board that exactly follows the outline of the crystals in the first exposure. Place the mask over the field lens, which is a conjugate plane to the specimen. This prevents the region exposed by the first exposure from being reexposed.

3. Expose the mountains, seas, sky, and more. To make "mountains," use recrystallized chemicals, including xanthin or HEPES. To make a "sky," place a low-intensity blue filter in the light path. Using bright-field optics, expose the remaining film. Short exposures make a dark blue "sky" and long exposures make a light blue "sky." "Skies" can also be made by using a sheet of polyethylene between

crossed polars or by defocusing a bead of epoxy resin using polarization optics and a full wave plate.

4. Expose a sun or moon. Close the field diaphragm until you get the size of the "sun" or "moon" you want. Move the substage condenser to put the sun or moon in the desired position. Lower the substage condenser to defocus the field diaphragm leaves until the "sun" or "moon" appears to be a circle. Place a color filter in the light path to make the "sun" orange, yellow, or red. If you insert a diffraction grating in the light path you will spread out the image and get "redder" colors. Use an exposed piece of Polaroid HC film for a diffraction grating.

5. Expose stars or clouds. To make "stars," place a sealed microscope slide that contains a solution of poly-benzyl-1-glutamate on the stage. Using a 10x objective lens, move the slide so that no "stars" are over the "sun" or "moon." Make sure the microscope is set up for Köhler illumination again and then expose the film. "Clouds" can be made by imaging defocused cibachrome bleach crystals under crossed polars for a long time to wash out any color. Image these crystals on the upper one-third of the slide. Most importantly, experiment! The photomicrographs can also be pieced together to form a panoramic landscape (Thurgood, 1995).

THE SIMILARITIES BETWEEN FILM AND THE RETINA

The retina is composed of rods and cones. The retinal rods make up a surface that is analogous to a sheet of black-and-white film, and the cones make up a surface that is analogous to a sheet of color film. The rods and cones, along with their accompanying neurons, are analogous to the silver halide grains in the film. The rods, which are used in dim light conditions (scotopic vision), can detect the presence of only 100 photons, which is only about 39 aJ of energy. The rods are used to detect differences in the brightness of an object in dim light, and the cones are used to distinguish the colors of an object in bright light (photopic vision). There are three kinds of cones: one senses violet-indigo-blue light (400–500 nm), one senses

green-yellow light (450–630 nm), and one senses orange-red–far-red light (500–700 nm).

James Clerk Maxwell, who had a deep interest in color vision, produced, along with his colleague Thomas Sutton, a red, a green, and a blue photograph of a colored tartan ribbon. In a lecture he gave at the Royal Institution in 1861, he superposed the three photographs to obtain a fully colored image of the tartan ribbon and thus proved that all the colors could be produced by the three primary colors, and these primary colors, as Thomas Young (Peacock, 1855) originally proposed, represent the three different color-sensors in our eyes (Glazebrook, 1896; Everitt, 1975; Niven, 2003). Other scientists who had a deep interest in color vision include Isaac Newton, Wolfgang Goethe, Ernst Mach, Hermann von Helmholtz, and Erwin Schrödinger.

WEB RESOURCES

Nikon's Small World Photomicrography Competition: http://www.nikonsmallworld.com/

Molecular Expressions: Images from the Microscope: http://micro.magnet.fsu.edu/micro/about.html

Georg N. Nyman's Website of Photomicrographs: http://www.gnyman.com/Photomicrography.htm

Stephen Durr's Website of Photomicrographs: http://www.btinternet.com/~stephen.durr/

Chapter 6

Methods of Generating Contrast

The resolving power attainable with the bright-field microscope is meaningless when we look at invisible, transparent, colorless objects typical of biological specimens. We can make such transparent specimens visible by closing down the aperture diaphragm; but when doing so, contrast is gained at the expense of resolving power. A goal of the light microscopist is to find and develop methods that increase contrast while maintaining the diffraction-limited resolving power inherent in the light microscope. In this chapter I will describe four methods (dark-field, Rheinberg illumination, oblique, and annular illumination) that can increase contrast in the light microscope by controlling the quality and/or quantity of the illuminating light when the microscope is set up for Köhler illumination. When using these four methods, the illumination is controlled by the aperture diaphragm situated at the front focal plane of the sub-stage condenser. I will also describe a method, known as phase-contrast microscopy, which can increase contrast and maintain resolving power by manipulating the light at the back focal plane of the objective lens as well as at the front focal plane of the sub-stage condenser (McLaughlin, 1977). Color plates 3 through 6 give examples of specimens observed with dark-field illumination, oblique illumination, and phase-contrast microscopy.

DARK-FIELD MICROSCOPY

All that is required for dark-field microscopy is to arrange the illuminating system so that the deviated (first- and higher order diffracted light) rays, but not the illuminating (zeroth-order diffracted light) rays enter the objective lens (Gage, 1920, 1925). Dark-field microscopy is as old as microscopy itself, and Antony von Leeuwenhoek, Robert Hooke, and Christiaan Huygens all used dark-field microscopy in the seventeenth century. Leeuwenhoek (in Dobell, 1932) wrote, "… I can demonstrate to myself the globules in the blood as sharp and clean as one can distinguish with one's eyes, without any help of glasses, sandgrains that one might bestrew upon a piece of black taffety silk."

Hooke wrote (in Martin, 1988), "If the flame of the candle were directly before the microscope, then all those little creatures appeared perfectly defined by a blackline, and the bodies of them somewhat darker than the water; but if the candle were removed a little out of the axis of vision all those little creatures appeared like so many small pearls or little bubbles of air, and the liquid in which they swam appeared dark."

Huygens wrote (in Martin, 1988), "I look at these animals not directly against the light but on turning the microscope a little which makes them appear on a black ground. One can best discover by this means the smallest animals living and can also distinguish best the parts of larger ones."

When dark-field illumination is desired, the specimen usually is illuminated with a hollow cone of light. In the absence of a specimen, the illuminating light does not enter the objective lens because the numerical aperture of the sub-stage condenser is larger than the numerical aperture of the objective (Lister, 1830; Reade, 1837; Queckett, 1848; Carpenter, 1856). Special sub-stage condensers are made for dark-field microscopy; however, for low magnifications, a clever and frugal person can create the hollow cone of light by inserting a "spider stop" or a black circular piece of construction paper in the front focal plane of a sub-stage condenser designed for the bright-field microscope. A clever or frugal person with a phase-contrast microscope can create a dark-field microscope by using the 100x phase-contrast annular ring in combination with a 10x or 20x objective. A high-contrast dark-field image can also be obtained by removing the undeviated, zeroth-order diffracted light by inserting an opaque stop in the central region of any plane that is conjugate with the aperture plane, including the back focal plane of the objective and the eye point (Figure 6-1). A dark-field image can also be produced by a bright-field microscope connected to a digital image processor that removes the low-frequency components of the Fourier spectrum in a process known as spatial filtering (Pluta, 1989; Chapter 14).

In order to ensure that the rays emanating from the sub-stage condenser are as oblique as possible, we must raise the sub-stage condenser as high as it goes and open the aperture diaphragm to its full capacity. We can also put water or immersion oil between the sub-stage condenser and the glass slide. Water is used for convenience; immersion oil is used for better resolution. The liquids reduce

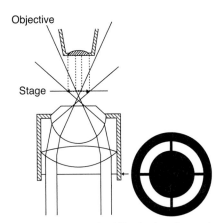

FIGURE 6-1 To achieve dark-field illumination, the specimen must be illuminated with a hollow cone of light that is too wide to enter the objective lens.

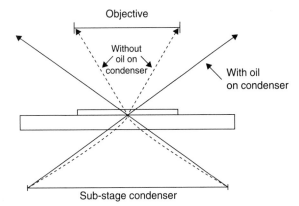

FIGURE 6-2 Oil placed between the condenser and the slide allows us to use an objective lens with greater numerical aperture when using dark-field illumination.

or eliminate the refraction that takes place at the air-glass interface. When refraction occurs at the air-microscope slide interface, the illuminating rays are refracted toward the optical axis of the light microscope. This means that an objective lens will capture more of the illuminating rays when there is air between the top lens of the sub-stage condenser and the microscope slide than when there is water or oil there. Thus when we place water or oil on top of the sub-stage condenser, we get better contrast with an objective lens with a low numerical aperture or can use an objective lens with a higher numerical aperture to get better resolution (Figure 6-2).

Sub-stage condensers that are especially made for dark-field microscopy are designed to transmit only the most oblique rays (Wenham, 1850, 1854, 1856). Dry dark-field condensers are made to be used with objectives with numerical apertures up to 0.75, and the oil immersion dark-field condensers are made to be used with objectives with numerical apertures up to 1.2. The dry dark-field condensers use prisms that are cut in such a manner that the light that enters the prism is reflected internally so that the light leaving the prism exits at a very oblique angle. Some oil immersion dark-field

sub-stage condensers use a convex mirror to bring the light to a concave parabolic mirror, which acts as the main lens (Figure 6-3). Mirrors, unlike glass lenses and prisms, have virtually no chromatic aberration.

Since the dark-field condition requires that the numerical aperture of the objective be smaller than the numerical aperture of the sub-stage condenser, an objective with a variable aperture or iris is very useful for dark-field microscopy. A variable iris in the objective lens lets us adjust the numerical aperture of the objective so that it is "just" smaller than the numerical aperture of the sub-stage condenser and thus obtain optimal resolution and contrast.

Once an object is inserted into the dark-field microscope, the illuminating light that interacts with the specimen is deviated by refraction and/or diffraction. The refracted and/or the first- and higher-order diffracted rays are the only rays that are able to enter the objective. These rays recombine to make the image, and the specimen appears bright on a dark background. Dark-field microscopy is best suited for revealing outlines, edges, and boundaries of objects. It is less useful for the revealing internal details of cells unless there are a lot of highly refractile bodies in a relatively transparent cytosol.

The more oblique the illuminating rays are, the easier it is to detect the presence of very minute objects. Of course the upper limit of obliquity is having the illuminating rays pass perpendicular to the optical axis of the microscope, Microscopes, known as ultramicroscopes, have the dark-field condenser set so that the illuminating rays emerge from the condenser at a 90 degree angle with respect to the optical axis. Typical dark-field microscopes can detect cilia, which are 250 nm in diameter, and single microtubules, which are 24 nm in diameter (Koshland et al., 1988); the ultramicroscopes can detect particles as small as 4 nm (Siedentopf and Zsigmondy, 1903; Hatschek, 1919; Chamot, 1921; Ruthmann, 1970). The volume of a particle is estimated by counting the number of particles in a solution containing a preweighed amount of substance of a known density (Kruyt and van Klooster, 1927; Weiser, 1939). The linear dimensions can then be ascertained by making an assumption about the shape of the particle.

Dark-field microscopes often are used to visualize particles whose size is much smaller than the limit of resolution. How is this possible? Doesn't the wavelength of light limit the size of a particle that we can see? No, this is a very important point; the limit of resolution is defined as the minimal distance between two object details that can just be recognized as separate structures. Resolving power is truly limited by diffraction. But the concept of the limit of resolution does not apply to the minimal size of a single particle whose presence can be detected because of its light scattering ability. The limit of detection in a dark-field microscope is determined by the amount of contrast attainable between the object and the background. To obtain maximal contrast, the rays that illuminate the object must be extremely bright, and the zeroth-order rays must not

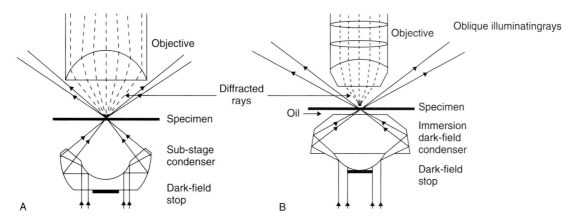

FIGURE 6-3   Two kinds of dark-field condensers.

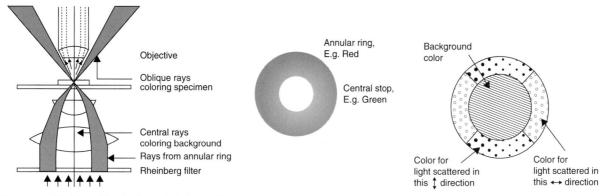

FIGURE 6-4   Rheinberg illumination and Rheinberg filters.

be collected by the objective lens. Moreover to obtain the maximal contrast it is important to have scrupulously clean slides, since every piece of dust will act as a glaring beacon of light. Conrad Beck (1924) says that dirt is "matter out of place" and glare is "light out of place."

We can infer the presence of minute objects using the naked eye. In a dark room, pierced by a beam of sunlight, we can detect the scattering of light by tiny motes of dust in the beam as long as we do not look directly into the beam. In fact, John Tyndall made use of this optical phenomenon to determine whether or not the room in which he was working while he was performing his experiments on the refutation of the theory of spontaneous generation was dust-free. For this reason, the phenomena of scattering by microscopic objects and the ability to detect minute objects by scattering is often referred to as the Tyndall Effect and Tyndall scattering, respectively (Gage, 1920, 1925).

RHEINBERG ILLUMINATION

Rheinberg illumination is a variation of dark-field microscopy first described by Julius Rheinberg in 1896. Rheinberg discovered this method when he accidentally placed colored glass filters in the sub-stage ring (Spitta, 1907). Rheinberg called this type of illumination "multiple or differential color illumination." It is also known as optical staining since when we use Rheinberg

illumination, the image of a normally colorless specimen is made to appear colored without the use of chemical stains. Zeiss introduced Rheinberg illumination under the name of Mikropolychromar in 1933.

When Rheinberg illumination is desired, the central opaque stop of the dark-field microscope is replaced with a transparent, colored circular stop inserted into a transparent ring consisting of a contrasting color (e.g., a red annulus surrounding a green circle). The Rheinberg stop is placed in the front focal plane of the sub-stage condenser. With Rheinberg illumination, the illuminating rays that pass through the annular ring are too oblique to pass through the objective lens and consequently, the background in the image plane is formed only from the illuminating rays that pass through the central area of the circular stop. In order to get the color of the annular ring in the image plane, the light originating from the annular ring must be deviated by refraction and/or diffraction so that it passes into the objective lens. Since the deviated rays originate from the illuminating light passing through the annular ring, the specimens appear to be the color of the annular ring. If we were to observe a protist like *Tetrahymena* with a microscope equipped with a Rheinberg filter with a green central stop inside a red annulus, the protist would appear red swimming in a green background (see Strong, 1968). If we were to use a yellow annulus around a blue central stop, the protist would appear yellow in a sea of blue (Figure 6-4).

Homemade Rheinberg filters are easy to make (Rheinberg, 1896; Needham, 1958; Taylor, 1984). To make Rheinberg filters, cut circles and rings from colored plastics used for theatre lighting with a cork borer. Then cement them with mounting medium to a piece of glass that fits in the filter holder at or near the front focal plane of the sub-stage condenser. For best results, the edges of the adjacent colors should be covered with an opaque border about 3 mm in width.

Tricolored Rheinberg filters can be made where the annulus is composed of four sectors (Hewlett, 1983). Contrasting colors are placed in the adjacent sectors. For example red, green, red, green. The central stop then is filled with a third color, perhaps blue. Tricolored Rheinberg illumination is particularly effective in objects that show striking periodicity, like diatoms and textiles, including silk, where the warp appears in one color and the woof in another. In designing Rheinberg filters, you are limited only by your imagination.

Contrast can be increased when we use Rheinberg illumination by adding several layers of the colored filter to the central portion or by adding a neutral density filter to the central portion. In this way a dim image of the specimen will not be lost in a bright background. Contrast can also be increased by putting a polarizer over the central stop, and then rotating another polarizer placed between the light source and the Rheinberg filter in order to decrease the brightness of the background and increase the contrast of the specimen (Strange, 1989; Chapter 7).

Rheinberg-like filters can also be made by putting together two semicircular sheets of colored material with contrasting colors. The sheets should slightly overlap.

This yields a kind of oblique illumination (Walker, 1971; Carboni, 2003).

OBLIQUE ILLUMINATION

Oblique or anaxial illumination, like dark-field illumination, is also as old as microscopy itself (Goring and Pritchard, 1832; Brewster, 1837; Naegeli and Schwendener, 1892). Oblique illumination is a great way to produce good contrast in images of transparent specimens. The contrast not only comes without a loss in resolving power, but with a gain in resolving power. With oblique illumination, the image of invisible colorless, microscopic specimens appears three-dimensional, although the image is actually a pseudo-relief image and does not represent the true three-dimensional structure of the specimen.

Oblique illumination is created by allowing light from only one portion of the light cone exiting the sub-stage condenser to illuminate the specimen. This is accomplished by blocking all but one portion of the oblique rays coming from the front focal plane of the sub-stage condenser. The unwanted rays can be blocked with your fingers or with a sector stop. Older microscopes were equipped with a translatable (i.e., decenterable) aperture diaphragm that was capable of illuminating the specimen with axial or oblique illumination (Figure 6-5).

A clever and frugal person with a phase-contrast microscope can readily produce oblique illumination by slightly moving the rotating turret of the phase-contrast condenser from the bright-field position to just slightly off-center and adjusting the aperture diaphragm to give the best pseudo-relief image.

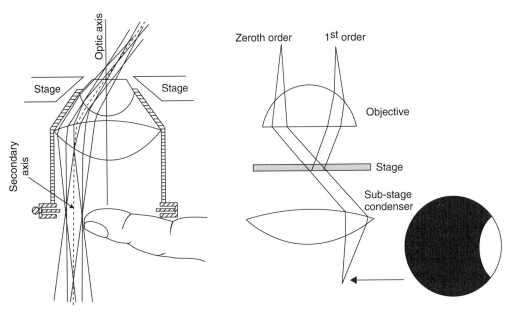

FIGURE 6-5 Two ways of producing oblique illumination.

I would like to emphasize that the image formed by oblique illumination appears three-dimensional; but the three-dimensional impression we get is just an introduced artifact and bears no relationship to real shadow effects. According to geometrical optics, the image plane is parallel to the object plane only when the illuminating rays are parallel to the optical axis. As the illuminating rays become more and more oblique, so does the image plane. The inclination of the image plane results in small objects being overfocused at one edge and underfocused at the other edge. This results in a pseudo-relief image (James, 1976). According to physical optics, the defocusing results in constructive interference on one side (the bright side) of an image and destructive interference on the other side (dark side).

There is currently no complete theory to explain the appearance of the image obtained using oblique illumination. Ellis (1978) has a theory that takes only diffraction into consideration, and Hoffman (1977) has a theory that only takes refraction into consideration. I will discuss the diffraction theory in this chapter after I discuss phase-contrast microscopy, and I will discuss the refraction theory in Chapter 10.

One advantage of using oblique illumination is that the resolving power of the light microscope is up to twice of what it would be with Köhler illumination (Figure 6-6). The only disadvantage is that the image must be viewed with caution since the (would-be) diffracted rays from one side of the object do not contribute to the formation of the image, and consequently a single image will not represent all asymmetrical specimens. On the other hand, oblique illumination makes it easy to discover and visualize asymmetries that may have gone undetected with bright-field illumination. In order to distinguish between asymmetries and symmetries in the specimen when using oblique illumination, it is important to rotate the specimen or the direction of illumination. The specimen can be rotated if the microscope is equipped with a rotating stage, and the

direction of illumination can be rotated by rotating the sector stop or the translatable aperture diaphragm.

In the nineteenth century, most microscopes were equipped with either a decenterable sub-stage condenser or a decenterable aperture diaphragm for doing oblique illumination. However, in the late nineteenth century, the optical attachments that made it possible to do oblique illumination became victims in the sub-stage condenser battle between the "axial-ists," like Abbe, and the "cone-ists," like Nelson and Koch. Microscope manufacturers took sides with the axial-ists, and oblique illumination was unintentionally lost in the process. Oblique illumination has been reinvented many times since then (Hoffman and Gross, 1975; Hoffman, 1977; Ellis, 1978; Hartley, 1980; Kachar, 1985; Inoué and Spring, 1997; Piekos, 1999). Snowflakes look sensational when they are photographed on a cold microscope slide using oblique illumination (Kepler, 1611; Nakaya, 1954; LaChapelle, 1969; Blanchard, 1998; Libbrecht, 2007). In 1991, almost 100 years after the battle between the cone-ists and the axial-ists, Zeiss introduced a type of oblique illumination known as variable relief contrast (Varel).

PHASE-CONTRAST MICROSCOPY

The human eye has the ability to detect differences in intensity and/or in color, and we see images when the points that make up those images vary in intensity or color compared with the background. The bright-field microscope can be considered an amplitude-contrast microscope, which can be used to visualize microscopic specimens that absorb light, thus reducing the amplitude of the waves and the intensity of the light. When a specimen differentially absorbs visible light of different wavelengths, the bright-field microscope renders a colored image. The human eye cannot, however, detect differences in the phase of waves. Transparent specimens with variations in refractive index,

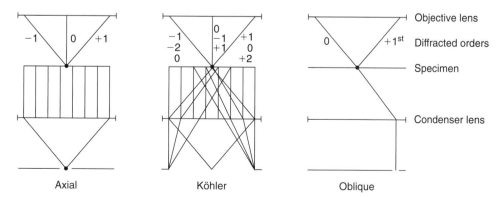

FIGURE 6-6 Comparison bet-ween axial, Köhler, and oblique illumination. The numbers on the diffracted rays indicate the diffraction order. With Köhler illuminate the zeroth-order rays that enter the left side of the objective lens, will coexist with the negative first-order diffracted rays from the axially-illuminated specimen and the negative second-order diffracted rays from the specimen illuminated with oblique rays that enter the right side of the objective lens.

or more correctly, variations in the optical path length (OPL), are able to differentially retard waves; but since we cannot detect the differences in phase with the naked eye, the object remains invisible. The phase-contrast microscope is able to convert differences in the optical path length between regions of a specimen into differences in intensity, thus making transparent specimens visible (Richards, 1950).

The phase-contrast microscope was invented by Fritz Zernike in 1934 (Zernike, 1942a, 1942b, 1946, 1948; Turner, 1982). Like many new discoveries (F. Darwin, 1887; C. Darwin, 1889; Planck, 1936, 1949; Cornford, 1966; Wayne and Staves, 1996), his invention was not immediately accepted by the powers that be. Zernike (1955) describes how his new invention was received by the Zeiss works:

> With the phase-contrast method still in the first somewhat prim-
> itive stage, I went in 1932 to the Zeiss works in Jena to dem-
> onstrate it. It was not received with as much enthusiasm as
> I had expected. This may be explained by the following facts.
> The great achievements of the firm in practical and theoretical
> microscopy were all the result of the work of their famous leader
> Ernst Abbe and dated from before 1890, the year in which Abbe
> became sole proprietor of the Zeiss works. After 1890 Abbe was
> absorbed in administrative and social problems, and partly also in
> other fields of optics. Indeed his last work on microscopy dates
> from that same year. In it he gave a simple reason for the dif-
> ficulties with transparent objects, which we now see was insuf-
> ficient. His increasing staff of scientific collaborators, evidently
> under the influence of his inspiring personality, formed the tradi-
> tion that everything worth knowing or trying in microscopy had
> already been achieved.

Eventually people realized the importance of the phase-contrast microscope for visualizing living, unfixed,

and unstained cells, and Zernike won the Nobel Prize for his discovery in 1953.

The optical path length is equal to the product of the refractive index and the thickness of the specimen (see Chapter 2). It often is given in units of nm. The optical path length can be expressed as the phase angle (in degrees or radians) by multiplying the optical path length by unity, where $1 = 360°/\lambda$ or $1 = 2\pi/\lambda$. These relations hold when we define the wavelength of a sinusoidal wave in terms of a circle. The optical path length can also be expressed as the phase change (in λ) by dividing the optical path length by λ, where represents the wavelength of the illuminating light.

An amplitude object (b) decreases the amplitude of the incident light waves (a) that propagate through it and a phase object (c) changes the phase of the incident light that propagates through it (Beyer, 1965; Figure 6-7). Mixed objects can influence both the amplitude and the phase of the incident light.

The optical path length of a phase object differs from the optical path length of the surround if there is a difference in the refractive index and/or the thickness. According to Maxwell's relation (see Chapter 3), the refractive index of a substance is a measure of an electrical property of a substance known as the dielectric constant or relative permittivity. The optical path length is a measure of the degree of interaction between the light and the electrons in the specimen compared to the degree of interaction between the light and the surround.

In most biological samples, the refractive index is related to the concentration of dry mass in the sample. The dry mass of biological samples is predominantly made

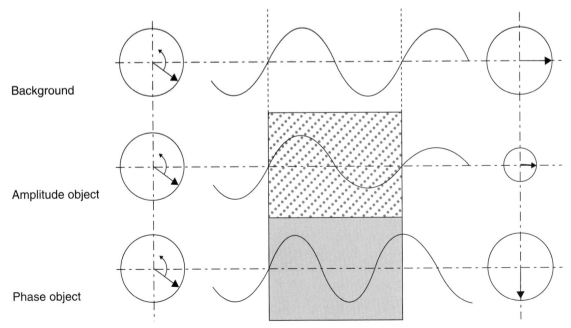

FIGURE 6-7 Amplitude objects reduce the amplitude of the incident wave, but phase objects influence the phase of the wave without changing the amplitude.

up of proteins, lipids, nucleic acids, and carbohydrates, and each of these macromolecules is composed primarily of light atoms, including carbon, oxygen, hydrogen, and nitrogen held together by single bonds. As a first approximation, all the electrons in these light atoms and in the bonds between them have similar electrical properties, and consequently they interact with light in a similar manner. The specific refractive increments of these macromolecules are similar because the electrical properties are similar. The specific refractive increment, which characterizes the degree of interaction of the macromolecules with light, is defined as the increase in the refractive index per one percent increase in the dry mass. Since biological specimens are predominantly composed of these macromolecules with similar specific refractive increments, the refractive index of a specimen is related to its dry mass.

The specific refractive increment is a characteristic of each molecule that describes how much the refractive index of an aqueous solution increases above the refractive index of water (n = 1.3330) for every one percent (w/v)

increase in dry mass. A one percent increase in dry mass represents an increase of 1 g/100 ml of solution. Table 6-1 lists the specific refractive increments of representative biological substances.

Most biological molecules have a specific refractive increment between 0.0017 and 0.0019, with an average of 0.0018. Tryptophan is the notable exception in that it has a somewhat higher than average specific refractive increment, due to the preponderance of double bonds. Lipids and carbohydrates, with their somewhat higher proportion of hydrogen atoms, have a somewhat lower than average specific refractive increment. For practical purposes, the living cell can be considered to be composed of protein. The specific refractive index of proteins is approximately 0.0018. Therefore a 5-percent solution of protein (in water) has a refractive index of 1.3330 + 5(0.0018) = , which is similar to the refractive index of a typical mammalian cell.

Understanding optical path difference (OPD) is the key to understanding phase-contrast microscopy. A region of a specimen and the background, or two regions in a

TABLE 6-1 The Specific Refractive Indices of Biological Molecules

Substance		Specific refractive increment (100 ml/g)
Protein		
	bovine serum albumin (BSA)	0.001854–00187
	horse serum albumin	0.001830–0.0018444
	human serum albumin	0.00181–0.001860
	egg albumin (chick)	0.001820
	γ-globulin (human)	0.00186
	serum globulin (horse)	0.00186
	lactoglobulin (bovine)	0.001818
	β_1 Lipoprotein	0.00171
	hemocyanine (*Helix*)	0.00179
	hemocyanine (*Carcinus*)	0.00187
	hemoglobin (human)	0.00194
Amino acids		
	glycine	0.00179
	alanine	0.00171
	tryptophan	0.0025
Nucleic acids		
	DNA	0.0016–0.0020
	RNA	0.00168–0.00194
Lipids (average)		0.0014
Carbohydrates		0.0013–0.0014

specimen, will have an optical path difference if they have different refractive indices or different thicknesses.

When light passes a homogeneous object that has a different refractive index than the surround, its velocity is altered, and its arrival at a given point is either advanced or retarded, depending on whether the refractive index of the object is less than or greater than, respectively, that of its surroundings. A difference in phase arises because the wave that goes through the specimen has a different optical path length than the wave that goes through the surround. If n_o is the refractive index of the object and n_s is the refractive index of the surround, the optical path difference (OPD) is given by:

$$OPD = (n_o - n_s)t$$

where t is the thickness of the specimen.

An OPD will also arise when the light passes through an object that has the same refractive index as the surround but has a different thickness. In this case:

$$OPD = n(t_o - t_s)$$

where n is the refractive index of the object and surround, t_o is the thickness of the object and t_s is the thickness of the surround. In general, $OPD = OPL_o - OPL_s$, where OPL_o is the optical path length through the object and OPL_s is the optical path length of the surround.

By convention an optical path difference of one wavelength is equal to 360 degrees, and in circular coordinates, the phase angle φ is given by:

$$\varphi = OPD\,(360°/\lambda) = (n_o - n_s)\,t\,[360°/\lambda]$$

where λ is the wavelength of the illuminating light.

The relative phase and amplitude of light waves can be represented by vectors, where the length of the vector represents the amplitude of the wave and the angle of the vector represents the relative phase of the wave (Zernike, 1942a, 1942b, 1946; Barer, 1952a, 1952b, 1953a, 1953b). We can draw the vector in a circle, whose circumference represents one wavelength of the incident light. We will consider that the illuminating wave, incident on the specimen, has an amplitude of 1, and a phase angle of zero degrees. With Köhler illumination, each point of the filament illuminates each and every point on the image plane, and consequently, each wave vector represents the sum of all the illuminating waves divided by the number of illuminating waves. The square of the amplitude of each vector gives the intensity of the image in the image plane. If the specimen is transparent, the vector that represents the waves that propagate through any given point in the specimen will also have an amplitude of one; but it will have a phase angle equal to φ, which will be related to the optical path difference between the specimen and the surround (Figure 6-8).

We will use this vector representation of light to predict with surprising accuracy the nature of the image that will be obtained in the phase-contrast microscope. I would like to emphasize, however, that although the mathematical models that describe phase-contrast microscopy are quantitative and useful, they are only first approximations to the truth, especially when dealing with complicated biological specimens (Zernike, 1946; Bennett et al., 1951; Barer, 1952a, 1952b, 1952c, 1955; Strong, 1958; Goldstein, 1990).

A light wave impinging on a given point in a transparent specimen will be diffracted by that point in the specimen, and the image of that point will be formed by the interference of the diffracted and nondiffracted light at the conjugate point in the image plane. The waves that contribute to the nondiffracted light at the image point originate from every point on the filament. The waves that contribute to the diffracted light at the image point originate only from the conjugate point of the specimen itself. I will call the zeroth-order diffracted light that contributes to an image point the undeviated light, and the sum of all the nonzero diffracted orders, the deviated light. The undeviated light that makes up a given image point in the image plane is represented by vector OA. The length of OA is proportional to the amplitude of the wave and the intensity is proportional to the square of the amplitude. The angle of OA represents the phase of the incident wave. If we define the phase of the undeviated light to be zero ($\varphi = 0$), we can represent the phase of other waves relative to OA.

The length of the vector (OP) that represents a point in the image plane that is conjugate to a perfectly transparent point in the specimen is equal in length to vector OA. Even if the point in the specimen is not perfectly transparent, the length of OP, in general, is indistinguishable from the length of OA. However, the phase angle of the vector that represents a point in the image plane conjugate to a point in a transparent specimen is nonvanishing and depends on the optical path difference between the point in the transparent specimen and the surround.

Vector OP is the sum of vector OA and vector AP. Since vector OP represents the sum of the amplitudes of the undeviated light and the deviated light and vector OA represents the undeviated light, then vector AP must represent the deviated light, which is the sum of the nonzero-order diffracted light (Figure 6-9).

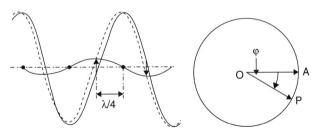

FIGURE 6-8 Wave and vector representation of light that does not interact with the specimen (A) and light that does interact with a point on the specimen (P).

In a typical transparent biological specimen, the phase angle is small (say $\varphi = 12°$), and the vector (AP) that represents the deviated or higher-order diffracted light is rotated about 90 degrees relative to vector OA that represents the zeroth-order diffracted light. A rotation of 90 degrees means that the higher-order diffracted light is out-of-phase with the undiffracted light by 90 degrees. As we learned in Chapter 3, the amplitude of the resultant that represents the interference of two waves with similar amplitudes but that are out-of-phase with each other by 90 degrees ($=\lambda/4$) is equal to the sum of the amplitudes of the two waves.

The light that makes up a point in the image plane is represented by vector OP, where OP = OA and $\angle AOP = \varphi$. Since the intensity of the light that makes the conjugate image point is represented by $(OP)^2$, and the intensity of the surround ($(OP)^2$) has the same value, the specimen would be invisible in a bright-field microscope. The phase-contrast microscope, however, has the ability to convert differences in phase into differences in intensity.

In order to make a transparent object visible, vector OP must differ in length from vector OA. This is accom-

plished by adding an additional phase of 90 degrees ($=\lambda/4 = \pi/2$) to either the direct or the deviated light. In positive phase-contrast microscopy, the direct wave is advanced 90 degrees relative to the deviated wave (Figure 6-10). The advancement is depicted in the vector diagram by a counter-clockwise rotation of the OA vector by 90 degrees to make vector OA'. The length and angle of the deviated beam remains unchanged. The image point now is represented by the vector sum of OA' and AP. In order to add the two vectors together, we translate vector AP to its new position at the terminus of vector OA', while keeping it parallel to vector AP. Because we keep the length and angle of the vector constant, vector A'P' equals vector AP. Then by summing vector OA' and vector A'P', we get vector OP', whose length differs from the length of OA' in a manner that depends on the optical path difference between the specimen and the surround. Depending on φ, the intensity of the image point may now be greater or less than the intensity of the background.

In negative phase-contrast microscopy, the direct wave is retarded 90 degrees relative to the deviated wave (Figure 6-11). The retardation is depicted in the vector diagram by a clockwise rotation of the OA vector by 90 degrees to make vector OA'. The length and angle of the deviated beam remains unchanged. The image point is now represented by the vector sum of OA' and AP. In order to add the two vectors together, we translate vector AP to its new position at the terminus of vector OA', while keeping it parallel to vector AP. Because we keep the length and angle of the vector constant, vector A'P' equals vector AP. Then by summing vector OA' and vector A'P', we get vector OP', whose length differs from the length of OA' in a manner that depends on the optical path difference between the specimen and the surround. Depending on φ, the intensity of the image point may now be greater or less than the intensity of the background.

Using the vector method, we can predict qualitatively how various specimens will appear in a positive

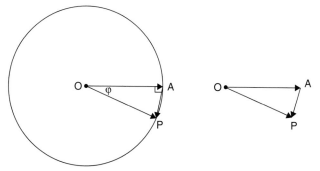

FIGURE 6-9   Vector representation OA represents the undiffracted light, AP represents the diffracted light, and OP represents the vector sum of the diffracted light and undiffracted light that makes up the image point.

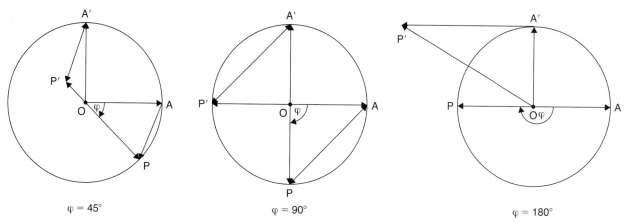

$\varphi = 45°$ $\varphi = 90°$ $\varphi = 180°$

FIGURE 6-10   Vector representations of specimens observed with positive phase-contrast microscopy.

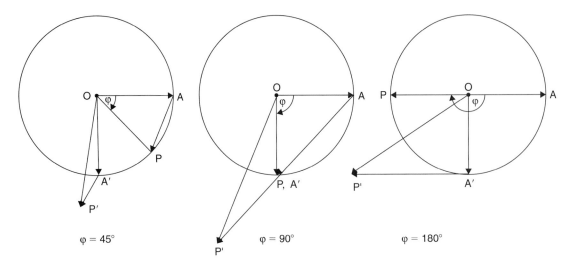

$\varphi = 45°$ $\varphi = 90°$ $\varphi = 180°$

FIGURE 6-11   Vector representations of specimens observed with negative phase-contrast microscopy.

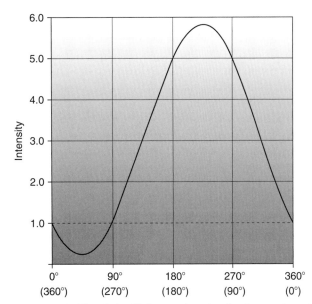

FIGURE 6-12   The relationship between intensity of an image point and phase angle. The phase angles for negative phase-contrast microscopy are given in parentheses.

phase-contrast microscope and a negative phase-contrast microscope. The relative lengths of vectors OP′ compared to the lengths of the background vectors OA′ for specimens with phase angles (φ) of 45, 90, or 180 degrees viewed in a positive phase-contrast microscope are given in Figure 6-10, and the relative lengths of vectors OP′ compared to the lengths of the background vectors OA′ for specimens with phase angles (φ) of 45, 90, or 180 degrees viewed in a negative phase-contrast microscope are given in Figure 6-11. The intensities of the specimen point and the background are given by the squares of OP′ and OA′, respectively. Figure 6-12 summarizes the intensity of the image point compared with the background with respect to phase angles. The relative intensities or the ratio of the intensities

of the specimen point compared to the surround are on the ordinate. The phase angles are presented on the abscissa. The phase angles without parentheses are used for specimens viewed with a positive phase-contrast microscope and the phase angles in parentheses are used for specimens viewed with a negative phase-contrast microscope.

The intensity of the image point is not linearly related to the phase angle, and contrast reversals do occur. In fact, the image points with phase angles of 0, 90, and 360 degrees will be invisible when viewed with a positive phase-contrast microscope and image points with phase angles of 0, 270, and 360 degrees will be invisible when viewed with a negative phase-contrast microscope. In a positive phase-contrast microscope, maximum darkness occurs when the specimen has a phase change of 45 degrees and maximum brightness occurs when the object has a phase change of 235 degrees. In a negative phase-contrast microscope, maximum brightness occurs when the phase angle is 135 degrees and maximum darkness occurs when the phase angle is 315 degrees. Figure 6-12 shows graphically the relative intensity (I) of the image point compared to the background. The relative intensities can also be computed analytically using the following equation:

$$I = 3 - 2\sin\varphi - 2\cos\varphi$$

We can put ourselves in Zernike's shoes and put our knowledge to work to design a phase-contrast microscope. Zernike's first phase-contrast microscope used axial illumination, where all the direct light is focused to a spot on the optical axis in the back focal plane of the objective. Zernike (1958) suggests doing an experiment that would help us realize that an image of a phase object is composed of the addition of the direct and diffracted light. Using a microscope in which the sub-stage condenser has been removed so as to ensure the specimen is illuminated with

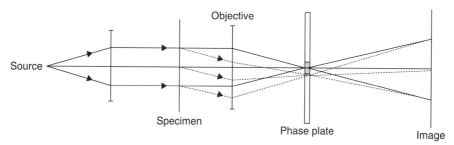

FIGURE 6-13 Phase-contrast microscope based on axial illumination.

parallel light, observe a specimen of India ink with bright-field illumination. Then place a piece of aluminum foil that contains a 0.5 mm pinhole in it on top of the last lens of a 10x objective. No matter what the relatively transparent specimen is, the image observed through 10x eyepieces will be that of a uniformly illuminated field, as if there were no specimen. This is the image of the direct light. Then, replace the pinhole with a glass disk with a 1 mm black dot in the center and look at the specimen through the eyepieces. The India ink will appear as white dots on a black field, indicating the reality of the diffracted light. The vector sum of the direct light and the diffracted light gives rise to a bright-field image.

To make a phase-contrast microscope, which converts differences in phase into differences in intensity, Zernike put a glass phase plate that is thinner in the region through which the direct light passes in the back focal plane of the objective. The region through which the direct light passes is called the conjugate region. Zernike made the conjugate area thin by etching the glass in the conjugate region with dilute hydrofluoric acid. Since the conjugate area is thinner than the rest of the phase plate known as the complementary area, the direct light is advanced relative to the deviated light. Zernike etched the conjugate area of the positive phase plate so that the direct light would be advanced $\lambda/4$ or 90 degrees relative to the deviated light, and he etched the complementary area of the negative phase plate so that the direct light would be retarded $\lambda/4$ or 90 degrees relative to the deviated light (Figure 6-13).

Since the resolving power of the light microscope depends on the obliquity of the light that enters the objective lens, Zernike opted for increased resolving power by inserting an annulus, also known as a phase ring, in the front focal plane of the sub-stage condenser and making the conjugate area in the phase plate an annulus too (Figure 6-14).

To turn a bright-field microscope into a phase-contrast microscope, we must add a phase ring at the front focal plane of the sub-stage condenser that allows a hollow cylinder of light to pass into the sub-stage condenser. A hollow cone of light illuminates the specimen. An inverted hollow cone of direct light is captured by the objective lens and is focused into a ring that coincides with the conjugate area of the phase plate at the back focal plane of the objective.

The contrast of the image formed in a phase-contrast microscope can be increased by coating the conjugate

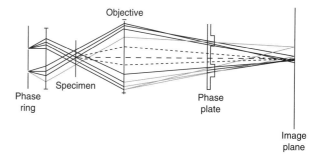

FIGURE 6-14 Phase-contrast microscope based on annular illumination.

region of the phase plate with MgF_2. The MgF_2 absorbs a portion of the direct light, thus decreasing its amplitude. In terms of the vector diagram, the MgF_2 coating results in a shorter vector OA′ while the vector A′P′ remains unchanged. Consequently vector OP′ becomes shorter relative to vector OA′. Thus the difference in the intensity of a point in the specimen to the intensity of the surround becomes greater as a result of the MgF_2 coating. The coating also increases the slope of the curve depicted in Figure 6-12 for specimens with small phase angles so that the reversal in phase occurs at smaller phase angles. The reversal of contrast takes place at 90, 54, and 12 degrees, for a conjugate area with 0, 75, and 99 percent absorption, respectively. At 100 percent absorption, the contrast reversal takes place at 0 degrees, and we end up with dark-field illumination (Figure 6-15).

Although the phase-contrast microscope is fundamentally a qualitative instrument for observing transparent cells (Huxley and Hanson, 1954), it can also be used quantitatively. It has been used quantitatively to measure the refractive index, density, and dry mass of cells. It has also been used to measure changes in the refractive index of cells during cell division, to measure hemoglobin and nucleic acid concentrations, to study the action of X-rays and drugs on lymphocytes, and to study the osmotic behavior of cells (Barer, 1952c, 1953c; Barer et al., 1953; Barer and Dick, 1957; Barer and Joseph, 1954, 1955, 1958; James and Dessens, 1962; Ruthmann, 1966; Ross, 1988; Wayne and Staves, 1991).

In order to measure the refractive index of a cell using a phase-contrast microscope, we observe cells that have

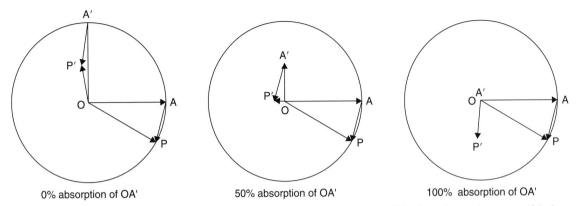

FIGURE 6-15 Vector representation of the influence of putting an absorbing layer on the annulus of the phase plate on the contrast of the image.

been placed in solutions that contain varying concentrations of a solute (e.g., bovine serum albumin) that is nontoxic and does not significantly influence the osmolarity of the solution. Bovine serum albumin (BSA) is nontoxic and has little influence on the osmolarity of a solution as a consequence of its large molecular mass. The specific refractive increment of bovine serum albumin is 0.0018 for every 1 percent increase in dry mass. When we immerse cells in various concentrations of BSA, we change the phase angle. As we decrease the phase angle, the contrast will typically decrease (we must known which portion of the phase change-intensity curve we are on) and will eventually disappear when the refractive index of the BSA solution equals the refractive index of the bulk of the cell. While doing this experiment we can see that if the cell is not homogeneous, certain organelles will disappear and reappear before others as a result of the fact that different organelles have different refractive indices.

The refractive index of a cell (n_o), obtained by finding the refractive index of the medium in which the cells disappear, is given by the following formula:

$$n_o = 1.3330 + \alpha C_{dm}$$

where 1.3330 is the refractive index of water, α is the specific refractive increment, and C_{dm} is the concentration of dry mass (in % w/v) in a cell. Since protein makes up the majority of the dry mass of a typical cell, C_{dm} could be used as an estimate of the concentration of protein in the cell. The concentrations of other macromolecules have been determined with quantitative phase-contrast microscopy. For example, the amount of DNA in a cell has been determined by measuring the refractive index before and after treatment with DNase. In Chapter 8, I will discuss how an interference microscope can be used more elegantly to quantitatively determine the dry mass of the cell or its parts.

It is a good exercise to observe your specimen in media of different refractive indices if you are going to use phase-contrast microscopy routinely. This is because the image of a given cellular structure may be obscured in one medium but made visible in another. In the old days, a

phase microscopist had a battery of phase-contrast objectives available, each with a different retardation, advance, and absorption to give optimal contrast when used with a certain specimen in a given medium. Now that we can buy only a couple of different kinds of phase-contrast objectives, we must be cleverer and vary the medium in order to vary the contrast. Remember each specimen exists in an environment, and usually, it is the relationship between the two that influences the image. This is a specific instance of the nature-nurture debate.

Phase-contrast microscopy allows us to observe transparent, living cells at high resolution, but it also generates some artifacts of its own. It is important to be able to recognize these artifacts and to minimize them if possible. Phase-contrast microscopy is limited to thin specimens. In thick specimens, the areas below and above the object plane will provide unwanted phase changes that cause out-of-focus images that are superimposed upon and obscure the focused image. Optimal phase-contrast is obtained with specimens, whose optical path differences do not exceed the depth of field of the objective. That means thick specimens must have low refractive indices and specimens with high refractive indices must be thin. As a rule of thumb, to obtain a good image in a phase-contrast microscope, the optical path difference between the specimen and the surround should be less than the depth of field of the objective.

Halos and shading-off are two artifacts that typically are introduced by a phase-contrast microscopes (Figure 6-16). Both of these artifacts result from the incomplete separation of the direct and deviated light at the phase plate. Since the deviated light is diffracted by the object in all directions, some of it passes through the conjugate area of the phase plate. Most of the light diffracted by objects with low spatial angular wave numbers is almost parallel to the undeviated light and consequently also goes through the conjugate region of the phase plate. Thus the diffracted light from coarse structures constructively interferes with the undeviated light and produces a bright spot in the image plane. In a positive phase-contrast

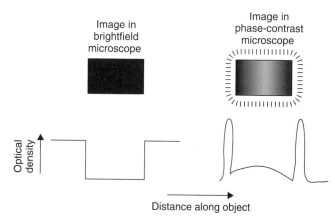

FIGURE 6-16 Shading-off effect in a phase-contrast microscope.

microscope, this results in bright halo around the image. The halo is an unresolved image with reversed contrast that is superimposed on the principle image (Ross, 1967). The darker the image, the brighter the halo and vice versa. Remember with a transparent specimen in a phase-contrast microscope, the dark images appear dark, not because of a decrease in the energy of the light due to absorption, but due to a redistribution of the energy in the image plane. If the energy in the reversed image and the principal image could be completely superimposed in register, the images would disappear completely.

A uniform object may not produce a uniform image in a phase-contrast microscope as a result of shading-off. Shading-off is when the contrast of the image decreases from the edge toward the center until at the center of the image, the intensity is the same as that of the background. Shading-off happens because objects with low spatial angular wave numbers send the majority of their deviated rays through the conjugate area of the phase plate. Thus the deviated light from these regions are advanced or retarded in the same manner as the direct light. Thus, there is no increase in contrast, and these regions have the same brightness as the background. In red blood cells, the biconcave shape enhances the shading-off effect, and the cells look like donuts in a phase-contrast microscope (Goldstein, 1990).

The halo effect and the shading-off effect can be reduced by decreasing the size of the annulus on the phase ring and the phase plate. Although reducing the nondiffracted light reduces the brightness of the image, this is no longer a problem when we use bright light sources and cameras with sensitive imaging chips. The unwanted introduction of halos can be minimized by surrounding both sides of the annular ring on the phase plate with neutral density filters in a process known as apodization, which literally means "removing the foot." By reducing the intensity of light diffracted at small angles, apodized lenses increase the contrast of small details at the expense of losing the contrast of large details. The halo

and shading-off effects can be completely eliminated by using an interference microscope in which the nondiffracted and diffracted beams are completely separated (see Chapter 8).

When given information about the specimen, the illuminating light, and the type of phase-contrast microscope, we can determine the nature of the image produced by the phase-contrast microscope using the following recipe:

1. Read the problem.
2. Identify the important pieces of information in the problem.
3. Convert the information into mathematical symbols (e.g., vectors).
   a. Determine the optical path length of the object (OPL_o).
   b. Determine optical path length of the surround (OPL_s).
   c. Determine the optical path difference ($OPD = OPL_o - OPL_o$).
      Since $OPL_i = n_i t_i$, A, B, and C require knowledge of the thicknesses and refractive indices.
   d. Determine the phase angle (φ). This requires knowledge of the OPD and the wavelength of light (λ).
   e. Plot the vector (OA) that represents the direct light ($\varphi = 0_o$, length = 1 by convention).
   f. Plot the image vector that represents the interference of the direct and deviated light (OP).
   g. By subtraction, determine the vector that represents the deviated light (AP).
   h. Determine the type of phase-contrast microscope. Is the direct light advanced or retarded? By how much? Is it reduced in amplitude? Plot vector OA′ to represent these changes.
   i. Since the deviated light represented by AP is unchanged, move the vector AP so that its tail is attached to the head of vector OA′ and the angle (relative to the horizontal) is unchanged. This new vector is called A′P′.
   j. Now find the vector sum of OA′, which represents the direct light and A′P′, which represents the deviated light. The sum of these two vectors gives the vector that represents the image.

For example: Describe the image of a 10,000 nm thick cell that contains 5 percent protein when it is immersed in water and viewed with 500 nm light in a positive phase-contrast microscope.

$$OPD = OPL_o - OPL_s$$
$$OPL_s = (1.333) 10,000 \, nm$$
$$OPL_o = [1.333 + 5(0.0018)] 10,000 \, nm$$
$$OPD = 5(0.0018) 10,000 \, nm = 90 \, nm$$
$$\varphi = (360°/500 \, nm) 90 \, nm = 64.8°$$

The image is dark on a bright background. How would the image look if we observed it with 360nm light?

There are many elegant and creative ways to design phase-contrast microscopes that let you visualize invisible specimens. Some of these designs utilize polarized light and some utilize interferometers (Pluta, 1989).

OBLIQUE ILLUMINATION RECONSIDERED

Details in a transparent biological specimen typically retard the deviated light by $\lambda/4$ relative to the undeviated light and consequently are invisible in a bright-field microscope (Zernike, 1955, 1958; Francon, 1961; Ellis, 1978; Kalchar, 1985). The vector that represents the $\lambda/4$ retardation, however, can be resolved into two component vectors, one that represents the light diffracted into the positive orders and one that represents the light diffracted into the negative orders (Figure 6-17).

The diffraction spectra of the positive and negative orders are not symmetrical because on one side of a vesicle, for example, the waves emitted by a particle will go through a medium with higher refractive index than the waves that go through the other side. Because the velocity of light is inversely proportional to the refractive index,

the waves that experience a higher refractive index will be shorter than the waves that experience a lower refractive index (Fresnel, 1827–1829). This can be easily visualized by using Huygens Principle.

We can take advantage of the fact that the deviated light is the sum of two component vectors by illuminating the specimen with oblique illumination and capturing either the positive orders or the negative orders from each side of a given detail, but not both diffraction orders. Consequently, the image of one side of a detail will be constructed from the interference of the positive orders and the undeviated light and the image of the other side of the detail will be formed from the interference of the negative orders and the undeviated light. The image from the middle of the detail will be constructed from some of the positive orders, some of the negative orders, and the undeviated light. Consequently, one side of the detail will be bright, the other side will be dark, and the middle will be invisible. Such an image will appear three-dimensional, but the pseudo-relief image does not represent the actual three-dimensional structure. The pseudo-relief image results from the fact that the gradient in optical path length is positive on one side of a detail and negative on the other side. In Chapters 9 and 10, I will discuss differential interference microscopy and modulation contrast microscopy, two other optical ways to obtain a pseudo-relief image. In

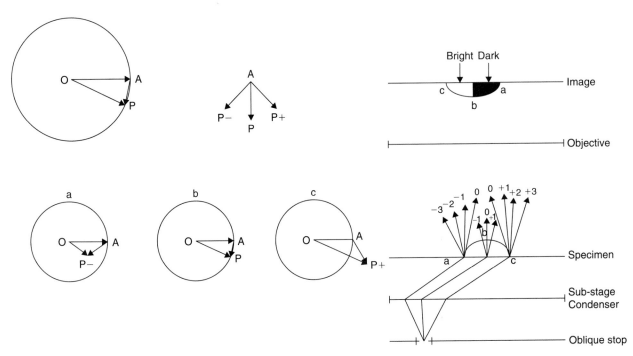

FIGURE 6-17 Vector representation of the generation of a pseudo-relief image by oblique illumination. Notice that the negative diffracted orders produce a dark image and the positive diffracted orders produce a bright image. The difference in refractive index between an object and the surround leads to a difference in the wavelength of Huygens wavelets passing through the object and the surround. Consequently, the diffraction pattern created by the interface of the object and surround is tilted left or right, depending on the direction of the difference in the refractive indices of the object and the surround. When using oblique illumination, the objective lens preferentially captures the positive diffraction orders from one side of the object and the negative diffraction orders from the other side of the object to produce a pseudo-relief image.

Chapter 14, I will discuss how a pseudo-relief image can be produced electronically and numerically.

ANNULAR ILLUMINATION

Another casualty of the great sub-stage condenser wars of the nineteenth century was the use of annular illumination. With this method, an annulus that allows only the most oblique rays to enter the objective is placed in the front focal plane of the sub-stage condenser (Shadbolt, 1850, 1851; Carpenter, 1883; Gordon, 1907; Spitta, 1907; Hallimond, 1947; Mathews, 1953). When using annular illumination, the numerical apertures of the sub-stage condenser and the objective lens should be identical. When using annular illumination, as opposed to Köhler illumination, the resulting image is formed only from the oblique light that produces images with the greatest resolution, and is not formed from the axial light that forms low resolution images.

Annular illumination lost out to Köhler illumination, in part, because the introduction of photomicrography and its slow films required the brightest possible light sources. However, by using annular illumination combined with monochromatic polarized light, the "mixture of a multitude of partial images" is further reduced, and the resolution of the final image is astonishing, although the image is somewhat low in contrast. With bright light sources, sensitive imaging chips, and contrast-enhancing digital image processors, there is no reason why all microscopes should not be equipped with annular illumination as a standard feature in order to maximize the resolving power of the light microscope. For a short time, the Unitron Corporation produced a phase microscope that had an annular illuminator and no phase plate.

Chapter 7

Polarization Microscopy

When we draw a picture of a light wave without thinking deeply, we usually draw a linearly polarized light wave because a linearly polarized wave is the simplest form of a wave. Because linearly polarized light is so simple compared with common nonpolarized light, it acts as a convenient and unencumbered tool to probe the physicochemical properties of matter. Polarized light microscopy has been used to determine the identity of molecules, their spatial orientation in biological specimens, and even their thermodynamic properties. Although polarized light is simpler than common light, there are many things to keep straight before we become comfortable with polarized light. For that reason, I present the information in the following chapter to my students in four 65-minute lectures, replete with hands-on demonstrations that give virtual witness to the experiments done by the masters who discovered the properties of polarized light and birefringent crystals. The purpose of the first lecture is to help the students become comfortable with linearly polarized light. The second lecture helps the students become comfortable with specimens that are able to convert linearly polarized light into elliptically polarized light as a consequence of having two indices of refraction. The third lecture helps students learn the elegant ways to quantify the degree of ellipticity of the light so they can make inferences about the chemical nature of the specimen and the orientation of the birefringent molecules. In the fourth lecture, I provide examples of how polarized light microscopy has been used thoughtfully and creatively to peek into the world of nanometer dimensions in living cells.

WHAT IS POLARIZED LIGHT?

Common or natural light is composed of sinusoidal waves of equal amplitude where all azimuths of the vibration of the electric field are represented equally. In end-view, the propagating light would look like Figures 7-1A and 7-1B. By contrast, when the light is polarized, all the azimuths of the vibration of the electric field are not represented equally. In partially polarized light, some azimuths are underrepresented

compared with others, or put another way, the amplitude of the electric field of some of the waves is less than the amplitude of the electric field of other waves. In end-view, the propagating partially polarized light may look like Figure 7-1C. In fully polarized light, coherent, sinusoidal oscillations propagate with a helical motion that appears as an ellipse in end-view. Polarized light propagating helically is known as elliptically polarized light (Figure 7-1E). In the special case, where the minor and major axes of the ellipse are equal, elliptically polarized light is called circularly polarized light (Figure 7-1E). In another special case where the minor axis of the ellipse vanishes, the light is known as linearly polarized light (Figure 7-1F, G). This is the form of polarized light we usually use to illuminate an object when doing polarized light microscopy

A linearly polarized light wave with an angular wave number k_x ($=2\pi/\lambda$, in 1/m), an angular frequency ω ($=2\pi\nu$, in rad/s), and a phase angle φ (in radians) propagates along the x-axis with the electric field (E) vibrating in the xy plane or the xz plane, perpendicular to the propagation vector, is described by the following equations:

$$\vec{E}_y\ (x,t) = E_{oy}\ \cos\theta_y\ (\sin\ (k_x x - \omega t + \varphi))$$

$$\vec{E}_z\ (x,t) = E_{oz}\ \sin\theta_y\ (\sin\ (k_x x - \omega t + \varphi))$$

where θ_y is the angle that the electric field makes in the yz plane with respect to the y-axis. The angles are positive for a counterclockwise rotation. These angles specify the azimuth of the linearly polarized light. When $\theta_y = 0°$, the plane of vibration is along the xy plane and the light wave is said to be linearly polarized in the y direction. When $\theta_y = 90°$, the plane of vibration is along the xz plane and the light wave is said to be linearly polarized in the z direction. Light that vibrates along any arbitrary plane at an angle relative to the xy plane can be formed by combining two orthogonal waves that are in-phase ($\varphi = 0$). When a wave linearly polarized along the xy plane combines with a wave with the same amplitude, but linearly polarized along the xz plane, a linearly polarized resultant wave whose plane of vibration is in a plane ±45 degrees relative to the xy-plane results.

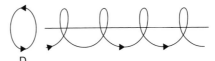

A   B   C   D

FIGURE 7-1 Common (A,B), partially-polarized (C) and elliptically- (D), circularly- (E) and linearly- (F,G) polarized light.

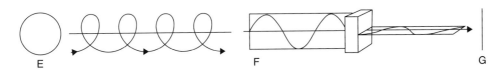

E   F   G

When the two orthogonal component waves have unequal amplitudes, the azimuth of the resultant will be at an angle relative to the xy plane, closer to the wave with the greatest amplitude. The azimuth of the resultant linearly polarized wave can be easily determined by adding vectors that represent the maximal amplitude of the two linearly polarized component waves. Here is a good time to stress that not only does the vector sum of two component waves give rise to the resultant, but every resultant can be viewed as being composed of two orthogonal components. This is the "double reality" of vectors and light (Figure 7-2).

In order to help us understand the various kinds of polarized light we can use a rope or long skinny spring. One person holds the rope or spring still and the other person moves his or her hand up and down in a straight vertical line (Figure 7-3). The wave generated in the rope or spring moves from one person's hand to the other person's hand. The energy in the wave causes each point of the rope or spring to move up and down. While the wave moves from one person to the other, the rope or spring does not move sideways, but moves only up and down, forming a linearly polarized wave whose azimuth of polarization is vertical.

If the second person moves his or her hand back and forth horizontally while the first person holds his or her hand still, the energy in the wave will cause each point of the rope or spring to move from side-to-side, forming a linearly polarized wave whose azimuth of polarization is horizontal or orthogonal to the vertically polarized wave.

If both people move their hands with equal amplitude at the same time, one vertically, and the other horizontally, a spiraling motion will move down the rope or spring, and each point of the rope or spring will move in a circular motion, forming a circularly polarized wave with two axes of equal length. We can decompose this wave by looking at the shadow of this circularly polarized rope or spring wave on the vertical wall. It will look just like the vertically polarized wave. Now look at the shadow on the horizontal floor. It will look just like the horizontally polarized wave. We can consider the circularly polarized waves to be a combination of two linearly polarized waves. Elliptically polarized light is also a combination of two linearly polarized waves, except that the amplitude of the up-and-down motion differs from the amplitude of the side-to-side

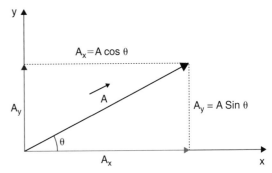

FIGURE 7-2 The resultant of two coherent waves is determined by finding the vector sum of the components. Moreover, any single wave can be viewed as being composed of two orthogonal components. This is the "double reality" of vectors and light.

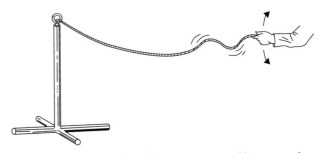

FIGURE 7-3 Demonstration of transverse waves with a rope or long spring.

motion. Professor Wheatstone designed a mechanical apparatus to demonstrate the interaction of two orthogonal linearly polarized waves (Pereira, 1854).

In biological polarization microscopy, elliptically polarized light usually is not produced by the combination of two orthogonal linearly polarized waves with different amplitudes, but by the combination of two orthogonal linearly polarized waves with different phases.

$$E_\theta(x,t) = E_y(x,t) + E_z(x,t) = y\, E_{oy}\, (\sin\,(k_x x - \omega t + \varphi_1)) + z\, E_{oz}\, (\sin\,(k_x x - \omega t + \varphi_2))$$

Letting $\Delta\varphi = \varphi_2 - \varphi_1$

$$\vec{E}_\theta\,(x,t) = y\, E_{oy}\, (\sin\,(k_x x - \omega t)) + z\, E_{oz}\, (\sin\,(k_x x - \omega t + \Delta\varphi))$$

$$\vec{E}_\theta \ (x,t) = E_o \ ((\sin (k_x x - \omega t)) + (\sin (k_x x - \omega t + \Delta\varphi)))$$

Since $\sin (A + B) = \sin A \cos B + \cos A \sin B$

$$\vec{E}_\theta(x,t) = E_o[(\sin (k_x x - \omega t))$$
$$+ (\sin(k_x x - \omega t) \cos (\Delta\varphi))$$
$$+ \sin\theta_y (\cos(k_x x - \varphi t) \sin (\Delta\varphi))]$$

When the phases of two orthogonal linearly polarized waves of equal amplitude differ by an even integral multiple (m) of $\pm m\pi$, the resultant will be linearly polarized. When the phases of two orthogonal linearly polarized waves of equal magnitude differ by an odd integral multiple (m) of $\pm m\pi$, the resultant will also be linearly polarized, but the vibrations of the electric field will be orthogonal to the vibrations produced by the result formed by two waves an even integral multiple of $\pm\pi$. When the phases of two orthogonal linearly polarized waves of equal amplitude differ by $\pm\pi/2$ ($=\pm 90°$), the resultant will be circularly polarized. As we will discuss in detail later, any other difference in phase angle gives rise to elliptically polarized light that describes a helix as the light propagates.

Sound waves, which are composed of longitudinal waves that vibrate parallel to the axis of propagation, cannot be linearly polarized. Since the concept of light waves developed out of the concept of sound waves (see Chapter 3), it was reasonable to assume *prima facie* that light waves were also longitudinally polarized. However, Thomas Young and Augustin Fresnel realized that light must be composed, at least in part, of transverse waves that vibrate perpendicular to the axis of propagation, because light, unlike sound, could be linearly polarized. Young's and Fresnel's radical proposal of the transverse nature of light was difficult for many to accept because of the implications it had for the mechanical nature of the luminous ether though which transverse were thought to propagate (Peacock, 1855; Cajori, 1916; Whittaker, 1951; Shurcliff and Ballard, 1964). Maxwell (1891) concluded that electromagnetic light waves were exclusively transversely polarized, when he obtained a solution for the propagation of electromagnetic waves in free space that had only transverse components. Maxwell's elimination of the longitudinal component of electromagnetic waves resulted from his assumption that the electromagnetic waves in free space had neither a source nor a sink.

FitzGerald (1896) and Roentgen (1899) tried to revive the idea that electromagnetic radiation might have a longitudinal component, but to a scientific community that had finally accepted the transverse nature of light, the reintroduction of a longitudinal component of electromagnetic waves was just too radical. The nature of the polarization of light is still mysterious. Quantum mechanics is unable to say whether a single photon is linearly polarized or circularly polarized (Dirac, 1958).

USE AN ANALYZER TO TEST FOR POLARIZED LIGHT

We can test whether or not light is linearly polarized by using an analyzer. One such analyzer is tourmaline, a colorful precious gem made out of various combinations of boron silicate. Tourmaline is considered dichroic, since it absorbs certain wavelengths of light vibrating in one direction but not in the orthogonal direction (Brewster, 1833b). Consequently, the color of tourmaline, like the colors of sapphires and rubies, depends on the azimuth of the incoming light (Pye, 2001). In one azimuth of polarization, sapphires and rubies pass only blue light and red light, respectively, while they pass white light in all other azimuths.

Edwin Land developed a synthetic crystal, known as a Polaroid, which has the ability to absorb all wavelengths of visible light that are vibrating in a certain azimuth. Light that is vibrating orthogonally to the azimuth of maximal absorption is transmitted by the Polaroid. Light that is vibrating at any other angle is transmitted according to the following relation, known as the Law of Malus:

$$I = I_o \ \cos^2 \theta$$

where I is the transmitted intensity, I_o is the incident intensity, and θ is the angle between the azimuth of transmission of the analyzer and the azimuth of the incident light. Thus, if the intensity of light transmitted through the analyzer varies according to this relation as one turns the analyzer, then the incident light can be said to be linearly polarized. When the axis of maximal transmission of the analyzer is parallel to the azimuth of polarization ($\cos^2 \theta = 1$), the interaction with the analyzer is minimal and light passes through the analyzer. When the axis of maximal transmission of the analyzer is orthogonal to the azimuth of polarization ($\cos^2 \theta = 0$), the interaction is maximal, and no light passes through the analyzer.

Edwin Land originally made Polaroids (of the J type) out of sulphate of iodo-quinine crystals or herapthite embedded in cellulose acetate. Herapathite was serendipitously discovered by the physician William Bird Herapath. Herapath and his student, Mr. Phelps, fed a dog quinine, which is a drug isolated from the bark of the fever-bark tree and is used to treat malaria. When they added iodine to the dog's urine as part of a microchemical test for quinine (Gage, 1891; Hogg, 1898), they noticed little scintillating green crystals. Herapath looked at the crystals in a microscope and noticed they were light or dark in some places where they overlapped and realized that they had discovered a new semi-synthetic dichroic material (Figure 7-4; Herapath, 1852; Herschel, 1876; Grabau, 1938; Land, 1951).

Land, while a nineteen-year-old undergraduate student at Harvard University, read the second edition of David Brewster's (1858) book on the kaleidoscope. Brewster

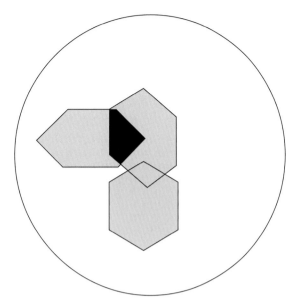

FIGURE 7-4 Herapath discovered that the crystals that formed in the iodine-stained urine of a dog that had been fed quinine were either bright or dark in the places where they overlapped. Since their transparency or opacity of the thin crystals depended on their mutual orientation, he realized that they produced polarized light and he had discovered that "the most powerful polarizing substance known … proved to be a new salt of a vegetable alkaloid."

wrote in this book that he dreamed of making plates of herapathite that were large enough and strong enough to use in kaleidoscopes because such kaleidoscopes would produce beautiful and intense interference colors from transparent crystals (see later), instead of the less-brilliant absorption colors produced by colored glass. Large flat herapathite crystals were too brittle to be used in the kaleidoscopes. Land had the idea of making large, stable, dichroic sheets by embedding millions of tiny crystals of herapathite in a gelatin medium. The medium was then subjected to mechanical stress in order to align the crystals so that their axes of maximal absorption were all coaligned. The H-type Polaroid sheets that are used today are made totally from synthetic crystals that are aligned by mechanical stress.

Land (1951) wrote about his motivation:

> Herapath's work caught the attention of Sir David Brewster, who was working in those happy days on the kaleidoscope. Brewster thought that it would be more interesting to have interference colors in his kaleidoscope than it would be to have just different-colored pieces of glass. The kaleidoscope was the television of the 1850's and no respectable home would be without a kaleidoscope in the middle of the library. Brewster, who invented the kaleidoscope, wrote a book about it and in that book he mentioned that he would like to use the herapathite crystals for the eye piece. When I was reading this book back in 1926 and 1927, I came across his reference to these remarkable crystals and that started my interest in herapathite.

Edwin Land left Harvard University after his freshman year so that he could spend his time developing his polarizer, and founded the Polaroid Corporation (Wensberg, 1987; McElheny, 1999).

PRODUCTION OF POLARIZED LIGHT

Any material that can analyze polarized light can also be used to produce polarized light. Thus a Polaroid will convert nonpolarized light into linearly polarized light by absorbing light of all but one azimuth. Any device that converts nonpolarized light into polarized light is called a polarizer. In microscopy, Polaroids typically are used to polarize light and to analyze it.

Crystals beside those found in Polaroids can be used to polarize light, and historically, calcite was the first crystal that was found to have the ability to polarize light. Before the discovery of calcite, there was not any reason to think that there were any hidden asymmetries in light that would allow it to be polarized, be it a corpuscle or a longitudinal wave. The discovery of calcite changed our view of light.

When nonpolarized light strikes a piece of calcite, it is resolved into two beams of light—the ordinary and the extraordinary light. The ordinary light experiences an index of refraction of 1.658 and the extraordinary light experiences an index of refraction from 1.486 to 1.658, depending on the relative orientation of the light wave and the crystal. Since calcite has two indices of refraction, it is known as a birefringent material. A birefringent material, if it is thick enough, has the ability to doubly refract light, causing a single beam of light to diverge into two beams. According to Newton (1730), "If a piece of this crystalline stone be laid upon a book, every letter of the book seen through it will appear double, by means of a double refraction."

The amazing properties of calcite were first discovered by Erasmus Bartholinus (1669). While playing with a piece of the newly discovered Iceland Spar (now called calcite), Bartholinus noticed that the transmitted light produced two images as if the incident light beam were split into two transmitted beams (Figure 7-5). One of the images precessed around the other image as he turned the crystal of calcite. The plane of the crystal that contains both the ordinary ray and the extraordinary ray is known as the principal section. Bartholinus named the rays that formed the stationary image, the ordinary rays since they acted like they passed though an ordinary piece of glass and obeyed the ordinary laws of refraction. He called the rays that made the precessing image, the extraordinary rays, since they behaved in an extraordinary manner. Bartholinus wrote,

> Everyone praises the beauty of the diamond, and indeed jewels, gems, and pearls can give many pleasures. Yet they serve only in idle display on finger and neck. I hope that those of my readers for whom knowledge is more important than mere diversion will derive at least as much pleasure from learning about a new substance, lately brought back to us from Iceland as transparent crystals.

Huygens (1690) explained the double refraction discovered by Bartholinus in terms of wave theory. He proposed that the atoms in the calcite acted asymmetrically

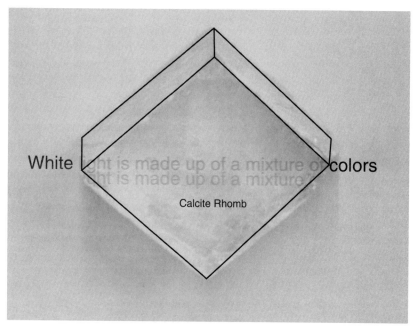

White light is made up of a mixture of colors

Calcite Rhomb

FIGURE 7-5 Bartholinus noticed that the light transmitted through a piece of calcite formed two images and must therefore be split into two beams.

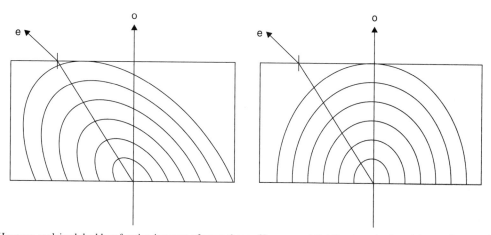

FIGURE 7-6 Huygens explained double refraction in terms of wave theory. He proposed that the atoms in the calcite acted asymmetrically upon the incoming light so that the waves that made up the ordinary beam were spherical and passed straight through the crystal, whereas the waves that made up the extraordinary beam were elliptical and traveled diagonally through the crystal. If he rotated the crystal while he viewed the double image through the top of crystal, the image produced by the ordinary ray (o) would remain stationary while the image produced by the extraordinary ray would precess around the ordinary image.

upon the incoming light so that the waves that made up the ordinary beam were spherical and the waves that made up the extraordinary beam were elliptical (Figure 7-6). This would result in an ordinary beam that was circular in outline when viewed end-on and an extraordinary beam that was elliptical when viewed end-on. Huygens, however, was at a loss to come up with anything more mechanistic and wrote, "But to tell how this occurs, I have hitherto found nothing which satisfies me." Newton (1730) explained the double refraction in terms of corpuscular theory, and concluded that the asymmetries in the atoms of the calcite acted upon corpuscles that had sides of various lengths and therefore were accelerated through the calcite in a manner

that depended on which side of the corpuscle interacted with the calcite.

Huygens (1690) and Newton (1730) performed experiments with two pieces of calcite that clearly demonstrated that a crystal of calcite could resolve common light into two components that differ in their polarity and that a second prism could recombine the two components to form common light. These experiments are analogous to those that were done by Newton, where he showed that one prism can resolve white light into its chromatic components and a second prism can recombine the chromatic components to form white light, and have become a classic way to experience the marvels and mysteries of polarized light (Brewster, 1933).

The description given here is lengthy so that anyone and everyone can personally repeat these illustrative experiments that demonstrate the conversion of common light to polarized light. These experiments can be done using a light source and two inexpensive pieces of optical calcite that can be obtained from the gift shop at a local science museum.

Make a pinhole in a piece of black poster board and place the pinhole on a convenient light source like a flashlight or a light table. The light coming through the aperture is common light. Place a piece of calcite on the aperture. The beam of common light is split into two beams, the distance between the two depending on the thickness of the calcite. We can tell the extraordinary beam from the ordinary beam by the way the extraordinary beam precesses around the ordinary beam. When the ordinary and extraordinary beams are aligned in a line that extends to the two obtuse angles of the crystal, the two beams are in a principal section that contains the optic axis of the crystal.

Place a second piece of calcite with a similar thickness on top of the first so that the two crystals have the same orientation. In this orientation, the principal sections are parallel, and each beam produced by the first crystal is not divided into two, but rather the two beams produced by the first crystal are further separated by the second crystal. The beam that underwent an ordinary refraction in the first crystal undergoes an ordinary refraction in the second crystal and the beam that underwent an extraordinary refraction in the first crystal undergoes an extraordinary refraction in the second crystal. The distance between the ordinary and extraordinary beams emerging from the second crystal is equal to the sum of the distances that would be produced by each crystal separately. Huygens (1690) felt "it is marvelous" why the beams incident from the air on the second crystal do not divide themselves the same as the beam that enters the first crystal. He first hypothesized that, in passing through the first crystal, the ordinary beam "lost something which is necessary to move the matter which serves for the irregular refraction …" and the extraordinary beam "lost that which was necessary to move the matter which serves for regular refraction."

When the top crystal is turned 90 degrees so that the principal sections of the two crystals are perpendicular, the two beams that emerged from the second crystal disappear and two new beams appear that are oriented diagonally with respect to the original beams and are closer together. This can be explained if the ordinary beam from the first crystal becomes the extraordinary beam in the second crystal and the extraordinary beam from the first crystal becomes the ordinary beam in the second crystal. Thus Huygens' first hypothesis was wrong; the ordinary beam that exits the first crystal is still able to undergo an extraordinary refraction and the extraordinary beam that exits the first crystal is still able to undergo an ordinary refraction.

When the top crystal is rotated from 0 to 90 degrees, the brightness of the beams decreases with the square of the cosine of the angle between the two principal sections. By contrast, the brightness of the newly emergent beams increases with the square of the sine of the angle between the two principal sections. The two beams that emerge when the principal sections of the crystals are perpendicular to each other are equally bright and have the same brightness as the two beams had when the principal sections were parallel to each other. Each of the four beams produced at 45 degrees is half as bright as each of the two beams produced at 0 and 90 degrees.

When the top crystal is rotated from 90 to 135 degrees, the brightness of the two beams present at 90 degrees decreases as the brightness of two new beams oriented diagonally relative to the two beams present at 90 degrees increases. The four beams present at 135 degrees are mirror images to the four beams present at 45 degrees.

When the top crystal is rotated a total of 180 degrees so that the principal sections of the two crystals are again parallel, the two original rays merge into a single beam. The single beam is twice as bright as each of the two beams present at 0 and 90 degrees and four times as bright as each beam present at 45 and 135 degrees. The single beam emerging from the two crystals oriented with their principal sections 180 degrees relative to each other has all the characteristics of common light.

Huygens concluded that the "waves of light, after having passed through the first crystal, acquire a certain form or disposition in virtue of which, when meeting the texture of the second crystal, in certain positions, they can move the two different kinds of matter which serve for the two species of refraction and when meeting the second crystal in another position are able to move only one of these kinds of matter."

Huygens realized that the waves had become polarized after passing through the first medium, but since he had no conception of the possibility of transverse waves, he was at a loss to explain how they became polarized. Huygens also realized that the calcite crystal itself had "two different kinds of matter which serve for the two species of refraction," but since he had no conception of the existence of electrons and their electrical polarization, he "left to others this research…."

According to David Brewster (1833b),

> The method of producing polarized light by double refraction is of all others the best, as we can procure by this means from a given pencil of light a stronger polarized beam than in any other way. Through a thickness of three inches of Iceland spar we can obtain two separate beams of polarized light one third of an inch in diameter; and each of these beams contain half the light of the original beam, excepting the small quantity of light lost by reflexion and absorption. By sticking a black wafer on the spar opposite either of these beams, we can procure a polarized beam with its plane of polarization either in the principal section or at right angles to it. In all experiments on this subject, the reader should recollect that every beam of polarized light, whether it is produced by the ordinary or the extraordinary refraction, or by positive or negative crystals, has always the same properties, provided the plane of its polarization has the same direction.

These experiments indicated that common light is composed of two types of polarized light with orthogonal planes of polarization. Calcite splits common light into two orthogonal beams, one that is polarized parallel to the principal section and one that is polarized perpendicular to the principal section. The plane of polarization in the minds of Huygens, Newton, and Brewster was not related to the plane of vibration in our minds, because they did not know of and/or accept the idea of transverse waves.

Until 1808, it seemed that double refraction was the only means by which transparent substances could polarize light. Then Ethienne Louis Malus, a proponent of the corpuscular theory of light, discovered that polarized light can also be produced by reflection from common transparent substances such as glass and water. While in his home in the Rue d'Enfer looking through a calcite crystal at the light from the sunset reflected from a window in Luxembourg Palace, Malus discovered that the light that was reflected from the castle windows was polarized. That is, instead of seeing two images of the reflected light when he looked through a calcite crystal, he saw only one. Depending on how he rotated the crystal, the image was made from the ordinary or the extraordinary beam. Before going to sleep that night, Malus showed that although candle light viewed directly through calcite produces two images, candle light viewed through calcite after it was reflected from a plate of glass or a pan of water produces only one (Arago, 1859a; Towne, 1988). Malus (in Whittaker, 1951) found that

> … light reflected by the surface of water at an angle of 52° 45' has all the characteristics of one of the beams produced by the double refraction of Iceland spar, whose principal section is parallel to the plane which passes through the incident ray and the reflected ray. If we receive this reflected ray on any doubly refracting crystal, whose principle section is parallel to the plane of reflection, it will not be divided into two beams as a ray of ordinary light would be, but will be refracted according to the ordinary law.

When Malus held the calcite crystal so that its principal section was parallel to the plane of incidence (which is the plane that includes the incident ray and the reflected ray), he saw only the ordinary ray and concluded that the ordinary ray was polarized parallel to the principal section. When he held the calcite crystal so that its principal section was perpendicular to the plane that passes through the incident ray and the reflected ray, he saw only the extraordinary ray and concluded that the extraordinary beam was polarized perpendicular to the principal section. Malus (in Herschel, 1876) concluded, "that light acquires properties which are relative only to the sides of the ray—which are the same for the north and south sides of the ray" (i.e., of a vertical ray), "using the points of a compass for description's sake only…." Malus defined the plane of polarization of the ordinary ray as being parallel to the principal section of calcite and the plane of polarization of the extraordinary ray as being perpendicular to the principal section (Knight, 1867).

Pereira (1853) confirmed that the azimuths of polarization of the ordinary and extraordinary beams are orthogonally polarized by using a plate of tourmaline to analyze the ordinary and extraordinary beam. He found that as he rotated the analyzer, the two beams were alternately extinguished, indicating that the two beams were linearly polarized perpendicularly to each other (Tyndall, 1873, 1887). By 1821, Fresnel, a proponent of the wave theory of light, was questioning the arbitrariness of the "plane of polarization" and wondered what the relationship was between the plane of vibration and the plane of polarization (Whittaker, 1951).

Although the experiments done by Brewster and Malus illustrate the differences between common light and polarized light, they are tricky to follow in the twenty-first century because the definition of the plane of polarization they used differs from our current definition (Brewster, 1833b; Stokes, 1852; Arago, 1859a, 1859b, 1859c; Preston, 1895; Schuster, 1909; Jenkins and White, 1937; Strong, 1959; Hecht and Zajac, 1974; Born and Wolf, 1980; Collett, 1993; Niedrig, 1999; Goldstein 2003). Malus arbitrarily defined the plane of polarization of light as the "plane which passes through the incident ray and the reflected ray." However, in the mid-to-late nineteenth century, after Maxwell showed that the refractive index depended on the electrical permittivity of a substance, it became clear that optical phenomena resulted from the action of matter on the vibrating electric field of light, and consequently the plane of polarization became redefined to mean the plane of vibration of the electric field. Arago (1859a; p. 153) wrote "… it was long a disputed question whether the vibrations of which they [light waves] consist, according to the wave theory, are actually performed in those planes or perpendicular to them; the later has now been shown to be the fact."

Using the current definitions, the electric field of the extraordinary wave vibrates parallel to the principal section of calcite and the electric field of the ordinary wave vibrates perpendicular to the principal section. Since the principal section contains the optic axis of the crystal, the electric field of the extraordinary wave vibrates parallel to the optic axis and the electric field of the ordinary wave vibrates perpendicular to the optic axis. Moreover, when light is reflected from a dielectric, the reflected wave is linearly polarized parallel to the surface of the dielectric.

David Brewster (1815a) quantified Malus' observations on the polarization of light by reflection. Brewster found that the degree of polarization in the reflected light is maximal when the angle of incidence (relative to the normal) is equal to the arc tangent of the refractive index of the reflecting surface:

$$\theta_{Brewster} = \text{arc tan } (n)$$

This angle is known as Brewster's angle (Figure 7-7). For glass, where the refractive index is 1.515, $\theta_{Brewster}$ is 56.57 degrees. Reflection is not a really useful way of

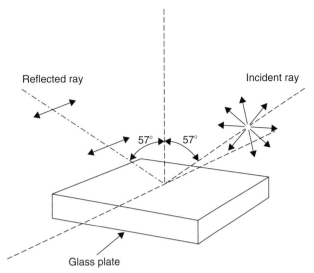

FIGURE 7-7 Production of polarized light by reflection at Brewster's angle. According to Malus, the "plane of polarization" was the plane that included the incident ray and the reflected ray.

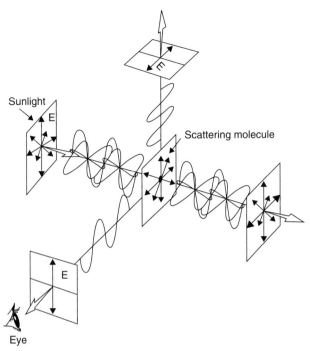

FIGURE 7-8 Production of polarized light by scattering sunlight from the molecules in the atmosphere.

obtaining polarized light for polarized light microscopy, but it is a useful way to determine the orientation of a laboratory polarizer. When we look through an analyzer at the glare reflected off a desktop or floor, and turn the analyzer to the position where the glare is maximal, we know that the azimuth of maximal transmission is parallel to the reflective surface. When we rotate the analyzer to the position that gives maximal extinction, we know that the azimuth of maximal transmission is perpendicular to the surface of the reflector. This is how Polaroid sunglasses work!

Polarized light can also be produced by scattering (Figure 7-8). When nonpolarized light interacts with atoms or small molecules compared with the wavelength of light, the atoms or molecules scatter the light in all directions (Rayleigh, 1870). This includes the molecules in the atmosphere. When one looks at the sky in a direction perpendicular to the sun's rays, the light coming toward one's eyes is linearly polarized (Können, 1985). The vibration is perpendicular to the plane made by the ray of the sun to the radiating molecule and the ray from the molecule to the eye. We can also use the polarization of skylight to tell the orientation of a laboratory analyzer. When the analyzer is rotated to the position where the sky appears maximally dark, the analyzer is oriented with it axis of maximal absorption vertical, and its axis of maximal transmittance horizontal.

We can detect polarized light with the rods in our eyes only under very specialized conditions, but typically we cannot detect the azimuth of polarized light (Minnaert, 1954). On the other hand, fish and bees can see and use the polarization of light to gather information about their world (Hawryshyn, 1992; Horváth and Varjú, 2004). Karl von Frisch (1950) showed that bees use the polarization of skylight to guide their flight and dances. Typically, when

bees return home after finding a good source of honey, they perform a dance that points in the direction of the honey to show the other bees in the hive where the honey is. When von Frisch experimentally altered the polarization of the skylight by using Polaroids, the dance, too, was altered and the bees no longer pointed in the direction of the honey.

As an aside on skylight, the intensity of the scattered light, compared to the light that passes straight through the atoms, molecules or particles is inversely proportional to the fourth power of the wavelength. Therefore, when we look at the sky anywhere except directly in the direction of the overhead sun, it appears blue (Rayleigh, 1870). This was first appreciated by Leonardo da Vinci in the 1500s. When we look directly at the sun at sunrise or sunset, it appears red since all the shorter wavelengths have been scattered away by the dust and/or water drops in the atmosphere by the time the light reaches our eyes.

Typically, there is little dust in the air in the morning, so the sky will be red only if there are enough water droplets in the air to scatter the light. Consequently, a red sky in the morning indicates rain. At the end of a typical day, there is a lot of dust in the air, so the sunset appears red. However, if it is raining in the west, the rain will have washed away most of the dust, which contributes to most of the scatter. Consequently, the shorter wavelengths will have not been scattered away and the sky will appear pale yellow, and there will not be a "red sky at night" (Minnaert, 1954).

We have discussed what linearly polarized light is, how to determine if light is linearly polarized, and how to produce linearly polarized light, as a result of dichrism, double

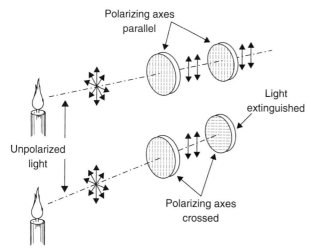

FIGURE 7-9 Extinguishing polarized light using a polarizer whose azimuth of maximal transmission is perpendicular to the azimuth of the linearly polarized light formed by the polarizer.

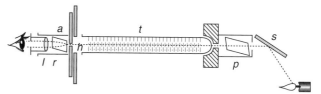

FIGURE 7-10 A polarimeter.

refraction, reflection, and scattering. Now we are ready to see how molecules that can produce linearly polarized light can influence the light that passes through crossed polars.

INFLUENCING LIGHT

All dichroic and birefringent materials that are able to polarize and analyze light also are able to influence the light that passes through crossed polars.

Two pieces of dichroic or birefringent material can be set up such that the first element, called the polarizer, produces linearly polarized light, and the second element, known as the analyzer, can block the passage of the linearly polarized light (Figure 7-9). The equation that describes the intensity of the light that passes through two polars is known as the Law of Malus:

$$I = I_o \cos^2 \theta$$

where θ is the angle made by the axes of maximal transmission of both polars.

In 1811, Francois Arago invented a polariscope to visualize the effect of a given substance on the polarization of light. A polariscope is a tube with a polarizer on one end and an analyzer at the other. (A polarimeter, by contrast, is a polariscope in which the analyzer can be turned in a graduated and calibrated manner.) In a polariscope, the polarizer and the analyzer are set orthogonally to each other so that no light exits the analyzer. Tongs, consisting of two pieces of tourmaline oriented orthogonally to each other, were designed by Biot to investigate the ability of a substance to influence polarized light. When a sample of tourmaline or calcite that is able to influence polarized light is inserted in the polariscope or in the tongs, light exits the

analyzer, even when in the crossed position. It turns out that tourmaline and calcite are able to influence the linearly polarized light. In fact, any dichroic or birefringent material that is capable of polarizing light and analyzing polarized light, when placed in a certain position between crossed polars, is capable of influencing linearly polarized light, so that light exits the analyzer. Thus crossed polars can be used to test whether or not a material is dichroic or birefringent.

Unlike thick crystals of calcite, thin crystals of calcite do not visibly separate laterally the extraordinary ray from the ordinary ray. The two rays are still orthogonally polarized and show another spectacular phenomenon that occurs when the rays partially overlap. Arago noticed that some transparent crystals appeared colored when viewed between crossed polars. Jean-Baptiste Biot discovered that, for a given crystal, there was a relationship between the thickness of the crystal and the colors produced. Moreover, different types of crystals (e.g., gypsum, quartz, and mica) with the same thickness produced different colors. Using the polarimeter, Biot also discovered that liquids, including turpentine, oil of lemon, and oil of laurel were able of influence linearly polarized light (Figure 7-10).

Biot proposed that the ability to influence polarized light was a property of the molecules that constituted the sample. In 1833, Biot showed that sucrose rotated the linearly polarized light to the right, but after it was heated with dilute sulfuric acid, the sugar solution rotated the linearly polarized light to the left—that is, the direction was inverted. Biot also studied the breakdown of starch into a compound he called dextrine, since it rotated the linearly polarized light to the right. Biot made the initial observations on the ability of tartaric acid to influence linearly polarized light. This set the foundation for the work of Pasteur (1860), Le Bel, and van't Hoff (1874, 1967) on stereochemistry (Pereira, 1854; Freund, 1906; Jones, 1913; Pye, 2001).

Louis Pasteur noticed that when he put tartar of wine, which is composed of tartaric acid, in his polarimeter, it rotated linear polarized light to the right. Oddly enough, racemic acid, which had the same chemical formula ($C_4H_6O_6$), was unable to rotate the azimuth of polarized light. Pasteur noticed that the sodium ammonium racemate was composed of two kinds of crystals that were mirror images of each other. One kind of crystal had right-sided tetrahedral faces and the other kind had left-handed tetrahedral faces. He separated the two kinds of crystals with tweezers and a magnifying glass. He then dissolved the two groups separately and determined that the class with the right-handed tetrahedral face rotated the linearly polarized light to the right and was like tartaric acid, and the class with a left-handed tetrahedral face rotated the linearly polarized light to the left. When he mixed the two in equal quantities, he reformed racemic acid, which no longer rotated the azimuth of linearly polarized light. Pasteur concluded that racemic acid was a mixture of tartaric acid and another chemical that was its mirror image. Le Bel and van't Hoff deduced that molecules that contained a carbon atom that bound four different atoms were asymmetrical and existed in a three-dimensional structure that could exist in two forms that were related by mirror symmetry.

If mineral and biological specimens are able to influence linearly polarized light; and, if the manner in which they influence linearly polarized light is a property of their molecular constituents, wouldn't it be reasonable to construct a polarizing light microscope to understand better, the molecular constituents of nature?

DESIGN OF A POLARIZING MICROSCOPE

Images of specimens taken with a polarized light microscope are shown in color plates 7 through 11. Henry Fox Talbot designed the first polarizing microscope, in which he could detect a change in the azimuth of polarization brought about by each point in the specimen. The polarizing microscope converts a change in the azimuth of polarization into a change in color and/or intensity. In designing the first polarizing light microscope, Talbot (1834a) wrote,

> Among the very numerous attempts which have been made of late years to improve the microscope, I am not aware that it has yet been proposed to illuminate the objects with polarized light.
>
> But as such an idea is sufficiently simple and obvious, it is possible that some experiments of this kind may have been published, although I am not acquainted with them. I have lately made this branch of optics a subject of inquiry, and I have found it so rich in beautiful results as entirely to surpass my expectations.
>
> As little else is requisite to repeat the experiments which I am about to mention than the possession of a good microscope, I think that in describing them I shall render a service to that numerous class of inquirers into nature, who are desirous of witnessing some of the most brilliant of optical phaenomena without the embarrassment of having to manage any large or complicated apparatus. And it cannot be without interest for the physiologist and natural historian to present him with a method of microscopic inquiry, which exhibits objects in so peculiar a manner that nothing resembling it can be produced by any arrangements of the ordinary kind.
>
> In order to view objects by polarized light, I place upon the stage of the microscope a plate of tourmaline, through which the light of the large concave mirror is transmitted before it reaches the object lens. Another plate of tourmaline is placed between the eyeglass and the eye; and this plate is capable of being turned round in its own plane, so that the light always traverses both the tourmalines perpendicularly.

The goal of a well set-up polarizing microscope is to generate contrast and to provide a bright image of an anisotropic substance against a black background at high resolution and high magnification. The degree to which the background can be darkened is expressed quantitatively as the extinction factor (EF):

$$EF = I_p/I_c$$

where I_p and I_c is the intensity of light that comes through the analyzer when the polarizer and analyzer are parallel and crossed, respectively. A typical polarizing microscope has an extinction factor of 1000. A good polarizing microscope, like the one designed by Shinya Inoué, has an extinction factor 10,000 or more (Inoué and Hyde, 1957; Inoué, 1986; Pluta, 1993). By convention, the azimuth of maximal transmission of the polarizer is oriented in the east-west direction on the microscope (which, by convention (ISO 8576), is left to right when facing the microscope; Pluta, 1993; Figure 7-11) and the azimuth of maximal transmission of the analyzer is oriented in the north-south orientation (which, by convention, is front to back. East, north, west, and south are considered to be 0, 90, 180, and 270 degrees, respectively, and when the polars are crossed, the polarizer is at 0 degrees and the analyzer is at 90 degrees.

In order to obtain maximal extinction, Talbot (1834a, 1834b) switched from using tourmaline for the polarizer and analyzer to using calcite prisms designed by William Nicol. According to Brewster (1833b), a single piece of calcite must be approximately three inches tall in order to separate the ordinary and extraordinary rays sufficiently to be practical to use as a polarizer and analyzer. William Nicol (1828, 1834, 1839) found an ingenious way to construct a prism that could separate the ordinary ray from the extraordinary using a minimal length of calcite. He constructed the prism out of a rhomboidal piece of calcite that he cut into two pieces. He cemented the two pieces together with Canada balsam, which has a refractive index intermediate between the refractive index experienced by the ordinary ray and the refractive index experienced by the extraordinary ray as they propagate through the calcite (Figure 7-12). Nonpolarized light striking the first part of the prism is resolved into the ordinary and extraordinary rays. When the ordinary ray, which had been experiencing a refractive index of 1.658, strikes the cement (n = 1.55), it is reflected away from the optical axis by total internal reflection. The extraordinary ray, which had been experiencing a refractive index of 1.486, is refracted through the cement interface and emerges through the far end of the prism, yielding linearly polarized light (Talbot, 1834b).

The Nicol prism gives extremely high extinction. Moreover, since it is transparent and thus does not absorb visible light, it is almost a perfect polarizer since it passes 50 percent of the incident light. Tourmaline and Polaroids,

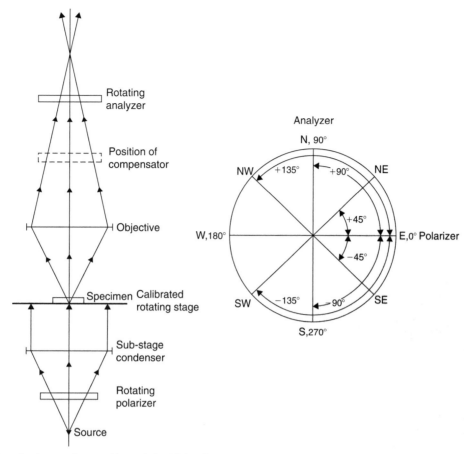

FIGURE 7-11   The design of and conventions used in a polarized light microscope.

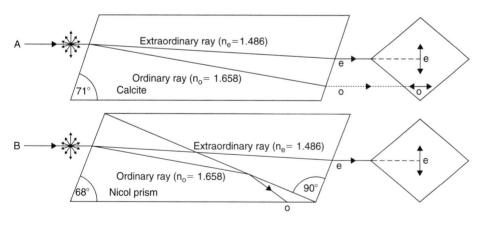

FIGURE 7-12   (A) Common light being split by calcite and (B) by a Nicol prism formed from two pieces of calcite cemented together with Canada balsam.

by contrast, pass less than 50 percent of the light because a small portion of the light that passes through the axis of maximal transmittance is absorbed. On the down side, Nicol prisms are very expensive, sometimes need rementing, and they have limited angular apertures because the maximal angle of incident light is determined by the critical angle that gives total internal reflection.

Many experimenters have designed polarizing prisms, made out of different materials that are cut and cemented in various ways in order to obtain the widest possible angular aperture (Thompson, 1905). The wide angular apertures are necessary in order to optimize resolution and extinction. Unfortunately, all the prisms pass the extraordinary beam, instead of the ordinary beam. Thus, the specimen is illuminated with spheroidal waves as opposed to spherical waves. This results in some astigmatism that can be corrected with cylindrical lenses.

The mechanical stress put on glass during the grinding process can introduce birefringence into the glass. Such "strain birefringence" reduces the extinction factor of the microscope and the contrast of the image by introducing elliptically polarized light in the absence of a specimen. The loss in extinction due to strain birefringence can be prevented by using strain-free sub-stage condenser and objective lenses. Often, a matched pair of strain-free objectives is used for the sub-stage condenser and objective lenses (Taylor, 1976).

The extinction factor in a polarizing microscope is much better with low numerical aperture objective lenses compared with high numerical aperture objective lenses. The reduced extinction factor is due to the rotation of the azimuth of polarization when polarized light strikes a surface at a large angle of incidence (Inoué, 1952b). An increase in the numerical aperture of 0.2 results in a tenfold increase in stray light. At one time the only way to get high extinction was to decrease the numerical aperture of both the sub-stage condenser and the objective lens to mitigate the decrease in contrast introduced by the depolarization of light as it strikes lenses at a large angle. In fact, it was suggested that only parallel light be used for illumination. Such a solution would compromise the resolving power of the light microscope. However we can get both a high extinction factor and good resolving power by using rectified optics, a method developed by Inoué and Hyde (1957; Inoué, 1961, 1986). It is also possible to obtain a high extinction factor and good resolving power using image processing techniques, including background subtraction (see Chapter 14).

The influence of a specimen on linearly polarized light depends on the orientation of the specimen with respect to the azimuth of polarization. Consequently, we should use a rotating stage in a polarizing microscope to accurately align the specimen relative to the azimuth of polarization. Talbot (1834a) describes the influence of specimen orientation on its appearance:

> The crystals, which were highly luminous in one position, when their axes were in the proper direction for depolarizing the light, became entirely dark in the opposite position, thus, as they rapidly moved onwards, appearing by turns luminous and obscure, and resembling in miniature the coruscations of a firefly. It was impossible to view this without admiring the infinite perfection of nature, that such almost imperceptible atoms should be found to have a regular structure capable of acting upon light in the same manner as the largest masses, and that the element of light itself should obey in such trivial particulars the same laws which regulate its course throughout the universe.

Talbot (1834a) noticed that the addition of a mica plate between the polarizer and the analyzer caused crystalline objects that were nearly invisible to become brightly colored. This formed the basis for the development of compensators. Compensators typically are inserted in a slot in the microscope body between the polarizer and the object or between the object and the analyzer. The concept of "compensation" was invented by Biot in order to get a hint

at the molecular architecture of a specimen. Talbot (1834a) describes the first use of a compensator in a microscope:

> When these miniature crystals [copper sulphate] are placed on the stage of the microscope, the field of view remaining dark as before, we see a most interesting phenomenon; for as every crystal differs from the rest in thickness, it displays in consequence a different tint, and the field of view appears scattered with the most brilliant assemblage of rubies, topazes, emeralds, and other highly coloured gems, affording one of the most pleasing sights that can be imagined. The darkness of the ground upon which they display themselves greatly enhances the effect. Each crystal is uniform in colour over all its surface, but if the plate of glass upon which they lie is turned round in its own plane, the colour of each crystal is seen to change and gradually assume the complementary tint. Many other salts may be substituted for the sulphate of copper in this experiment, and each of them offers some peculiarity, worthy of attention, but difficult to describe. Some salts, however, crystallize in such thin plates that they have not sufficient depolarizing power to become visible upon the dark ground of the microscope. For instance, the little crystals of sulphate of potash, precipitated by aether, appear only faintly visible. In these circumstances a contrivance may be employed to render evident their action upon light. It must be obvious that if a thin uniform plate of mica is viewed with the microscope, it will appear coloured (the tint depending on the thickness it may happen to have), and its appearance will be everywhere alike, in other words it will produce a coloured field of view. Now if such a plate of mica is laid beneath the crystals, or beneath the glass which supports them, these crystals, although incapable of producing any colour themselves, are yet frequently able to alter the colour which the mica produces; for instance, if the mica has produced a blue, they will, perhaps, alter it to purple, and thus will have the appearance of purple crystals lying on a blue ground.

Following the invention of the polarizing microscope by Talbot (1834a, 1834b, 1836a, 1836b, 1837, 1839a), chemists, physicists, geologists, engineers, and biologists (Brewster, 1837; Quekett, 1852; Pereira, 1854; Valentin, 1861; Carpenter, 1883; Naegeli and Schwendener, 1892; Hogg, 1898; Spitta, 1907; Reichert, 1913; Chamot, 1921; Schaeffer, 1953; Chamot and Mason, 1958; Bartels, 1966; Goldstein, 1969; see Inoué, 1986 for many references) have been able to study birefringent or dichroic specimens with this elegant tool. The polarizing microscope can do far more than introduce color and contrast to make transparent birefringent specimens visible. But before I describe the other miracles a polarizing microscope is capable of performing, I must first describe the molecular basis of birefringence.

WHAT IS THE MOLECULAR BASIS OF BIREFRINGENCE?

In order to understand the interaction of light with matter, we must understand both the nature of matter and the nature of light. While Malus, Arago, Biot, Wollaston, Talbot, and Brewster were experimenting with the interaction of polarized light with matter, it was not clear whether light was a particle or a wave. Therefore it was difficult to come up with a general theory of polarization. Newton (1730) wrote "for both must be understood, before the reason of their Actions upon one another can be known." In Chapter 3, I described what was known about the nature of light at the turn of the nineteenth century. Here, I will briefly describe a little bit about what was known about the nature of crystalline matter at the same time.

Christiaan Huygens (1690) knew that two crystals, Iceland spar and quartz, were birefringent. He proposed that the corpuscles in matter that had one refractive index were spherical; and these spherical corpuscles radiated spherical wavelets of light. Then he went on to say that the corpuscles that constituted birefringent crystals were spheroidal, and they gave rise to spheroidal wavelets of light. Charles François Dufay, who discovered the two types of electricity, known as vitreous and resinous, proposed that all crystals, except those that were cubic, may end up being birefringent.

Rene-Just Haüy (1807) accidentally dropped a beautiful specimen of Iceland spar on the floor and noticed that it broke into tiny pieces with a regular rhomboidal shape. He found that other crystals could be broken, with a sharp tool, into pieces with regular and characteristic shapes, including rhomboids, hexagonal prisms, tetrahedrons, and such. He believed that characteristic shapes of the "integrant particles" were related to the arrangement of the moleculae that that made up the crystal. The hardness of a crystal depended on the cohesive force between the moleculae, and if the hardness was not symmetrical, then the cleavage planes were not symmetrical. This meant that the cohesive forces between moleculae in asymmetric crystals like Iceland spar were asymmetric. The cleavage planes represent the planes in which the bonding between the moleculae is relatively weak.

If the asymmetry of the constituent moleculae of Iceland spar were the reason behind the birefringence, then it should be possible to reduce the asymmetry by heating up the Iceland spar and randomizing the constituent parts. Indeed Mitscherlich found that heating a crystal of Iceland spar differentially affects the sides of the rhomb and causes the rhomb to approach a cube. *Pari passu*, the double refraction diminishes (Brewster, 1833b).

David Brewster (1815b) and Thomas Seebeck independently showed that they could induce birefringence in glass, a substance that usually is not birefringent, by heating the glass and cooling it rapidly. Similarly, Brewster (1815c) and Fresnel independently found that they could induce birefringence in glass and other substances by compressing or stretching it. These experiments support the contention that birefringence is a result of the asymmetry of attractive forces between the particles that make up a given crystal (Brewster, 1833b). Stress-induced strain-birefringence is particularly spectacular in molded transparent plastics, including Plexiglas and clear plastic tableware.

The refractive index is a measure of the ability of a transparent substance to decrease the speed of light as a result of interactions between the substance and the light. Based on the dynamic wave theory of light, the smaller the ratio of the elasticity to the density of the aether within a transparent substance, the slower light propagates through the substance.

Elasticity characterizes the ability to resist deformation or to restore a deformed object so that it takes up its original position. In terms of the electromagnetic theory, electrons held tightly in place by the electromagnetic forces of the nucleus or nuclei would not be easily polarized; whereas electrons held in place less tightly by the electromagnetic forces of the nucleus or nuclei would be easily polarized. Substances that have readily polarizable, deformable electron clouds have a higher refractive index than substances that have tightly held rigid electron clouds. Thus, according to the dynamical electromagnetic theory of light, the speed of light though a substance depends on the elasticity of the spring that holds an electron in the substance. The less deformable the spring, the faster the light propagates through the substance, the more deformable the spring, the slower the light propagates through the substance.

The electrons are not held in place by a restoring force transmitted through mechanical springs, but by analogous electromagnetic forces that depend on the electric permittivity and magnetic permeability. According to Maxwell's (1865) electromagnetic wave theory of light, the square of the index of refraction is approximately equal to the dielectric constant or relative permittivity (K_e) according to the following equation:

$$n^2 = K_e = \varepsilon/\varepsilon_o$$

where ε is the frequency-dependent electric permittivity of the substance and ε_o is the electric permittivity of the vacuum. The relative permittivity thus must be a measure of the deformability of a substance. The ability of external electric fields ($\vec{E}$), including the oscillating electric fields of light, to deform or separate positive and negative charges in a substance is characterized by the electric polarization ($\vec{P}$).

$$\vec{P} = (\varepsilon - \varepsilon_o)\vec{E}$$

If we divide both sides of the equation by ε_o, we get

$$\vec{P}/\varepsilon_o = (\varepsilon/\varepsilon_o - \varepsilon_o/\varepsilon_o)\vec{E} = (\varepsilon/\varepsilon_o - 1)\vec{E}$$

and since $\varepsilon/\varepsilon_o = n^2$, then

$$n^2 = 1 + \vec{P}/\varepsilon_o\vec{E} = 1 + \vec{P}/\vec{D}$$

where the product of ε_o and E is equal to the electric flux density (D, in N/Vm). In general, an electron vibrating in a bond can be displaced, deformed, or polarized parallel to the bond, more than it can be displaced, deformed, or polarized perpendicular to the bond. Consequently, the refractive index along the bond will be greater than the refractive index perpendicular to the bond.

If such bonds are randomly distributed throughout the molecule, light will be slowed down in a manner that is independent of the azimuth of polarization. Such a substance is called isotropic, and it will be invisible in a polarizing light microscope. Gases, liquids, most solids, and glasses are isotropic because the bonds are arranged randomly in these substances. However some cubic crystals, like NaCl, are isotropic because the electrons are arranged symmetrically with respect to every azimuth. If a biological specimen is isotropic, we can assume that the bonds in the substance are distributed randomly with respect to all axes since cubic crystals are rarely found in biological material.

If the bonds in a substance are not random, but coaligned, the substance is said to be anisotropic. Since the electron clouds can be deformed more readily by electric fields vibrating parallel to a bond than electric fields vibrating perpendicular to a bond, light whose electric fields are linearly polarized parallel to the bond will be slowed down more than light whose electric fields are linearly polarized perpendicular to the bond. Such a birefringent specimen will be visible in a polarized light microscope.

The interaction of polarized light with electrons can be studied very effectively using microwave radiation (Figure 7-13). As opposed to visible light that has wavelengths in the nanometers or micrometer range, microwave radiation has wavelengths in the centimeter range. The microwaves are produced and received by a diode and are measured with an electrical field meter. A wire grating is used as a polarizer. Microwave radiation linearly polarized perpendicular to the slits of the wire grating pass through the grating, but microwave radiation linearly polarized parallel to the wire grating interacts with the grating. The radiation linearly polarized parallel to the wires interacts with the electrons in the wire grating because the electrons can be readily moved back and forth through a conductor. Consequently, the radiant microwave energy is converted into kinetic energy, which eventually is converted to heat by the electrons moving along the length of the wire. The wire grating has a large cross-section for the microwave radiation linearly polarized parallel to the wires, and removes it from the radiation that passes through the grating.

On the other hand, the electrons cannot move across the air gaps between the wires of the grating, so the microwave radiation linearly polarized perpendicular to the wires interacts minimally with the electrons and consequently propagates right through the grating.

The interaction between electromagnetic radiation and a material can be characterized by the refractive index and the extinction coefficient of the material for electromagnetic radiation with a given angular frequency. Refraction takes place when the natural characteristic angular frequency of the vibrating electrons is much greater than the angular frequency of the incident light (Jenkins and White, 1950; Wood, 1961; Hecht and Zajac, 1974). The natural angular frequencies of glasses for example are in the ultraviolet range. When the angular frequency of light approaches the natural angular frequency of a substance, the substance no longer refracts the light, but absorbs it, which is why ordinary glass cannot be used to transmit ultraviolet light. The extinction coefficient, given in units of area/amount of substance, characterizes the ability of the substance to reduce the output light captured by an axial detector by converting the incident light into chemical energy or heat, while the refractive index characterizes the ability of a substance to reduce the direct output light captured by an axial detector by bending the incident light away from the detector. Anisotropic substances have two refractive indices or two extinction coefficients for light with a given angular frequency. Some anisotropic crystals, including calcite and quartz are birefringent and some, including tourmaline, sapphires, and rubies, are dichroic.

In general, there are two kinds of birefringence, positive and negative—quartz is an example of a positively birefringent substance and calcite is an example of a negatively birefringent substance. The two classes of birefringence are distinguished geometrically. In positively birefringent substances, the axis with the greater refractive index is parallel to the optic axis whereas in negatively birefringent substances the axis with the greater refractive index is perpendicular to the optic axis.

When we view an aperture illuminated with common light through a birefringent crystal like calcite, we will typically see two images of the aperture, and the separation of the two images will depend on the orientation of the crystal. There is one particular orientation that will produce only a single image. In this orientation, we are looking down the optic axis. The optic axis of the calcite is coparallel with the imaginary line through the crystal that connects the aperture to the eye.

The propagation of linearly polarized light through a birefringent crystal will depend on the sign of birefringence, the angle that the propagation vector makes with the optic axis, and the azimuth of polarization. In general,

1. Nonpolarized light composed of electric fields vibrating in all azimuths, propagating through a positively

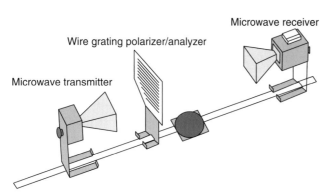

FIGURE 7-13 Studying the interaction of polarized light with electrons using microwave radiation.

Microwave receiver

Wire grating polarizer/analyzer

Microwave transmitter

birefringent or a negatively birefringent crystal with its propagation vector parallel to the optic axis, will experience only one index of refraction, and thus will form only one image. The light will travel through the crystal as a spherical wave (Figure 7-14). In this case, the electric vectors experience the ordinary index of refraction (n_o).

2. Nonpolarized light composed of electric fields vibrating in all azimuths, propagating through a negatively birefringent crystal perpendicular to the optic axis will experience two indices of refraction. The components of the electric field whose azimuths are perpendicular to the optic axis will form the ordinary wave and the components of the electric field whose azimuths are parallel to the optic axis will form the extraordinary wave. In negatively birefringent substances, the extraordinary wave experiences a smaller index of refraction (n_e) than the ordinary wave (n_o), and thus two images will be formed on top of each other: the image formed by the ordinary wave, which experiences a larger index of refraction than the extraordinary wave will be closer to us than the image formed by the extraordinary wave. This is because the difference in refractive index between the negatively birefringent crystal and air is greater for the ordinary wave than for the extraordinary wave. When we trace back the rays refracted from the crystal-air interface, the ordinary waves appear to originate from a spot closer to us than the extraordinary waves. The ordinary wave propagates through the crystal as a spherical wave and the extraordinary wave propagates as a spheroidal wave.

3. Nonpolarized light composed of electric fields vibrating in all azimuths, propagating through a positively birefringent crystal perpendicular to the optic axis will experience two indices of refraction. The components of the electric field whose azimuths are perpendicular to the optic axis will form the ordinary wave and the components of the electric field whose azimuths are parallel to the optic axis will form the extraordinary wave. The extraordinary wave will experience a larger index of refraction (n_e) than the ordinary wave (n_o), and thus two images will be formed on top of each other: the image formed by the ordinary wave, which experiences a smaller index of refraction than the extraordinary wave, will appear farther from us than the image formed by the extraordinary wave. This is because the difference in refractive index between the positively birefringent crystal and air is greater for the extraordinary wave than for the ordinary wave. When we trace back the rays refracted from the crystal-air interface, the ordinary rays appear to originate from a spot farther from to us than the extraordinary rays. The ordinary wave propagates through the crystal as a spherical wave and the extraordinary wave propagates as a spheroidal wave.

4. Nonpolarized light composed of electric fields vibrating in all azimuths, propagating through a positively birefringent crystal at any angle between 0 and 90 degrees relative to the optic axis will experience two indices of refraction. Whereas the ordinary wave will be constituted only from the components of the electric field whose azimuths are perpendicular to the optic axis, the extraordinary

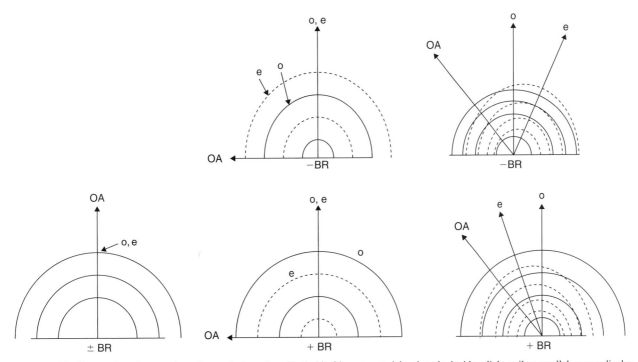

FIGURE 7-14 Propagation of waves through negatively and positively birefringent materials when the incident light strikes parallel, perpendicular, and oblique to the optic axis. Ordinary waves (——), extraordinary waves (- - - -).

wave will be constituted from the components of the electric field that are perpendicular and parallel to the optic axis. Consequently, the extraordinary wave will be spheroidal and will produce an astigmatic image that is laterally displaced from the image made by the ordinary ray. An astigmatic image is one where a circle in the object is represented as a spheroid in the image. The image made by the extraordinary wave will tend away from the optic axis. The distance between the two images will depend on the thickness of the crystal. The ordinary wave propagates through the crystal as a spherical wave and the extraordinary wave propagates through the crystal as a spheroidal wave.

5. Nonpolarized light composed of electric fields vibrating in all azimuths, propagating through a positively birefringent crystal at any angle between 0 and 90 degrees relative to the optic axis will experience two indices of refraction. Whereas the ordinary wave will be constituted only from the components of the electric field whose azimuths are perpendicular to the optic axis, the extraordinary wave will be constituted from the components of the electric field that are perpendicular and parallel to the optic axis. Consequently, the extraordinary wave will produce an astigmatic image that is laterally displaced from the image made by the ordinary ray. The image made by the extraordinary wave will tend toward the optic axis. The distance between the two images will depend on the thickness of the crystal. The ordinary wave propagates through the crystal as a spherical wave and the extraordinary wave propagates through the crystal as a spheroidal wave.

Light vibrating perpendicular to the optic axis is always refracted ordinarily and consequently, the refractive index perpendicular to the optic axis is known as n_o. By contrast, light vibrating parallel to the optic axis can be refracted in an extraordinary manner and thus the refractive index parallel to the optic axis is called n_e. Birefringence is defined as $n_e - n_o$. Thus when $n_e > n_o$, the specimen is positively birefringent and when $n_e < n_o$, the specimen is negatively birefringent. Birefringence (BR) is an intrinsic or state quantity, which can be used to identify a given substance because it is independent of the amount of substance. That is birefringence, like density is an intrinsic or state quantity in contrast to mass and volume, which are extrinsic qualities that vary with the amount of the substance.

Birefringence in a specimen can be detected with polarization microscopy, by placing the specimen between crossed polars and rotating the stage 360 degrees. Upon rotation, a birefringent specimen will alternate between being bright and dark. The brightness depends on the orientation of the optic axis of the specimen relative to the crossed polars. If the azimuth of the axis of maximal transmission of the polarizer is defined as 0 degrees, then the brightness increases as the optic axis of the specimen is rotated from 0 to 45 degrees, then decreases from 45 to 90 degrees, then increases from 90 to 135 degrees, then decreases from 135 to 180 degrees, and so on. If we determine that the substance is birefringent, we can deduce that the bonds in that substance are anisotropic or asymmetrical.

INTERFERENCE OF POLARIZED LIGHT

When nonpolarized light strikes a relatively thick piece of calcite or quartz, the image is duplicated since the incident light is split into two laterally displaced beams of polarized light that vibrate perpendicularly to each other. Since one of the beams travels faster than the other one, the phases are also different. When the crystals are placed on the stage of a polarizing microscope so that the optic axis is parallel to the stage and oriented $\pm 45°$ relative to the polarizer, the incident linearly polarized light is split laterally into two beams that are linearly polarized perpendicular to and out-of-phase with each other.

In the case of a very thin crystal or a birefringent biological specimen, the incident light is divided into two beams that vibrate orthogonally to and out-of-phase with each other. However, when the specimen is thin enough, there is almost no lateral separation, and both beams, which originate from the same point in the specimen and are coherent, are close enough to interfere with each other after they pass through the analyzer. The difference in phase, which is known as the retardation (Γ) in the polarization literature, depends on two things:

• The difference in the refractive indices between the extraordinary beam and the ordinary beam (n_e–n_o)
• The thickness of the specimen (t).

The retardation (in nm) is given by the following equation:

$$\Gamma = (n_e - n_o)t = (BR)t$$

The retardation, which is reminiscent of the optical path difference, can also be expressed in terms of a phase angle. The phase angle (in degrees and radians) is given by the following equation:

$$\varphi = \Gamma (360°/\lambda) = [(n_e - n_o)t] (360°/\lambda)$$
$$= [(n_e - n_o)t] (2\pi/\lambda)$$

and the phase change in terms of the wavelength of the incident light λ_i is given by:

$$\text{Phase change} = [(n_e - n_o)t)] (\lambda/\lambda_i)$$

In order to understand how an image is formed in a polarized light microscope, I am going to give several examples of imaginary specimens and a method that can be used to determine the nature of their images. Then I will present a general method to understand the nature of any image obtained with a polarized light microscope. Imagine putting a positively birefringent specimen ($n_e = 1.4805$,

$n_o = 1.4555$, t = 5,000 nm) on a rotating stage in a polarized light microscope and orienting it so its optic axis is +45 degrees (NE-SW) relative to the azimuth of maximal transmission of the polarizer (0°, E-W). Illuminate the specimen with linearly polarized light with a wavelength of 500 nm (λ = 500 nm). The birefringent specimen resolves the incident linearly polarized light into two orthogonal linearly polarized waves, the ordinary wave and the extraordinary wave. The extraordinary wave vibrates linearly along the NE-SW axis and the ordinary wave vibrates linearly along the SE-NW axis. The ordinary wave will be ahead of the extraordinary wave by 125 nm, and the extraordinary wave will be retarded relative to the ordinary wave by 125 nm. This is equivalent to the extraordinary wave being retarded by 90 degrees, $\pi/2$ radians or $\lambda/4$ of 500 nm light. Because the extraordinary wave propagates slower than the ordinary wave, the axis of electron polarization that gives rise to the extraordinary wave is called the slow axis, and the axis of electron polarization that gives rise to the ordinary wave is known as the fast axis. The optic axis of a positively birefringent specimen is the slow axis and the axis perpendicular to the optic axis is known as the fast axis.

Only coherent waves whose electric fields are coplanar or whose electric fields have components that are coplanar can interfere with each other. Thus orthogonal waves, whose electric fields are perpendicular to each other, cannot interfere with each other. However, if we allow the coherent, orthogonal, linearly polarized waves leaving the birefringent specimen to pass through an analyzer, whose azimuth of maximal transmission is oriented at a 90-degree (N-S) angle relative to the azimuth of maximal transmission of the polarizer, a component of each of the two orthogonal waves can pass through the analyzer. The out-of-phase components of the ordinary wave and the extraordinary wave that pass through the analyzer are coherent, linearly polarized, and coplanar. Consequently they will interfere with each other.

We can use a vector method to model the interaction of light with a point in an anisotropic specimen, and to predict how bright the image of that point will be. Consider the specimen just described. First, draw the relative phases of the ordinary wave and the extraordinary wave (Figure 7-15). Second, make a table of the amplitudes of each wave at various convenient times during the period (Table 7-1).

Third, create a Cartesian coordinate system where the azimuth of polarization of the incident light goes from E to W, and the azimuth of light passed by the analyzer goes from N to S (Figure 7-16). The linearly polarized light that strikes the specimen, whose optic axis is oriented at +45 degrees (NE to SW) will be converted into two linearly polarized waves that vibrate perpendicularly to each other. The extraordinary wave vibrates parallel to the optic axis (NE to SW); the ordinary wave vibrates perpendicular to the optic axis (SE to NW). Draw the vectors that represent the extraordinary waves on the NE-SW axis and the

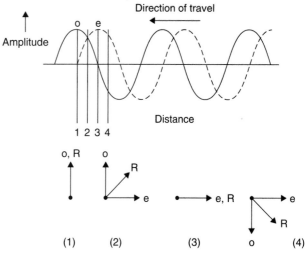

Remember the two waves are orthogonal, and since $n_e > n_o$, the o-wave travels faster than the e-wave.

FIGURE 7-15 Determination of the resultant (R) of two orthogonal linearly polarized waves that propagate through a point in the specimen 90° out-of-phase with each other.

TABLE 7-1 The Amplitudes of the Extraordinary (e) Wave and the Ordinary (o) Wave at Various Time Points for a Birefringent Specimen (n_e = 1.4805, n_o = 1.4555, t = 5,000 nm) Illuminated with 500 nm Light

Time	Position (in degrees)	Relative amplitude of e wave	Relative amplitude of o wave
1	0	0	1
2	45	0.707	0.707
3	90	1	0
4	135	0.707	−0.707
5	180	0	−1
6	225	−0.707	−0.707
7	270	−1	0
8	315	−0.707	0.707
9	360	0	1

vectors that represent the ordinary waves on the SE-NW axis.

Plot the amplitudes of the ordinary wave and the extraordinary wave at each convenient time. The resultant wave is obtained by adding the vectors that represent the electric fields of the ordinary and the extraordinary waves. The amplitude of the resultant wave at a given time point is represented by the length of the vector and the azimuth of the wave is represented by the angle of the vector. In this

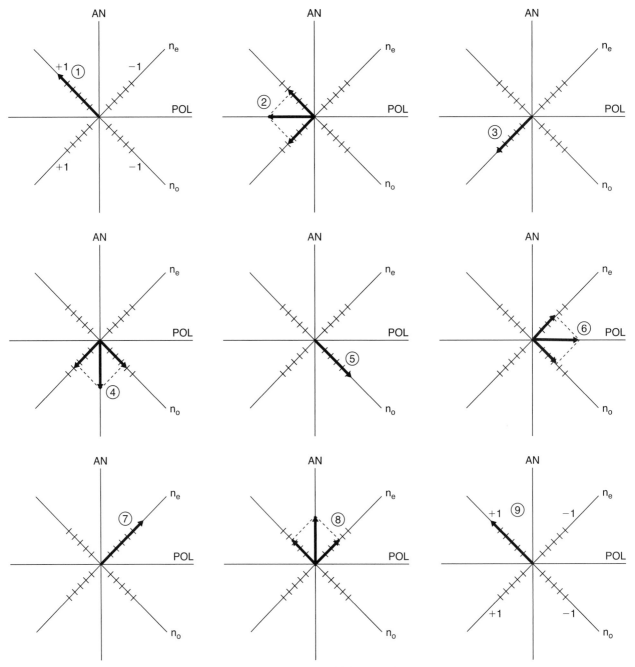

FIGURE 7-16 Determination of the resultant wave from its components at individual points in time.

case, the resultant wave is circularly polarized and going in a counterclockwise direction (Figure 7-17). Circularly polarized light can be considered to be composed of two orthogonal, linearly polarized waves of equal amplitude, 90 degrees out-of-phase with each other: one vibrating parallel to the azimuth of maximal transmission of the polarizer and one vibrating parallel to the axis of maximal transmission of the analyzer. Consequently, half of the intensity of the incident light will pass through the analyzer and the specimen will appear bright on a black background.

If the same specimen were placed so that its optic axis were −45 degrees (SE-NW) relative to the polarizer, the

resultant wave would also be circularly polarized, but in the clockwise direction (Figure 7-18). The specimen would still appear bright on a dark background.

Imagine putting a positively birefringent specimen ($n_e = 1.4805$, $n_o = 1.4555$, $t = 10,000\,nm$) on a rotating stage in a polarized light microscope and orienting it so its optic axis is +45 degrees (NE-SW) relative to the azimuth of maximal transmission of the polarizer (0°, E-W). Illuminate the specimen with linearly polarized light with a wavelength of 500 nm ($\lambda = 500\,nm$). The birefringent specimen resolves the incident linearly polarized light into two orthogonal linearly polarized waves, the ordinary wave and the extraordinary wave. The extraordinary wave vibrates linearly along the NE-SW axis and the ordinary wave vibrates linearly along the SE-NW axis. The ordinary wave will be ahead of the extraordinary wave by 250 nm, and the extraordinary wave will be retarded relative to the ordinary wave by 250 nm. This is equivalent to the extraordinary wave being retarded by 180 degrees, π radians or $\lambda/2$ of 500 nm light. Again we can use the vector method to model the interaction of light with a point in an anisotropic specimen, and to predict how bright the image of that point will be. Repeat the process described earlier. First, draw the relative phases of the ordinary wave and the extraordinary wave (Figure 7-19). Second, make a table of the amplitudes of each wave at various convenient times during the period (Table 7-2).

Third, create a Cartesian coordinate system where the azimuth of polarization of the incident light goes from E to W, and the azimuth of light passed by the analyzer goes from N to S (Figure 7-20). The linearly polarized light that strikes the specimen, whose optic axis is oriented at +45 degrees (NE to SW), will be converted into two linearly polarized waves that vibrate perpendicularly to each other. The extraordinary wave vibrates parallel to the optic axis (NE to SW); the ordinary wave vibrates perpendicular to the optic axis (SE to NW).

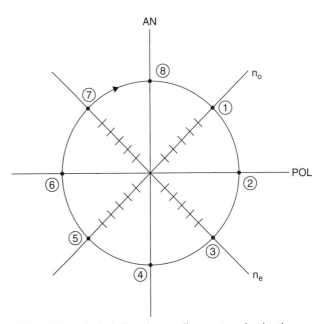

FIGURE 7-18 A single Cartesian coordinate system showing the propagation of a clockwise circularly polarized wave.

TABLE 7-2 The Amplitudes of the Extraordinary (e) Wave and the (o) Ordinary Wave at Various Time Points for a Birefringent Specimen ($n_e = 1.4805$, $n_o = 1.4555$, t = 10,000 nm) Illuminated with 500 nm Light

Time	Position (in degrees)	Relative amplitude of e wave	Relative amplitude of o wave
1	0	0	0
2	45	0.707	−0.707
3	90	1	−1
4	135	0.707	−0.707
5	180	0	0
6	225	−0.707	0.707
7	270	−1	1
8	315	−0.707	0.707
9	360	0	0

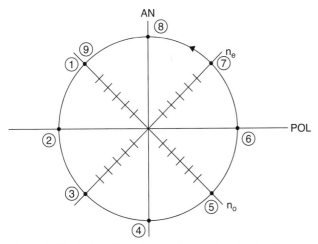

FIGURE 7-17 A single Cartesian coordinate system showing the propagation of a counterclockwise circularly polarized wave.

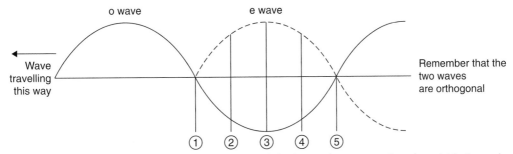

FIGURE 7-19 Determination of the resultant of two orthogonal linearly polarized waves that propagate through a point in the specimen 180° out-of-phase with each other.

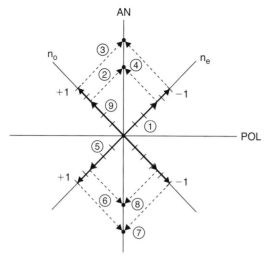

FIGURE 7-20 Determination of the resultant wave from its components at individual points in time.

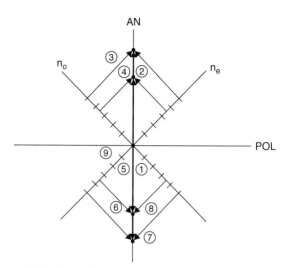

FIGURE 7-21 Linearly polarized light parallel to the azimuth of maximal transmission of the analyzer.

Next, plot the amplitudes of the ordinary wave and the extraordinary wave at each convenient time. The resultant wave is obtained by adding the vectors that represent the electric fields of the ordinary and the extraordinary waves. The amplitude of the wave is represented by the length of the vector and the azimuth of the wave is represented by the angle of the vector. In this case, the resultant wave is linearly polarized in the N-S direction, along the axis of maximal transmission of the analyzer. The specimen will appear bright on a black background (Figure 7-21). In this case, the image would look the same if its optic axis were placed −45 degrees relative to the polarizer.

Imagine putting a negatively birefringent specimen ($n_e = 1.4555$, $n_o = 1.4805$, $t = 5,000\,nm$) on a rotating stage in a polarized light microscope and orienting it so its optic axis is +45 degrees (NE-SW) relative to the azimuth of maximal transmission of the polarizer (0°, N-S). Illuminate the specimen with linearly polarized light with a wavelength of 500 nm ($\lambda = 500\,nm$). The birefringent specimen resolves the incident linearly polarized light into two orthogonal linearly polarized waves, the ordinary wave and the extraordinary wave. The extraordinary wave vibrates linearly along the NE-SW axis and the ordinary wave vibrates linearly along the SE-NW axis. The extraordinary wave will be ahead of the ordinary wave by 125 nm, and the ordinary wave will be retarded relative to the extraordinary wave by 125 nm. This is equivalent to the ordinary wave being retarded by −90 degrees, −π/2 radians or −λ/4 of 500 nm light. Because the ordinary wave propagates slower than the extraordinary wave, the axis of electron polarization that gives rise to the ordinary wave is called the slow axis, and the axis of electron polarization that gives rise to the extraordinary wave is known as the fast axis. The optic axis of a negatively birefringent specimen that produces the extraordinary wave is the fast

axis and the axis perpendicular to the optic axis is known as the slow axis.

We can use the vector method to model the interaction of light with a point in an anisotropic specimen, and to predict how bright the image of that point will be. First, draw the relative phases of the ordinary wave and the extraordinary wave (Figure 7-22). Then make a table of the amplitudes of each wave at various convenient times during the period (Table 7-3).

Third, create a Cartesian coordinate system where the azimuth of polarization of the incident light goes from E to W, and the azimuth of light passed by the analyzer goes from N to S. The linearly polarized light that strikes the specimen, whose optic axis is placed at +45 degrees (NE to SW), will be converted into two linearly polarized beams that vibrate perpendicularly to each other. The extraordinary wave vibrates parallel to the optic axis (NE to SW), which is the fast axis; the ordinary wave vibrates perpendicular to the optic axis (SE to NW), which is the slow axis. The amplitude of the wave is represented by the length of the vector and the azimuth of the wave is represented by the angle of the vector. Plot the amplitudes of the ordinary wave and the extraordinary wave at each convenient time on a Cartesian coordinate system as I explained above.

The specimen produces circularly polarized light that travels in the clockwise direction. The specimen appears bright on a black background. If the specimen were placed so that its optic axis was −45° relative to the polarizer, the specimen would produce circularly polarized light that travels in the counterclockwise direction.

Imagine putting a negatively birefringent specimen ($n_e = 1.402$, $n_o = 1.452$, $t = 10,000\,nm$) on a rotating stage in a polarized light microscope and orienting it so its optic axis is +45° (NE-SW) relative to the azimuth of

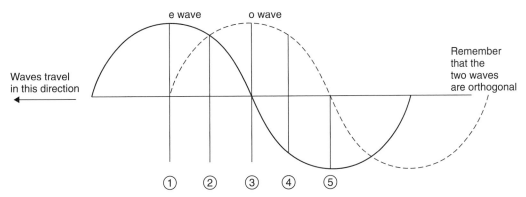

FIGURE 7-22 Determination of the resultant of two orthogonal linearly polarized waves that propagate through a point in the specimen 90° out-of-phase with each other.

TABLE 7-3 The Amplitudes of the Extraordinary (e) Wave and the Ordinary (o) Wave at Various Time Points for a Birefringent Specimen ($n_e = 1.4555$, $n_o = 1.4805$, t = 5,000 nm) Illuminated with 500 nm Light

Time	Position (in degrees)	Relative amplitude of e wave	Relative amplitude of o wave
1	0	1	0
2	45	0.707	0.707
3	90	0	1
4	135	−0.707	0.707
5	180	−1	0
6	225	−0.707	−0.707
7	270	0	−1
8	315	0.707	−0.707
9	360	1	0

the relative phases of the ordinary wave and the extraordinary wave (Figure 7-23). Second, make a table of the amplitudes of each wave at various convenient times during the period (Table 7-4).

Third, create a Cartesian coordinate system where the azimuth of polarization of the incident light goes from E to S, and the azimuth of light passed by the analyzer goes from N to S. The linearly polarized light that strikes the specimen, whose optic axis is placed at +45 degrees (NE to SW), will be converted into two linearly polarized beams that vibrate perpendicularly to each other. The extraordinary wave vibrates parallel to the optic axis (NE to SW); the ordinary wave vibrates perpendicular to the optic axis (SE to NW). Next, plot the amplitudes of the ordinary wave and the extraordinary wave at each convenient time.

The resultant wave is linearly polarized along the azimuth of maximal transmission of the polarizer and perpendicular to the azimuth of maximal transmission of the analyzer, and consequently the specimen is invisible on a black background (Figure 7-24). Although a birefringent specimen whose retardation is equal to the wavelength of the monochromatic illuminating light is invisible, it can be made visible by using white light, since for all other colors, the retardation will not equal an integral number of wavelengths.

Birefringent specimens influence the incident linearly polarized light and form two linearly polarized waves that are out-of-phase and orthogonal to each other. The direction and ellipticity of the resultant wave depends on the magnitude and sign of birefringence and the thickness of the specimen. It also depends on the position of the specimen with respect to the polarizer and the wavelength of light used to illuminate the specimen. When using monochromatic light, a birefringent specimen that introduces a phase angle greater than 0 degrees and less than 90 degrees produces elliptically polarized light whose major axis is parallel to the azimuth of maximal transmission of the polarizer. When the phase angle is between 90 and 180 degrees, the specimen produces elliptically polarized light whose major axis is parallel to azimuth of maximal transmission of the analyzer. The greater the component of

maximal transmission of the polarizer (0°, N-S). Illuminate the specimen with linearly polarized light with a wavelength of 500 nm ($\lambda = 500$ nm). The birefringent specimen resolves the incident linearly polarized light into two orthogonal linearly polarized waves, the ordinary wave and the extraordinary wave. The extraordinary wave vibrates linearly along the NE-SW axis and the ordinary wave vibrates linearly along the SE-NW axis. The extraordinary wave will be ahead of the ordinary wave by 500 nm, and the ordinary wave will be retarded relative to the extraordinary wave by 500 nm. This is equivalent to the ordinary wave being retarded by −360 degrees, −2π radians or −λ of 500 nm light.

We can use the vector method again to model the interaction of light with a point in a birefringent specimen, and to predict how bright the image of that point will be. First, draw

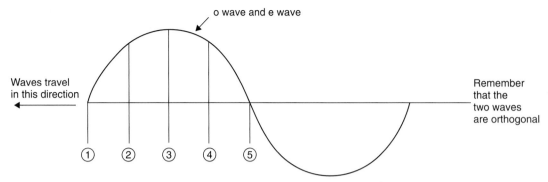

FIGURE 7-23 Determination of the resultant of two orthogonal linearly polarized waves that propagate through a point in the specimen 0° out-of-phase with each other and through the azimuth of maximal transmission of the analyzer.

TABLE 7-4 The Amplitudes of the Extraordinary (e) Wave and the Ordinary (o) Wave at Various Time Points for a Birefringent Specimen ($n_e = 1.402$, $n_o = 1.452$, $t = 10,000$ nm) Illuminated with 500 nm Light

Time	Position (in degrees)	Relative amplitude of e wave	Relative amplitude of o wave
1	0	0	0
2	45	0.707	0.707
3	90	1	1
4	135	0.707	0.707
5	180	0	0
6	225	−0.707	−0.707
7	270	−1	−1
8	315	−0.707	−0.707
9	360	0	0

TABLE 7-5 The Appearance of a Specimen with a Given Phase Angle in a Polarized Light Microscope Using Monochromatic Illumination

Phase change, degrees	Brightness
0	Dark
±45	Somewhat bright
±90	Half of the light incident on the specimen
±135	Pretty bright
±180	As bright as the light incident on the specimen
±225	Pretty bright
±270	Half of the light incident on the specimen
±315	Somewhat bright
±360	Dark

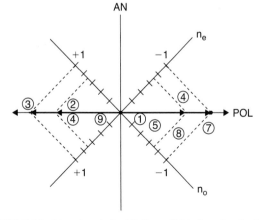

FIGURE 7-24 Linearly polarized light parallel to the azimuth of maximal transmission of the polarizer.

elliptically polarized light that is parallel to the azimuth of transmission of the analyzer, the brighter the specimen will be (Table 7-5).

We can determine how any point on any specimen will appear in a polarizing light microscope by using the following rules.

1. Identify the important pieces of information in the problem.

2. Determine the retardation: $\Gamma = (n_e - n_o)t$.

3. Determine the phase angle: $\varphi = \Gamma(360°/\lambda) = [(n_e - n_o)t](360°/\lambda)$. Plot the amplitudes of the ordinary and extraordinary waves.

4. Make a table of the amplitudes of the ordinary wave and the amplitude of the extraordinary wave at various convenient time points.

5. Draw a coordinate system and place the azimuth of maximal transmission of the polarizer in the E-W direction and the azimuth of maximal transmission of the analyzer in the N-S position. Then orient the optic axis of the specimen in either a +45 or a −45 degree angle relative to the azimuth of transmission of the polarizer. Label the optic axis n_e, since it represents the extraordinary refractive index and the axis perpendicular to the optic axis n_o, since it represents the ordinary refractive index. Draw a vector representation of the extraordinary wave parallel with the optic axis (n_e) and a vector representation of the ordinary wave perpendicular to the optic axis (n_o).

6. Add the amplitudes of the ordinary and extraordinary wave vectors at each convenient time point to get the resultant. Label each time point.

7. Connect the dots (in order) and determine the direction of the resultant light as it moves in time from the paper toward you.

8. If the resultant has any component along the azimuth of the analyzer, the object will be visible. The greater the component along the azimuth of transmission of the analyzer is, the brighter the image will be.

THE ORIGIN OF COLORS IN BIREFRINGENT SPECIMENS

In the examples just given, I have assumed that the specimens were illuminated with monochromatic light. What happens when the specimens are illuminated with white light? The phase angle of the light that leaves a birefringent specimen depends on the wavelength of light. A birefringent specimen with a given retardation will introduce a different phase for light of each wavelength. The wavelengths that undergo a phase angle of 180 degrees (or $m \pm \lambda/2$, where m is an integer) will come through crossed polars as the brightest. The wavelengths that undergo a phase angle of 0 degrees (or $m\lambda$, where m is an integer) will not pass through the analyzer in the crossed position. The wavelengths that undergo intermediate phase angles will come through intermediately.

A birefringent specimen that introduces a phase angle of 360 degrees for 530 nm light will appear lavender because the greenish-yellow 530 nm light leaving the specimen will be linearly polarized parallel to the azimuth of maximal transmission of the polarizer and perpendicular to the azimuth of maximal transmission of the analyzer. In the crossed position, no 530 nm light will pass through the analyzer, but both reddish and bluish wavelengths will, since they will not be linearly polarized, but elliptically polarized. The specimen will appear lavender, which is a mixture of blue and red, and which is the complementary color to greenish-yellow (530 nm).

The color of a specimen viewed between crossed polars will be the complementary color to the color whose wavelength is retarded one whole wavelength by the specimen. Consequently, the color of a specimen in a polarizing microscope provides information about the retardation introduced by the specimen. Since the retardation is equal to the product of the birefringence and the thickness; the birefringence can be estimated from the color of the specimen if the thickness is known. If the birefringence is known, the thickness can be estimated from the color of the image. A Michel-Lévy color chart, first presented by Auguste Michel-Lévy in 1888 to identify minerals, is used to simplify this identification. Many published Michel-Lévy color charts have been compiled by John Delly (2003) and are available online at http://www.modernmicroscopy. com/main.asp?article=15&page=1.

If the analyzer were turned so that it was parallel to the polarizer, the specimen described earlier would appear greenish-yellow, which is the complementary color of lavender. A complementary color is defined as the color of light, which when added to the test color, produces white light. The color of a birefringent specimen placed between parallel polars and illuminated with white light is, in essence, the color of the wavelength that gives a 360 degree phase change. The color of a birefringent specimen placed between crossed polars and illuminated with white light is, in essence, the complementary color of the wavelength that gives a 360 degree phase change.

USE OF COMPENSATORS TO DETERMINE THE MAGNITUDE AND SIGN OF BIREFRINGENCE

We are now in a position to utilize the experimental observations and deep thinking of Bartholinus, Huygens, Newton, Arago, Malus, Wollaston, Biot, Fresnel, Talbot, Herschel, and Brewster to understand much about the material nature of the specimens we observe with a polarizing microscope. We can use a polarizing microscope to determine the birefringence of a specimen quantitatively. The magnitude of birefringence of a sample characterizes the asymmetry of its molecular bonds. The birefringence is also a state quantity, so we can use it to identify the molecules in the sample. In addition, if we already know the identity of a birefringent substance, we can determine the orientation of the birefringent molecules in the specimen. Lastly, we can interpret any natural or induced changes in the birefringence of a cell to indicate physicochemical changes in cellular organization and/or chemistry (Johannsen, 1918; Bennett, 1950; Oster, 1955; Pluta, 1993).

As Talbot (1834a) noticed, many specimens do not appear colored in a polarizing microscope, but appear bright white on a black background. This is because these specimens do not introduce a retardation that is large enough to eliminate any wavelengths of visible light that would result in the specimen appearing as the complementary

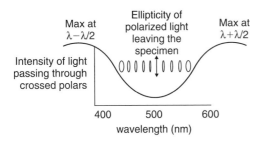

FIGURE 7-25 The intensity and ellipticity of the light passing through a first-order wave plate is wavelength dependent.

color. Experience shows that specimens with retardations between about 50 nm and 350 nm appear white.

Talbot (1834a) showed that specimens illuminated with white light in a polarizing microscope appeared colored when he placed a piece of mica between the crossed polars. Nowadays we insert a crystal known as a first-order red plate (a.k.a. red I plate or full wave plate) between the crossed polars in order to produce color contrast (Bennett, 1950; Pluta, 1993). The first-order red plate usually is made from a sheet of birefringent gypsum. The first-order red plate is inserted in a known orientation so that its slow axis, usually marked with a γ or n_γ, is fixed at ± 45 degrees relative to the transmission azimuth of the polarizer. Depending on the manufacturer, the first-order red plate introduces a retardation of between 530 and 590 nm. Thus, when the first-order red plate is inserted between crossed polars, at an orientation ± 45 degrees relative to the polarizer, it retards greenish-yellow light one full wavelength. The greenish-yellow light is linearly polarized parallel to the maximal transmission azimuth of the polarizer and will not pass the analyzer. Everything but greenish-yellow light will be elliptically polarized and will have a component that will go through the analyzer (Figure 7-25). Consequently, the background will appear lavender—a mixture of reddish and bluish colors, between crossed polars and greenish-yellow between parallel polars.

When the first-order wave plate is inserted in a polarizing microscope in the presence of a specimen, each point in the image depends on two factors: the elliptically polarized light produced by each point in a birefringent specimen and the elliptically polarized light produced by the compensator. Depending on the relative orientation of the slow axis of the specimen and the slow axis of the compensator, the elliptically polarized light that reaches the analyzer is either added together or subtracted from one another.

When a birefringent specimen that introduces a retardation of approximately 100 nm is placed on the stage, it will appear white between crossed polars. It will appear bluish, yellow-orangish, or both when the first-order red plate is inserted. I will call the first-order red plate, or any birefringent material we insert into the microscope, to determine the magnitude and sign of birefringence, the compensator. The compensator can be made out of a positively or negatively birefringent material. In order to explain this miraculous production of color, let me review some conventions. For a compensator made from a positively birefringent crystal, n_e and n_o are the slow axis and fast axis, respectively. For a compensator made from negatively birefringent crystal, n_e and n_o are the fast axis and slow axis, respectively. For a positively birefringent specimen, n_e and n_o are the slow axis and fast axis, respectively. For a negatively birefringent specimen, n_e and n_o are the fast axis and slow axis, respectively.

Each point in the image depends on the elliptically polarized light produced by each point in the specimen and the elliptically polarized light produced by the compensator. The sum of or difference between these two contributions of elliptical polarized light will also be elliptically polarized light. If the resultant light of a given wavelength is linearly polarized in the azimuth of maximal transmission of the polarizer, none of that color will contribute to the image point. If the resultant light of a given wavelength is linearly polarized parallel to the azimuth of maximal transmission of the analyzer, that color will contribute greatly to the image point. The contribution of any other wavelength to an image point will depend on the relative ellipticity of the resultant.

When the slow axis of a region of the specimen is parallel to the slow axis of the compensator, the image of this region will appear bluish. Blue is known as an "additive color" because the linearly polarized light that passes through the slow axis of the specimen is additionally retarded as it passes through the slow axis of the compensator. If the fast axis of a region of the specimen is parallel to the slow axis of the compensator, the object will appear yellowish-orange. Yellowish-orange is known as a "subtractive color" since the linearly polarized light that passes through the slow axis of the specimen is advanced as it passes through the fast axis of the compensator (Figure 7-26).

Let us start from the beginning and assume that we have a microscope equipped with crossed polars, in which the azimuth of maximal transmission of the polarizer is along the E-W axis and the azimuth of maximal transmission of the analyzer is along the N-S axis. The slow axis of the first-order wave plate compensator is inserted +45 degrees (NE-SW) relative to the polarizer. Let's further assume that light passing through the slow axis of the compensator will be retarded 530 nm relative to the light passing through the fast axis of the compensator. The light bulb on the microscope produces nonpolarized white light, which becomes linearly polarized after it passes through the polarizer. When the linearly polarized white light passes through a specimen, which is oriented ± 45 degrees relative to the polarizer, the linearly polarized white light is resolved into two linear components that

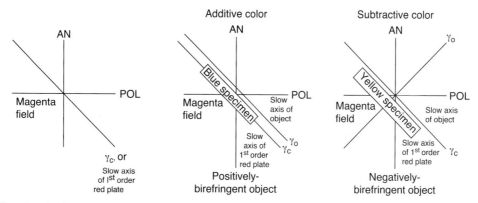

FIGURE 7-26 The color of a birefringent object in a polarized light microscope with a first-order wave plate depends in part on the orientation of the fast and slow axes of the specimen relative to the slow axis of the compensator.

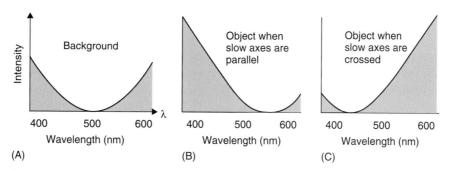

FIGURE 7-27 The colors produced by a first-order wave plate, in the absence (A) and presence (B) of a specimen with its slow axis oriented parallel to the slow axis of the compensator (B) and perpendicular to the slow axis of the compensator (C).

differ in phase and vibrate perpendicularly to each other. Let's assume that the light that passes through the slow axis of the specimen is retarded 100 nm relative to the light that passes through the fast axis of the specimen.

If the slow axis of the specimen is parallel to the slow axis of the compensator, the light that passes through these two slow axes will be retarded 630 nm. Therefore, 630 nm light will be retarded one full wavelength. Since 630 nm light is orange-ish, orange-ish light will be linearly polarized parallel to the azimuth of maximal transmission of the polarizer and will not pass through the analyzer. Light of all other wavelengths will be elliptically polarized. Elliptically polarized light with wavelengths close to 630 nm will have a very small minor axis parallel to the maximal transmission azimuth of the analyzer and wavelengths further away from 630 nm will have a larger one. Wavelengths longer than about 700 nm are invisible to us and therefore we will see many more wavelengths shorter than 630 nm than wavelengths longer than 630 nm. Consequently, the specimen will appear bluish, which is the complementary color of orange-ish (Figure 7-27).

If the slow axis of the specimen is parallel to the fast axis of the compensator, the light that passes through these two axes will be advanced 430 nm. Therefore, 430 nm light will be advanced one full wavelength. Since 430 nm light is bluish, bluish light will be linearly polarized parallel to the maximal transmission azimuth of the polarizer and will not pass through the analyzer. Light of all other wavelengths will be elliptically polarized. Wavelengths close to 430 nm will have a very small minor axis parallel to the maximal transmission azimuth of the analyzer and wavelengths further away from 430 nm will have a larger one. Wavelengths shorter than about 400 nm are invisible to us and therefore we will see many more wavelengths longer than 430 nm than wavelengths shorter than 430 nm. Consequently, the specimen will appear yellow-orange-ish, which is the complementary color of bluish.

The color of individual specimens scattered in all orientations throughout the field of a polarized light microscope depends on the orientation of its slow axis relative to the orientation of the slow axis of the first-order red plate.

Many birefringent specimens are not linear, but spherical or cylindrical and thus may show both additive and subtractive colors in different regions of the specimen. In these cases, it is helpful to think of the Cartesian coordinate system as being capable of translating over the specimen to any desired point. We also have to make the assumption that the slow axis is parallel to the physical axis in positively

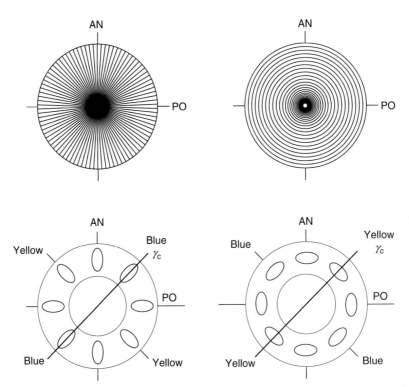

FIGURE 7-28 The colors of a specimen composed of positively birefringent molecules is a function of the orientation of the molecules. What colors would we observe in the specimens above if they were composed of negatively birefringent molecules?

birefringent molecules and perpendicular to the physical axis in negatively birefringent molecules.

When a specimen is composed of positively birefringent molecules that are radially arranged, the molecules will appear bluish in the regions where the physical axes of the positively birefringent molecules are parallel to the slow axis of the compensator, and yellowish-orange in the regions where the physical axes of the positively birefringent molecules are perpendicular to the slow axis of the compensator.

When a specimen is composed of positively birefringent molecules that are tangentially arranged, the molecules will still appear bluish in the regions where the physical axes of the positively birefringent molecules are parallel to the slow axis of the compensator; and yellowish-orange in the regions where the physical axes of the positively birefringent material are perpendicular to the slow axis of the compensator. However, in this case the arrangement of colors will be different than the arrangement of colors produced by a specimen in which the positively birefringent material is radially arranged (Figure 7-28). We can then distinguish radially arranged positively birefringent molecules from tangentially arranged positively birefringent molecules from the color pattern.

When the specimen is composed of negatively birefringent molecules that are radially arranged, the specimen will appear yellowish-orange in the regions where the physical axes of the negatively birefringent molecules

are parallel to the slow axis of the compensator; and bluish in the regions where the physical axes of the negatively birefringent molecules are perpendicular to the slow axis of the compensator. Such a specimen has the same color pattern as a specimen composed of tangentially arranged positively birefringent molecules. To distinguish between the two, you must know beforehand whether you are looking at a specimen composed of positively birefringent or negatively birefringent molecules.

When the specimen is composed of negatively birefringent molecules that are tangentially arranged, the specimen will appear yellowish-orange in the regions where the physical axes of the negatively birefringent molecules are parallel to the slow axis of the compensator; and bluish in the regions where the physical axes of the negatively birefringent molecules are perpendicular to the slow axis of the compensator. Such a specimen has the same color pattern as a specimen composed of radially arranged positively birefringent molecules. To distinguish between the two, you must know beforehand whether you are looking at a specimen composed of positively birefringent or negatively birefringent molecules.

When we know the sign of birefringence of the molecules that make up the specimen, we can determine the orientation of those molecules in the specimen. On the other hand, when we know the orientation of molecules from independent experiments (e.g., electron microscopy), we can determine the sign of birefringence of the molecules that make up the

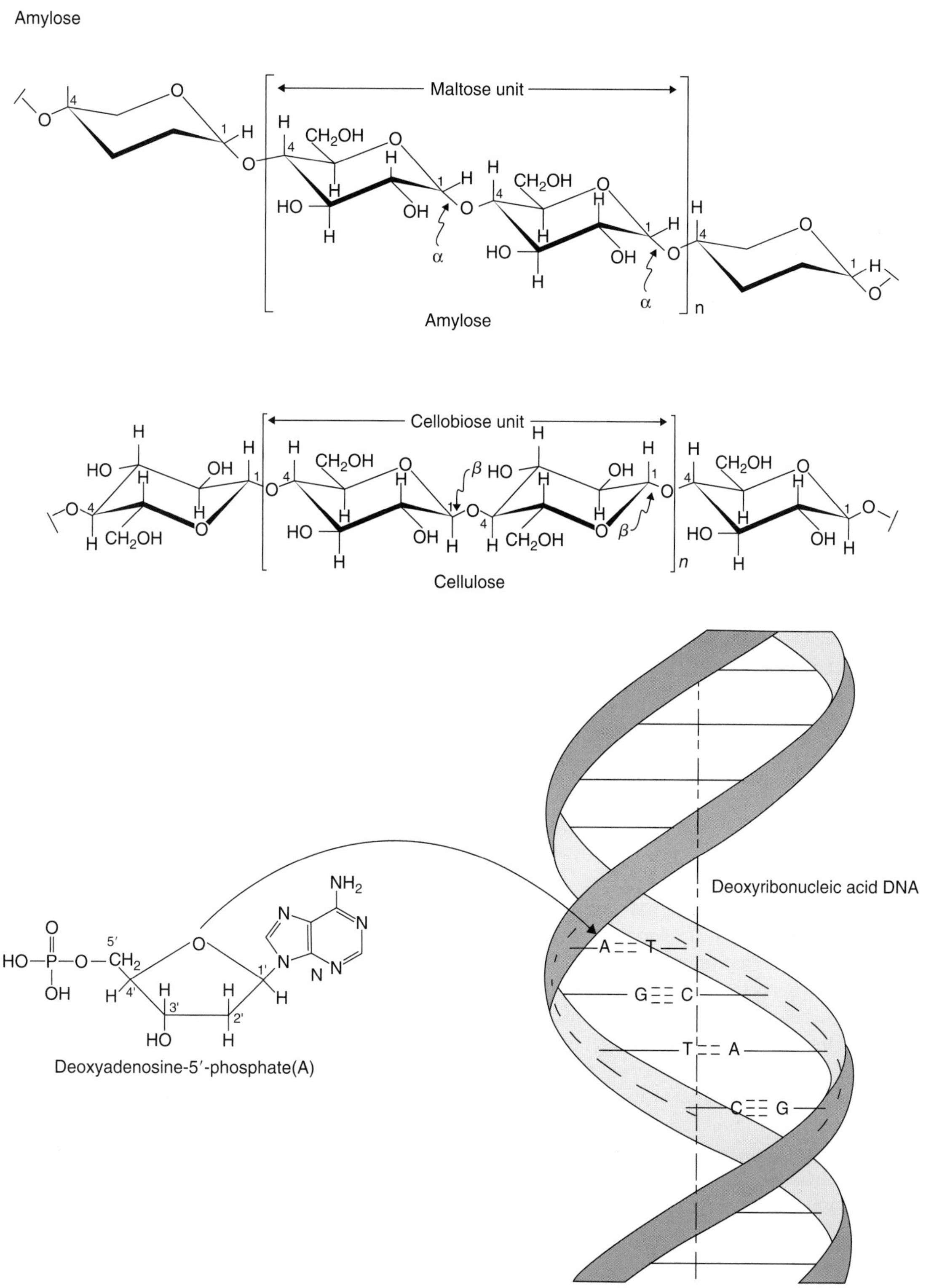
Amylose
Maltose unit
Amylose
Cellobiose unit
Cellulose
Deoxyadenosine-5′-phosphate(A)
Deoxyribonucleic acid DNA

FIGURE 7-29 The structure of positively birefringent molecules (e.g., starch and cellulose) and a negatively birefringent molecule (DNA).

specimen from the pattern of colors that appear in a polarizing microscope equipped with a first order red plate.

In biological specimens, we often do not know where the optic axis of a given specimen is. Thus we have to make an operational definition of n_e and n_o. We usually assume that the molecule is uniaxial and that n_e is parallel to the physical axis of the specimen and n_o is perpendicular to the physical axis of the specimen. Biaxial crystals, which have two optic axes and three indices of refraction, typically are not found in biological materials. Immobilized molecules, like amylose (in starch), cellulose, or fatty acyl-containing lipids, where the majority of the electrons vibrate parallel to the physical axis of the molecule, are considered positively birefringent. Molecules, like DNA, where the majority of the electrons vibrate perpendicular to the physical axis of the molecule, are considered negatively birefringent (Figure 7-29).

As I mentioned earlier, if a specimen placed on a microscope between crossed polars has a great enough retardation to eliminate one visible wavelength, the specimen will appear colored and we can determine its retardation directly by comparing the color of the specimen with the Newton colors that appear on a Michel-Lévy chart. If we know the thickness of the specimen by independent means (e.g., interference microscopy), we can read the magnitude of birefringence of the specimen from the Michel-Lévy chart. However, the retardations caused by most biological specimens are too small to give interference colors without a compensator. We can determine the retardation of these specimens by comparing the color of the specimen viewed with a first-order red plate, with the Newton colors that appear on a Michel-Lévy chart. Where we find a match in the colors, we can read the retardation due to the specimen and the first-order red plate directly from the Michel-Lévy chart and then subtract from this value the retardation due to the compensator. The retardation of the specimen is equal to the absolute value of this difference. Again, if we know the thickness of the specimen by independent means, we can determine the magnitude of birefringence (BR) of the specimen from the following equation:

$$BR = n_e - n_o = \Gamma/t.$$

The magnitude of retardation can also be rapidly estimated and the sign of birefringence determined with the aid of another fixed azimuth compensator known as the quartz wedge compensator (Bennett, 1950; Pluta, 1993). Quartz is a positively birefringent crystal with $n_e = 1.553$ and $n_o = 1.544$ (BR = +0.009). Since the retardation is equal to the product of birefringence and thickness, and since the thickness varies along the compensator, the retardation varies from zero to about 2000 nm. When the quartz wedge is slid into the microscope so that the slow axis of the compensator, often denoted with a γ or n_γ, is fixed at ±45 degrees relative to the crossed polars and illuminated with white light, a series of interference fringes, identical in order to the colors in the Michel-Lévy chart, appears in the field.

When using a quartz wedge, the specimen is placed on the stage so its slow axis is ±45 degrees relative to the azimuth of maximal transmission of the polarizer and illuminated with white light. In order to get complete extinction, the aperture diaphragm must be closed all the way. When the slow axis of the specimen and compensator are perpendicular, there will be one position of the compensator where the retardation of the compensator equals the retardation of the specimen

and the specimen will be extinguished. At this position, the compensator introduces an ellipticity that is equal in magnitude but opposite in sense to the ellipticity introduced by the birefringent specimen. If you cannot bring the specimen to extinction, then its slow axis is coaligned with the slow axis of the compensator or its retardation exceeds 2000 nm.

We find the position of extinction by gradually sliding the quartz wedge into the compensator slot. As this is done the color of the specimen changes from its initial color toward the "subtractive colors" and eventually it turns black. When the specimen is brought to extinction, look at the adjacent background color produced solely by the elliptical light introduced by the quartz wedge. Then match this color with the identical color on the Michel-Lévy color chart and read the retardation of the compensator directly from the color chart.

The quartz wedge will bring the specimen to extinction only if the slow axis of the compensator is perpendicular to the slow axis of the specimen. Therefore, we can use the quartz wedge compensator like the first-order wave plate to determine the orientation of molecules as long as we know the sign of birefringence of those molecules. Double wedge or Babinet compensators work the same way as the quartz wedge compensators, but are more accurate for estimating smaller retardations (Bennett, 1950; Pluta, 1993).

There are many compensators that can introduce a range of retardations by tilting or rotating the birefringent crystals in the compensator (Pluta, 1993). The Berek or Ehringhaus compensators are examples of tilting compensators—that is, compensators whose ability to produce elliptically polarized light is increased by tilting the compensator from a position where its optic axis is parallel to the microscope axis to an angle where the compensator's optic axis is at an angle relative to the microscope axis (Figure 7-30). The Berek compensator consists of a MgF_2 or calcite plate that is cut with its plane surfaces perpendicular to its optic axis. Therefore, when the Berek compensator is placed in the microscope horizontally, it acts like an isotropic material and cannot compensate a birefringent object. The MgF_2 or calcite plate can then be tilted with respect to the optical axis of

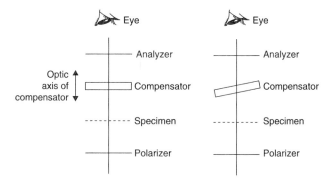

FIGURE 7-30 Compensation with a Berek or Ehringhaus compensator. The compensator is tilted from the horizontal position to the vertical position to compensate the specimen.

the microscope. The MgF_2 or calcite plate provides maximal compensation when it is oriented in a vertical position.

Tilting compensators introduce maximal retardations of 2000 to 3000 nm. At any intermediate tilt, the compensator retards the light from the specimen an intermediate amount. The tilting angle of the compensator that brings the specimen to extinction is read directly from the compensator tilting knob. We then look at a table supplied by the compensator manufacturer that relates the retardation to the compensator angle and the wavelength of illuminating light. In order to increase the accuracy of the measurement, the compensator is tilted in both directions and the average of the two tilting angles is used to calculate the retardation. Maximum accuracy is obtained by using monochromatic light and by making several measurements on a single sample and calculating the average. Retardation is obtained from the following formula:

$$\log \Gamma = \log c + \log f(i)$$

where c is the constant of the individual compensator plate for the wavelength of light used, i is the difference between the two compensation readings and f(i) is the function of i given in the manufacturer's table. Since tilting compensators have a logarithmic calibration curve, they are able to compensate specimens with a wide range of retardations. However, although tilting compensators have a large range,

their sensitivity is very low: on the order of 22 nm per degree (Figure 7-31).

Although the first order wave plate, quartz wedge, Babinet, Berek, and Ehringhaus compensators are useful for estimating the retardation of specimens with large retardations, we have to use rotary compensators such as a Brace-Köhler compensator or a de Sénarmont compensator in order to measure the small retardations introduced by many biological specimens. In general, the greater the sensitivity of a compensator, the smaller the range in which it is useful; therefore it is important to understand something about both your specimen and the various compensators in order to chose the best compensator.

A Brace-Köhler compensator is a rotary compensator made out of a mica plate (Bennett, 1950; Pluta, 1993). The mica plate can be rotated ±45 degrees around the microscope axis to introduce elliptically polarized light. Depending on the sensitivity of the compensator, the 45 degree rotation corresponds to a λ/30, λ/20, or λ/10 of retardation. Initially, the slow axis of the compensator is parallel to the azimuth of maximal transmission of the polarizer or the analyzer. At either of these positions, the compensator introduces no ellipticity to the light. When the slow axis of the compensator is rotated away from one of these positions, the Brace-Köhler compensator produces elliptically polarized light. The ellipticity of the light increases as the slow axis of the compensator is rotated

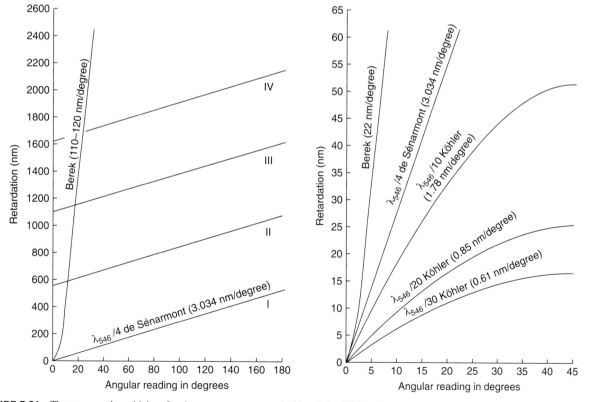

FIGURE 7-31 The ranges and sensitivies of various compensators used with polarized light microscopes.

between 0 and ±45 degrees, and it introduces maximum ellipticity to the light when its slow axis is ±45 degrees. In order to determine the retardation of the birefringent specimen, the compensator is rotated until the specimen is brought to extinction. The specimen is brought to extinction, or compensated, at the position where the compensator produces elliptically polarized light that is equal in magnitude, but opposite in sense to that of the object.

To use the Brace-Köhler compensator, center the stage and the objectives on a polarizing microscope, and then place the specimen on a rotating stage, focus and illuminate the specimen with Köhler illumination, using a green interference filter of the appropriate wavelength (see later). Then cross the polarizer and the analyzer to get maximum extinction. Close down the field diaphragm as much as you can to minimize stray light, and to get the highest extinction factor. To this end, also close down the aperture diaphragm as much as you need to optimize extinction and resolution. Rotate the specimen so that it is maximally bright. Or better yet, since our vision is more accurate in determining the maximal darkness on a black background than maximally bright on a black background, find the position where the specimen is maximally dark and then rotate the stage exactly ±45 degrees. At this position, the slow axis of the specimen will be ±45 degrees relative to the polarizer. Make sure that you are in a dark room and your eyes are dark-adapted.

Next, insert the compensator and turn the knurled knob until the background is maximally black. In this position, the slow axis of the compensator, denoted with a γ or n_γ, will be parallel to the azimuth of maximal transmission of the polarizer or the analyzer. The compensator should read ±45 degrees. The initial angle, read off the compensator, is called c_1. Now rotate the compensator knob until the specimen is brought to extinction, and note c_2, the new angle of the compensator. Depending on whether the specimen is positively or negatively birefringent, the Brace-Köhler compensator will have to be turned clockwise or counterclockwise. As you turn the knob, the compensator will move from a position where the slow axis of the compensator is parallel to the azimuth of maximal transmission of the polarizer or analyzer to a position where there is maximal retardation and the slow axis is ±45 degrees relative to the polarizer, and the angle that the compensator was rotated can be read directly from the compensator. A $\lambda/10$ Brace-Köhler compensator gives a maximum of $\lambda/10$ retardation; a $\lambda/20$ Brace-Köhler compensator gives a maximum of $\lambda/20$ retardation; and a $\lambda/30$ Brace-Köhler compensator gives a maximum of $\lambda/30$ retardation. The wavelength of the compensator should match the wavelength of the interference filter used to illuminate the specimen. The retardation of the specimen is obtained by inserting the angle read from the compensator into the following formula:

$$\Gamma_{specimen} = -(\Gamma_{compensator})\sin(2\theta)$$

where $\theta = c_1 - c_2$ and $\Gamma_{compensator}$ is equal to $546\,nm/30 = 18.2\,nm$ for $\lambda/30$ Brace-Köhler compensator, $546\,nm/20 = 27.3\,nm$ for a $\lambda/20$ Brace-Köhler compensator, and $546\,nm/10 = 54.6\,nm$ for a $\lambda/10$ Brace-Köhler compensator. The minus sign indicates that the slow axis of the compensator and the slow axis of the specimen lie on opposite sides of the azimuth of maximal transmission of the polarizer. Notice, that when $\theta = 45°$, the $\sin 2\theta = 1$ and $\Gamma_{specimen} = -(\Gamma_{compensator})$.

Let's look at stress fibers in cheek epithelial cells for an example. Insert the compensation in the SE-NW direction. When the $\lambda/30$ Brace-Köhler compensator is set for 45 degrees, its slow axis is parallel with the analyzer. The background is maximally dark and we are able to see bright stress fibers going from NE to SW. The stress fibers are brought to extinction when we rotate the compensator to 59 degrees. The retardation of the specimen is $8.5\,nm$ according to the following calculation (Figure 7-32):

$$\Gamma_{specimen} = -(\Gamma_{compensator})\sin(2\theta)$$
$$\Gamma_{specimen} = -18.2\,nm\,(\sin 2(45° - 59°))$$
$$\Gamma_{specimen} = -18.2\,nm\,(\sin 2(-14°))$$
$$\Gamma_{specimen} = -18.2\,nm\,(\sin(-28°))$$
$$\Gamma_{specimen} = -18.2\,nm\,(-0.4695)$$
$$\Gamma_{specimen} = 8.5\,nm$$

Consider the same sample that has stress fibers oriented from NE to SW. Our first reading, c_1, will still be 45 degrees, but our second reading, c_2, will be 31 degrees. The retardation of the specimen is $-8.5\,nm$ according to the following calculations:

$$\Gamma_{specimen} = -(\Gamma_{compensator})\sin(2\theta)$$
$$\Gamma_{specimen} = -18.2\,nm\,(\sin 2(45° - 31°))$$
$$\Gamma_{specimen} = -18.2\,nm\,(\sin 2(14°))$$
$$\Gamma_{specimen} = -18.2\,nm\,(\sin(28°))$$
$$\Gamma_{specimen} = -18.2\,nm\,(0.4695)$$
$$\Gamma_{specimen} = -8.5\,nm$$

The sign of the retardation is an indication of the relationship between the slow axis of the compensator and the slow axis of the specimen. If you know the orientation of the specimen, you can determine the sign of birefringence; if you know the sign of birefringence, you can determine the orientation of the birefringent molecules relative to the orientation of the slow axis of the compensator.

When using this method of Brace-Köhler compensation, the background lightens as the object is brought to extinction. Thus, it is a little tricky to determine the absolute compensation angle that brings the specimen to maximal extinction. In order to overcome this difficulty, Bear and Schmitt introduced a new method in which they determined the angle of the compensator when the object and the isotropic background were equally gray (c_3), and

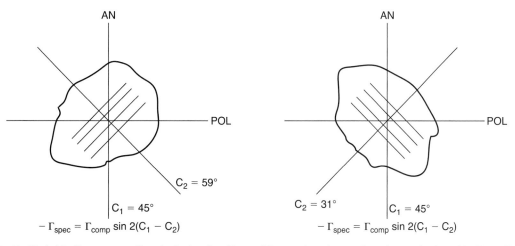

FIGURE 7-32 Positively birefringent stress fibers in cheek cells with two different orientations are brought to extinction with a Brace-Köhler compensator.

TABLE 7-6 Sensitivity and Maximal Compensation of Brace-Köhler Compensators Using Different Methods of Compensation

	Traditional method		Bear-Schmitt method	
Compensator	Maximum comp*	Sensitivity**	Max comp*	Sensitivity**
λ/30	18.33 nm	0.41 nm/degee	36.6 nm	0.81 nm/degee
λ/20	27.5 nm	0.61 nm/degree	55 nm	1.22 nm/degee
λ/10	55 nm	1.22 nm/degree	110 nm	2.44 nm/degee

*Assuming $\lambda = 550\,nm$.
**Obtained by dividing the maximum compensation by 45°.

subtracted this angle from c_1. The difference between these two angles is α. The retardation of the specimen is given by the following formula:

$$\Gamma_{specimen} = -2(\Gamma_{compensator})\sin(2\alpha)$$

This method is half as sensitive as the original method but more accurate, and the gain in accuracy may more than balance the loss of sensitivity. Using this method, we find that the stress fibers in the SE-NW direction are just as gray as the isotropic background when the compensator is set at 52 degrees. The retardation of the specimen is 8.8 nm, according to the following calculations:

$$\Gamma_{specimen} = -2(\Gamma_{compensator})\sin(2\alpha)$$
$$\Gamma_{specimen} = -2(18.2\,nm)\sin(2(c_1 - c_3))$$
$$\Gamma_{specimen} = -2(18.2\,nm)\sin(2(45° - 52°))$$
$$\Gamma_{specimen} = -2(18.2\,nm)\sin(2(-7°))$$
$$\Gamma_{specimen} = -2(18.2\,nm)\sin(-14°)$$
$$\Gamma_{specimen} = -2(18.2\,nm)\sin(-0.2419)$$
$$\Gamma_{specimen} = 8.8\,nm$$

Identical values of retardation would have been obtained by both methods if c_3 could be read precisely to 51.786. The maximum compensation and sensitivity of the various Brace-Köhler compensators are given in Table 7-6.

With the traditional method, Brace-Köhler compensators cannot be used to compensate objects with retardations greater than their maximum. However, using the Bear-Schmitt method, the maximum retardation that can be measured can be increased by a factor of two.

In general, we use compensators, including a first-order wave plate, a quartz wedge, a Berek compensator, an Ehringhaus compensator, and a Brace-Köhler compensator to extinguish the specimen by introducing an ellipticity to the linearly polarized light coming from the polarizer that is equal in magnitude but opposite in sense to that introduced by the specimen. Upon extinguishing the birefringent specimen, the retardation introduced by the specimen is obtained from either the Michel-Lévy chart or from formulas that relate the retardation to the angle the compensator was tilted or rotated.

The de Sénarmont compensation method is a little different from the methods described earlier, in that the

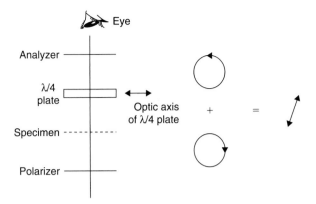

FIGURE 7-33 A de Sénarmont compensator typically utilizes a λ/4 plate and a rotating analyzer.

specimen is brought to extinction by rotating the analyzer (or polarizer) instead of moving the compensator (Figure 7-33). de Sénarmont compensation requires a polarizer and an analyzer, one of which can be rotated accurately 180 degrees. The birefringent specimen is placed with its slow axis at a ±45-degree angle relative to the polarizer. A λ/4 plate is then inserted into the compensator position with its slow axis, denoted with a γ or n_γ, parallel to the fixed polarizer (or if the analyzer is fixed it must be parallel to the fixed analyzer). Monochromatic light must be used, and the wavelength of the illuminating light must match the wavelength of the λ/4 plate.

To do de Sénarmont compensation, set the rotating polar so that the background is maximally black and make a note of that angle, which is most likely, but not always, 0 degrees. Then turn the rotating polar until the birefringent specimen is brought to extinction and note this angle. The difference between the two angles is called the compensation angle (a_c). The retardation of the specimen is given by the following formula:

$$\Gamma_{specimen} = (2a_c)(\lambda/360°) + m\lambda.$$

The sensitivity of the de Sénarmont compensator is 3.034 nm per degree and it normally is used for retardation up to one wavelength. It can be used for objects that have many wavelengths of retardation; however, it is capable of giving only the fraction of the wavelength above one order. The value of m can be deduced from a rough measurement made with a quartz wedge compensator. The sign of birefringence can be obtained since the rotating polar must be turned in opposing directions to extinguish positively or negatively birefringent specimens oriented with their physical axes in the same direction.

Consider that the slow axis of a specimen is oriented +45 degrees (NE-SW) relative to the fixed polarizer (E-W) and the slow axis of the λ/4 plate (E-W). The de Sénarmont compensator works because linearly polarized light passing through a specimen is resolved into two

orthogonal linearly polarized components that are out-of-phase with each other. Instead of considering the wave emerging from the specimen as being elliptically polarized, as we did when discussing the other compensation methods, consider the emerging light as consisting of two orthogonal, linearly polarized waves of equal amplitude. Each wave passes through the λ/4 plate after it emerges from the specimen. The slow wave emanating from the specimen passes through the λ/4 plate and makes an angle of +45 degrees (NE-SW) relative to the slow axis of the λ/4 plate. This is optically equivalent to the slow axis of the λ/4 plate making an angle of −45 degrees relative to the incoming linearly polarized light. Thus the slow linearly polarized wave is turned into circularly polarized light with a clockwise rotation.

Now, consider the fast component that emerges from the birefringent specimen as a linearly polarized wave that transverses the λ/4 plate at an angle of −45 degrees (SE-NW) relative to the slow axis of the λ/4 plate. The fast linearly polarized wave emanating from the specimen is converted into circularly polarized light that travels with a counterclockwise sense of direction. Thus two coherent circularly polarized waves have equal amplitudes and frequencies, but opposite senses when they emerge from the λ/4 plate. Each of these two waves has a component that is coplanar with the azimuth of maximal transmission of the analyzer, and consequently, they can interfere in that plane.

In order to determine the resultant azimuth of these two circularly polarized waves we must add the amplitudes of each vector at various time intervals. Two coherent circularly polarized waves with equal magnitudes but opposite senses recombine to give linearly polarized light. The azimuth of the linearly polarized wave depends on the phase angle between the two circularly polarized waves. The azimuth of the resultant linearly polarized wave will not be parallel to the azimuth of maximal transmission of the rotating analyzer. The rotating analyzer is then turned from its original position where the background was maximally dark, to a position where the specimen is brought to extinction. The angle through which the rotating analyzer is turned is known as the compensation angle.

I will describe another vector method for describing the brightness of the specimen when the analyzer is rotated through any angle (Figure 7-34). Consider a polarizing microscope with a fixed polarizer and a rotating analyzer. The slow axis of the λ/4 plate is parallel to the azimuth of maximal transmittance of the polarizer. Consider a birefringent specimen that causes a phase change of 45 degrees oriented so that its slow axis is +45 degrees relative to the polarizer. The slow axis of the λ/4 plate is −45 degrees relative to the linearly polarized wave that originates from the slow axis of the specimen. Thus this light wave will be circularly polarized by the λ/4 plate and will travel in the clockwise direction. The slow axis of the λ/4 plate is +45 degrees relative to the linearly polarized

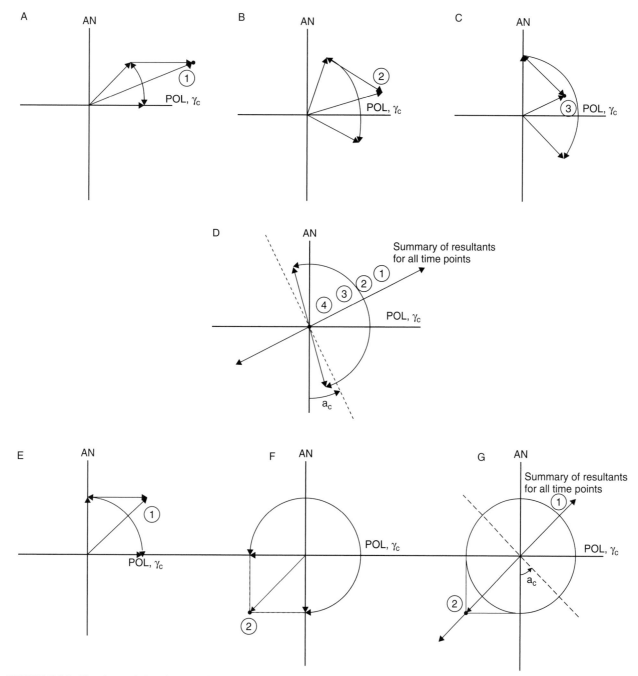

FIGURE 7-34 The phase relations between the clockwise and counterclockwise circularly polarized waves that leave the quarter wave plate for a specimen with a phase angle of 45 degrees (A-D) and a specimen with a phase angle of 90 degrees (E-G). When the counterclockwise circularly polarized wave is 45° ahead of the clockwise circularly polarized wave, the resultant linearly polarized wave is rotated 22.5° counterclockwise relative to the axis of maximal transmission of the polarizer. To extinguish the resultant, the analyzer also has to be rotated 22.5° counterclockwise. When the counterclockwise circularly polarized wave is 90° ahead of the clockwise circularly polarized wave, the resultant linearly polarized wave is rotated 45° counterclockwise relative to the axis of maximal transmission of the polarizer. To extinguish the resultant, the analyzer also has to be rotated 45° counter clockwise.

wave that originates from the fast axis of the specimen. Thus this light wave will be circularly polarized by the λ/4 plate and will travel in the counterclockwise direction.

Draw a Cartesian coordinate system so that the abscissa (E-W) represents the azimuth of maximal transmission of the polarizer and the ordinate (N-S) represents the azimuth of maximal transmission of the analyzer. Draw the path of the two circularly polarized light beams on this coordinate system. The wave originating from the fast axis of the specimen will be ahead of the wave originating from the slow axis of the specimen by an amount equal to the phase angle. Here are two examples: In the first example

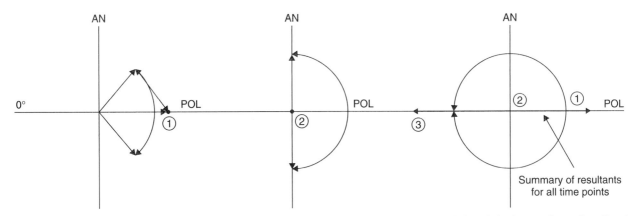

FIGURE 7-35 When the counterclockwise circularly polarized wave is 0° ahead of the clockwise circularly polarized wave, the resultant linearly polarized wave is rotated 0° clockwise relative to the axis of maximal transmission of the polarizer. To extinguish the resultant, the analyzer also has to be rotated 0°.

(like the example earlier), the specimen introduces a 45-degree phase angle, linearly polarized light is produced by the λ/4 plate, and the analyzer must be rotated 22.5 degrees counterclockwise to extinguish the specimen. In the second, the specimen introduces a 90-degree phase angle, linearly polarized light is produced by the λ/4 plate, and the analyzer must be rotated 45 degrees counterclockwise to extinguish the specimen. In both case, the analyzer must be rotated through an angle that is equal to half the fractional phase change of the specimen.

$$\text{fractional phase change} + n\lambda = (360°/\lambda)\,\Gamma_{specimen} = 2a_c.$$

Following is an example of a specimen that introduces no difference in phase or a phase angle equivalent to an integral number of wavelengths of the monochromatic illuminating light (Figure 7-35).

A clever person can make his or her own de Sénarmont compensator by attaching a Polaroid to a circular protractor to make a rotating polarizer and by making a λ/4 plate by layering six to seven layers of Handi-Wrap together. In order to eliminate any wrinkles, cut a hole in a piece of cardboard and tape each layer of Handi-Wrap to the cardboard. A clever person can also make a λ/4 plate out of mica $(H_2Mg_3(SiO_3)_4)$ that is split into thinner and thinner pieces with a fine needle. Make several of these pieces, and then test them to find which ones are λ/4 plates for the wavelength of interest. Mount the λ/4 plate in Canada Balsam between two cover glasses (Spitta, 1907).

The retardations measured with all compensators are based on trigometric functions and thus the retardation data may need to be transformed when doing statistics in order to meet the assumptions of linearity (Neter et al., 1990).

I have discussed a number of methods we may use to measure the retardation of birefringent specimens. However, to know the magnitude of birefringence, we must be able to measure the thickness of the specimen. The best

way to measure the thickness is to use interference microscopy (see Chapter 8). There are two other methods that may be acceptable.

With the focusing micrometer method, we use a lens with a high numerical aperture and minimal depth of field. First, calibrate the fine focusing knob (see Chapter 2; Clark, 1925; McCrone et al., 1984). Second, focus on the top surface of the specimen and read the measurement on the fine focusing knob. Third, focus on the bottom of the object and again read the measurement on the fine focusing knob. Since the greater the refractive index of the specimen, the thinner it appears, this method is accurate only when we already know the average refractive index of the specimen.

We can also estimate the thickness of a spherical or cubic specimen by using the ocular micrometer method, in which we measure the length of the specimen in the horizontal dimension and assumes that it is the same as the length of the specimen in the vertical dimension.

CRYSTALLINE VERSUS FORM BIREFRINGENCE

Some molecules are not intrinsically birefringent themselves, but they can be aligned in such a way that linearly polarized light will be retarded more when it passes through the specimen in one azimuth compared to another. Such oriented molecules are visible when viewed between crossed polars. This is known as form birefringence. Form birefringence can result from natural biological processes that align macromolecules (e.g., spindle formation during mitosis). Form birefringence can also result from the alignment of molecules by cytoplasmic streaming (Ueda et al., 1988). This is also known as flow birefringence.

Birefringence in solids, induced by mechanical stress, is known as strain birefringence (Brewster, 1833b; Lintilhac and Vesecky, 1984). The quality of glass used to make lenses is monitored, in part, through measurements of strain birefringence (Musikant, 1985). Photo-elastic studies use polarized light and polarized light microscopy to determine the strain in materials by looking at the pattern of polarization colors produced by a specimen under stress.

Form, but not intrinsic birefringence, disappears when a permeable specimen is immersed in media of increasing refractive indices and reappears when the refractive index of the medium surpasses the refractive index of the specimen (Bennett, 1950; Frey-Wyssling, 1953, 1957; Colby, 1971; Taylor, 1976). Intrinsic birefringence is not affected by immersion media because the molecules that make up the immersion media are two big to penetrate the bonds of the anisotropic molecules. When selecting immersion fluids it is important to use fluids that will not change the physicochemical nature of the specimen itself. Patzelt (1985) compiled a useful list of immersion fluids with refractive indices between 1.3288 (methanol) and 1.7424 (methyl iodide) that can be used to distinguish between form and intrinsic birefringence.

ORTHOSCOPIC VERSUS CONOSCOPIC OBSERVATIONS

Up until now I have been discussing orthoscopy, which means that I have been discussing what the specimen looks like at the image plane. Mineralogists also find it useful to observe the diffraction pattern produced by the specimen in the reciprocal space of the aperture plane. This is known as conoscopy. When doing conoscopy, it is convenient to be able to position the specimen in three dimensions by using a goniometer, which is a rotating stage that can rotate the specimen in the specimen plane as well as tilt the specimen ±45 degrees relative to the optical axis of the microscope. By observing the diffraction plane, conoscopy provides an effective way of determining whether a mineral is uniaxial or biaxial; that is, if it has one optic axis and two refractive indices or if it has two optic axes and three indices of refraction (Patzelt, 1985).

REFLECTED LIGHT POLARIZATION MICROSCOPY

Polarized light microscopes can be used in the reflected light or epi-illumination mode. In epipolarization microscopy, the light passes through the objective before it strikes the specimen and then the reflected light is captured by the same objective lens. Epipolarization microscopes are used for metallurical work and have also been used in biological work to localize antibodies that are conjugated to colloidal gold (Hughes, 1987; Hughes et al., 1991; Gao and Cardell, 1994; Gao et al., 1995; Stephenson et al., 1998; Ermert et al., 1998, 2000, 2001).

USES OF POLARIZATION MICROSCOPY

Polarization microscopy is used in geology, petrology, and metallurgy to identify minerals (McCrone et al., 1979, 1984). Polarization microscopy is also used in art identification in order to determine the age, and thus the authenticity of a painting (McCrone, 1992; Brouillette, 1990). Each pigment used to make paint usually has a metallic base. Throughout history new pigments have been invented. The pigments can be identified in a "pinprick" of paint with a polarizing microscope. White pigment made out of rutile ($TiO_2/CaSO_4$) was not introduced until 1957, so if you find them in a painting supposedly done in 1907, the painting is probably a forgery. Walter McCrone (1990) estimates that 90 percent of paintings are fakes.

Polarization microscopy also has been used extensively in biology to study the structure of DNA, membranes, the microfibrils in cell walls, wood, cell walls, microtubules in the mitotic apparatus and the phragmoplast, chloroplast structure, muscle fibers (the I band and the A band stand for isotropic band and anisotropic band, respectively), starch grains and biological crystals composed of calcium oxalate, calcium carbonate and silica (Quekett, 1848, 1852; Carpenter, 1883; Naegeli and Schwendener, 1892; Schmidt, 1924; Noll and Weber, 1934; Hughes and Swann, 1948; Swann, 1951a, 1951b, 1952; Inoué, 1951, 1952a, 1953, 1959, 1964; Inoué and Dan, 1951; Preston, 1952; Frey-Wyssling, 1953, 1957, 1959, 1976; Inoué and Bajer, 1961; Forer, 1965; Ruch, 1966; Salmon, 1975a, 1975b; Taylor, 1975; Frey-Wyssling, 1976; Palevitz and Hepler, 1976; Hiramoto et al., 1981; Wolniak et al., 1981; Inoué, 1986; Cachon et al., 1989; Schmitt, 1990; Inoué and Salmon, 1995). Perhaps the most elegant use of the polarized light microscope in biology was to determine the entropy and enthalpy of the polymerization reaction of the mitotic spindle fibers by Shinya Inoué.

The mitotic spindle is a dynamic structure that reversibly breaks down when exposed to elevated hydrostatic pressures and microtubule-depolymerizing drugs, including colchicine (Inoué, 1952a). While observing the mitotic spindle of *Chaetopterus* with polarization microscopy, Shinya Inoué (1953, 1959, 1964) noticed a remarkable and completely unexpected phenomenon. He observed that as he increased the temperature of the stage, the amount of anisotropy of the mitotic spindle increased (Figure 7-36). Since a completely random specimen is isotropic, and anisotropy typically indicates that the specimen is ordered—an increase in the anisotropy of the specimen means an increase in order. This was surprising because typically, increasing the temperature increases the disorder in a system.

This unusual effect of temperature on protein had been observed before with muscle protein and the tobacco mosaic virus protein. It turns out that protein monomers have a shell of water around them, and this bound water prevents the protein monomers from interacting among

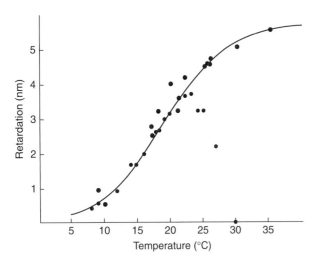

FIGURE 7-36 The retardation of the mitotic spindle is temperature-dependent.

themselves and forming a polymer. The removal of the bound water requires an input of energy, which comes from raising the temperature. As the temperature increases, the shell of water surrounding the protein monomers is removed and hydrophobic interactions between the protein monomers themselves can take place. Inoué proposed that the fibers in the spindle were made out of proteins, and the fibers formed when the protein subunits polymerized.

Inoué quantified his work in the following way: He assumed that the maximum anisotropy occurred when all the subunits were polymerized. Inoué called the total concentration of the protein subunits A_0 and the concentration of polymerized subunits B, where B is proportional to the amount of anisotropy (Γ, in nm) measured with the polarizing microscope. When all the subunits are polymerized $B = A_0$, and when all the subunits are depolymerized, $B = 0$. The free subunits can be calculated as $A_0 - B$. Inoué assumed that polymerization occurred according to the following reaction:

$$A_0 - B \underset{k_{off}}{\overset{k_{on}}{\rightleftharpoons}} B$$

where the equilibrium constant $K_{eq} = [B]/[A_0-B]$

Inoué determined the concentration of polymerized subunits at a given temperature by measuring the retardation of the spindle at that temperature. He assumed that the maximal value of retardation was an estimate of the total concentration of subunits. He calculated the concentration of free subunits at a given temperature by subtraction.

The degree of polymerization is determined by the equilibrium constant (K_{eq}), which represents the ratio of polymerized to free subunits ($[B]/[A_0-B]$). The molar

standard free energy (G^{std}) relative to the energy that occurs at equilibrium ($G^{eq} = 0$) can be calculated from the following equations:

$$G^{eq} - G^{std} = RT \ln K_{eq}/K_{std}$$
$$= RT \ln\{([B]/[A_0 - B])/(1/1)\}$$

$$G^{std} - G^{eq} = -RT \ln K_{eq}/K_{std}$$
$$= -RT \ln\{([B]/[A_0 - B])/(1/1)\}$$

At standard pressure, the standard free energy is composed of a standard enthalpy term (J/mol) and a standard entropy term (in $J\ mol^{-1}\ K^{-1}$) according to the following equation:

$$\Delta G^{std} = \Delta H^{std} - T\Delta S^{std}$$

Therefore,

$$\Delta G^{std} = \Delta H^{std} - T\Delta S^{std} = RT\ln([B]/[A_0 - B])$$

This can be rewritten as a linear equation:

$$-\ln([B]/[A_0 - B]) = (\Delta H^{std}/R)(1/T) - \Delta S^{std}/R$$

or

$$\ln([B]/[A_0 - B]) = -(\Delta H^{std}/R)(1/T) + \Delta S^{std}/R$$

Inoué plotted $\ln([B]/[A_0-B])$ versus $(1/T)$ to get a van't Hoff plot, where the slope is equal to $(\Delta H^{std}/R)$ and the Y intercept is equal to $\Delta S^{std}/R$ (Figure 7-37). Thus at a given temperature he could estimate $\Delta H^{std}/R$ and $T\Delta S^{std}$.

Inoué determined the standard enthalpy of polymerization at room temperature to be 34 kcal/mol and the product of the standard entropy and the absolute temperature to be 36.2 kcal/mol. Since $T\Delta S^{std}$ is greater than ΔH^{std}, polymerization of the subunits is an entropy-driven process. Thus Inoué concluded that the subunits of the spindle fibers were surrounded by bound water and the removal of the bound water caused an increase in entropy. This increase in entropy provided the free energy necessary to drive the polymerization reaction. Moreover, he proposed that the controlled depolymerization of the spindle fibers provided the motive force that pulled the chromosomes to the poles during anaphase. These experiments are remarkable in light of the fact that they were done approximately 20 years before the discovery of tubulin. After the discovery of microtubules with the electron microscope, Inoué and Sato (1967) recast the dynamic equilibrium hypothesis in terms of microtubules (Sato et al., 1975; Inoué, 1981).

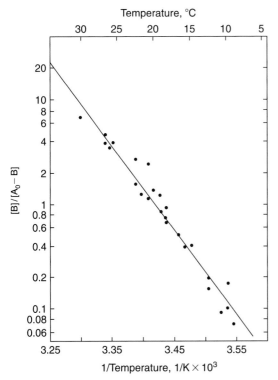

FIGURE 7-37 The entropy and enthalpy of polymerization of spindle fibers can be determined by drawing a van't Hoff plot with ln ([B]/[A$_o$–B]) vs 1/T.

OPTICAL ROTATORY (OR ROTARY) POLARIZATION AND OPTICAL ROTATORY (OR ROTARY) DISPERSION

E Pluribus Unum. E Unum Pluribus. It is a general truth that, when all factors are accounted for, the whole is equal to the sum of its parts. In explaining how polarized light interacts with birefringent specimens, I have freely resolved vectors into their components and summed the two components into a resultant. In order to emphasize the "double reality" of vectors and their components, I will end this chapter where I began, discussing the rotation of linearly polarized light. Here we will discuss how linearly polarized light can be resolved into two circularly polarized waves that rotate with opposite senses. We will use this knowledge to understand how, in a molecule, electron vibrations are not always linear but also helical. Polarized light allows us to visualize the three-dimensional electronic structure of molecules.

Francois Arago discovered in 1811 that quartz is unusual. He passed linearly polarized yellow light produced by a sodium flame and a Nicol prism polarizer through a 1 mm section of quartz cut perpendicular to the optic axis, and he discovered that the yellow light passed right through the crossed Nicol prism that he used as an

analyzer. The light that passed through the analyzer was still linearly polarized since he could extinguish the light by rotating the analyzer 22 degrees. The quartz appeared to rotate the azimuth of polarization in a process Arago named optical rotatory polarization. Substances that are capable of rotating linearly polarized light are called optically active. In 1817, Jean-Baptiste Biot found that when he passed white light through the quartz plate, the plate appeared colored, and the color changed as the analyzer was rotated. That meant that the degree of rotation was wavelength-dependent. The wavelength-dependence of optical rotatory polarization is known as optical rotatory dispersion (Wood, 1914; Slayter, 1970). Substances that rotate the azimuth of polarization to the right are known as dextro-rotatory and substances that rotate the azimuth of polarization to the left are known as laevo-rotatory. In 1822 John Hershel found that quartz molecules are dextro-rotatory and others are laevo-rotatory. Lenses used to transmit ultraviolet light are made of quartz. These must have equal contributions of dextro-rotatory and laevo-rotatory quartz to the optical paths to ensure that the lens itself does not introduce rotatory polarization (Wood, 1914).

Optical rotatory polarization was explained by Augustin-Jean Fresnel (Wood, 1914; Robertson, 1941). He proposed that the incident linearly polarized light can be considered to be composed of two in-phase components of circularly polarized light that propagate with opposite senses with differing speeds. An optically active substance can be considered to have circular birefringence.

By looking at the angle the light bends when it passes a quartz-air interface, Fresnel showed that a circularly polarized wave with clockwise rotation experiences a different refractive index than the circularly polarized wave with counterclockwise rotation as they spiral though a substance capable of optical rotatory polarization. Upon recombination in the azimuth of maximal transmission of the analyzer, the linearly polarized resultant passed through the analyzer placed in the crossed position because the two circularly polarized waves recombine to make linearly polarized light whose azimuth of polarization was changed as a consequence of the phase angle introduced by the specimen. The analyzer is then turned either clockwise or counterclockwise to find the position where the light is again extinguished. This position, combined with the direction of rotation, gives the value for the optical rotatory polarization of a given thickness of a pure liquid or solid, or at a given thickness, the concentration of a liquid.

Biot accidentally found that the oil of turpentine that he used to mount crystals in also had the ability to rotate the azimuth of polarization. Solutions of randomly arranged molecules, including sugars, amino acids, and organic acids, are also able to rotate the azimuth of linearly polarized light. In 1848 to 1851, Louis Pasteur found that when a solution of laboratory-synthesized racemic acid, which was identical to naturally occurring tartaric acid in every way except for its inability to rotate linearly polarized light, was allowed to crystallize, two types of crystals formed. The two crystals were mirror images of each other. Using a microscope, Pasteur separated the two types of crystals into right-handed crystals and left-handed crystals. When the right-handed crystals were dissolved in solution, they rotated the linearly polarized light to the right. Solutions of the left-handed crystals rotated polarized light to the left. Thus racemic acid did not rotate polarized light because it consisted of an equal mixture of two optical isomers of tartaric acid. Pasteur concluded that there must be a chemical asymmetry in the two optical isomers, but he did not know what caused the asymmetry (Jones, 1913).

By considering molecules to be three-dimensional rather the two-dimensional structures that were drawn on paper, Jacobus van't Hoff and Joseph Le Bel independently proposed in 1874 that the carbon atom was tetrahedral and that the asymmetries in molecules were caused by a tetrahedral carbon atom that made bonds with four different functional groups. Optical rotatory polarization and optical rotatory dispersion can be used to identify molecules and deduce their structure (Djerassi, 1960). Proteins with helical structures also tend to exhibit optical rotatory polarization. According to Paul Drude, electrons in substances that show optical rotatory polarization vibrate in a helical manner instead of back and forth (Wood, 1914). In principle, both optical rotation and optical rotatory dispersion can be done through the microscope to identify small quantities of material in a sample.

If the two circularly polarized components of the incident light are differentially absorbed, then the resultant will be elliptically polarized and will pass through crossed polars. This is known as circular dichroism. The Cotton Effect, named after A. Cotton, results from a combined effect of circular rotatory dispersion and circular dichroism.

Optical rotation, or circular birefringence, is different than linear birefringence in that a solution that has no preferred directionality associated with it can rotate the plane of linearly polarized light as a consequence of three-dimensional asymmetry in the structure of the solutes randomly arranged in the solution (Robertson, 1941). To visualize this, hold a slinky in your hand—no matter which way you turn it, its sense of rotation will be the same to you.

WEB RESOURCES

Polarized Light

Hyperphysics: http://hyperphysics.phy-astr.gsu.edu/hbase/phyopt/polarcon.html#c1

Polarized Light Microscopy

Olympus Microscopy Resource Center: http://www.olympusmicro.com/primer/techniques/polarized/polarizedhome.html

Nikon Microscopy U:http://www.microscopyu.com/articles/polarized/polarizedintro.html

Molecular Expressions Optical Microscopy Primer: http://micro.magnet.fsu.edu/primer/techniques/polarized/polarizedhome.html

Polarized light has been utilized by Austine Wood Comarow to make *Polages*, which are "paintings without pigments." http://www.austine.com/

Chapter 8

Interference Microscopy

Interference microscopy is similar to phase-contrast microscopy in that both types of microscopes turn a difference in phase into a difference in intensity, and for this reason, a chapter on interference microscopy would naturally follow Chapter 6. However, since many interference microscopes utilize polarized light, I present interference microscopy to my students after they become fully comfortable with polarized light. Color plates 12 and 13 provide examples of images taken using image duplication-interference microcopy. Moreover, since interference microscopes are not commonly found in laboratories, to make the process of interference more familiar to my students, I begin by taking them outside to blow bubbles while we discuss the optics of thin films. It also gives them a chance to picture Isaac Newton blowing bubbles, to realize the truth in Albert Szent Györgyi's words that, "Discovery consists in seeing what everyone else has seen and thinking what no one else has thought."

GENERATION OF INTERFERENCE COLORS

Sometimes less is more, and it is really pretty amazing how transparent objects such as a solution of soap or gasoline become colored when they become thin enough. Newton (1730) wrote, "It has been observed by others, that transparent Substances, as Glass, Water, Air, &c. when made very thin by being blown into Bubbles, or otherwise formed into Plates, do exhibit various Colours according to their various thinness, altho' at a greater thickness they appear very clear and colourless." The observation of colors produced by thin plates was first observed by Robert Hooke (1665), and published in his book, *Micrographia*. Newton, however, quantitatively studied the generation of colors by thin plates, and we usually speak of the colors generated by interference as Newton's colors. Newton (1730) wrote,

> If a Bubble be blown with Water first made tenacious by dissolving a little Soap in it, 'tis a common Observation, that after a while it will appear tinged with a great variety of Colours.... As soon as I had blown any of them I cover'd it with a clear Glass, and by that means its Colours emerged in a very regular order, like so many concentrick Rings encompassing the top of the Bubble. And as the Bubble grew thinner by the continual subsiding of the Water, these Rings dilated slowly and overspread the whole Bubble, descending in order to the bottom of it, where they vanished successively. In the mean while, after all the Colours were merged at the top, there grew in the center of the Rings a small round black Spot.

Interference is not only responsible for the colors of thin plates, but can also account for the iridescent colors of some algae, the blue leaves of *Selaginella* (Fox and Wells, 1971), as well as some insects, butterflies, and birds, including peacocks (Fox, 1936). Newton (1730) wrote, "The finely colour'd Feathers of some Birds, and particularly those of Peacocks Tails, do, in the very same part of the Feather, appear of several Colours in several Positions of the Eye, after the very same manner that thin Plates were found to do...."

Newton found that the colors produced by a thin film formed by a given substance were dependent on the thickness of the thin film. He also found that the color of the thin film depends on the angle of observation since the length through which light propagates through the thin film to get to our eyes changes when we look at it from different angles. He also found that, for films of identical thickness and viewed at identical angles, the color depends on the refractive index of the substance that makes up the thin film. All these observations can be summed up by the following sentence: The color of a thin film depends on the difference in the optical path lengths of the light waves that reflect off the top surface of the thin film and the light waves that reflect off the bottom surface of the thin film.

Newton tried to explain the generation of colors in terms of the corpuscular theory of light, but he had to introduce an *ad hoc* hypothesis that the aetherial medium through which the corpuscles travel underwent "fits." Thomas Young (1801, 1807) thought that perhaps the light itself underwent the fits and was thus wave-like. He then suggested that light waves that were reflected from the bottom surface of the thin film combined with the light waves that were reflected from the top surface and destroyed each other in a process he called interference. By the 1820s, Augustin Fresnel (Fresnel, 1827, 1828, 1829; Arago, 1857; Mach, 1926) came up with a complete theory of the generation of interference colors for thin films of both isotropic and birefringent material. I discussed the colors formed by birefringent material in Chapter 7. Here I will discuss the

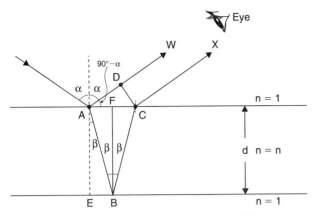

FIGURE 8-1 Interference of light reflected from a thin film in air.

generation of colors by isotropic substances that have only one refractive index.

Consider the following situation: Light, traveling through air, which has a refractive index of 1, strikes a thin film with a refractive index n, at A, at an angle α relative to the normal. The light then is refracted toward the normal and travels to B. The light then is reflected in such a way that the angle of reflection equals the angle of incidence and thus both angles equal $\angle\beta$. The reflected light then strikes the film-air interface at C, where it is refracted away from the normal, and travels parallel to wave W. Wave W and wave X are able to interfere with each other because they are coherent; i.e. generated from the same source, and close together. The fact that waves that are close enough together are able to interfere indicates that waves have a wave width as well as a wave length. The optical path difference (OPD) between the two waves is given by the following formula (Figure 8-1):

$$OPD = n(AB + BC) - AD$$

As written, this formula is not very useful to us. However, a little geometry, a little trigonometry, and the Snell-Descartes Law will make this equation more useful.

Since AE and BF are parallel, $\angle ABF = \angle EAB = \beta$. From the definition of tangent and cosine, we get the following relations:

$$\tan\beta = AC/2d \quad AC = 2d\tan\beta$$

$$\cos(90 - \alpha) = AD/AC$$

$$\cos\beta = d/AB = d/BC \quad AB = BC = d/\cos\beta$$

The following is a trigonomic identity:

$$\sin\alpha = \cos(90° - \alpha)$$

Thus,

$$\cos(90° - \alpha) = AD/AC$$

and

$$\sin\alpha = AD/AC$$

Since $\tan\beta = (AC/2)/d$, $AC = 2d\tan\beta$ and $\sin\alpha = AD/(2d\tan\beta)$, solve for AD:

$$AD = 2d\tan\beta\sin\alpha$$

Since $\tan\beta = \sin\beta/\cos\beta$:

$$AD = 2d(\sin\beta/\cos\beta)\sin\alpha$$
$$= 2d(\sin^2\beta/\cos\beta)(\sin\alpha/\sin\beta)$$

Use the Snell-Descartes Law (for air). Since $\sin\alpha/\sin\beta = n$, then

$$AD = 2dn(\sin^2\beta/\cos\beta)$$

Since the optical path length is the product of the refractive index and the distance, AD is the optical path that the reflected light takes in air.

$$OPL_1 = 2dn(\sin^2\beta/\cos\beta)$$

Now we must determine the optical path length that the wave takes to and from the second surface. Since $\cos\beta = d/(AB) = 2d/2AB$:

$$2AB = 2d/\cos\beta$$

The optical path length of the wave reflected at the second surface is the product of the refractive index and the distance traveled and is given by

$$OPL_2 = 2ABn = (2dn/\cos\beta)$$

The optical path difference (OPD) is

$$OPD = OPL_2 - OPL_1$$
$$= (2dn/\cos\beta) - 2dn(\sin^2\beta/\cos\beta)$$
$$OPD = (2dn/\cos\beta)(1 - \sin^2\beta)$$

Since, $1 - \sin^2\beta = \cos^2\beta$

$$OPD = (2dn/\cos\beta)(\cos^2\beta)$$

and

$$OPD = (2dn\cos\beta)$$

This equation can also we written in terms of the angle of incidence using the following form of the Snell-Descartes Law: Since $\sin\alpha/\sin\beta = n$,

$$n = \sin\alpha/\sqrt{(1 - \cos^2\beta)}$$

and

$$\cos\beta = \sqrt{(1 - ((\sin^2\alpha)/n^2))}$$

or

$$OPD = 2dn\sqrt{(1 - ((\sin^2\alpha)/n^2))}$$
$$= 2d\sqrt{(n^2 - (\sin^2\alpha))}$$

This equation, if it is true, suggests that when the thickness of the thin film is zero, there is no optical path difference between the wave reflected from the top surface and the wave reflected from the bottom surface, and consequently, all the wavelengths of light should constructively interfere to form a white light reflection. Unfortunately, experience shows that this is not always true, and that often there is no reflection when the thickness is zero and consequently the thin film appears black. Both Young and Fresnel were able to model reality better by postulating that the reflection that takes place at the surface between the rarer and the denser medium introduces a $\lambda/2$ change in phase. In Thomas Young's (1802) words,

> In applying the general law of interference to these colours, as well as to those of thin plates already known, I must confess that it is impossible to avoid another supposition, which is a part of the undulatory theory, that is, that the velocity of light is the greater, the rarer the medium; and that there is also a condition annexed to the explanation of the colours of thin plates, which involves another part of the same theory, that it, that where one of the portions of light has been reflected at the surface of a rarer medium, it must be supposed to be retarded one half of the appropriate interval, for instance, in the central black spot of a soap-bubble, where the actual lengths of the paths very nearly coincide, but the effect is the same as if one of the portions had been so retarded as to destroy the other. From considering the nature of this circumstance, I ventured to predict, that if the two reflections were of the same kind, made at the surfaces of a thin plate, of a density intermediate between the densities of the mediums containing it, the effect would be reversed, and the central spot, instead of black, would become white; and I now have the pleasure of stating, that I have fully verified this prediction, in interposing a drop of oil of sassafras [n = 1.536] between a prism of flint glass [n = 1.583] and a lens of crown glass [n = 1.52]; the central spot seen by reflected light was white, and surrounded by a dark ring.

Thus, according to Young, the colors of reflected light seen at a given angle depend not only on the thickness and refractive index of the material that makes up the thin film, but also on the difference in refractive indices at both surfaces (Young, 1807; Fresnel, 1827, 1828, 1829; Baden-Powell, 1833; Lloyd, 1873; Mach, 1926). Two reflecting surfaces can also be made from three different media.

According to Young's and Fresnel's rules, when the difference between the refractive indices of the incident medium and the transmitting medium is positive, there is no change in the phase of the reflected light. When the difference is negative, there is a half-wavelength change in phase in the reflected light. For example, for a thin film soap bubble in air, n_1-n_2 is negative and n_2-n_3 is positive. As a result, there is a half-wavelength change in phase for all wavelengths introduced at the first surface.

$$OPD = OPL_1 - OPL_2 = 2dn \cos\beta - \lambda/2$$

The optical path difference between the refracted and reflected wave vanishes when $2dn \cos\beta = \lambda/2$, and light with a wavelength equal to $2dn \cos\beta$ is not reflected.

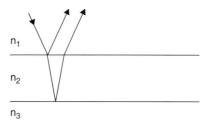

FIGURE 8-2 Interference of light reflected from a thin film when $n_1 < n_2 < n_3$. This is the case for the thin films of metal oxides that are responsible for the iridescent colors of glass.

Wavelengths close to this wavelength are partially reflected. By conservation of energy, the transmitted light represents the portion of each wavelength that is not reflected and the color of the transmitted light is complementary to the color of the reflected light (Young, 1807; Dunitz, 1989). The observed color depends on the refractive index of the thin film, its thickness, and the angle of the observation (which is related to β). When the thickness of the film (d) approaches 0, the optical path differences for all wavelengths approach $-\lambda/2$. Consequently, all wavelengths destructively interfere and there is no reflected light causing the thin film to look black.

In the case described earlier, $n_1 < n_2 > n_3$; however, when the refractive indices are such that $n_1 < n_2 < n_3$ such as it would be at an air/oxide/glass interface of a air/metal oxide/metal interface, then there is a $\lambda/2$ change in phase of the reflected light at both interfaces and the optical path difference is given by (Figure 8-2):

$$OPD = (2dn \cos\beta + \lambda/2) + \lambda/2 = 2dn \cos\beta + \lambda$$

The optical path difference of a given wavelength vanishes when $2dn \cos\beta = \lambda$, and light with a wavelength equal to $2dn \cos\beta$ is completely reflected and the complementary color is transmitted. When the thickness approaches zero, all wavelengths are reflected and the film appears white. Thin films of metal oxides are responsible for the iridescent colors of glass made by Frederick Carder of Steuben Glass and Louis Comfort Tiffany of the Tiffany Glass Company and of some jewelry available today.

Lenses often are coated with thin films to minimize reflections. The idea for antireflection coatings came to Lord Rayleigh when he realized that the old optical glass that had tarnished transmitted more light than an untarnished piece of glass. Thin films made of magnesium fluoride replace the tarnish in antireflection coatings used on lenses today. The thin films act as antireflection coatings when the unidirectional optical path length of the film is $\lambda/4$ and the light reflected from the two interfaces destructively interferes. The antireflection coating reduce the percentage of reflected light from 4 percent to 1 to 2 percent.

Before I discuss how the knowledge of how interference colors produced by thin films can be used to our advantage in microscopy, I want to briefly discuss the generation of interference colors by diffraction (see Chapter 3).

In 1611, Maurolico wrote:

> If you view the light of a candle, placed not too far away, through a white feather from a dove or from some other bird, when placed opposite the eye you will see, between the lines on the feather and those branches, a certain distinct cross with a wonderful variety of colors, such as are seen in the rainbow. This can happen only through the light being received between the small grooves of the feather tufts, and there multiplied, continually incident and by turns reflected.

The colors of some bird feathers and of many brightly colored beetles result from the process of diffraction. The importance of diffraction in the coloration of beetles was first suggested by Thomas Young (1802). In order to artificially produce colored spectra, David Rittenhouse (1786) and Joseph Fraunhöfer (1821; see Meyer, 1949) independently discovered that gratings were able to separate light into its constituent colors. Fraunhöfer etched extremely precise gratings on glass and illuminated them with a slit of white light. He found that in the center, a white image of the slit is produced, only slightly darker than the image would have been in the absence of a grating. Both sides of the bright band were dark. On the outsides of the dark bands were colored bands of violet, blue, green, yellow, orange, and red, in that order. Fraunhöfer found that the angular position of each color in the first spectrum could be found with the following formula:

$$\sin \theta = \lambda/d$$

where d is the distance between the etched lines.

Since each wavelength is diffracted to a different position, a colored spectrum is produced when white light strikes a diffraction grating. John Barton decided to take advantage of the production of colors by multiple slits and developed a machine that was able to cut grooves 1/2000 to 1/10,000 of an inch apart in steel with a diamond knife. Mr. Barton made colorful buttons and ornaments known as "iris ornaments" (Brewster, 1833b; Baden-Powell, 1833).

Many animals and some plants produce vivid structural colors as a result of thin film interference, diffraction, or a combination of the two (Fox, 1936; Rossotti, 1983; Lee, 2007). In general, however, animal and plant colors result from the differential absorption of light by chemical pigments, and the reflection of colors that are not absorbed. The beauty of jewelry made from minerals results, in part, from the optical properties of the minerals, including their index of refraction, their dispersion, their transparency, and the wavelength-dependent extinction coefficient of their pigments (Boyle, 1664, 1672).

THE RELATIONSHIP OF INTERFERENCE MICROSCOPY TO PHASE-CONTRAST MICROSCOPY

A phase-contrast microscope turns a difference in phase into a difference in intensity. In a phase-contrast microscope, contrast from a transparent object arises from the constructive or destructive interference that takes place at the image plane between the nondeviated and deviated waves that emerge from each point on the specimen. The incomplete separation of the nondeviated and deviated waves going through the phase plate in the back focal plane of the objective results in the introduction of undesirable artifacts, such as halos and shading-off, in the image (Osterberg, 1955; Tolansky, 1968).

An interference microscope turns a difference in phase into a difference in intensity, or more marvelously, into a difference in color. In an interference microscope, contrast is generated by the interference of the light wave, which passes through each point of the specimen, and a reference wave that is coherent with the light that passes through the specimen. Since the reference wave does not pass through the specimen, the separation of the two interfering waves is complete, and, as a result, halos and shading-off are prevented.

Consider a point on a transparent biological specimen, with a refractive index (n_o) of 1.33925 and a thickness of 10,000 nm, mounted in water ($n_s = 1.3330$). When the specimen is observed with green light (500 nm), it introduces an optical path difference (OPD) and phase angle (φ) of 62.5 nm and 45 degrees, respectively, according to the following formulas:

$$OPD = (n_o - n_s)t$$
$$\varphi = OPD\,(360°/\lambda)$$

In a bright-field microscope, the image is formed from the interference of the nondiffracted (U) and the diffracted (D) waves. Using a vector method to illustrate the appearance of an invisible, transparent specimen, we see that p, the resultant of U and D, is the same length as U and thus the object is equal in intensity to the background and is thus invisible (Figure 8-3).

In a phase-contrast microscope, we either advance the U wave (positive phase-contrast) or retard the U wave (negative phase-contrast) to transform a difference in phase into a difference in intensity compared with the background (Figure 8-4). In a positive phase-contrast microscope, such a specimen will appear darker than the background; with a negative phase-contrast microscope, this specimen will appear brighter than the background.

Now consider what happens in an interference microscope when we introduce a reference wave R that is coherent with, and capable of interfering with, both the U wave and the p wave. Assume that the U wave and the p wave are equal in amplitude so that the object is invisible (a). Now introduce the R wave so that it is retarded 90 degrees relative to the U wave. A new resultant (U′) will occur from the interference of the U wave and the R wave (b). Furthermore, another new resultant (P′) will occur from the interference of the p wave and the R wave (c). Notice that P′ is longer than U′, thus the object will appear bright against a less bright, but not dark, background (Figure 8-5).

Imagine that we can vary the phase of the reference wave (R) until so that the background appears maximally dark. That is, when the U wave and the R wave destructively interfere, the amplitude of the resultant wave (U′) will be zero (Figure 8-6).

Imagine that we can also vary the phase of reference wave R, so that the specimen appears maximally dark. That is, when the p wave and the R wave interfere, the amplitude of the resultant P′ will be zero (Figure 8-7).

Imagine that we can introduce a phase change of 0 to 360 degrees to the reference wave. If, when we vary the phase of the reference (R) wave we can find one place where the reference wave is exactly 180 degrees out-of-phase with the nondeviated (U) wave, the background will be maximally dark. If the interference microscope is equipped with a compensator, we can then read the angle (θ_1) of compensator when the background is maximally dark. Next turn the compensator until the particle is maximally dark. This is where the reference (R) wave is exactly 180 degrees out of phase with the particle (p) wave.

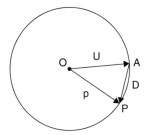

FIGURE 8-3 Vector representation of an image point (P) produced by the nondiffracted (U) and diffracted waves (D) in a bright-field microscope.

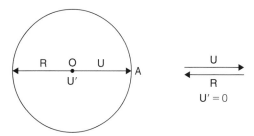

FIGURE 8-6 By varying the phase of the reference beam (R), we can bring the background to extinction.

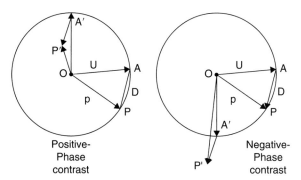

Positive-Phase contrast   Negative-Phase contrast

FIGURE 8-4 Vector representations of an image point (P′) produced by the nondiffracted (U) and diffracted waves (D) in positive and negative phase-contrast microscopes.

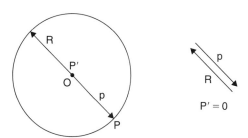

FIGURE 8-7 By varying the phase of the reference beam (R), we can bring a point in the image to extinction.

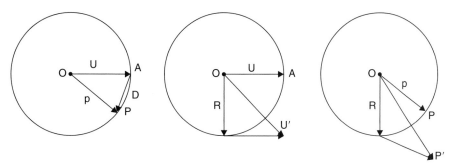

FIGURE 8-5 Vector representation of a background point (U′) produced by the vector sum of the nondiffracted light (U) and the reference beam (R). Vector representation of an image point (P′) produced by the vector sum of the diffracted (D), nondiffracted (U), and the reference beam (R) in an interference microscope. The vector sum of the diffracted (D) and nondiffracted (U) light is equal to the vector (p) that represents all the light that comes from a point in the specimen.

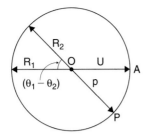

FIGURE 8-8 The phase angle of a point in the image compared to the background can be determined from the difference in the angles that bring the image point and the background to extinction.

Then we read this angle (θ_2). The difference in the compensation angles ($\theta_1 - \theta_2$) is equal to the phase angle (ϕ) introduced by the specimen, since the phase angle is equal to the optical path difference (OPD, in nm) times 360 degrees divided by the wavelength of incident light (Figure 8-8). That is,

$$\varphi = (OPD)(360°/\lambda) = (\theta_1 - \theta_2)$$

Implicit in this formula is the assumption that a 360-degree turn of the compensator compensates 1 λ. However, when we use an interference microscope with a de Sénarmont compensator, a half-turn (180°) of the analyzer compensates 1 λ. Thus the following formula applies:

$$\varphi = (OPD)(360°/\lambda) = 2(\theta_1 - \theta_2)$$

Zeiss Jena produced a microscope that lets us do both phase-contrast microscopy and interference microscopy with the same apochromatic objectives of any magnification (Beyer, 1971; Schöppe et al., 1987; Pluta, 1989). In this microscope, a focusable image transfer system projects the diffraction pattern from the back focal plane of the objective lens onto an interferometer that splits the nondiffracted light from the diffracted light. In the phase-contrast mode, a phase plate that advances the nondiffracted light to produce positive phase-contrast images or a phase plate that retards the nondiffracted light to produce negative phase-contrast images is inserted into the interferometer. To convert from phase-contrast microscopy to interference microscopy, the phase-contrast insert is replaced with an interference insert that includes a compensator.

Although the unification of Germany was a wonderful event in the world, and it still brings tears to my eyes when I think of the old men chipping away at the Berlin wall, it also meant that all microscope manufacturers would be driven by a market economy as opposed to being driven by the craftsmanship that meant so much to Ernst Abbe. I realized that this meant that the elegant but complicated microscopes made by the East German Zeiss craftsmen would no longer be made. I quickly bought a Jenalpol Interphako. I love that microscope because it illustrates to my students how there are alternative ways to solve technological problems. Moreover, we can look at the optics above the

objective lens as an analog image processor that serves as an inspiration of what we can do with digital cameras and image processors. I will discuss this microscope more in Chapter 9. So before I describe various interference microscopes, I will describe how they can be used quantitatively.

QUANTITATIVE INTERFERENCE MICROSCOPY: DETERMINATION OF THE REFRACTIVE INDEX, MASS, CONCENTRATION OF DRY MATTER, CONCENTRATION OF WATER, AND DENSITY

With a quantitative interference microscope, we can accurately measure the phase angle that we must introduce into the reference wave to bring either the background or the specimen to maximal extinction. Once we know the phase angle, we can easily determine the optical path difference with the following formula:

$$OPD = \varphi \, (\lambda/360°)$$

If we know the thickness of the specimen from independent measurements, and the refractive index of the medium (n_s), we can calculate the refractive index of the object (n_o) from the following formula:

$$OPD = (n_o - n_s)t$$

or

$$n_o = [(OPD)/t] + n_s$$

or more explicitly:

$$n_o = (\theta_1 - \theta_2)(\lambda/360°)(1/t) + n_s$$

As I discussed in Chapter 6, the refractive index of a biological specimen is related to the concentration of dry mass in the specimen. The refractive index of an aqueous specimen is equal to the refractive index of water (1.3330) plus the refractive index due to the concentration of dry mass (C_{dm}, in g/100 ml). The refractive index due to the concentration of dry mass is equal to the product of the concentration of dry mass (C_{dm}, in g/100 ml) and the specific refractive increment (α; in 100 ml/g). For most macromolecules that make up biological specimens, the refractive index increases by 0.0018 for every 1 percent (w/v) increase in dry mass. The refractive index of an aqueous specimen is given by the following formula:

$$n_o = 1.3330 + \alpha C_{dm}$$

Once we know the refractive index of the specimen, we can calculate its concentration of dry mass by rearranging the formula above:

$$C_{dm} = (n_o - 1.3330)/\alpha = (n_o - 1.3330)/0.0018$$

We can combine the following two equations to find the relationship between the optical path difference and the concentration of dry mass (in g/100 ml) in the cell. Assuming $n_s = 1.3330$,

$$OPD = (n_o - 1.3330)t$$

and

$$C_{dm} = (n_o - 1.3330)/0.0018$$

Solve both equations for $(n_o - 1.3330)$:

$$OPD/t = C_{dm}(0.0018)$$

or

$$OPD = C_{dm}(0.0018)t$$

or

$$C_{dm} = OPD/[(0.0018)t]$$

The optical path difference is equal to the concentration of dry mass (C_{dm}) times the specific refractive increment (α) times the thickness (t). The concentration of dry mass can be conveniently calculated with the following equation:

$$C_{dm} = OPD/(\alpha t) = (\theta_1 - \theta_2)(\lambda/360°)(1/\alpha t)$$

The thickness is usually difficult to measure accurately, making it inaccurate to calculate the concentration of dry mass, but we can accurately determine the mass per unit area in the following manner:

$$C_{dm} = (\theta_1 - \theta_2)(\lambda/360°)(1/\alpha t)$$

$$C_{dm} = mass/volume$$

$$C_{dm} = mass/(area\ t)$$

$$mass/(area\ t) = (\theta_1 - \theta_2)(\lambda/360°)(1/\alpha t)$$

$$mass/area = (\theta_1 - \theta_2)(\lambda/360°)(1/\alpha)$$

The area can be measured accurately with a microscope, particularly with the aid of an image processor (see Chapter 14). When we know the area of the specimen, $(\theta_1-\theta_2)$, λ and α, we can then determine the dry mass of the specimen.

$$mass = (area)(\theta_1 - \theta_2)(\lambda/360°)(1/\alpha)$$

Although the area is relatively easy to measure accurately, the thickness is more difficult to measure accurately, and the percent error in measuring thickness is much greater than the percent error in measuring area. We can measure the concentration more precisely if we can measure it independently of the thickness.

This can be done by measuring the optical path difference twice, each time using a medium with a different refractive index. For example, first measure the optical path difference between the specimen and the surround in water (OPD_w) and then measure the optical path difference between the specimen and the surround in a 5 percent

aqueous solution of bovine serum albumin (OPD_b). It is important to select media that will not cause any osmotic swelling or shrinkage, since any change in volume will introduce an unwanted change in the concentration of dry mass. As a consequence of the colligative properties of matter, which depend on the concentration of particles and not on their individual chemical properties, the solute of choice must have a high molecular mass. Solutions of high molecular mass have high densities and refractive indices yet low osmotic pressures. Observing the specimen in two media allows us solve two simultaneous equations with two unknowns, t and n_o

$$OPD_w = (n_o - n_w)t$$

and

$$OPD_b = (n_o - n_b)t$$

where OPD_w is the optical path difference when the cells are in water, OPD_b is the optical path difference when the cells are in 5% BSA, n_w is the refractive index of water (1.333), n_b is the refractive index of 5% BSA (1.333 + 5 (0.0018)) = 1.3420, and n_o is the refractive index of the object. Solving these equations for t, we get,

$$t = OPD_w/(n_o - n_w) = OPD_b/(n_o - n_b)$$

After rearranging, we get:

$$OPD_w(n_o - n_b) = OPD_b (n_o - n_w)$$
$$OPD_w n_o - OPD_w n_b = OPD_b n_o - OPD_b n_w$$
$$OPD_w n_o - OPD_b n_o = OPD_w n_b - OPD_b n_w$$
$$n_o(OPD_w - OPD_b) = OPD_w n_b - OPD_b n_w$$
$$n_o = (OPD_w n_b - OPD_b n_w)/(OPD_w - OPD_b)$$

The concentration of dry mass can then accurately be determined with the following formula:

$$C_{dm} = (n_o - n_w)/0.0018$$

Furthermore, once we know n_o, we can accurately calculate the thickness of the specimen with the following formula:

$$t = OPD_w/(n_o - n_w) = OPD_b/(n_o - n_b)$$

If a cell is composed of a single substance, we can determine the molarity of this substance by dividing its concentration (in kg/m^3) by its molecular mass (in kg/mol). Even if a cell is composed of many substances, we can determine the DNA content, for example, by measuring the concentration before and after the specimen is treated with DNase. Trypsin, lipase, and such can also be used to measure the concentration of protein, lipid, and other macromolecules. (Davies et al., 1954, 1957).

We can also determine the percentage of water in a cell by assuming that the specific volume of protoplasm is the same as the specific volume of proteins ($V_{sp} = 75$ ml/100 g $= 0.75$ ml/g). The concentration of water (C_w, in g/100 ml or %) is obtained from the following formula:

$$C_w = 100\% - V_{sp}C_{dm}$$

Thus, a solution composed of 10 percent (w/v) protein contains 92.5 percent (w/v) water. The molarity of water (M_w) in a cell can be determined using the following formula:

$$M_w = [C_w \text{ (in percent)}/100\%] \, 55.55 \, M$$

where 55.55 M is the molarity of pure water. Thus a solution composed of 10 percent (w/v) protein will contain 51.34 M water.

The density of a cell or organelle (in kg/m^3) can be determined with the following formula:

$$\text{Density} = (C_w/100\%)D_w + C_{dm}$$

where C_w is the concentration of water (in %), D_w is the temperature-dependent density of water (~ 1000 kg/m^3), and C_{dm} is the concentration of dry matter in the specimen (in kg/m^3).

I have described how it is possible to measure refractive index, mass, concentration of dry mass, concentration of water, density, and thickness with an interference microscope. In order to see how sensitive an instrument an interference microscope is, I will calculate the minimum amount of mass an interference microscope can measure.

With a good interference microscope it is easy to measure an optical path difference of about $\lambda/100$. This translates to about 5.5 nm if we illuminate the specimen with 550 nm light. By combining the following two formulas derived earlier:

$$\text{mass} = (\text{area})(\theta_1 - \theta_2)(\lambda/360°)(1/\alpha)$$

$$\varphi = (\text{OPD})(360°/\lambda) = (\theta_1 - \theta_2)$$

we get:

$$\text{mass} = (\text{area})(\text{OPD})(1/\alpha) = (\text{area})$$
$$(5.5 \times 10^{-9}\text{m})(1/1.8 \times 10^{-7}\text{m}^3/\text{g})$$

where 1.8×10^{-7} m^3/g is the specific refractive increment in units of m^3/g. It is equal to 0.0018/g/100 ml.

Since we can resolve a unit area of about 0.2×10^{-6} m $\times$ 0.2×10^{-6} m or 4×10^{-14} m^2 with an apochromatic objective lens (NA $= 1.4$), the limit of detection of mass with an interference microscope is equal to 1.2×10^{-15} g (about a femtogram), which is approximately the mass of a single band on a chromosome.

SOURCE OF ERRORS WHEN USING AN INTERFERENCE MICROSCOPE

There are a number of possible sources of error when using an interference microscope for quantitative measurements (Davies and Wilkins, 1952; Barer, 1952a, 1952b, 1953c, 1966; Barer et al., 1953; Mitchison and Swann, 1953; Barer and Joseph, 1955, 1958; Ross, 1954; Davies et al., 1954, 1957; Barer and Dick, 1957; Ingelstam, 1957; Hale, 1958, 1960; Francon, 1961; Ross and Galavazi, 1964; Beneke, 1966; Bartels, 1966; Chayen and Denby, 1968; Wayne and Staves, 1991). These include errors due to geometry, errors in the determination of optical path difference, errors due to differences in the specific refractive index, errors due to inhomogeneous specimens, and errors due to birefringence of the specimen.

Sometimes we cannot exactly determine the shape of the specimen, we can only approximate the area. However, using image processors to determine the area of a specimen, errors in geometry can be virtually eliminated.

Since we cannot always determine what is optimally dark, different observers may choose a different dark level in order to obtain the different optical path differences. This error in the determination of the optical path difference is typically between 0.5 and 3 percent, and can be minimized by using an image processor to find the maximally dark position.

The specific refractive increment varies among various compounds and the specific refractive increment used to determine the mass of a specimen may differ from the actual specific refractive increment of the macromolecules in the specimen. Errors due to differences in the specific refractive increment can be as much as 5 percent.

Most specimens are inhomogeneous and thus one part of the specimen is brought to extinction before another part. Errors due to the inhomogeneity of the specimen can be minimized by determining the maximal darkness of as large an area as possible; that is, by integrating. However when doing this, spatial resolution is sacrificed for a decrease in the inhomogeneity error.

Many interference microscopes used polarized light and consequently, the refractive index of the specimen will vary with the orientation of the optic axes of the molecules in the specimen. When measuring birefringent specimens, use an interference microscope that is not based on polarized light.

I have described the theory behind how interference microscopes can be used as a very sensitive balance to weigh a cell or organelle. Although the first interference microscope was built by J. L. Sirks in 1893, companies were not interested in developing interference microscopes until the 1950s when Barer (1952) and Davies and Wilkins (1952) figured out how interference microscopes could be used quantitatively. Companies then began to develop interference microscopes based on a variety of interesting optical principles, and it was a dream of the interference microscopists of the 1950s to be able to put a cell in a microscope, press a button, and read of the mass of the cell directly. However, it is the winner-take-all nature of science that the introduction of a newer technique with yet

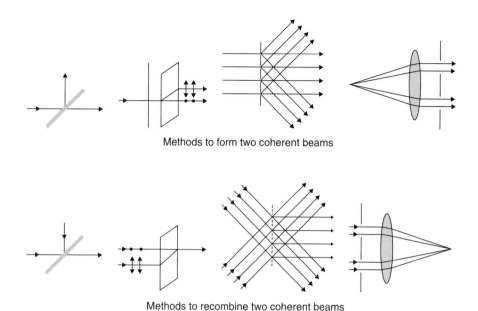

Methods to form two coherent beams

Methods to recombine two coherent beams

FIGURE 8-9 Methods to form and to recombine two coherent beams.

unknown errors and limitations kills the development and use of a technique with known errors and limitations. Thus, in the 1950s, the introduction of the electron microscope, with its potential for atomic resolution, brought a temporary stop to the development of light microscopes.

However, in the 1970's, the resurgence of the light microscope slowly began as several independent cell biologists, including Robert Allen, Andrew Bajer, Arthur Forer, Paul Green, Peter Hepler, Shinya Inoué, Eiji Kamitsubo, R. Bruce Nicklas, Barry Palevitz, Jeremy Pickett-Heaps, Ted Salmon, and D. Lansing Taylor, among others, who were both admired and emulated, showed that light microscopes could be used to observe the molecular structure, motility, and physicochemical properties of living cells without perturbing the natural state of the cell; whereas the electron microscope was able to obtain its high resolving power (Slayter, 1970) only at the expense of killing the cell.

The resurgence of light microscopy was helped by some bio-organic chemists, including Alan Waggoner and Roger Tsien, who developed probes that could be used to visualize the physicochemical properties of living cells and to localize macromolecules using fluorescence microscopy (Tsien et al., 2006; Chapter 11). The microscope manufacturers began to redesign microscopes for the study of living cells, and imaging centers in academia and industry have a variety of microscopes that will perform many fantastic tasks (Webb, 1986). Although the functions of the interference microscope have not been replaced by the other techniques, the poor interference microscope is relatively unknown and unused. Especially when combined with digital image processing, the interference microscopy is still a very useful method to measure such fundamental quantities as mass and thickness in a living cell (Dunn, 1991).

MAKING A COHERENT REFERENCE BEAM

In a phase-contrast microscope, the two coherent waves that interfere in the image plane arise from the same point in the specimen. In an interference microscope, we must generate two coherent waves (Svensson, 1957; Koester, 1961; Krug et al., 1964). A single wave can be split into two coherent waves with a variety of methods (Figure 8-9), including reflection, refraction, diffraction, and by using two holes in an opaque barrier. A wave can be split by reflection by passing it through a half-silvered mirror, where half of the amplitude of the wave is reflected and half of the amplitude of the wave is transmitted. A wave can also be split by double refraction where the incident wave is split into two orthogonal, in-phase, linearly polarized waves. A wave can also be split by a diffraction grating where the incident wave is split into the various orders, and the $+1$ and -1 orders (for example) can be used as the two coherent waves. A wave can also be split into two waves by passing it through two holes in an opaque surface. Optical processes are reversible and the same four methods can be used to recombine two coherent waves into one resultant. An interferometer consists of a beam (or wave) splitter and a beam (or wave) combiner.

The closer two waves with finite widths are to each other, the more likely they are to be coherent and interfere. In describing how interference microscopes work, I will be considering the interference of nearly parallel waves. The amplitude of the resultant ($\Psi(x)$) of two interfering waves with equal amplitude (Ψ_o), but that differ in phase by φ, is given by the following time-independent version of a formula given in Chapter 6:

$$\psi(x) = 2\psi_o \cos(\varphi/2)$$

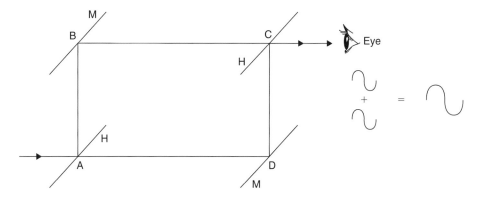

FIGURE 8-10 A diagram of a Mach-Zehnder interferometer composed of fully silvered mirrors and half-silvered mirrors in the absence of a specimen.

The phase angle is equal to $(360°/\lambda)(OPD)$, where OPD is the optical path difference between the reference wave and the specimen wave. The interference can occur between two waves (or beams) or between multiple waves (or beams).

DOUBLE-BEAM VERSUS MULTIPLE-BEAM INTERFERENCE

When a specimen is placed on the stage between the wave splitting and wave recombining portions of the interferometer so that only one set of waves propagates through the specimen and the other set acts as a reference wave, the interference pattern in the corresponding portion of the image plane is altered in a specimen-dependent manner. In white light, the image is spectacular!

Light passing through an interferometer consisting of a wave (or beam) splitter and a wave (or beam) combiner in a microscope interferes in the image plane (Michelson, 1907). If all the elements in the interferometer are parallel, the interference between two coherent waves produces a homogeneous background color. The color will be the complementary color to the one whose phase angle is 0 degrees. If one of the elements is tilted, then waves of different wavelengths will have a phase angle of 0 degrees at different places in the field. As a result, colored fringes will occur across the field, identical to those found in a Michel-Lévy color chart. When monochromatic light is used, the pattern in the image plane consists of a uniformly monochromatic background, when the wave fronts are parallel or a pattern of alternating dark and monochromatic light bands when the wave fronts arrive at a small angle.

When the background is uniform, the difference between the specimen and the surround is characterized by a process known as double-beam interferometry. We compensate the specimen by noting the color of the surround and then turning the compensator until the specimen becomes the same color. At this point, we read the angle from the compensator. We must remember the color of the

original surround since the background color also changes as additional optical path length are introduced into the reference wave in order for the specimen to attain the color of the original background. Double-beam interferometry is good for biological specimens (Pluta, 1989, 1993).

When a specimen is placed in a microscope whose field has multiple fringes, the background fringes are displaced laterally in the position the specimen occupies. This is because, in the presence of the specimen, the optical path length changes so that the position, where the phase angle for a given wavelength is 0 degrees, is displaced laterally to a position where the tilted element of the compensator gives a phase angle of 0 degrees for that wavelength in the presence of a specimen. Multiple-beam interferometry is valuable for accurately characterizing linear elements like fibers (Pluta, 1989, 1993; Barakat and Hamza, 1990).

INTERFERENCE MICROSCOPES BASED ON A MACH-ZEHNDER TYPE INTERFEROMETER

Interference microscopes, unlike bright-field microscopes, include an interferometer. A Mach-Zehnder interferometer (Figure 8-10), independently designed in the 1890s by Ludwig Mach, the son of Ernst Mach, and Lugwig Zehnder, is a "round the square" interferometer. In this interferometer, the incident light passes through a half-silvered mirror (A). At the half-silvered mirror, half of the light is reflected 90 degrees relative to the incoming beam and half of the light passes straight through the half-silvered mirror. Both waves propagate until they hit two different fully-silvered mirrors (B and D). Each wave is then reflected by one of two fully-silvered mirrors so that the wave that was perpendicular to the incident wave becomes parallel to the incident wave and the wave that was parallel to the incident wave becomes perpendicular to it. Then the two reflected waves travel to another half-silvered mirror (C), where the wave from mirror B propagates straight through it and the wave from mirror D is reflected

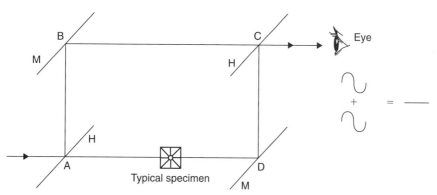

FIGURE 8-11 A diagram of a Mach-Zehnder interferometer in the presence of a specimen whose phase angle is 180°. The specimen appears dark. Fully-silvered mirror (M), half-silvered mirror (H).

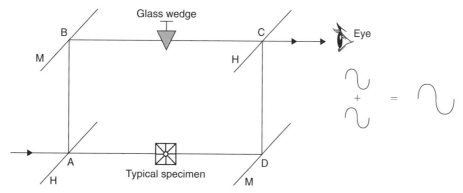

FIGURE 8-12 Compensation of a specimen in a Mach-Zehnder interferometer. The specimen appears bright. Fully-silvered mirror (M), half-silvered mirror (H).

perpendicular to the incident wave. The final half-silvered mirror recombines the two waves. When the mirrors are placed correctly, the distances the two waves travel are equal and the two waves exit in phase and constructively interfere to make a bright resultant wave.

The wave that takes the path ABC represents the reference wave, and the wave that takes the path ADC represents the specimen wave. Imagine placing a pure phase specimen that introduced a phase change of λ/2 between mirrors A and D. The two waves that exit mirror C would then be 180 degrees out-of-phase and would destructively interfere to create blackness (Figure 8-11).

Now imagine that we put a transparent glass wedge compensator of a given refractive index between mirrors B and C and slide it into the optical path until the specimen becomes maximally bright. Then we have introduced an identical increase in optical path length to path ABC as the specimen added to path ADC. At this point, the optical path difference (OPD) vanishes. If we could read the increase in optical path length we introduced from a knob on the compensator, then we would know the optical path length introduced by the specimen (Figure 8-12).

In practice it is easier to discern differences between variations in blackness than variations in brightness, so we first find the position where the background is maximally black (θ_1) and then find the position where the specimen is maximally black (θ_2). The difference between these two

readings ($\theta_1 - \theta_2$) is equal to the phase angle introduced by the specimen (φ). From the measured phase angle, we can determine the refractive index, thickness, mass, density, and percent water of the specimen using the equations I discussed earlier.

The Leitz interference microscope uses a Mach-Zehnder interferometer. It is really a double microscope. The illuminating light is split by a specially designed prism consisting of a half-silvered mirror following the principles of reflection. The prism separates the two waves a distance of 6.2 cm. The two waves then pass through perfectly matched condensers. Each wave then goes through separate objectives. The object is placed under one of the objectives. The optical path length of the two light beams can be adjusted with the aid of compensators. The light waves that pass through the two matching objectives are then recombined with a prism, similar to the one that separated the original wave. This microscope can be used for almost any biological specimen since the two waves are separated by such a large distance (Figure 8-13).

The Dyson (1961) interference microscope is another commonly found microscope that is based on the Mach-Zehnder interferometer, except that the "round the square" is slightly distorted into a rhombus. The Dyson interference system is mounted in front of an ordinary objective. The specimen (O) is mounted under a cover glass and on a glass slide. The slide and cover glass are fully immersed in

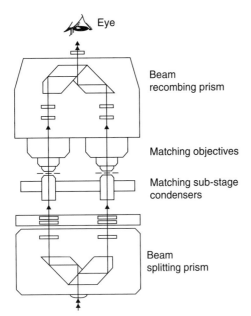

FIGURE 8-13 Diagram of a Leitz interference microscope based on a Mach-Zehnder interferometer.

immersion oil, which, on the bottom, comes in contact with a half-silvered surface (H_1) on the top of a slightly wedged plate. The bottom of the wedge has a fully silvered mirrored spot. The plate (H_1) can be adjusted with two screws. Above the object, there are two wedge-shaped plates with two half-silvered surfaces (H_2 and H_3). Above the two halfsilvered mirrors is a glass block with a fully-silvered concave surface (C) that acts as a reflection lens with a magnification of 1. The lens has a transparent hole at which the image of the specimen appears. The ordinary objective lens of the microscope is used to view this image. In comparison to the Leitz microscope, the object must be very small to allow the reference beam to go around it (Figure 8-14).

INTERFERENCE MICROSCOPES BASED ON POLARIZED LIGHT

In the Jamin-Lebedeff interference microscope, the illuminating beam is linearly polarized with a Polaroid filter. The linearly polarized light is then passed through either a calcite or a quartz crystal that is oriented with its optic axis 45 degrees relative to the azimuth of maximal transmission of the polarizer. This crystal functions as a wave (or beam) splitter. The birefringent plate separates the linearly polarized light into an ordinary wave and an orthogonally-polarized extraordinary wave that are laterally separated. The degree of lateral separation of the ordinary and extraordinary waves depends on the birefringence and thickness of the crystal. In the Jamin-Lebedeff microscope, the degree of separation is 546 μm for the 10X objective-condenser pair; 175 μm for the 40X condenser-objective pair; and 54 μm for the 100X condenser-objective pair.

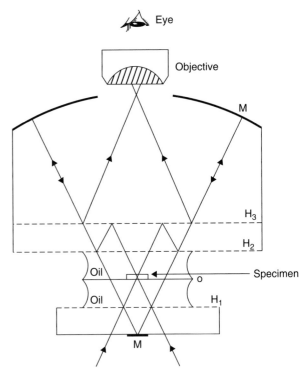

FIGURE 8-14 Diagram of a Dyson interference microscope based on a modified Mach-Zehnder interferometer.

The ordinary wave and the extraordinary wave then pass through a half-wave plate so that their azimuths of vibration are rotated 180 degrees. Now the ordinary wave vibrates in the same azimuth that the extraordinary wave vibrated before it reached the halfwave plate and the extraordinary wave vibrates in the same direction as the ordinary wave did before it reached the half-wave plate. Thus, at the half-wave plate, the ordinary wave and the extraordinary wave interchange their azimuth of polarization.

In the absence of a specimen the two waves enter a second birefringent crystal that is cut and oriented exactly the same as the first wave splitting crystal.

In the case of the positively birefringent quartz prisms, where the extraordinary wave is retarded relative to the ordinary wave, the extraordinary wave that exited the wave splitter enters the fast axis of the wave (or beam) combiner where it experiences the refractive index of the ordinary wave (n_o) and thus acts as the ordinary wave and passes straight through the wave (or beam) combiner. The ordinary wave that exits the wave splitter enters the slow axis of the beam combiner where it experiences the refractive index of the extraordinary wave (n_e) and thus acts as the extraordinary wave and experiences the anomalous refraction. In this way, the two waves are recombined into the same axial ray although they still are vibrating perpendicularly to each other (Figure 8-15).

In the case of the negatively birefringent calcite, where the ordinary wave is retarded relative to the extraordinary wave, the ordinary wave enters the fast axis of the beam combiner where it experiences the refractive index of the

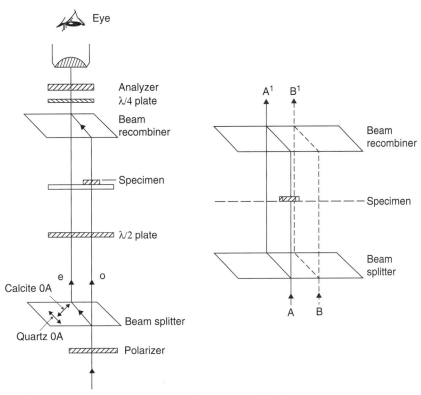

FIGURE 8-15   Diagram of a Zeiss interference microscope based on a Jamin-Lebedeff interferometer.

extraordinary wave (n_e) and thus acts as the extraordinary wave and experiences the anomalous refraction. The extraordinary wave that enters the wave splitter enters the slow axis of the wave combiner where it experiences the refractive index of the ordinary wave (n_o) and thus acts as the ordinary wave and goes straight through the wave combiner. In this way, the two waves are recombined into the same axial ray although they still are vibrating perpendicularly to each other.

The recombined ordinary and extraordinary waves then strike a quarter wave plate whose slow axis is parallel to the polarizer. The λ/4 plate then turns each of the two orthogonal, linearly polarized waves into two circularly polarized waves that rotate with opposite senses. The light then travels to the analyzer, which is crossed relative to the polarizer. In the absence of a specimen the background is maximally dark since there is no optical path difference introduced into the light path. The resultant of the extraordinary wave and the ordinary wave, in the absence of a specimen, is linearly polarized in the azimuth of maximal transmission of the polarizer.

When a specimen is inserted into the microscope it introduces a retardation of the ordinary wave relative to the extraordinary wave that leaves the wave splitter. This will cause the two circularly polarized waves that leave the λ/4 plate to be out-of-phase. The two put-of-phase circularly polarized waves produce linearly polarized light whose azimuth is determined by the phase angle between the ordinary and extraordinary waves. The change in azimuth can then be determined with the de Sénarmont compensator.

When using a de Sénarmont compensator, we must use monochromatic light to measure the optical path difference introduced by the specimen. First, bring the background to extinction by turning the analyzer and read the angle on the analyzer knob (θ_1). Then rotate the analyzer until the part of the specimen that is of interest is extinguished, and read the angle on the analyzer knob (θ_2). The optical path difference introduced by the specimen is then calculated from the following formula:

$$OPD = 2[(\theta_1 - \theta_2)\lambda/360°]$$

When we look at an image in an interference microscope based on a Jamin-Lebedeff interferometer, we see two images that are laterally displaced from each other. One image is in focus and the other one is astigmatic and out-of-focus. When I spoke about the operation of this microscope I talked as if only the ordinary waves go through the specimen. However extraordinary waves that come from other regions of the wave splitter may also pass through the specimen. We obtain an image A^1 from the ordinary waves that pass through the specimen and an image B^1 from the extraordinary waves that pass through the specimen. The image made by the extraordinary waves that pass through a specimen is an astigmatic image since the refractive index that these waves experience depends on the angle of incidence. The two images are of complementary colors.

One type of interference microscope designed by Smith uses a Wollaston prism to produce two coherent

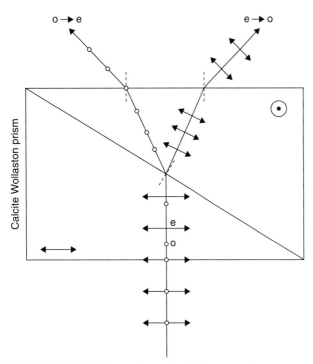

FIGURE 8-16 A Wollaston prism splits light into two coherent waves. The waves that strike the interface between the two prisms at the midway point in the prism leave the prism in phase. An optical path difference between the ordinary wave and the extraordinary wave can be introduced by sliding the Wollaston prism. The ordinary wave emerges ahead of the extraordinary wave in the beam splitter when the prism is shifted to the left, and the extraordinary wave emerges ahead of the ordinary wave in the beam splitter when the prism is shifted to the right.

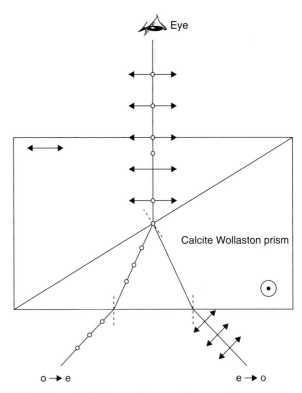

FIGURE 8-17 A Wollaston prism can also recombine two coherent waves. An optical path difference between the recombined ordinary wave and the extraordinary wave can be introduced by sliding the Wollaston prism of the recombining prism. The beam that enters the recombining prism as the ordinary wave emerges ahead of the extraordinary wave when the prism is shifted to the right and the beam that enters the recombining prism as the extraordinary wave emerges ahead of the ordinary wave when the prism is shifted to the left.

and orthogonal, linearly polarized waves. In these microscopes, the nonpolarized, monochromatic light from the lamp is linearly polarized by passage through a Polaroid below the sub-stage condenser. The linearly polarized light is split into two orthogonal linearly polarized waves by a Wollaston prism (Figure 8-16).

A Wollaston prism is made from two wedges of a birefringent material that are positioned so that their optic axes are perpendicular to each other. Linearly polarized light that is vibrating at an azimuth ±45 degrees relative to the optic axis of the first wedge is resolved into two linearly polarized components by the lower wedge: the component vibrating parallel to the optic axis experiences only n_e, whereas the component vibrating perpendicular to the optic axis experiences only n_o. For a negatively birefringent crystal like calcite, $n_e < n_o$ and thus $v_e > v_o$, and the ordinary wave that vibrates perpendicular to the optic axis is retarded relative to the extraordinary wave that vibrates parallel to the optic axis. Since each wave experiences only one refractive index, only spherical wavelets are formed and both the extraordinary wave and the ordinary wave pass through the lower crystal without being bent.

Once the extraordinary wave and the ordinary wave strike the interface, the ordinary wave enters a medium

with a lower index of refraction, so it is bent away from the normal and becomes an extraordinary wave. By contrast, the extraordinary wave enters a medium with a higher refractive index and is bent toward the normal and becomes an ordinary wave. On entering air they both separate further by bending away from the normal. When the Wollaston prism is inverted, it acts as a wave (or beam) recombiner (Figure 8-17).

In a Smith interference microscope (Figure 8-18), the ordinary wave is the image forming wave that passes through the specimen and the extraordinary wave passes beside the specimen. A second Wollaston prism recombines the two rays. The second Wollaston prism is oriented opposite to the first one so that in the absence of a specimen, the second prism exactly cancels the phase angle introduced by the first prism.

The recombined beam then passes through a $\lambda/4$ plate whose slow axis is parallel to the azimuth if maximal transmission of the polarizer. The $\lambda/4$ plate turns the ordinary and extraordinary waves into circularly-polarized waves with opposite senses of rotation. Together, the analyzer and the $\lambda/4$ plate act as a de Sénarmont compensator and the analyzer can then be rotated to bring the background and then the specimen to extinction.

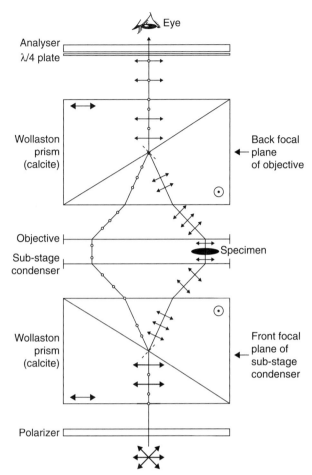

FIGURE 8-18 A Smith interference microscope. Note the antiparallel orientation of the cement line of the two Wollaston prisms. In most publications, the cement lines in the two Wollaston prisms are shown with a parallel orientation—an arrangement that could not work. I used to tell my students that I thought that the orientation of the two prisms should be antiparallel, even though the majority of publications, including the technical report put out by Zeiss (Lang, 1968) show parallel Wollaston prisms, and since I may be crazy, they were free to go with the majority opinion. After many years of saying this, I finally called Zeiss and told them that I think that their technical publication was incorrect and that the Wollaston prisms must have an antiparallel orientation to split and recombine the beams. Ernst Keller of Zeiss graciously called me back, saying, "You have the whole building upside down" and indeed "I was right." I use this as an example for my students, to base their conclusions on first principles and not on what the majority says–the majority can be right, but it is not always right.

Smith invented another kind of interference microscope, produced by Baker and by American Optical (Richards, 1963, 1964, 1966), that has a sensitivity of $\lambda/100$ or about 5 nm. In this interference microscope, the birefringent prisms that make up the wave splitter and wave combiner are placed on the top lens of the sub-stage condenser and bottom lens of the objective, respectively (Figure 8-19).

The Smith interference microscope is adjusted by setting up Köhler illumination using white light. Then an ocular is replaced with a centering telescope to view the colored fringes in the back focal plane of the objective. The lower

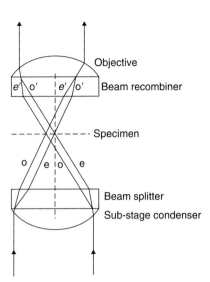

FIGURE 8-19 A diagram of a Smith-Baker interference microscope based on polarized light.

birefringent prism is tilted and rotated until the first-order red fringe is centered and then spread uniformly across the rear focal plane of the objective. The iris diaphragm is then closed enough to eliminate other interference colors in the back focal plane of the objective but not so much as to decrease resolution. The ocular is replaced and an image appears in strikingly brilliant colors. The colors can be varied by rotating the analyzer.

The measurement of the optical path difference introduced by the specimen requires monochromatic light. The green (546 nm) filter is inserted and the analyzer is turned until the region of the surround near the specimen is brought to extinction. Once the angle of the analyzer (θ_1) is read, the analyzer is rotated until the part of the specimen that is of interest is brought to extinction. The angle of the analyzer (θ_2) is read again and the optical path difference (OPD) is then calculated from the following formula:

$$\text{OPD} = 2[(\theta_1 - \theta_2)\lambda/360°]$$

With the Smith interference microscope, the specimen must be smaller than the separation of the two waves at the object plane. On the image plane, the region where the astigmatic image forms must be clear of primary images of other specimens. The size limits vary with the objective-condenser pairs: With the 10X objective-condenser pair, the beams are separated by 330 μm; with the 40× objective-condenser pair, the beams are separated by 160 μm; and with the 100× objective-condenser pair, the beams are separated by 27 μm.

For specimens that are too large for the lateral shearing method described earlier, the interference microscope comes with another set of condenser-objective pairs known as the double focus system. Unlike the condenser-objective pairs that permit shearing interference microscopy, where the reference wave and the specimen wave are laterally displaced, in the double focus system, the reference wave is axially displaced so that the microscope forms an out-of-focus image underneath the real image. This microscope is easy to set up, good for qualitative microscopy as a method to generate contrast, but cannot be used for quantitative interference microscopy.

In interference microscopes that use birefringent beam splitters, the numerical aperture of the condenser and the objective should not be too high; otherwise the optical path difference of the light that strikes the specimen from different angles will be too different and the optical path difference that is calculated will differ from the actual optical path difference introduced by the specimen. This is known as the obliquity error and it is approximately 0.06 percent for a lens with a NA of 0.10 and about 10 percent for a lens with a NA of 0.70 (Ingelstam, 1957; Richards, 1966).

THE USE OF TRANSMISSION INTERFERENCE MICROSCOPY IN BIOLOGY

Interference microscopy has been used to study pollen development and mitosis in *Tradescantia* stamen hairs (Davies et al., 1954), muscle contraction (Huxley and Niedergerke, 1954, 1958; Huxley and Hanson, 1957; Huxley, 1974), mitosis in endosperm cells (Richards and Bajer, 1961), sea urchin egg development (Mitchison and Swann, 1953), the growth of yeast cells (Mitchison, 1957), osmotic behavior and density of chick fibroblasts (Barer and Dick, 1957), spore development in fungi (Barer et al., 1953), the mass of red blood cells (Barer, 1952a,b), the mass of fibroblast nuclei (Hale, 1960) and sperm nuclei (Mellors and Hlinka, 1955) as well as the biochemical composition (Davies et al., 1954), the refractive index of the Nebenkern of sperm (Ross, 1954), the mass of an onion nucleolus (Svensson, 1957), chick nucleolus (Merrium and Koch, 1960), and myoblast nucleolus (Ross, 1964), the thickness of the wall of *Nitella* (Green, 1958, 1960), the thickness of a bacterium (Ross and Galavazi, 1964), the mass of cartilage (Goldstein, 1964; Galjaard and Szirmai, 1964), the mass of the mitotic apparatus before and after isolation (Forer and Goldman, 1972), the density of the endoplasm of *Nitellopsis* and *Chara* (Wayne and Staves, 1991), and auxin-regulated wall deposition and degradation (Baskin et al., 1987; Bret-Harte et al., 1991).

One interesting experiment using interference microscopy was done by Hugh Huxley and Jean Hanson (1957) using a Cooke-Dyson interference microscope. In developing the sliding-filament theory of muscle contraction, they wanted to know where myosin was localized in the sarcomere. Was it in the anisotropic (A) band or in the isotropic (I) band? They used the interference microscope to measure the optical path differences introduced by the dry matter in the A band and I band of a sarcomere. Then they treated the muscle with 0.6 M KCl, which selectively

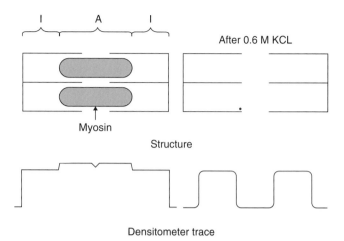

FIGURE 8-20 Diagrams and densitometer tracings of a sarcomere before and after myosin extraction from Huxley and Hanson (1957).

dissolves and removes the myosin. Then they measured the optical path differences again. They found that treating the sarcomere with 0.6 M KCl selectively decreased the optical path difference of the A band without affecting the I band. This demonstrated that myosin is localized in the A band and not in the I band of sarcomeres (Figure 8-20).

Following is a diagram of Andrew Fielding Huxley's (1952, 1954) interference microscope based on Smith's design using Wollaston prisms in the interferometer to split and combine the waves. In his low-power microscope, the Wollaston prisms were placed at the front focal plane of the sub-stage condenser and at the back focal plane of the objective. However, the Wollaston prisms cannot get close enough to the back focal plane of the water-immersion objective (NA 0.9) he needed to localize myosin. Therefore Huxley (1954) built a "corrector," which is placed above the second Wollaston prism. It is made out of four pieces of calcite placed such a distance and oriented in such a way as to recombine the two rays that emerge from the second prism (Figure 8-21).

REFLECTION-INTERFERENCE MICROSCOPY

The principles used for transmitted-light interference microscopy can also be used for reflected-light interference microscopy, or incident-light interference microscopy as it is often called. A number of microscopes including the Sangnac's interference microscope, Linnik's interference microscope, and the Zeiss (Raentsch) interference microscope are based on the principle of a Michelson interferometer, where the incident wave is split into two orthogonal waves by a wave splitter. One wave, which is reflected by the wave splitter, is incident on the specimen; the other wave, which is transmitted by the wave splitter, strikes a mirror. The specimen and reference waves are reflected by the specimen and mirror, respectively, and then pass

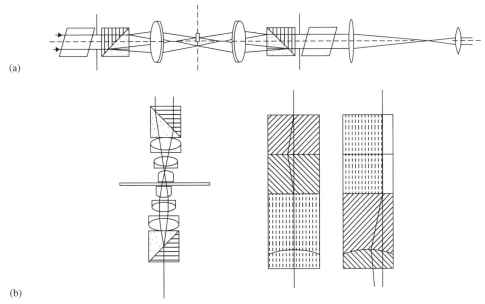

(a)

(b)

FIGURE 8-21 Diagrams of the optical arrangements designed by Andrew Fielding Huxley using various prisms made from quartz and calcite to observe interference images of muscle (Huxley, 1952, 1954).

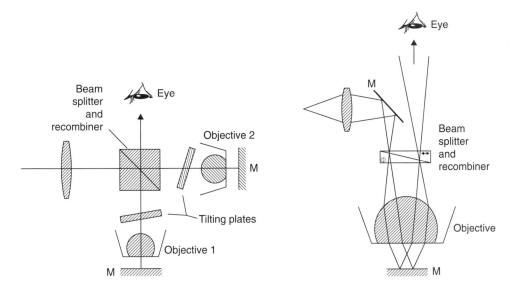

FIGURE 8-22 Diagram of two possible arrangements used in reflection-interference microscopes.

through the wave recombiner where the specimen wave is recombined with the reference wave and interfere in the image plane (Figure 8-22). Other reflection interference microscopes are based upon polarized light and use a single birefringent prism to split and recombine the specimen and reference waves to make an interference image.

USES OF REFLECTION-INTERFERENCE MICROSCOPY IN BIOLOGY

Reflection-interference microscopy can be used to measure small distances, and has been used in biology to determine the distance between a cell and its substrate (Curtis, 1964;

Opas, 1978). Knowledge of the distance helps to characterize the attractive forces between the cell and the substrate. Izzard and Lochner (1976) studied the movement of fibroblasts on glass with reflection-interference microscopy. The image obtained in a reflection-interference microscope results from three reflections that interfere to make the image: A reflection from the cover glass-medium interface, one from the top of the cell-medium interface, and a third from the bottom of the cell-medium interface (Figure 8-23).

Izzard and Lochner found that by using the widest possible numerical aperture, which gives the smallest depth-of-field, they could focus on the top of the cell and capture the reflections only from the glass-medium interface and the medium-top of the cell interface. In this way, the

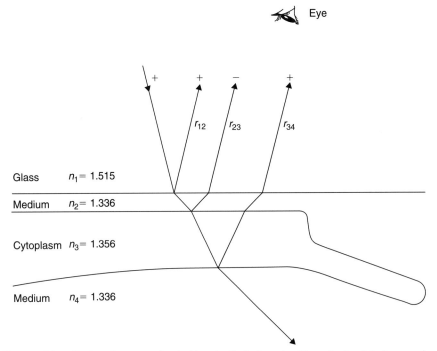

FIGURE 8-23 A ray diagram of the reflections that occur when looking at cell attachments with a reflection-interference microscope.

medium between the cell and the glass acts as a thin film with a refractive index smaller than the refractive index of the material above or below the medium.

There is a $\lambda/2$ phase change in the light reflected at the medium-cell interface (r_{23}) since the cell has a higher refractive index than the medium. On the other hand, there is no phase change at the glass-medium interface (r_{12}) since the glass has a higher refractive index than the medium. If the cell attached directly to the glass, the thickness of the medium would be zero and the reflected white light would destructively interfere, giving a black image on a bright background.

As the distance between the cell and the glass increases, the thickness of the medium increases, and the color of the cell changes in the same order as Newton's colors.

In this way, Izzard and Lochner (1976) were able to measure the thickness of the thin film and showed that cells attach to the substrate at focal contacts (0.25–0.5 μm wide × 2–10 μm long), which are separated from the glass surface by 10 to 15 nm. The focal contacts are not located at the edge, but near the edge of the advancing lamellapodia, which are separated from the glass by 100 nm or more.

Chapter 9

Differential Interference Contrast (DIC) Microscopy

In Chapter 8, I discussed image-duplication interference microscopy, in which the image results from the interference between a wave that propagates through a point in the specimen and a reference wave that is laterally displaced from that point in the specimen. In this chapter, I will discuss differential interference contrast microscopy (DIC). Color plates 14 through 18 provide examples of images obtained using differential interference microscopy.

In a differential interference contrast microscope, the two waves are laterally displaced a distance that is smaller than the resolving power of the objective lens. By producing two laterally displaced coherent waves, a differential interference contrast microscope is able to convert a gradient in the optical path length into variations in light intensity. Steep gradients in optical path length appear either bright or dark, depending on the sign of the gradient, whereas areas that have a uniform optical path length appear gray. A differential interference contrast microscope can also be used to change gradients in optical path lengths in transparent objects into spectacular differences in color. If the two laterally separated images produced by a differential interference microscope were brought together in perfect alignment, the introduced contrast would disappear and the image of a transparent object would be invisible.

Gradients in optical path length are mathematically equivalent to the first derivative of the optical path length (Figure 9-1). Consequently, a differential interference contrast microscope can be considered to be an analog computer that gives the first derivative of the optical path length of a specimen. By contrast, an image duplication interference microscope can be considered to be an analog computer that gives the integral of the optical path length.

DESIGN OF A TRANSMITTED LIGHT DIFFERENTIAL INTERFERENCE CONTRAST MICROSCOPE

Differential interference contrast microscopes were designed and developed throughout the 1950s and beyond by F. Smith, M. Francon, G. Nomarski, and H. Beyer (Pluta, 1989). The Zeiss Jena interference microscope uses a Mach-Zehnder type interferometer to separate and combine the specimen wave and the reference wave (Figure 9-2). The lateral separation of the specimen and reference waves can be controlled so that the same microscope can function as an image duplication interference microscope, or as a differential interference microscope with minimal adjustment. Most differential interference contrast microscopes, however, utilize birefringent prisms to split and recombine the specimen and reference waves.

In the Zeiss Jena interference microscope, the specimen is illuminated by a slit placed in the front focal plane of the substage condenser. The image of the specimen appears in a field plane just in front of the interferometer, and an image of the slit is coaligned with another slit placed in an aperture plane in one arm of the Mach-Zehnder interferometer. The light passing through this arm functions as the reference wave. The light passing through the other arm functions as the specimen wave. There is a tilting plate in the reference arm that is used to laterally separate the reference wave from the specimen wave at the image plane, and a rotating plate that is able to add additional optical path length into the reference arm, which results in a colored image.

The differential interference microscope based on polarized light has a certain similarity with the polarizing

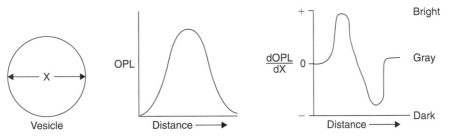

FIGURE 9-1 A diagram of a vesicle, a graph of the optical path length across a vesicle, and a graph of the first spatial derivative of the optical path length. In a differential interference microscope, the first spatial derivative of the optical path length gives rise to shades of gray or colors that represent the first spatial derivative.

microscope and the image duplication interference microscope based on polarized light (Lang, 1968; Allen et al., 1969; Figure 9-3). In a polarizing microscope, a single beam of linearly polarized light that is produced by a polarizer passes through a birefringent specimen that is oriented so that its slow axis is at a ±45-degree angle relative to the azimuth of maximal transmission of the polarizer. The specimen splits a linearly polarized wave into orthogonal linearly polarized waves. The phase difference between the two waves is a function of the retardation introduced by the specimen. The two out-of phase orthogonal linearly polarized waves can be considered as a single elliptically polarized wave. The component of the elliptically polarized wave parallel to the azimuth of maximal transmission of the analyzer passes through the analyzer. The brightness of

an image point depends on the degree of anisotropy in the bonds that make up the conjugate point in the specimen.

In an image duplication interference microscope based on polarized light, the linearly polarized light from the polarizer is split into an ordinary wave and an extraordinary wave that are laterally separated from each other by 10 to 500 μm. One wave passes through the specimen while the other passes through the surround. The two waves are recombined in the wave combiner and are turned into elliptically polarized light. The component of the elliptically polarized wave parallel to the azimuth of maximal transmission of the analyzer passes through the analyzer. The brightness of the image point depends on the phase difference between the wave that goes through the conjugate point in the specimen and the reference wave.

In a differential interference contrast microscope, the linearly polarized light from the polarizer is acted upon by a prism that laterally separates the ordinary wave and the extraordinary wave by only 0.2 μm to about 1.3 μm. The two orthogonal waves that propagate through two nearby points in the specimen are recombined in the wave recombiner and are turned into elliptically polarized light. The component of the elliptically polarized wave parallel to the azimuth of maximal transmission of the analyzer passes through the analyzer. The brightness of the image point depends on the phase difference between the waves that propagate through the two nearby points in a specimen.

In differential interference microscopes based on polarized light, the linearly polarized light from the polarizer is split into two orthogonal laterally displaced waves and recombined into an elliptically polarized wave by Wollaston prisms or modified Wollaston prisms, known as Nomarski prisms (Figure 9-4).

Wollaston prisms are constructed from two wedges of either calcite or quartz, which are cemented together so that the optic axes of the two birefringent crystals that make up a prism are perpendicular to each other. Wollaston prisms are often too thick to be placed in front of the substage condenser lens or behind the objective lens so that

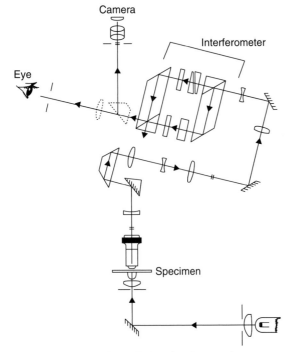

FIGURE 9-2 Diagram of the Zeiss Jena interference microscope.

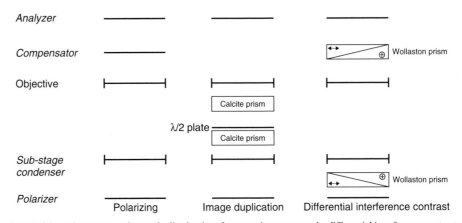

FIGURE 9-3 A comparison of a polarizing microscope, an image-duplication interference microscope, and a differential interference contrast microscope.

the center of the prism interface is located at the front focal plane of the substage condenser lens and the back focal plane of the objective lens, respectively. This problem can be overcome by using Nomarski prisms, in which the optic axis in the first crystal in a prism is oblique relative to the second. Because of this arrangement, the waves are split and recombined outside the prism. The Nomarski prisms are placed in the microscope such that the plane outside the prism, where the waves are split, is placed at the front focal plane of the substage condenser and the plane, in which the waves are recombined outside the second prism, is placed at the back focal plane of the objective.

INTERPRETATION OF A TRANSMITTED LIGHT DIFFERENTIAL INTERFERENCE CONTRAST IMAGE

In order to understand how contrast is produced with a differential interference contrast microscope based on polarized light; consider four pairs of waves produced by the lower Wollaston or Nomarski prism (Figure 9-5). The interface of the prism is placed at the front focal plane of the substage condenser so that the ordinary wave and the extraordinary wave appear to diverge from the front focal plane of the condenser and exit the substage condenser as a parallel beam of in-phase, orthogonal ordinary and extraordinary waves. Imagine that one pair of waves passes through the surround (A), one pair of waves passes along an edge of an object (B), one pair of waves passes through the center of the object (C), and the last pair of waves passes along the other edge of the object (D). The first and third pairs of waves (A and C) experience no optical path differences, whereas one of the waves of a pair in the second and fourth pair of waves (B and D) will experience an optical path difference relative to the other wave of the pair.

Suppose that the object, like most biological objects, introduces a $\lambda/4$ phase retardation between the ordinary and extraordinary waves of a pair. Also suppose that the second prism is set so as to introduce another $\lambda/4$ phase change between the members of a pair and so that the ordinary ray is retarded relative to the extraordinary ray (Figure 9-6).

Pairs where both the ordinary and extraordinary waves go through the surround, or pairs where both the ordinary and extraordinary waves pass through regions where there are no differences in the optical path length will be $\lambda/4$ out

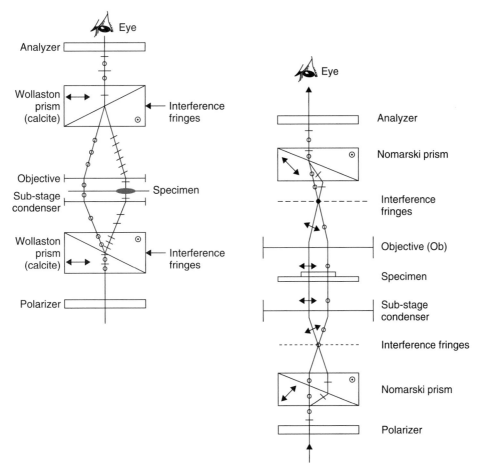

FIGURE 9-4 Differential interference microscopes comprised of negatively birefringent calcite Wollaston prisms (left) or positively birefringent quartz Nomarski prisms (right).

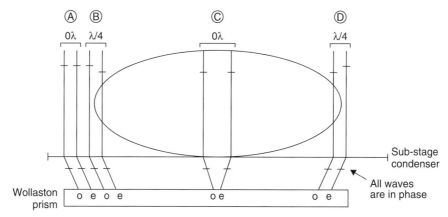

FIGURE 9-5   Formation of a differential interference contrast image. The two members of each pair of waves have been laterally separated but not axially separated.

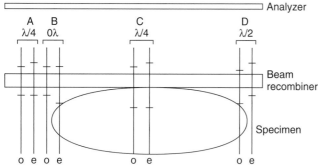

FIGURE 9-6   Formation of a differential interference contrast image. The two members of each pair of waves have been laterally separated and axially separated.

of phase after they are recombined by the second prism (e.g., A and C). The resultant wave will be circularly polarized and these regions will appear gray in the image. On the other hand, pairs of waves whose ordinary wave experiences a phase shift as it passes through a specimen will be $\lambda/2$ out-of-phase after being recombined by the second prism (D). The resultant wave will be linearly polarized parallel to the azimuth of maximal transmission of the analyzer and this point in the image will be bright. A pair of waves whose extraordinary wave experiences a phase shift as it passes through the specimen will be $0\ \lambda$ out-of-phase after being recombined by the second prism B. The resultant wave will be linearly polarized parallel to the azimuth of maximal transmission of the polarizer and this image point will be dark. The contrasts would be white-black reversed if the specimen were phase-advancing relative to the medium instead of phase-retarding. As a result of the conversion of gradients of optical path into intensity differences, the image appears as if the specimen were a three-dimensional object illuminated from the side. However, the three-dimensional appearance of the image, like the appearance of a specimen that is obliquely-illuminated (see Chapter 6), is only an illusion (Rittenhouse, 1786; Hindle and Hindle, 1959).

Only gradients in optical path length that are approximately the same size as the distance that the two waves are laterally separated show up in relief. The microscopic objects that have characteristic lengths approximating the distance laterally separating the two waves include membranes, organelles, chromosomes, and vesicles in the cell. To get an optimum image, the specimen should be rotated so that the azimuth of separation between the two waves, known as the azimuth of shear, maximally enhances the particular structure we want to observe. The azimuth of shear is the azimuth that contains both the E-ray, which is perpendicular to the extraordinary wave front, and the O-ray, which is perpendicular to the ordinary wave front, as they emerge from the first Wollaston or Nomarski prism.

A single image obtained with a differential interference microscope will not represent all asymmetrical specimens. On the other hand, differential interference microscopy, like oblique illumination, makes it easy to discover and visualize asymmetries that may have gone undetected with bright-field illumination. In order to distinguish between asymmetries and symmetries in the specimen when using differential interference microscopy, it is important to rotate the specimen. The specimen can be rotated easily if the microscope is equipped with a rotating stage.

Imagine that a specimen in a differential interference contrast microscope based on polarized light is illuminated with white light. There will be an infinite number of wave pairs that pass through each point of the specimen, each pair representing a different wavelength. Each wavelength will experience the same retardation as the other wavelengths going through a given point, but since the wavelengths differ, the phase angle introduced for each wavelength will differ. Consequently, each wavelength will become elliptically polarized to a different extent and each wavelength will pass through the analyzer to a degree dependent upon its ellipticity.

The phase angles introduced by typical biological specimens are usually not large enough to produce interference

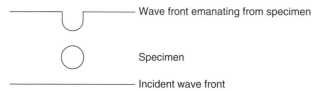

FIGURE 9-7 Wave fronts approaching and leaving a transparent specimen.

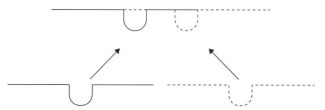

FIGURE 9-8 Allow a laterally separated pair of waves to propagate through specimen.

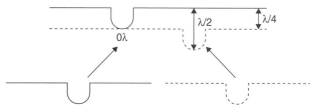

FIGURE 9-9 Axially separate the pair of laterally separated waves that propagate through the specimen.

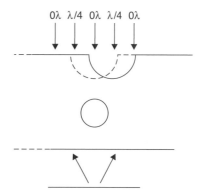

FIGURE 9-10 Without axial separation of the two waves, the contrast in the image of a vesicle will be symmetrical.

colors. A first-order wave plate is either included with the wave-combining prism or added as an accessory so that the small retardations introduced by the specimen can be added or subtracted from the 530 nm retardation introduced by the fist-order wave plate. All microscope manufacturers provide a different wave splitting prism for each objective lens. The correct prism is inserted into the light path by turning a turret on the sub-stage condenser to the position where the prism matches the objective. Some microscope manufacturers provide a single wave recombiner that work for all objectives. In this case the wave recombiner is inserted just under the analyzer. Other manufacturers provide a different wave recombiner for each objective. In this case the wave recombiners are inserted in the objective mounts.

Earlier, I described image formation in terms of four point pairs along the azimuth of shear. Now I would like to use the concept of wave fronts to give an alternative description of how contrast is generated with a differential interference microscope. Although this treatment also applies to waves split by a birefringent prism, I will specifically describe waves that are produced by a Mach-Zehnder interferometer in the Zeiss Jena differential interference contrast microscope as an example because this microscope can easily be adjusted for use as a bright-field microscope; an image duplication interference microscope, in which two widely separated wave fronts emanate from the specimen; and a differential interference contrast microscope, in which two minutely separated wave fronts emanate from the specimen.

Consider a spherical object that introduces a λ/4 increase in phase relative to the background in a bright-field microscope. Imagine that the object retards a plane wave that is moving in an upward direction (Figure 9-7).

Now consider what happens when the wave coming from the real image produced by the objective is split into two coherent laterally displaced plane waves. This is what happens in an image duplication interference microscope

(Figure 9-8). Now consider what happens when these two coherent waves are recombined to produce an interference image and the reference wave is axially displaced until it is in phase with the specimen wave coming from a given position in the specimen (Figure 9-9). The specimen will be bright in regions where there are no phase differences between the two waves and dark in regions where the phase change is λ/2. Regions that are λ/4 out of phase will appear gray. Two images of the specimen with opposite contrast are formed—one where the phase difference is 0λ and one where the phase difference is λ/2.

Now consider what happens when the wave coming from the real image produced by the objective is split into two coherent plane waves that are minutely separated laterally, as occurs in a differential interference contrast microscope (Figure 9-10). Without any axial separation between the two waves, a vesicle will appear like a doughnut with a gray ring around a bright center. Now consider when one of the laterally displaced waves is axially retarded relative to the other. In the regions where the two waves are displaced by 0 wavelengths, the image will be bright. In the regions where the two waves are displaced by λ/2, the image will be dark. Regions that are λ/4 out of phase will appear gray (Figure 9-11). Consequently, the vesicle will appear in pseudo-relief as if it were illuminated from the side.

When the specimen is illuminated with white light, there are an infinite number of wave pairs, each with the same optical path difference, but the phase angle for each wavelength will be different. Thus when one wavelength constructively interferes and produces a given color on one side of an object, the other side will appear as the complementary color. Additional color can be added in the Zeiss Jena differential interference microscope by introducing an additional phase to the reference wave by inserting a glass wedge into the reference wave path.

In an image duplication interference microscope based on polarized light, the amount of light transmitted through the analyzer also depends on the phase of the two wave fronts. In the description of a differential interference contrast microscope based on polarized light given earlier, I described image formation in terms of four point pairs along the azimuth of shear. To use the concept of wave fronts to give an alternative description of how contrast is generated, imagine that all the extraordinary rays of a given wavelength are connected together in a group to form a wave front and all the ordinary rays of that wavelength are connected together in another group to form another wave front. The first group would represent the extraordinary wave front and the second group would represent the ordinary wave front that leaves the specimen. In regions where the phase change between the extraordinary wave front

and the ordinary wave front is zero, the resultant is linearly polarized in the azimuth of maximal transmission of the polarizer and thus no light is transmitted by the analyzer and the image in this region will be black. In regions where the phase change between the extraordinary wave front and the ordinary wave front is $\lambda/2$, the resultant is linearly polarized in the azimuth of maximal transmission of the analyzer and the most light will be transmitted by the analyzer and the image in this region will be bright. In regions where the phase change is $\lambda/4$, the resultant will be circularly polarized and an intermediate amount of light will pass through the analyzer and the image in this region will be gray.

In a differential interference microscope based on polarized light, the amount of lateral separation is determined by the wave splitting prism and the axial separation is set by the adjustable wave recombining prism. The wave recombiner is adjusted until the two laterally displaced waves are in-phase on one side of the object of interest and $\lambda/2$ out-of-phase on the other side of the object of interest. The differential contrast will be expressed only along the azimuth of shear.

A differential interference contrast microscope introduces contrast into transparent objects and produces a pseudo-relief image. The pseudo-relief image does not relate to the topography itself but to the first derivative of the optical path length of the specimen. Since details in a real specimen introduce a variety of different phase changes, the wave recombiner can be adjusted to give the maximal contrast for a given specimen detail.

Consider what an image of a plant cell would look like if we have gradients in the optical path length that are opposite in sign (Figure 9-12). For example, consider a cell with a

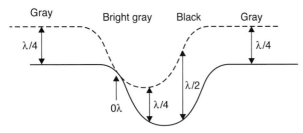

FIGURE 9-11 With axial separation of the two laterally displaced waves, the image of the specimen will be asymmetrical and appear to have relief.

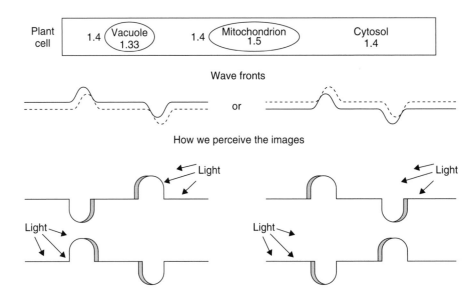

FIGURE 9-12 Representation of waves going through a plant cell containing organelles with high refractive index (e.g., mitochondria) and organelles with low refractive index (e.g., vacuole). Whether the organelles appear as hills or valleys depends on the direction from which we imagine the organelles are illuminated. Since they are illuminated from the bottom, the image is only a pseudo-relief image, not based on topography, but on the first spatial derivative of the optical path length.

vacuole (n = 1.33) and a mitochondrion (n = 1.5) in a cytoplasm with a refractive index of n = 1.4. We can advance or retard one of the waves relative to the other so that in one case the illumination appears to be coming from the top right and in the other case the illumination appears to be coming from the top left. In one case, the vacuole appears as a hill and the mitochondrion as a valley and in the other case, the vacuole appears as a valley and the mitochondrion as a hill. The apparent relief of the image depends on where we imagine the light to be coming from (Rittenhouse, 1786). In reality, it is being transmitted from the bottom.

The image seen in a differential interference contrast microscope can be described by the following rules (Allen et al., 1969).

1. The optical property of a microscopic object that generates differential interference contrast is the gradient of optical path length across the object in the direction of shear.

2. Contrast varies proportionally with the cosine of the angle made by the azimuth of the object with the direction of shear.

3. The pseudo-relief effect is emphasized when one slope of the image in the direction of shear is brought to extinction by varying the wave recombiner. This setting also yields the highest possible contrast and the most faithful geometric image of the object.

4. Gradients of optical path length of opposite sign produce shadows in opposite directions.

5. In a differential interference microscope, the contrast is generated independently of the aperture diaphragm. Consequently, we can keep the aperture diaphragm maximally opened to get maximal resolution and a very shallow depth of field that allows us to optically section.

Differential interference contrast microscopes are better than phase-contrast microscopes when viewing objects whose optical path differences are greater than the depth of field of a given objective lens. Differential interference contrast microscopes that are based on polarized light are not good for studying tissue cultured cells that grow on birefringent plastic culture plates. In this case, we may choose to view the specimens with a differential interference contrast microscope based on a Mach-Zehnder interferometer, oblique illumination, or Hoffman Modulation Contrast microscopy (HMC; see Chapter 10).

DESIGN OF A REFLECTED LIGHT DIFFERENTIAL INTERFERENCE CONTRAST MICROSCOPE

Reflected light differential interference contrast microscopy reveals surface contours by changing differences in height into variations in amplitude or color. It is used in biology, medicine, microelectronics, and other disciplines. Reflected

light differential interference microscopy is based on the same principles as transmitted light differential interference microscopy (Figure 9-13). In reflected light differential interference contrast microscopes based on polarized light, linearly polarized light is generated by an epi-illuminator, which is then directed to a half-silvered mirror and reflected through a Wollaston or Nomarski prism (where the beam is split), then through an objective lens and onto the specimen. The light is then reflected back through the objective lens and Wollaston or Nomarski prism (where the beams are recombined), through the half-silvered

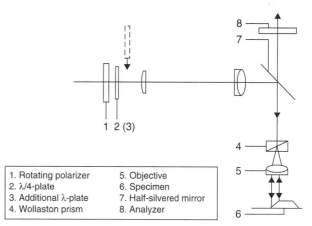

1. Rotating polarizer	5. Objective
2. λ/4-plate	6. Specimen
3. Additional λ-plate	7. Half-silvered mirror
4. Wollaston prism	8. Analyzer

FIGURE 9-13 Diagram of a reflected light differential interference microscope.

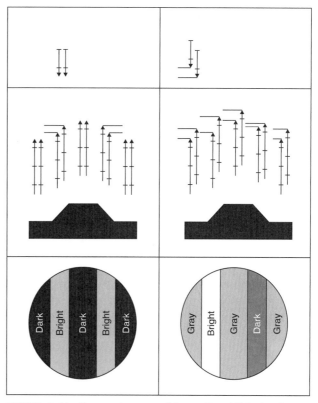

FIGURE 9-14 Contrast is generated in a reflected light interference contrast microscope as a result of differences in microtopography.

mirror and through the analyzer placed in the crossed position. A full wave or a λ/4 plate can be inserted after the polarizer in order to vary the retardation between the ordinary ray and the extraordinary ray and thus vary the color of the image.

The Zeiss Jena interference microscope can also be used for reflected light differential interference microscopy by changing from the transmitted light illuminator to the epi-illuminator and by changing the objectives from those designed for use with transmitted light to those designed for use with reflected light.

INTERPRETATION OF A REFLECTED LIGHT DIFFERENTIAL INTERFERENCE CONTRAST IMAGE

The illuminating linearly polarized light that passes through the Wollaston or Nomarski prism is split into two laterally separated orthogonal linearly polarized waves. Both waves then are reflected from the surface of the object and recombined in the original prism. When a pair of waves strikes a horizontal surface, there is no phase change introduced between the two, and the surface appears black.

However, when two waves of a pair strike an inclined surface, a phase change will be introduced and, as a result, the image will appear bright. The magnitude of the phase change introduced by the incline depends on the slope of the surface. The brightest spots in the image will appear where the height difference between the two waves of a pair equal λ/2 (Figure 9-14).

A λ/4 plate can be introduced after the polarizer so that the horizontal areas are gray, and the gradients in topology along the plane of shear appear white or black, depending on the direction of the slope. If we introduce a first-order wave plate after the polarizer, the horizontal areas will appear lavender and the slopes along the plane of shear will show additive (bluish) or subtractive (yellow-orangish) colors, depending on the slope.

Chapter 10

Amplitude Modulation Contrast Microscopy

According to Ernst Abbe, image formation by microscopes can be understood in terms of diffraction (see Chapter 3; Abbe, 1876−1878, 1889; Wright, 1907; von Rohr, 1936). His thinking went like this: A microscopic specimen diffracts the illuminating light. The diffracted light captured by the lens forms a diffraction pattern in the back focal plane of the lens. This diffraction pattern is an imperfect representation of the actual diffraction pattern produced by the specimen because the lens cannot capture the highest orders of diffraction. The spots of the diffraction pattern in the back focal plane of the objective act as sources of spherical waves that interfere in the image plane to form the image. Consequently, the image is related directly to the diffraction pattern and indirectly only to the specimen itself.

Abbe's theory is to date the most complete and useful theory of microscopical vision and, along with G. Johnstone Stoney's (1896) application of Fourier's Theorem to image formation, provides a strong mathematical basis for using digital image processors to determine the real structure of a specimen from knowledge of the optical properties of the microscope that convolve the image by introducing artifacts like low pass filtering. However, Abbe's diffraction theory was developed from the study of high-contrast amplitude objects and not transparent phase objects, and consequently, it does not take into consideration differences in the refractive indices of the points in a specimen that cause lens-like refractions. Such considerations, although complex, would yield a more complete theory of microscopic vision (Nelson, 1891).

The word diffraction is derived from the Latin words *dis* and *frangere*, which mean "apart" and "to break," respectively. Diffracted light, according to G. Johnstone Stoney (1896), is "light which advances in other directions than those prescribed by geometrical optics." Although diffraction theory is ultimately more complete than geometric optical theory, geometrical optics can be considered to be a reasonably approximate theory of image formation in the microscope when the ratio of object length to wave length is large. Under this condition, the majority of the diffracted light is almost indistinguishable from the nondiffracted light.

It is likely that few of us, if any, have the ability or the computational power to apply diffraction theory in a rigorous enough way to perform a complete characterization of a complex biological specimen (Meyer, 1949); consequently, many simplifying assumptions about the specimen must be introduced. It is possible, however, that in making the simplifying assumptions, elements that may be necessary to relate points in the image to points in the specimen become unintentionally discounted or obscured.

The simultaneous or occasional application of geometric optical theory to image formation may provide us with a certain degree of insight that may help in the construction of a complete and universal theory of image formation. For this reason, I present a few simple examples that will remind us of the importance of refraction in image formation. After that, I will use refraction theory to describe how images are formed in an amplitude modulation contrast microscope (Hoffman and Gross, 1975a, 1975b; Hoffman, 1977). Gordon Ellis (1978) has used diffraction theory to describe image formation in his version of an amplitude modulation contrast microscope, which he calls a "single-sideband edge enhancement" or SSEE microscope. Color plate 19 provides an example of an image obtained with a Hoffman modulation contrast microscope.

The importance of refraction in understanding the images of complex biological specimens formed by a microscope can be seen by doing the following experiment. Mix together water, oil, and air and place the mixture on a microscope slide (Figure 10-1). The air bubbles, which have a refractive index less than that of water, will act as diverging lenses, and each air bubble will produce a virtual image of the illuminating light beneath the bubble. That is,

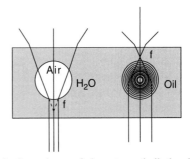

FIGURE 10-1 In a mixture of air, water, and oil, the oil drops act as converging lenses and air bubbles act as diverging lenses. They project a bright image of the aperture diaphragm above and below the lens-like object, respectively.

we will see a bright spot, which is the image of the aperture diaphragm, when we move the stage up. On the other hand, the oil droplets, which have a refractive index greater than that of water, will act as converging lenses, and each oil droplet will produce a real image of the illuminating light above the bubble. That is, we will see a bright spot, which is the image of the aperture diaphragm, when we move the stage down.

When this same experiment is done with a compound microscope using oblique illumination, the air bubble, which acts as a diverging lens, appears with the bright side on the opposite side as the illuminating light beneath the condenser. The oil droplet, which acts as a converging lens, appears with the bright side on the same side as the illuminating light beneath the condenser (Figure 10-2).

This little demonstration should always be kept in mind when interpreting the structure of biological specimens since the cytoplasm can be considered to be a sample that contains many spherical vesicles with refractive indices greater than and less than the refractive index of the cytoplasm (Naegeli and Schwendender, 1892; Wright, 1907; Gage, 1917). Thus each vesicle and organelle may act as a converging or diverging lens.

HOFFMAN MODULATION CONTRAST MICROSCOPY

The Hoffman modulation contrast microscope makes transparent specimens visible by converting gradients in optical path length into differences in intensity. According to

Hoffman, the direct light is responsible for image formation and contains information about the specimen. Specifically, it contains information about the gradients in optical path length that exist in the specimen. The modulation contrast microscope, like a differential interference contrast microscope, and a microscope that uses oblique illumination, produces pseudo-relief images. The images of transparent specimens produced by a modulation contrast microscope are high in contrast and resolution. Differential interference microscopes based on polarized light cannot be used for birefringent objects or specimens growing in culture dishes made of birefringent plastic, whereas modulation contrast microscopy is not limited to isotropic specimens.

The Hoffman modulation contrast attachments are made by Modulation Optics, Inc. to fit any bright-field microscope. The system includes a set of objective lenses that have been modified by adding a modulator to the back focal plane, a turret condenser that has a rotating slit whose size is specific for each objective lens, and a polarizer (Figure 10-3). Olympus markets this technique under the name of "relief contrast" and Zeiss markets a similar system under the name of "variable relief contrast" (Varel).

The Hoffman modulation contrast modulator is placed asymmetrically in the back focal plane of the objective. It has three regions that differ in their transmittance. The region at one edge has a transmission of less than 1 percent, the next region has a transmission of 15 percent, and the other side has a transmission of 100 percent. The modulator introduces only changes in amplitude without any change in phase.

The slit is placed in the front focal plane of the sub-stage condenser and is off-center so it will produce oblique

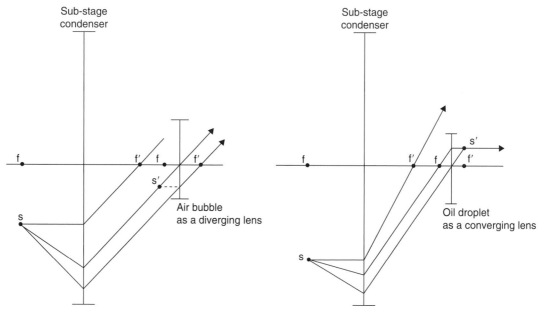

FIGURE 10-2 With oblique illumination, the air bubble, which acts as a diverging lens, appears with the bright side on the opposite side as the illuminating light after passing through the objective, and the oil droplet, which acts as a converging lens, appears with the bright side on the same side as the illuminating light after passing through the objective.

illumination. An image of the slit is produced by the sub-stage condenser-objective pair, on the modulator in the back focal plane of the objective. Part of the slit is covered with a polarizer. When the polarizer that is placed above the field diaphragm is crossed relative to the polarizer in the slit, the slit effectively is reduced to one-half its maximal width.

When setting up a Hoffman modulation contrast microscope, the slit, which is placed in the front focal plane of the condenser, is imaged on top of and aligned with the 15 percent transmittance portion of the modulator in the back focal plane of the objective. Therefore, in the absence of a specimen, the image plane will be filled with light that is about 15 percent of the intensity of the light that passes through the slit. Consequently, the background in the image plane will appear gray (Figure 10-4).

In order to understand how an image is formed in a Hoffman modulation contrast microscope, consider a gradient in optical path length to be equivalent to a prism made from isotropic material. When the prism is placed in the light path, in the position where the specimen would be, the incident light will be refracted by the prism and the image of the slit will be displaced laterally. That is, the slit will be imaged on top of either the 100 percent (Figure 10-5) or the 1 percent (Figure 10-6) transmittance portion of the modulator, depending on the orientation of the prism. If the refractive index of the prism is less than the refractive index of the surround, then the incident light will be refracted to the opposite side of the modulator (Figure 10-7). Figure 10-8 shows how a cell or a vesicle within the cell can be modeled as a series of prisms with different orientations.

Thus the intensity of a given point in a Hoffman modulation contrast image is determined by the gradient in optical path lengths in the object. Contrast is generated only by gradients in the optical path lengths that are parallel to the short axis of the slit. Consequently, the orientation of the specimen determines how the pseudo-relief image appears and a single image will not represent all asymmetrical specimens. Like oblique illumination and differential interference contrast microscopy, Hoffman modulation contrast microscopy makes it easy to discover and visualize asymmetries that may have gone undetected with bright-field illumination. In order to distinguish between asymmetries and symmetries in the specimen when using Hoffman modulation microscopy, it is important to rotate the specimen using a rotating stage.

All the nonrefracted light goes through the 15 percent transmission portion of the modulator when the azimuth of maximal transmission of the polarizer that covers part of the slit is perpendicular to the azimuth of maximal transmission of the first polarizer. This gives maximal contrast, but a grainy image. Some of the incident light also goes

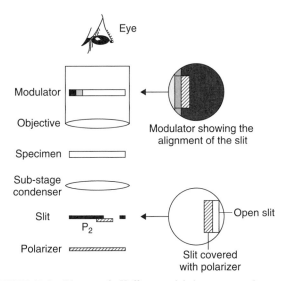

FIGURE 10-3   Diagram of a Hoffman modulation contrast microscope.

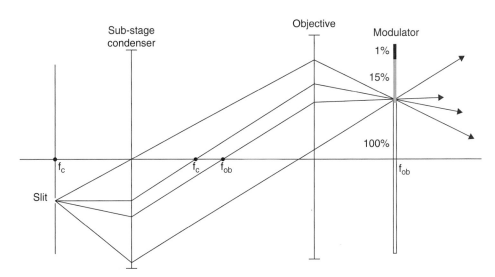

FIGURE 10-4   Ray diagram of a Hoffman modulation contrast microscope in the absence of a specimen.

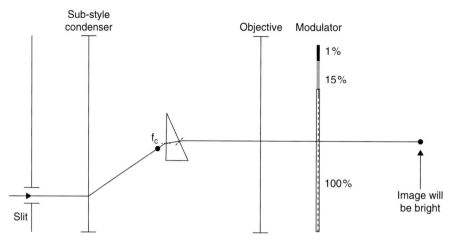

FIGURE 10-5   Ray diagram of a Hoffman modulation contrast microscope in the presence of a prism-like specimen of a given orientation.

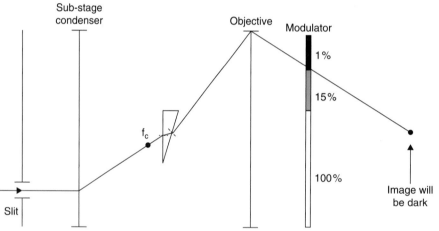

FIGURE 10-6   Ray diagram of a Hoffman modulation contrast microscope in the presence of a prism-like specimen with the opposite orientation as that in Figure 10-5.

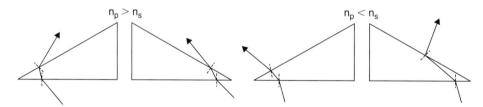

FIGURE 10-7   Ray diagrams of prism-like specimens with a refractive index greater than or less than the surround.

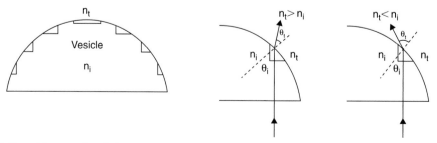

FIGURE 10-8   Model of a vesicle as a series of prisms.

through the 100 percent transmission portion of the modulator when the azimuth of maximal transmission of the polarizer covering part of the slit is parallel to the azimuth of maximal transmission of the first polarizer. This reduces the contrast, but gives a smooth metallic-like image. Thus the contrast of the image can be adjusted by rotating the first polarizer. The percent contrast of the image is described by the following formula:

$$\text{percent contrast} = (I_b - I_i)/I_b \times 100\%$$

where I_i is the intensity of the image and I_b is the intensity of the background.

Hoffman modulation contrast microscopy lets us obtain good resolution without sacrificing the resolving power of the objective lens. Since the full aperture of the objective lens is used for Hoffman modulation contrast microscopy, not only is the full resolving power of the microscope utilized, but the depth of field is minimized. Thus Hoffman modulation contrast optics, like differential interference contrast optics, lets us optically section the specimen.

Hoffman modulation contrast microscopy offers several advantages:

● Good contrast is obtained at high resolution since the illumination is oblique and the full numerical of the sub-stage condenser and objective is utilized.

● Since the numerical aperture is high, the depth of field is small and we can optically section.

● Since the image is in black and white, we can use relatively inexpensive achromats or plan achromats with a monochromatic green filter.

● Birefringent specimens do not degrade the image as they do in differential interference contrast microscopy, and in Hoffman modulation contrast microscopy, the specimens can be observed in birefringent plastic Petri plates, making the observation of transparent culture cells possible. This is possible because in Hoffman modulation contrast microscopy, the two polarizers are beneath the specimen.

● We can do epifluorescence microscopy and modulation contrast microscopy sequentially without having to remove a Wollaston or Nomarski prism as we would have to do when using differential interference contrast optics.

● Hoffman modulation contrast optics is relatively inexpensive compared to differential interference contrast optics since we do not have to buy the relatively expensive Wollaston or Nomarski prisms.

● A type of dark-field microscopy can be done by aligning the slit with the 1 percent transmission region of the modulator.

The disadvantages of Hoffman modulation contrast microscopy are:

● We must be careful in interpreting an image of a vesicle as a hill or as a valley. The pseudo-relief image does not necessarily represent actual three-dimensional objects, but only gradients in optical path lengths.

● In interpreting the image, we must remember that maximal image contrast arises from objects oriented perpendicular to the length of the slit.

● Images are not rendered in color, although optical staining can be achieved in a homemade system by replacing the 1, 15, and 100 percent regions of the modulator with colored filters. For example, if the 1 percent area were blue, the 15 percent area gray, and the 100 percent area magenta, then gradients of optical path length in one direction will appear blue, gradients in the other direction will appear magenta, and areas without gradients will appear gray.

REFLECTED LIGHT HOFFMAN MODULATION CONTRAST MICROSCOPY

Hoffman modulation contrast optics can be used with reflected light to convert variations in surface contours into variations in image brightness (Hoffman and Gross, 1970). In reflected light Hoffman modulation contrast microscopy, just like reflected light differential interference contrast microscopy, the three-dimensional image is a true representation of the surface contour of the specimen.

The reflected light Hoffman modulation contrast microscope is set up in an optically similar way as the transmitted light version except that in the reflected light version, an epi-illuminator and half-silvered mirror are added (Figure 10-9).

In interpreting an image generated by reflected light Hoffman modulation contrast microscopy, the specimen is considered to be composed of reflecting surfaces, and the direction of the reflected ray is described strictly by the law of reflection given by geometric optical theory (see Chapter 2). The reflected rays pass through the brighter or darker region of the modulator, depending on the slope of the surface of the specimen (Figure 10-10).

THE SINGLE-SIDEBAND EDGE ENHANCEMENT MICROSCOPE

A single-sideband edge enhancement (SSEE) microscope is a transmitted light microscope that is able to change a gradient in optical path length into variations in intensity (Ellis, 1978). It is a special case of oblique illumination. It is not commercially available, but it can be built very easily (Figure 10-11).

The incident light in a single-sideband edge enhancement microscope passes through a collector lens and a field diaphragm and is focused onto an aperture diaphragm equipped with an adjustable semicircular half stop that is placed at the front focal plane of the sub-stage condenser. The adjustable half-stop is set to occlude one-half of the

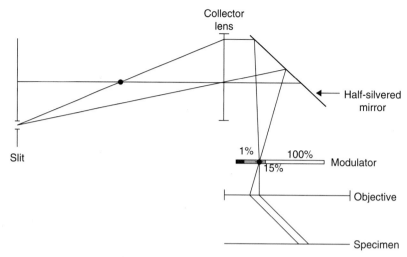

FIGURE 10-9 Diagram of the illuminating rays in a reflected light Hoffman modulation contrast microscope.

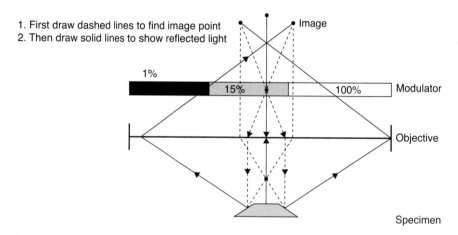

FIGURE 10-10 Ray diagram showing that contrast in a Hoffman modulation contrast microscope represents microtopography.

aperture diaphragm. The oblique light passes from the sub-stage condenser to the specimen. The light then passes into an ordinary objective. A relay lens then is used to project the back focal plane of the objective lens onto a carrier attenuation filter. The relay lens also moves the image of the specimen up the optical tube. There are two crossed polars, one above and one below the carrier attenuation filter.

The carrier attenuation filter is in a conjugate plane with the aperture diaphragm and the back focal plane of the objective. It is aligned so that the nondiffracted wave, which Ellis calls the carrier wave, passes through one half, and the positive or negative orders of diffracted waves, which Ellis calls the single sideband, passes through the other half. Ellis (1978) has made a variety of carrier attenuation filters that are used to enhance contrast. One filter, which is placed between parallel polars, is made up of two Polaroids oriented perpendicular to each other. The filter

can be rotated so as to differentially absorb either the carrier or the single-sideband light, thereby enhancing or reducing the contrast of various gradients in optical path length in the specimen.

The intensities of the image points formed by differential interference contrast and modulation contrast microscopes reflect the first derivative of the optical path lengths along a given azimuth. Computers excel in performing mathematical operations like taking derivatives, integrals, and Fourier transforms rapidly and inexpensively. In Chapter 14, I will discuss microscopes that include a video or digital camera so that an electrical or numerical signal that represents intensities and colors in the image can be passed into a computer. The computer can then perform a number of image processing functions, including taking the first derivative of the brightness in order to produce a pseudo-relief image.

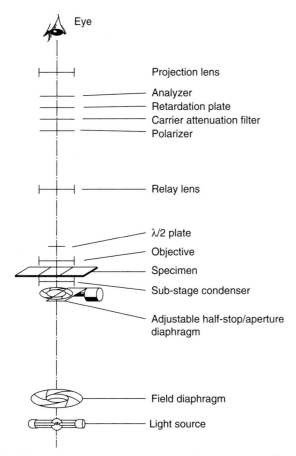

Eye

Projection lens

Analyzer
Retardation plate
Carrier attenuation filter
Polarizer

Relay lens

λ/2 plate
Objective
Specimen
Sub-stage condenser

Adjustable half-stop/aperture
diaphragm

Field diaphragm

Light source

FIGURE 10-11 A diagram of a single-sideband edge enhancement microscope.

I have shown you that the various properties of light (amplitude, phase, polarization, and wavelength) can be used effectively by a variety of microscopes to produce high contrast images of transparent biological specimens with great resolving power. The correct interpretation of these images provides us with quantitative information about the structure and physical properties of the specimen (e.g., mass, birefringence, entropy and enthalpy, thickness, absorption spectrum, density). This ends our ability to use geometric and physical optics to discuss image formation. In Chapter 11, in which I will discuss fluorescence microscopy, we will have to introduce the quantum theory of radiation (Einstein, 1905a, 1909, 1917; Dirac, 1927; Heitler, 1944; Finkelstein, 2003; Loudon, 2003; Zajonc, 2003).

Fluorescence Microscopy

The popular revival of light microscopy was due, in part, to the development of reflection fluorescence microscopy with its ability to localize enzymes, substrates, and genes, and its ability to characterize physicochemical properties of the cell, including membrane potentials, viscosity, pH, Ca^{2+}, Mg^{2+}, Na^+, and Cl^- with high resolution and contrast. While classical optics have been very useful in understanding the formation and interpretation of microscopic images formed by bright-field, dark-field, phase-contrast, polarization, interference, differential interference, and modulation contrast microscopes, it will fail us in understanding the formation of images in a fluorescence microscope. In order to understand image formation is fluorescence microscopes, we will have to explore the quantum nature of light. Color plates 20 and 21 provide examples of specimens observed with fluorescence microscopy.

DISCOVERY OF FLUORESCENCE

Perhaps fluorescence was first noticed by Nicolo Monardes, the physician from Seville who had published a book on the newly discovered medicinal plants from America (Boyle, 1664; Priestley, 1772; Harvey, 1957). In 1575, Monardes noticed that the wood of *Lignum nephriticum* (*Eysenhardtia polystachya*), when hollowed out into cups and filled with water, appeared to emit a bluish light. An extract of the wood, which was used to treat kidney diseases, also emitted the light. This was so spectacular that these cups were given to the royalty and visiting dignitaries for the next 100 years.

A century later, Robert Boyle (1664) noted that adding vinegar (or other acids) to the extract decreased the amount of fluorescence, whereas adding basic solutions (oil of tartar, solution of alum, spirit of hartshorn, or urine) restored the fluorescence. Boyle concluded that the color of the solution of *Lignum nephriticum* can be used to discern the acidity or alkalinity of a substance. This set the scene for using fluorescent compounds for measuring physiological properties of the living cell (e.g., pH).

Throughout history, philosophers have thought about how the color of a body is related to its fundamental composition or structure. While experimenting with a prism and illuminating objects with monochromatic light, Newton (1730) showed that the color of an object was not an absolute property of the object itself, but depended on the color of the illuminating light. He noticed that if an object looked red when illuminated with white light, it looked black when illuminated with anything but red light, indicating that the color of these objects was due to the color of light that was reflected from the object. Newton thought about this relationship and proposed that, "The bigness of the component parts of natural Bodies may be conjectured from their Colours." Sir David Brewster (1833a) continued to study the cause of natural colors and extracted chlorophyll from the Laurel (*Prunus Lauro-cerasus*) and 19 other plants with alcohol. He wrote:

> In making a strong beam of the sun's light pass through the green fluid, I was surprised to observe that its colour was a brilliant red, complementary to the green.... I have observed the same phenomenon in various other fluids of different colours, that it occurs almost always in vegetable solutions.... One of the finest examples of it which I have met with may be seen by transmitting a strong pencil of solar light through certain cubes of bluish fluor-spar. The brilliant blue colour of the intromitted pencil is singularly beautiful.

Brewster concluded that the absorption of rays by the atoms of a substance must play some role in the change in color. He wrote:

> The true cause of the colours of natural bodies may be thus stated: When light enters any body, and it is either reflected or transmitted to the eye, a certain portion of it, of various refrangibilities, is lost within the body; and the colour of the body, which evidently arises from the loss of part of the intromitted light, is that which is composed of all the rays which are not lost; or, what is the same thing, the colour of the body is that which, when combined with that of all the rays which are lost, compose the light. Whether the lost rays are reflected or detained by a specific affinity for the material atoms of the body, has not been rigorously demonstrated. In some cases of opalescence, they are either wholly or partly reflected; but it seems almost certain, that in all transparent bodies, and in that great variety of substances in which no reflected tints can be seen, the rays are detained by absorption.

Perhaps even more amazing than a green solution giving off red light, was to find a colorless solution that gave off blue light when irradiated with invisible ultraviolet light. John Herschel (1845a,b) observed a solution of quinine sulphate and found:

> Though perfectly transparent and colourless when held between the eye and the light, or a white object, it yet exhibits in certain aspects, and under certain incidences of the light, an extremely vivid and beautiful celestial blue colour, which from the circumstances of its occurrence, would see to originate in those strata which the light first penetrates in entering the liquid….

George Gabriel Stokes (1852, 1854) repeated Herschel's observation with sulphate of quinine. Stokes wrote, "It was certainly a curious sight to see the tube instantaneously lighted up when plunged into the invisible rays: it was literally darkness visible. Altogether the phenomenon had something of an unearthly appearance." Stokes irradiated the solution with variously colored light obtained by passing sunlight through a prism. He noticed that the emitted light always had a longer wavelength than the incident light. He wrote (1885):

> Perhaps the most striking feature in this phenomenon is the change in refrangibility of light which takes place in it, as a result of which visible light can be got out of invisible light, if such an expression may be allowed: that is, out of radiations which are of the same physical nature as light, but are of higher refrangibility than those that affect the eye; and in the same way light of one kind can be got out of light of another, as in the case for instance of an alcoholic solution of the green colouring matter of leaves, which emits a blood red light under the influence of the indigo and other rays. Observation shows that this change is always in the direction of a lowering.

Stokes called this phenomenon, where specimens absorb light of one wavelength and reemit it at a longer wavelength, *fluorescence*, after the mineral fluor-spar, which shows the same phenomenon. The phenomenon that the light emitted by fluorescent objects always has a longer wavelength than the light absorbed is now known as Stokes' Law. Frances Lloyd (1924) described the working situation where Stokes came up with the great law that bears his name:

> To grasp clearly the nature of fluorescence was the work of Sir George Stokes, who, my friend and colleague Professor A. S. Eve, tells me, was given to working in the 'back scullery and a small one at that', using the leaves of laurel and other plants which grew in his garden; and thus was led to the establishment 'of a great principle with accommodation and apparatus which would fill the modern scientific man with dismay'.

Stokes also postulated that fluorescence was related to phosphorescence. The only difference is that light given off by specimens that showed fluorescence stopped immediately after the incident light was shut off, whereas phosphorescent specimens continued to glow for relatively long periods of time after the incident light was removed. Indeed, with fluorescence, light emission stops almost immediately (within 10^{-8} s) after the cessation of the activating (or actinic) radiation, whereas with phosphorescence the emitted light persists for seconds, minutes, hours, days, or even months, depending on the material, after the cessation of the actinic radiation (Dake and De Ment, 1941).

Stokes (1852) tried to come up with a physical mechanism to describe how short wavelength light could turn into long wavelength light after it interacted with the fluorescent molecules. He weakly proposed that the incident light sent the atoms in a fluorescent molecule into a vibration and the light emitted from this vibration was of a longer wavelength. He did not like this conclusion, and believed that his explanation made no physical sense since it was physically impossible, according to classical wave theory, to get a short wavelength wave to give rise to a long wavelength wave. A better explanation had to await the development of quantum theory.

PHYSICS OF FLUORESCENCE

In Chapters 2 and 3, I discussed how Huygens (1690) and Newton (1730) realized that the various phenomena of light required taking into consideration that light and the ether through which it traveled, had the complementary properties of particles and waves. The distinction between Huygens' and Newton's ideas is that Huygens believed that light was a wave that traveled through a particulate ether, while Newton believed that light consisted of particles that traveled through a vibrating ether.

Up until the mid-nineteenth century, prominent scientists supported Newton's conception that light was particulate, and forgot about his need to include the wave nature of the ether for a full description of light (Anonymous, 1803, 1804; Young, 1804b). However, after Foucault's (1850) demonstration that the speed of light was faster in air than it was in water, Maxwell's (1865) miraculous equations that unified electricity and magnetism, and Heinrich Hertz's (1893) demonstration that electromagnetic waves can be transmitted, received, reflected, refracted, focused, polarized, and interfere with each other, the electromagnetic wave theory of light became the widely accepted theory of light. The need to incorporate Huygens' particulate nature of the ether, in general, was neglected and forgotten.

As Maxwell's equations were enjoying success in describing optical phenomena, the laws of thermodynamics were being applied to the study of radiation from hot bodies or incandescence. Initially, incandescent bodies served as model systems to understand solar radiation (Herschel, 1800a, 1800b; Leslie, 1804; Carson, 2004). However, Kirchhoff realized that there must be a universal function, dependent only on wavelength and temperature, and independent of the material in which an incandescent body is made, which describes the radiation (Kirchhoff, 1860, 1861; Draper, 1878; Houstoun, 1930). Black body radiators are black because they absorb all wavelengths of radiation, and when a black body radiator is heated to incandescence, it emits all wavelengths of radiation in a temperature-dependent manner (Figure 11-1).

Wilhelm Wien (1893, 1896, 1911; Drude, 1939) was interested in finding the universal function that described black body radiation and turned to Maxwell's (1897) kinetic theory of gases, which described the effect of temperature on the mean velocity and the distribution of velocities of molecules in a gas to use as an analogy to describe the effect of temperature on black body radiation (Figure 11-2). Wien made the following assumptions:

1. Maxwell's kinetic theory of gases, which describes the influence of temperature on the mean velocity and

distribution of velocities of gas molecules, should apply to the mean velocity and distribution of velocities of the atoms that make up the solid wall of a black body radiator.

2. The wavelength of light emitted by an atom depends only on the velocity of vibration of the atom. The wavelength is inversely proportional to the velocity and therefore the higher the velocity, the shorter the wavelength emitted.

3. The intensity of the light emitted at a given wavelength is proportional to the number of atoms vibrating at the velocity necessary to emit light at that wavelength.

Using these assumptions, Wien came up with an equation, which related the energy density (e_λ in J/m^3) of any given wavelength in a black body cavity to the temperature of the black body and Wien's equation was similar in form to Maxwell's velocity distribution equation. The formula, known as Wien's radiation law, required two constants to be dimensionally correct.

$$e_\lambda = (c_1/\lambda^5)(\exp(-c_2/\lambda T))$$
$$= (c_1/\lambda^5)(1/\exp(c_2/\lambda T))$$

where c_1 (in $J\,m$) and c_2 (in mK) were constants that could be determined empirically. Lummer and Pringsheim (1899a, 1899b, 1900) built a black body radiator that was optimal enough to test Wien's radiation law and found that it described the energy density of short wavelengths, but was inadequate to describe the energy density of long wavelengths (Figure 11-3; Kangro, 1976; Kuhn, 1978).

Other aspects of Maxwell's thermodynamics were used to find a theoretical foundation to the black body radiation curve. Lord Rayleigh (1900, 1905a, 1905b, 1905c), helped by Sir James Jeans (1905a, 1905b, 1905c, 1924) applied the principle of equipartition, which states that each particle in a group of identical particles collides with each other elastically, and as a consequence has a nearly identical energy. According to the wave theory of light, the energy of a wave

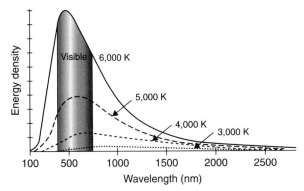

FIGURE 11-1 The emission spectrum of a black body radiator.

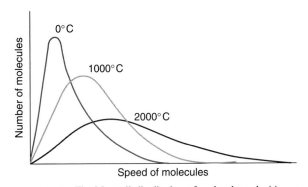

FIGURE 11-2 The Maxwell distribution of molecular velocities as a function of temperature.

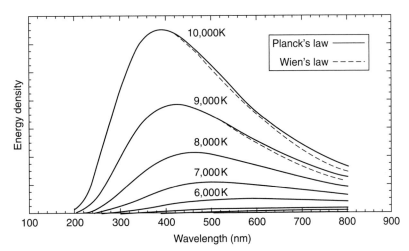

FIGURE 11-3 Wien's law describes the distribution of short wavelengths radiated from a black body but deviates from the experimental results at long wavelengths.

Number of half
wavelengths

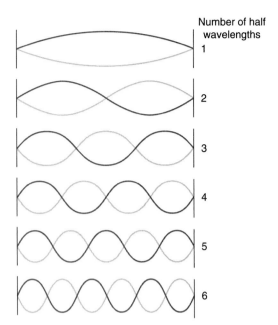

FIGURE 11-4 How waves of different lengths can be packed into a space. According to the wave theory of light, the energy contributed to the black body radiation by each wavelength is a function of the number of half-wavelengths in the space. The number of modes equals the number of half-wavelengths minus one.

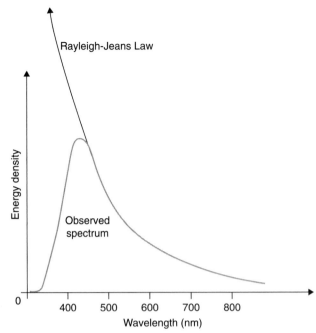

FIGURE 11-5 The "Ultraviolet catastrophe" that was predicted by the classical electromagnetic wave theory of light.

is proportional to the square of the amplitude. Following the principle of equipartition, Rayleigh assumed that the amplitude of waves of all wavelengths in the cavity would be the same. However, waves with shorter wavelengths would have more regions of maximal amplitude that fit in a given volume than waves with longer wavelengths. Usually, regions of zero amplitude within the cavity, called modes, are counted instead of regions with maximal amplitude. The number of modes equals the number of regions of maximal amplitude minus one (Figure 11-4). According to the boundary conditions of electromagnetic theory, standing electrical waves must have zero amplitude at the edge of the cavity. Thus if one half-wave with a wavelength equal to twice the length of the cavity would fit, two half-waves with a wavelength half as long would fit and three half-waves with a wavelength one-third as long would fit in the cavity. This trend goes on such that an infinite number of waves with an infinitely short wavelength would fit in the cavity.

Using the principle of equipartition of energy, Rayleigh and Jeans came up with an equation that related the energy density (e_λ in J/m^3 per given wavelength) to the temperature:

$$e_\lambda = (8\pi/\lambda^4)kT$$

where k is Boltzmann's constant (1.38×10^{-23} J/K), kT describes the energy in the cavity at a given temperature T (in Kelvin), and ($8\pi/\lambda^4$) describes the number of modes for a given wavelength. The Rayleigh-Jeans equation was able to explain the distribution of long wavelengths, but

failed in predicting the distribution of short wavelengths in that it predicted that a black body should give off an infinite amount of short wavelength light (Figure 11-5). Lord Kelvin (1904) described the inability of classical theory to describe black body radiation as one of the two "Nineteenth Century Clouds over the Dynamical Theory of Heat and Light," and in 1911, with 20/20 hindsight, Paul Ehrenfest gave the moniker, the "ultraviolet catastrophe" to the poor prediction.

In 1899, Max Planck saw that there was an element of truth in Wien's radiation law for short wavelengths and an element of truth in the Rayleigh-Jeans law for long wavelengths, but neither of them alone was capable of giving a theoretical foundation for understanding black body radiation or predicting what the spectrum would be at a given temperature. Planck (1949a, 1949b) reluctantly but courageously questioned the statistical assumptions upon which both laws were based. What if, he thought, all wavelengths do not share the thermal energy of the cavity equally and that shorter wavelengths are underrepresented? And what if the atoms in the wall did not give off radiation continuously, but in discreet packets he called quanta?

According to Planck's new hypothesis, black body radiation would be governed by two effects, one that was described by Wien's radiation law and one that was described by the Rayleigh-Jeans radiation law. First, since a greater number of short wavelength light can fit in a given volume compared with long wavelength light, there will be a tendency to fill the cavity with shorter wavelength light than longer wavelength light. Second, Planck postulated that the probability of the atoms in the walls radiating

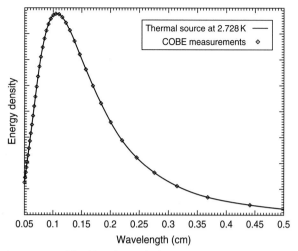

FIGURE 11-6 The black body distribution of the cosmic microwave background radiation. Note the identity of the experimental (◇) and theoretical (——) results.

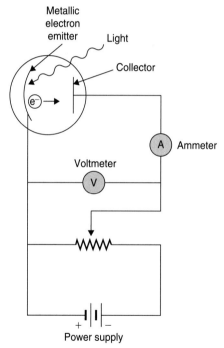

FIGURE 11-7 Diagram of the apparatus used to measure the photoelectric effect.

shorter wavelength light was less than the probability of radiating longer wavelength light. Planck proposed that an increase in the temperature increased the probability that short wavelength light would be radiated.

In order for the probability of radiation to be wavelength-dependent, Planck had to unwillingly assume that the energy did not flow continuously but was emitted in discrete quantities. This required using sums rather than integrals to account for all the wavelengths of the spectrum and gave a solution in the form of $1/(\exp[x]-1)$ instead of

$1/\exp[x]$ used by Maxwell and Wien. Planck visualized the oscillating atoms in the wall of the cavity as having discrete energy levels. That is, as the oscillating atom went from an excited state to a ground state, it lost energy in steps of one quantum, and at the same time, emitted the quantum of energy into one quantum of radiation. In order to achieve the curve generated by black body radiation, Planck assumed that the energy (E, in J) in one quantum of radiation was related to the frequency (ν) and wavelength (λ) of the radiated light by the following formula:

$$E = h\nu = hc/\lambda$$

where h is known as Planck's constant (6.626×10^{-34} Js) and c is the speed of light (2.99792458×10^8 m/s). Planck's assumption contrasts with the assumption of the wave theory of light where the energy is independent of the wavelength and proportional to the square of the amplitude.

Planck created an equation that related the energy density of a given wavelength to the temperature of the black body radiator. That equation, as follows, exactly fit the experimental data at the time and also fits the observational data currently being obtained for the cosmic microwave background radiation, which is a remnant of the incandescence of the universe that occurred shortly after the big bang (Figure 11-6). Planck's radiation law is:

$$e_\lambda = (8\pi/\lambda^4)(hc/\lambda)[1/(\exp(hc/(\lambda kT)) - 1)]$$
$$= (8\pi hc/\lambda^5)[1/(\exp(hc/(\lambda kT)) - 1)]$$

Notice that this equation is similar to Wein's except that $[1/(\exp(hc/\lambda kT) - 1)]$ replaces $[1/\exp(c_2/\lambda T)]$ and Wein's constants are replaced with the fundamental constants, h, c, and k. Wein's c_1 became $8\pi hc$ and Wein's c_2 became hc/k. Planck's radiation law, which accounts for the whole black body spectrum, reduces to the Rayleigh-Jean radiation law for long wavelengths, and to Wien's radiation law for short wavelengths (Richtmyer, 1928; Richtmyer and Kennard, 1942, 1947; Richtmyer et al., 1955, 1969; Ter Haar, 1967). The energy density can be converted into intensity in $J\,m^{-2}\,s^{-1}$ by multiplying the energy density by $c/4$.

Planck solved the problem of black body radiation by introducing quantization and assuming that the emitted energy was inversely proportional to wavelength. This really interested Albert Einstein since it implied that light itself was both particulate and wave-like. Einstein (1905a) proposed that not only is light emitted in quanta but also it is absorbed in quanta and travels in quanta that have energies equal to hc/λ. Einstein (1905a) asserted that this dual nature of light, which really goes back to Newton and Huygens, would be useful in understanding fluorescence. He also showed that the relationship between energy and wavelength was useful in understanding the photoelectric effect, which recently had been discovered by Heinrich Hertz and elucidated by Philipp Lenard in 1902 (Figure 11-7).

The photoelectric effect is the phenomenon where light causes the release of electrons from the surface of metals.

Although the classical theory predicted that the kinetic energy ($\frac{1}{2}mv^2$) of the released electrons should be proportional to the square of the amplitude of the incident light, Lenard found that the kinetic energy of electrons did not depend on the intensity (amplitude squared) of the incident light, but depended on its wavelength (Figure 11-8). Moreover, there was a threshold wavelength above which electrons were not ejected. At the threshold wavelength, electrons were ejected but virtually without any kinetic energy. As the light intensity increased more electrons were released but they had virtually no kinetic energy. The kinetic energy could be increased only by decreasing the wavelength of the light. As long as the wavelength was below the threshold, there was an immediate release of photoelectrons no matter how low the intensity of the light was (Serway et al., 2005).

If the wavelength of the incident light is short enough, the quantum of light has enough energy to overcome the electromagnetic binding forces that keep the electron in the metal. This amount of energy is known as the work function (w, in J) of the metal. If the wavelength is too long, the quantum does not have enough energy to overcome the work function. So no matter how many quanta arrive at the metal at one time, no electrons will be ejected unless the wavelength is short enough. Einstein suggested the following formula to explain the relationship between the kinetic energy (KE, in J) of the photoelectron and the wavelength of light:

$$KE = \tfrac{1}{2}mv^2 = hc/\lambda - w$$

Lenard measured the kinetic energy of the electron by making the electrical potential (V, in volts) of the collector plate more and more negative until the photoelectrons from the metallic emitter no long had enough energy to overcome the potential energy barrier created by the electric field. The kinetic energy was obtained by equating KE to eV, where e is the elementary charge of an electron (1.6×10^{-19} C). Robert Millikan (1917, 1924, 1935; Kargon, 1982) used this relationship to determine the value of Planck's constant.

Einstein (1917) stressed that particles of light have momentum equal to h/λ, and soon afterward, Arthur Compton (1929a, 1929b) observed that the wavelength of light scattered from electrons increased in a manner we would predict if light were composed of particles with energy ($E = \hbar\omega = h\nu$) and momentum ($p = \hbar k = h/\lambda$). The change of the wavelength upon scattering is known as the Compton Effect. Compton's experiment provided what was considered experimental proof for the existence of the particulate nature of light. Quanta are now known as photons, a name proposed by Gilbert Lewis. Interesting enough, Lewis (1926a, 1926b) gave the name photon to something that was not quite light itself. Lewis (1926b) took "the liberty of proposing for this hypothetical new atom, which is not light but plays an essential part in every process of

radiation, the name photon." Could he actually have given that name to the light-ether relation? Eventually, the ether could no longer be reconciled with the theory of special relativity and a photon came to be synonymous with a mathematical point of light. Lorentz (1924) argued that the photon had lost its spatial extension that gave light its wave-like properties.

Louis de Broglie (1922, 1924) and S. N. Bose (1924) independently realized that there was a logical inconsistency in the derivation of Planck's radiation law in that assumptions based on classical wave hypotheses, such as the number of degrees of freedom of the ether, were required to obtain the temperature-independent prefactor in Planck's quantum radiation law. De Broglie and Bose then worked out a strictly quantum derivation by characterizing the momentum of photons using a six-dimensional phase space. They assumed that the total number of cells in the phase space was equal to the number of possible ways of placing a quantum of light in the phase volume and then multiplied the answer by two since there were two azimuths of polarization. By doing so they obtained the prefactor of Planck's radiation law. The prefactor can also be derived in the following reasonable, although nontraditional way.

A photon is moving through a vacuum at velocity c, has an equivalent mass (m) equal to hc/λ, which is obtained by equating the photon's relativistic mass-energy to its light energy ($mc^2 = hc/\lambda$). Once the photon's equivalent mass is established, it is possible to combine classical and quantum theory to extend the mechanical idea of a photon and determine its radius (r) from its angular momentum (L). According to quantum mechanics, the angular momentum of all photons is $\hbar$ ($= h/2\pi$) and according to classical physics, the angular momentum is equal to mvr, where m is the mass of a particle, v is its angular velocity, and r is its radius. Since $v = \omega r$, then $\hbar = m\omega r^2$. Substituting $2\pi\hbar/(c\lambda)$ for m, we get $r = 1/k$, since $k = \omega/c = 2\pi/\lambda$. That means that the radius of a photon is equal to the inverse of its angular wave number or its wavelength divided by 2π.

As mentioned in Chapters 3 and 8, the phenomena of diffraction and interference suggest that light has a wave width as well as a wavelength. Given the radius of the photon given earlier, the cross-section of a photon would be $\pi(\lambda/2\pi)^2$. If the photon oscillated longitudinally with a length between 0 and λ (see Appendix II), the average length of a photon would be $\lambda/2$, and the average volume (V_λ) of a photon with wavelength λ would be:

$$V_\lambda = \pi(\lambda/2\pi)^2 \, \lambda/2 = \lambda^3/8\pi$$

The reciprocal volume of a photon would be:

$$1/V_\lambda = 8\pi/\lambda^3$$

The energy density of a photon with an energy of hc/λ would be $8\pi hc/\lambda^4$ and the energy density per given wavelength would be $8\pi hc/\lambda^5$. According to this calculation, for

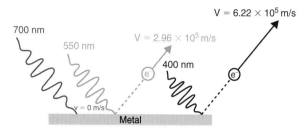

FIGURE 11-8   In the photoelectric effect, the kinetic energy of the emitted electron is a function of the frequency of the incident light.

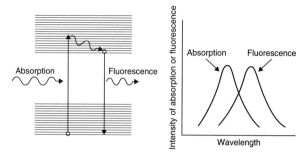

FIGURE 11-9   Energy diagram of a fluorescent molecule. The absorption and fluorescent emission spectrum of the molecule are shown on the right.

a given energy in a black body cavity, the prefactor represents how many photons of a given wavelength can fit into the volume of the cavity. The greater the temperature in the cavity, the greater the number of photons that are created, the smaller the volume of those photons, and the greater the energy density. Assuming that the radiant energy density per unit wavelength distributed within a black body cavity is quantized into photons with a real volume, and assuming the probability of producing those photons is $[1/(\exp(hc/\lambda kT) - 1)]$, the energy density per unit wavelength of the cavity would be:

$$e_\lambda = (8\pi hc/\lambda^5)[1/(\exp(hc/\lambda kT) - 1)]$$

which is the Planck radiation law.

In the early years of the quantum theory, many physicists proposed models of the structure and properties of the photon, yet the photon still remains an enigma (Lamb, 1995), being considered alternately as a mathematical point and an infinite plane wave. Some models propose that the photon is not an elementary particle, but a compound structure composed of two complementary elementary particles (Bragg, 1907a, 1907b, 1907c, 1911; Bragg and Madsen, 1908; de Broglie, 1924, 1932a, 1932b, 1932c, 1933, 1934a, 1934b, 1934c, 1946; Jordan, 1928, 1935, 1936a, 1936b, 1936c; Jordan and de L. Kronig, 1936; de Broglie and Winter, 1934; Born and Nagendra Nath, 1936a, 1936b; Nagendra Nath, 1936; de L. Kronig, 1935a, 1935b, 1935c; Dvoeglazov, 1999; Valamov, 2002; Appendix II). Currently, the most complete description of the interaction with light and matter is given by the theory of quantum electrodynamics (Feynman, 1988).

When the light emitted by an excited atom is passed through a prism or a diffraction grating, the light is split into a series of discrete bands known as a line spectrum (Figure 11-8; Nernst, 1923; Sommerfeld, 1923; Pauling and Goudsmit, 1930; Herzberg, 1944; Hund, 1974; Bohm, 1979). In 1885 Balmer came up with a formula that fit the observed wavelengths of light emitted from hydrogen (656.210, 486.074, 434.01, and 410.12 nm). His formula, in modern notation, is:

$$1/\lambda = R[1/n_f{}^2 - 1/n_i{}^2]$$

where n_i stands for an integer that represents the initial position, n_f stands for an integer that represents the final position,

and R is the Rydberg constant ($1.0973732 \times 10^7 \mathrm{m}^{-1}$). By applying Planck's quantization ideas to the quantization of angular momentum of electrons in orbitals, Einstein's concept of the photoelectric effect, in which the energy of a photon is related to the binding energy of an electron, and select principles of classical mechanics, Niels Bohr (1913) derived Balmer's formula, replacing the empirical Rydberg constant with fundamental constants:

$$1/\lambda = (m_e e^4)/(8\varepsilon_o{}^2 h^3 c) \, [1/n_f{}^2 - 1/n_i{}^2]$$

where e is the elementary charge ($1.6 \times 10^{-19}\mathrm{C}$), m_e is the mass of an electron ($9.1 \times 10^{-31}\mathrm{kg}$), ε_o is the electrical permittivity of a vacuum ($8.85 \times 10^{-12}\mathrm{F/m}$), h is Planck's constant, and c is the speed of light ($3 \times 10^8\mathrm{m/s}$).

Atomic absorption results in the transfer of an electron from a low energy ground state (n_i) to a higher energy excited state ($n_f > n_i$) in a process that takes about one period of light vibration ($\sim 10^{-15}\mathrm{s}$). The negative value of $1/\lambda$ indicates that energy is added to the atom in the form of a photon. Emission occurs when an electron falls from the excited state (n_i) to the ground state ($n_f < n_i$). The absorption spectrum and the emission spectrum of a gaseous atom are identical. The wavelength of emitted light gives a signature of the energy differences between electrons in the ground and excited states (Figure 11-9).

$$\lambda = hc/[E_m - E_{ground}]$$

The emitted wavelength depends on the energy difference between the excited state and the ground state according to the following formula:

$$E_{excited} - E_{ground} = \hbar kc = \hbar\omega = h\nu = hc/\lambda$$

When gaseous atoms are combined together into gaseous molecules, the electrons are shared between nuclei of different atoms and form molecular orbitals. The nuclei that share an electron can vibrate and rotate relative to each other. Consequently, complex molecules have many vibrational and rotational states and form band spectra instead of line spectra. Gaseous atoms give relatively clean spectra that correspond to transitions in the ultraviolet and visible region, and gaseous molecules give relatively clean band spectra that correspond to transitions that correspond to ultraviolet, visible, and infrared wavelengths. The spectra of liquids or solids, on the other hand, become broadened because a range of transition energies result from the interactions between molecules. The various lines and bands become overlapping and the spectrum appears as a continuous spectrum. In solids, the spectrum appears as a continuous band as described by Planck's radiation law.

A flexible molecule has many vibrational states and rotational states (Figure 11-9). Consequently, the excited state of a flexible molecule can dissipate energy in a variety of ways, which takes 10^{-15} to 10^{-9} s. Initially, the electronic energy can be conserved within the molecule, in a process known as internal conversion or radiationless transfer, where the electronic energy is converted to kinetic energy, which accompanies the vibrational and rotational movement of the molecule. Eventually, the kinetic energy is completely lost to the surround through collisions or as thermal energy with wavelengths of $2.5 - 100 \times 10^{-6}$ m for each vibrational transition and wavelengths of $50 - 350 \times 10^{-6}$ m for each rotational transition.

Once an electron reaches the lowest vibrational or rotational level of the excited state, it can return to the ground state only by emitting a photon in a process known as fluorescence, which takes about 10^{-8} s. Because some of the original radiant energy is converted to kinetic energy, the wavelength of the emitted photon is greater than the wavelength of the absorbed photon. This is the reason behind Stokes' Law and the basis for fluorescence microscopy.

The ratio between the number of photons absorbed and the number of photons emitted by fluorescence is called the fluorescence quantum yield (Q_{FL}). The fluorescence quantum yield ranges between 0 and 1, and depends on the chemical nature of the fluorescing molecule, fluorophore or fluorochrome, the excitation wavelength, and other factors, including pH, temperature, hydrophobicity, and viscosity. The greater the fluorescence quantum yield, the better the dye is for fluorescence microscopy.

The fluorescence quantum yield can be decreased by quenching or photobleaching. Quenching is the decrease in fluorescence as a result of the transfer of energy from the excited state to the ground state through a radiationless pathway between molecules. This occurs when the excitation and emission peaks overlap, and the fluorescent light given off by one molecule is reabsorbed by another. Bleaching results from the destruction of the molecule directly by light or though the light-induced production of free radicals. Bleaching can be minimized by adding antifade agents like propylgallate, (0.05%) phenylenediamine, or (0.05%) other antioxidants, including ascorbic acid to the preparation.

Understanding the physics of fluorescence helps in the design of fluorescent dyes that are both selective and bright as well as high resolution fluorescence microscopes that introduce enough contrast to allow the detection of a single fluorescing molecule.

DESIGN OF A FLUORESCENCE MICROSCOPE

In 1903, Siedentopf and Zsigmondy (1903) developed the ultraviolet microscope in order to "beat" the limit of resolution set by visible light and visualize colloidal particles (Siedentopf, 1903; Cahan, 1996). In 1911, Heinrich Lehmann and Stanislaus von Prowazek used this as the basis for the first fluorescence microscope (Rost, 1995).

The first obstacle to overcome in developing a fluorescence microscope was to find a light source that was intense enough in the ultraviolet range. Initially carbon arcs were used, but they were replaced with mercury vapor lamps, xenon lamps, and lasers. In a fluorescence microscope, the excitation light is passed through an excitation filter that should match the excitation spectrum of the fluorochrome. The excitation filter can be a colored absorption filter, which passes the light that is not absorbed by the dyes in the filter or an interference filter, which passes the wavelengths that constructively interfere and reflect the ones that destructively interfere.

There are three methods of illuminating the specimen found in fluorescence microscopes (Figure 11-10): (1) the bright-field method, which was originally made by Zeiss in 1911; (2) the dark-field method, which was originally made by Reichert in 1911; and (3) the reflected light method, which was first made commercially by Zeiss around 1930. The bright-field fluorescence microscope does not block enough of the excitation light, so the image contrast is low. The contrast can be increased by using a dark-field type of sub-stage condenser, but the numerical aperture of the objective must be reduced, thus limiting resolution. The reflected light or epi-illumination fluorescence microscope permits high contrast and high resolution without compromise (Ploem and Tanke, 1987; Oldfield, 1994).

Since fluorescent light is often extremely weak compared to the excitation light, a good fluorescence microscope must filter out the excitation light completely in order to distinguish the fluorescent light. This is relatively easy since the wavelengths of fluorescent light are typically longer than the excitation wavelengths, as described earlier. In epi-fluorescence microscopes, the light filtering is done by a chromatic (or dichromatic) beam splitter and a barrier filter. The chromatic beam splitter works on the principle of constructive and destructive interference. The chromatic beam splitter is a typical interference filter tipped at a 45-degree angle. At this angle, 90 percent of the light with wavelengths shorter than a certain value is reflected and 10 percent is transmitted. Similarly, 90 percent of the fluorescent light, which is longer than the cut off is transmitted

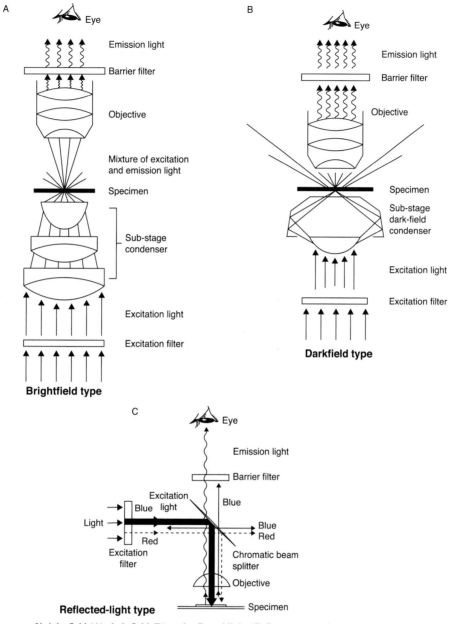

FIGURE 11-10 Diagrams of bright-field (A), dark-field (B), and reflected light (C) fluorescence microscopes.

through toward the ocular, and 10 percent is reflected back to the light source. Because of this property, a chromatic beam splitter is more efficient than a standard half-silvered mirror (Mellors and Silver, 1951; Ploem, 1967). Figure 11-11 is a ray diagram that shows the aperture and field planes in an epi-fluorescence microscope.

The fluorescent light that passes through the chromatic beam splitter is then filtered again by a barrier filter that further removes the excitation light. This filter can be a long pass filter, a wide bandpass, or a narrow bandpass interference filter. The more selective the filter, the dimmer the image will be. Consequently, we must select filters with bandwidths based on the tradeoff between brightness and selectivity.

The original fluorescence microscopes used ultraviolet light for excitation and these microscopes required fluorite objectives. However, ultraviolet photons are energetic and are often deadly to cells. Consequently, dyes, whose excitation spectrum falls in the visible range, have been developed, making high numerical aperture apochromatic objective lenses more desirable. Moreover, the numerical aperture of the objective lens is especially important in the epi-illumination mode, since the objective lens is used both as the condenser and the objective, and the brightness of the image depends on the fourth power of the numerical aperture divided by the square of the magnification.

Mary Osborn and Klaus Weber (1982) have developed a unique technique to make stereo pairs of immunofluorescence

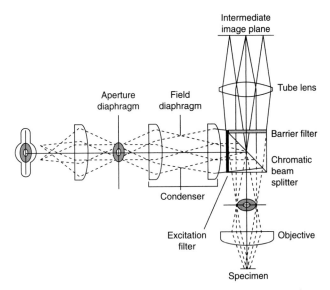

FIGURE 11-11 Ray diagram of a reflected light fluorescence microscope showing the conjugate aperture and field planes.

micrographs that combines oblique illumination with epi-fluorescence. It requires a commercially available Zeiss intermediate ring that is used to increase the tube length for a given objective. It is inserted between the objective and the nosepiece. It also requires a stereo insert, which can be constructed by you by welding a half moon diaphragm into a Zeiss differential interference contrast slider. Put the slider into the intermediate ring so it covers half the field in the rear of the objective at a time. In this way the specimen is illuminated with oblique illumination and only half the diffraction orders are collected to make the image. After photographing the specimen with the moon in one position, the half moon is put in the opposite position and the specimen is photographed again. Rotate the two micrographs by 90 degrees and then mount them side by side. When viewed with a 2x stereo viewer, they give a three-dimensional image.

FLUORESCENCE PROBES

Many cells have native molecules that are fluorescent. For example, chlorophyll fluorescence can be used to study the development of fern spores (Scheuerlein et al., 1988) and lignin fluorescence can be used to study the formation of secondary walls. However, the real effectiveness of the fluorescence microscope in understanding the dynamics of cell structure and function is due in large part to the development of sensitive and selective artificial dyes, in the form of free molecules or a quantum dots (Taylor et al., 1992; Tsien et al., 2006).

With bio-organic chemists developing dyes with engineered properties that relate to fluorescence and binding, the sky is almost the limit in finding a dye that will perform a desired function. Fluorescent dyes that accumulate

only in living cells have been developed as a quick way of determining cell viability. Other fluorescent dyes are able to selectively stain certain cell organelles and cytoplasmic fibers (Haigler et al., 1980; Morikawa and Yanagida, 1981; Wick et al., 1981; Wolniak et al., 1981; Wang et al., 1982; Terasaki et al., 1984, 1986; Pagano and Sleight, 1985; Parthasarathy, 1985; Parthasarathy et al., 1985; Upsky and Pagano, 1985; Matzke and Matzke, 1986; Lloyd, 1987; Wu, 1987; Pagano, 1988, 1989; Chen, 1989; Mitchison, 1989; Pringle et al., 1989; Quader et al., 1989; Terasaki, 1989; Wang, 1989; McCauley and Hepler, 1990; Zhang et al., 1990). Dyes, whose fluorescent excitation spectra, emission spectra, and/or quantum yield depend on their local environment, have also been developed to measure viscosity, hydrophobicity, membrane potentials, pH, Ca^{2+}, Cl^-, Na^+, K^+, or Mg^{2+}, and such (Saunders and Hepler, 1981; Grotha, 1983; Liu et al., 1987). High molecular mass fluorescent probes can be used to measure permeability (Luby-Phelps et al., 1986, 1988; Luby-Phelps, 1989; Hiramoto and Kaneda, 1988). Even higher molecular mass dyes that are unable to permeate the plasma membrane have been used as an assay for endocytosis (Ginzburg et al., 1999). Enzymatic substrates that are conjugated to dyes that become fluorescent only after an enzyme acts on the substrate can be used to visualize enzymatic reactions in the cell.

Many marine organisms, including the jellyfish (*Aquorea victoria*), the anemone (*Discosoma striata*), and the anthrozoans (*Entacmaea quadricolor* and *Anemonia majano*) are luminescent, in part because they contain proteins that fluoresce (Harvey, 1920, 1940; Barenboim et al., 1969; Morise et al., 1974; Prendergast and Mann, 1978; Zimmer, 2005; Pieribone and Gruber, 2006). The gene sequences that code the fluorescent proteins have been cloned (Prasher et al., 1992). The DNA sequence for the fluorescent protein can be inserted into a gene of interest so that the engineered gene produces a fluorescent chimerical protein. In this way the dynamics of the chimerical protein, which hopefully represents the dynamics of the native protein, can be visualized in a fluorescence microscope (Chalfie et al., 1994). Originally only green fluorescent protein from the jellyfish was used, but now many colored proteins having emission spectra that span the visible spectrum have been isolated and varied by mutagenesis (Shaner et al., 2005). This makes it possible for cells to express many fluorescent proteins at a time. Genetically targetable fluorescent proteins can be used to measure intracellular calcium (Palmer and Tsien, 2006). Again, the sky is the limit (Tsien, 2003, 2005; Giepmans et al., 2006)!

Quantum dots are semiconductors made out of silicon- or carbon-based crystals that are 2 to 10nm in diameter and are composed of 100 to 100,000 atoms. Quantum dots are almost like artificial "designer" atoms with electrons confined to specific orbitals whose sizes depend on the size of the quantum dots (Reed, 1993). The more an electron is confined in a space like an atom or a quantum dot, the higher its energy is. When an excited confined electron returns to the ground state, it gives off light whose color and energy are determined by the size of the quantum dot. Quantum dots can be targeted to specific protein in the cell to localize those proteins (Howarth et al., 2005).

PITFALLS AND CURES IN FLUORESCENCE MICROSCOPY

When doing fluorescence microscopy, it is important always to test for autofluorescence under the same conditions you use to observe an introduced fluorochrome. This control lets you make the claim that the observed fluorescence is due to the added fluorochrome. Moreover, since fluorescence microscopy often requires an intense light source, molecules in the cell can be altered, and this may alter the autofluorescence in a time-dependent manner. We can minimize autofluorescence problems by using more selective excitation and/or barrier filters or by using a dye that fluoresces in a different region of the spectrum than the autofluorescent molecules. Autofluorescence is not always a problem. van Spronsen et al. (1989) used the autofluorescence of chlorophyll to image the structure of the grana in living chloroplasts.

Richard Williamson (1991) pointed out a very interesting and important point when it comes to resolution in the fluorescence microscope. We can localize proteins that are smaller than the limit of resolution, but resolving where they are and how they are arranged is still limited by diffraction. For example, using antibodies or green fluorescent protein, we can fluorescently stain microtubules that are normally 24 nm in diameter. The image is inflated through diffraction to the limit of resolution of the lens, which is about 200 nm. That is, if two microtubules in reality are about 100 nm apart (4 diameters), they will appear as one microtubule. Therefore, unconnected microtubules may appear as branched microtubules, or overlapping microtubules may appear as one very long one. So fluorescence allows the detection but not resolution of 24 nm microtubules. We need an electron microscope to resolve them.

The image in a fluorescence microscope is often dim. This can be rectified by using a brighter lens with a higher numerical aperture and/or a lower magnification. It may be rectified by using a fluorite lens, if we are using ultraviolet excitation. The dimness problem can also be rectified by using lasers in a confocal microscope (see Chapter 12) and/or a more sensitive digital imaging system (see Chapter 13). If photographic images are underexposed, set the exposure meter in the camera to the setting that is appropriate to photograph scattered bright objects in a dark field (see Chapter 5).

Quenching of the dye can be remedied by decreasing the dye concentration, and bleaching of the dye can be minimized by using antifade agents.

If there is too much background light, reduce the dye concentration or focus the excitation light for critical illumination by focusing the arc at the level of the specimen. If traditional methods fail, use confocal microscopy (see Chapter 12) and/or digital image processing (see Chapter 14).

WEB RESOURCES
Fluorescent Microscopy

Molecular Expressions web site: http://micro.magnet.fsu.edu/primer/techniques/fluorescence/fluorhome.html
Nikon Microscopy U: http://www.microscopyu.com/articles/fluorescence/fluorescenceintro.html
Olympus Microscopy Resource Center: http://www.olympusmicro.com/primer/techniques/fluorescence/fluorhome.html
Leica Microsystems: http://www.fluorescence-microscopy.com

Fluorescent Dyes

Tsien Laboratory: http://www.tsienlab.ucsd.edu/Default.html
Invitrogen: http://www.probes.comwww.probes.com
Fluorophores.org: http://www.fluorophores.org
Green Fluorescent Protein Applications Page: http://www.yale.edu/rosenbaum/gfp_gateway.html
Fluorescence Microscopy and Fluorophores: http://www.micro-scope.de/fluoro.html

Interference Filters

Semrock: http://www.semrock.com/Catalog/BrightlineCatalog.html
Omega Optical: http://www.omegafilters.com
Chroma: http://www.chroma.com
Zeiss: http://www.micro-shop.zeiss.com/us/us_en/spektral.php?cp_sid=&f=db

Chapter 12

Various Types of Microscopes and Accessories

Microscopes have allowed us to peer into the world of small things and also have allowed us to see the microscopic building blocks of the macroscopic world. In this chapter, I describe how optical microscopes can be combined with lasers and centrifuges to study better the microscopic nature of the world. I will describe how other electromagnetic waves from X-rays through radio waves can be used to probe the structure of biological specimens. I will also describe how the Fresnel diffraction pattern, as opposed to the Fraunhöfer diffraction pattern, can be captured to form a high resolution image that beats Abbe's diffraction limit. I will also describe how longitudinal sound waves can be used in an acoustic microscope to image the viscoelastic properties of biological specimens. I will close by describing microscope accessories that help us study the microscopic world. The descriptions of these techniques are cursory, by necessity because my experience is limited. Even though this chapter borders on the limits of my knowledge, I present this information to my students to give them an idea of the unlimited potential of microscopy in forming a variety of images, each of which captures a different aspect of the reality of the specimen.

CONFOCAL MICROSCOPES

The first confocal microscope was invented and built by Marvin Minsky (1988) in 1955. Minsky was interested in taking a "top-down" approach to understand the brain, but when he looked at the whole brain, he saw nothing:

> And here was a critical obstacle: the tissue of the central nervous system is solidly packed with interwoven parts of cells. Consequently, if you succeed in staining all of them, you simply can't see anything. This is not merely a problem of opacity because, if you put enough light in, some will come out. The serious problem is scattering. Unless you can confine each view to a thin enough plane, nothing comes out but a meaningless blur. Too little signal compared to the noise: the problem kept frustrating me.

Minsky had a chance to overcome this problem as a Junior Fellow at Harvard:

> This freedom was just what I needed then because I was making a change in course. With the instruments of the time so weak,

there seemed little chance to understand brains, at least at the microscopic level. So, during those years I began to imagine another approach.... In the course of time, that new top down approach did indeed become productive; it soon assumed the fanciful name, Artificial Intelligence.... Artificial Intelligence could be tackled straight away—but my ideas about doing this were not yet quite mature enough. So (it seems to me in retrospect) while those ideas were incubating I had to keep my hands busy and solving that problem of scattered light became my conscious obsession. Edward Purcell ... obtained for me a workroom in the Lyman laboratory of Physics.... (That room had once been Theodore Lyman's office. Under an old sheet of shelf paper I found a bit of diffraction grating that had likely been ruled, I was awed to think, by the master spectroscopist himself.) One day it occurred to me that the way to avoid all that scattered light was to never allow any unnecessary light to enter in the first place. An ideal microscope would examine each point of the specimen and measure the amount of light scattered or absorbed by that point. But if we try to make many such measurements at the same time then every focal image point will be clouded by aberrant rays of scattered light deflected [by] points of the specimen that are not the point you're looking at. Most of those extra rays would be gone if we could illuminate only one specimen point at a time. There is no way to eliminate every possible such ray, because of multiple scattering, but it is easy to remove all rays not initially aimed at the focal point; just use a second microscope ... to image a pinhole aperture on a single point of the specimen. This reduces the amount of light in the specimen by orders of magnitude without reducing the focal brightness at all. Still, some of the initially focused light will be scattered by out-of-focus specimen points onto other points in the image plane. But we can reject those rays, as well, by placing a second pinhole aperture in the image plane that lies beyond the exit side of the objective lens. We end up with an elegant, symmetrical geometry: a pinhole and an objective lens on each side of the specimen. (We could also employ a reflected light scheme by placing a single lens and pinhole on only one side of the specimen-and using a half-silvered mirror to separate the entering and exiting rays.)

In the 30 years in which confocal microscopes have been produced commercially, microscopists all over the world have continued blazing the trail initiated by Minsky by using their knowledge of the physical nature of light to increase the resolving power and application of the confocal microscope (Shack et al., 1979; Hall et al., 1991; Denk et al., 1990; Pawley, 1990; Wilson, 1990; Hell et al., 1994).

The ability to illuminate a single point on the specimen with sufficient intensity was facilitated by the introduction of lasers as illuminating sources for microscopes. Lasers provide high intensity, coherent, monochromatic light. The laser light in a confocal microscope, in contrast to a wide-field microscope set up with Köhler illumination, is focused on the specimen or field plane as opposed to the aperture or Fourier plane. In this way, the illumination used in confocal microscopes is similar to critical illumination (see Chapter 4). However, in confocal microscopes, only a tiny region of the specimen, close to the limit of resolution, is illuminated at a given time. In a wide-field microscope with either Köhler or critical illumination, all points in all the parts and levels of a specimen are simultaneously illuminated so that each point gives off light that obscures the neighboring details, but in confocal microscopes, the obscuring light is rejected in the formation of each image point. This characteristic of confocal microscopes is particularly important when viewing thick specimens. Images taken with a confocal microscope seem to be perfectly thin optical sections (Jovin and Arndt-Jovin, 1989). A series of images taken at different axial levels can be combined by a digital image processor to reconstruct a true three-dimensional image.

The term confocal relates to the fact the plane containing the illumination pinhole, the plane containing the specimen, and the plane containing the detector pinhole are all conjugate planes (Figure 12-1). Consequently, the image of the illumination pinhole is in focus at the specimen plane and at the detector plane. Although confocal microscopes can be used in the transmission mode, usually they are used in the reflected light mode. Moreover, any kind of microscopy can be done with a confocal microscope, but in practice, confocal microscopes typically are used for fluorescence microscopy. Using a confocal microscope to do fluorescence microscopy is so widely used because the pinhole eliminates much of the out-of-focus fluorescence that reduces the contrast in the image (Figure 12-2).

In a confocal scanning fluorescence microscope, light from a laser passes through a pinhole, which is at the front focal plane of a converging lens. Axial light emerges from the lens and is reflected by a chromatic beam splitter through the objective lens to irradiate a point in the specimen, which is placed at the focus of the objective lens. The fluorescent light that is emitted from that point passes back through the objective lens as parallel light. It then passes straight through the chromatic beam splitter and through a barrier filter. After leaving the barrier filter, the light passes through a projection lens where it is focused on a pinhole adjacent to the light sensitive region of a photomultiplier tube. The output voltage of the photomultiplier tube then is digitized and stored in a computer memory along with the x,y,z coordinates of the specimen point.

The image is created point by point from the scanned object, and the scan can be accomplished in two general ways. Either the specimen can be moved through the beam, as was done in Minsky's original confocal microscope, or the beam can be moved over the specimen. Either way, the x,y,z coordinates of the specimen point have to be stored along with the intensity information. The specimen

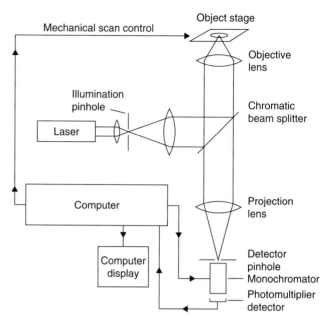

FIGURE 12-1 Diagram of a confocal microscope showing the conjugate field planes.

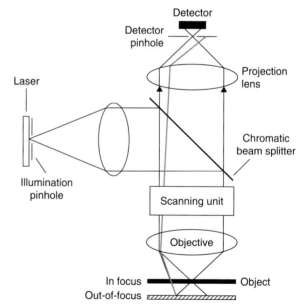

FIGURE 12-2 In a confocal microscope, the out-of-focus fluorescence is blocked by the pinhole.

is moved through the beam by stepper motors connected to the stage with a reliability of 10 to 20 nm, about one-tenth of the expected optical resolution. The advantage of moving the specimen is that the excitation and fluorescent light is propagated on-axis through the objective lens and light going to and coming from each point in the specimen experiences the same optical conditions (Brakenhoff et al., 1989). Moreover, the field of view in an on-axis confocal microscope is limited only by the mechanical scanning

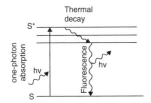

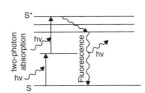

FIGURE 12-3 Energy diagram illustrating one- and two-photon absorption.

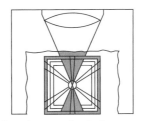

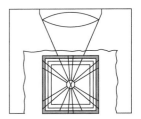

One-photon confocal microscope   Two-photon confocal microscope

FIGURE 12-4 Diagrams illustrating the depths to which a one- and two-photon confocal microscope excites a specimen. In the two-photon microscope there are only enough photons to excite the fluorochrome at the x.

distance of the stage. The disadvantage of a moving stage is that the stage must be moved relatively slowly in order not to deform the specimen. The slow movement limits the ability of a moving stage confocal microscope to follow rapid cellular processes at high resolution.

It has proved difficult to keep the specimen stationary and move the optical system so that rapid on-axis imaging can be accomplished without deforming the specimen. Consequently, most manufacturers build confocal microscopes that scan the specimen by moving the light spot using a series of moving mirrors, or slits in a rotating disk. This is known as off-axis imaging. In off-axis imaging, the incident and fluorescent light going to and coming from each point in the specimen experiences a different optical path. Moreover, the size of the field is limited by the objective lens. The advantages of off-axis imaging are that it allows fast acquisition times and does not deform the specimen.

Denk et al. (1990) developed a two-photon confocal laser scanning fluorescence microscope that uses a colliding-pulse, mode-locked dye laser that produces a stream of pulses that have a duration of 100 fs and a repetition rate of 80 MHz. With this laser, it is possible to pump photons into a molecule so quickly that a molecule will absorb two long wavelength photons nearly simultaneously, thus providing an electron with the equivalent energy of a single short wavelength photon (Figure 12-3). The idea for the two-photon confocal microscope came from knowledge of studies on two-photon absorption in photosynthesis (Shreve et al., 1990).

The two photon excitation technique has many advantages. The first is that the ability to optically section is improved because two photons must be captured nearly simultaneously in order to excite the fluorochrome. This means that the probability of fluorescence emission is proportional to the square of the excitation light intensity instead of the intensity itself. In general, the probability of emission is given by the following equation:

$$\text{probability of emission} = \Phi_{Fl}(\text{excitation intensity})^n$$

where Φ_{Fl} is the fluorescent quantum yield, and n is the number of photons needed to excite the molecule. The intensity of the laser is set so that the only plane where the intensity is great enough to excite the fluorochrome is at the very focus of the laser beam, thus minimizing any

out-of-focus images that would lower the contrast and obscure the image (Figure 12-4). The tight focusing of the excitation light can be seen in experiments set up to bleach dyes. The area bleached by a typical one-photon confocal microscope is enormous compared to the area bleached by a two-photon confocal microscope. Another advantage of the two- or multiphoton excitation process is that the multiphoton excitation spectra of dyes are wider than they are for dyes excited by single photons (Xu and Webb, 1996). This makes it possible to excite a number of fluorochromes at the same time with the same infrared laser.

Interestingly, in a multiphoton confocal microscope, the light used to excite the fluorochromes has a longer wavelength than the emission light, and consequently, a chromatic beam splitter that reflects long wavelength light and transmits short wavelength light must be used.

The resolving power of the confocal microscope is being improved by a technique known as stimulated emission depletion microscopy (STED). Stimulated emission depletion microscopy reduces the size of the excited region by following the excitation pulse with a ring-shaped depletion pulse, which surrounds the excitation pulse and is tuned to the emission wavelength of the dye. The depletion pulse causes most of the electrons except those in the very center of the excitation pulse to fall from the excited state to the ground state, giving off fluorescent light by stimulated emission. Whereas the fluorescence from the excitation pulse enters the aperture, the stimulated emission is focused as a ring around the aperture, and does not contribute to the image. In this way, the size of the imaged spot is decreased and the resolving power of the confocal microscope is increased (Hell and Wichmann, 1994; Klar et al., 2000, 2001).

A confocal microscope can be used to image single molecules interacting with a surface using a technique known as total internal reflectance fluorescence microscopy (TIRF; Figure 12-5). As a consequence of binding kinetics, the presence of many molecules is required for a single molecule to bind to a receptor surface. If the molecules are fluorescent, the fluorescence of the bound molecule will be overwhelmed by the fluorescence of the free molecules. In order to increase the contrast of bound fluorescent

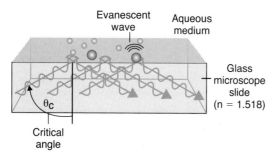

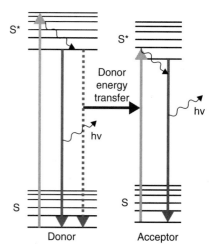

FIGURE 12-5 Diagram of the illumination of a specimen in a total internal reflection fluorescence microscope.

FIGURE 12-6 Energy diagram for fluorescence resonance energy transfer between two fluorochromes.

molecules, the surface of the specimen is placed about 100 nm from the surface of the microscope slide. The excitation light hits the surface of the slide at the critical angle so that the light is reflected away from the specimen by total internal reflection (see Chapter 2). However, if the fluorescent molecules are close enough to the surface, the evanescent wave will excite them and they will fluoresce while the free fluorochromes, which are too far from the surface for the evanescent wave to excite, will remain in the ground state (Axelrod et al., 1982; Axelrod, 1984, 1990; Tokunaga and Yanagida, 1997). With the contrast generated, one can visualize proteins translating across a strand of DNA (Gorman et al., 2007). In order to determine the axial location of a single molecule, a second reflecting surface can be introduced so that interference microscopy can be done with the fluorescent light that comes from a single molecule. This technique is known as spectral self-interference fluorescence microscopy (SSFM; Moiseev et al., 2006).

A confocal microscope can be used to determine the distance between two interacting molecules by making use of the fluorescence (or Förster) resonance energy transfer (FRET) technique (Figure 12-6). With this technique, one molecule is labeled with a donor fluorochrome and the other is labeled with an acceptor fluorochrome, and then the sample is irradiated with the excitation wavelength of the donor fluorochrome and observed at the emission wavelength of the acceptor fluorochrome. The two fluorochromes, which are typically color variants of green fluorescent protein, are inserted into proteins of interest using genetic engineering techniques. When the two molecules are within 1 to 10 nm of each other, energy can be transferred between the donor fluorochrome and the acceptor fluorochrome and subsequently, the acceptor fluorochrome gives off the excitation energy absorbed by the donor fluorochrome as fluorescence. The efficiency of resonance energy transfer is due to a dipole-dipole interaction that is inversely proportional to the sixth power of the distance. Consequently, if the two target molecules are not in close proximity, the excitation energy is given off as fluorescence from the donor fluorochrome. Protein conformational changes can also be monitored by putting the donor and acceptor fluorochromes on the same protein (Periasamy and Day, 2005).

A confocal microscope can also be used to localize specific genes in chromosomes or chromatin using a technique known as fluorescent in-situ hybridization (FISH). With this technique, the gene of interest is stained with a fluorescent probe made from fluorescent nucleotides arranged in an order that is complementary to the order of nucleosides in the gene of interest (Zhong et al., 1996, 1998; Jackson et al., 1998). Fluorescent in-situ hybridization can also be done with a wide-field fluorescence microscope and often is used in *in vitro* fertilization clinics for determining which alleles exist in an embryo.

A confocal microscope can be used to measure the rate of movement of molecules tagged with fluorescent probes using a technique known as fluorescence recovery or redistribution after photobleaching (FRAP). With this technique, the fluorescence in a given area is measured and then the fluorescent molecules are bleached with an intense light source so that the fluorescence disappears. The intense beam is shut off and as fluorescent molecules from other regions diffuse into the bleached area, the fluorescence recovers. The diffusion coefficient of the moving molecules can be determined from the rate in which the fluorescence recovers (Axelrod et al., 1976; Edidin et al., 1976; Wang, 1985; Baron-Epel et al., 1988; Gorbsky et al., 1988; Luby-Phelps et al., 1986).

For further information and practical guidance on how to use a confocal microscope, consult *Fundamentals of Light Microscopy and Electronic Imaging*, by Douglas B. Murphy (2001).

LASER MICROBEAM MICROSCOPE

Lasers also are used in wide-field microscopes to perform laser microsurgery (Figure 12-7; Aist and Berns, 1981; Berns et al., 1981, 1991; Koonce et al., 1984; Aist et al., 1991; Bayles et al., 1993), the principles of which are the

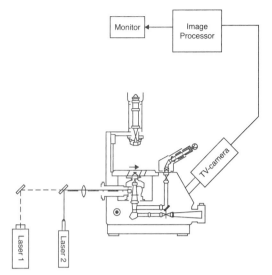

FIGURE 12-7 Diagram of a microscope with a laser microbeam and a laser optical trap.

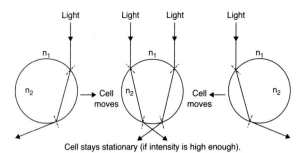

Cell stays stationary (if intensity is high enough).

FIGURE 12-8 Ray diagrams of light passing through a specimen. The specimen (n_2) recoils as the light passes out of it and into the medium (n_1 where $n_2 > n_1$).

same as those used in laser surgery of humans (Berns, 1991). In a microscope, the laser light is routed though the epi-fluorescence port. The laser does selective damage to a subcellular component in one of three ways:

- The damage may be due to classical absorption of the intense, coherent, monochromatic light by natural or added chromophores with the subsequent generation of heat.
- The damage may be due to the light stimulated addition of a chemical, for example, the light stimulated binding of psoralen to nucleic acids.
- When the photon density is high enough, heat can be generated by multiphoton absorption.

A clever and frugal person can do microsurgery using the ultraviolet light source on an epi-fluorescence microscope (Forer, 1965, 1991; Leslie and Pickett-Heaps, 1983; Wayne et al., 1990; Forer et al., 1997).

OPTICAL TWEEZERS

Photons have both linear momentum ($\hbar k = h/\lambda$) and angular momentum ($\hbar$); (Nichols and Hull, 1903a, 1903b; Poynting, 1904, 1910; Einstein, 1917; Schrödinger, 1922; Dirac, 1924; Beth, 1936; Friedberg, 1994; Kleppner, 2004) and can be used to move or rotate macroscopic objects (Lebedew, 1901; Nichols and Hull, 1903a, 1903b; Gerlach and Golsen, 1923), cells, organelles, and proteins (Ashkin and Dziedzic, 1987, 1989; Ashkin et al., 1987, 1990; Block et al., 1990; Svoboda and Block, 1994; Neuman and Block, 2004) and even stop atoms (Frisch, 1933; Ashkin, 1970a, 1970b, 1978; Bjorkholm et al., 1975; Chu, 1997; Cohen, 1997; Phillips, 1997; Johnson et al., 2007). By routing the light from an infrared laser through the epi-fluorescence, we can add "optical tweezers" to a microscope (Greulich, 1992; Ashkin and Dziedzic, 1987, 1989; Ashkin et al.,

1987, 1990; Block et al., 1990; La Porta and Wang, 2004; Moffitt et al., 2006).

The optical tweezers can be used to measure force or resistance by determining the light intensity (in J m^{-2} s^{-1}) necessary to move a stationary object or to hold a moving object stationary. The force (in N) exerted on an object can be determined with the following equation:

$$F = \eta(\text{light intensity})(\text{area of object})/c$$

where η is a dimensionless number between 1 for totally absorbing objects and 2 for totally reflecting objects and c is the speed of light. Optical tweezers can hold a moving cell still because the photons that leave the object cause the object to rebound in the opposite direction consistent with Newton's Third Law that states that for every action there is an equal and opposite reaction (Figure 12-8).

LASER CAPTURE MICRODISSECTION

Laser capture microdissection is a method that allows us to isolate and capture pure cells and their contents, particularly RNA, from heterogeneous tissue slices under the microscope (Emmert-Buck et al., 1996; Bonner et al., 1997; Nakazono et al., 2003; Woll et al., 2005; Cai and Lashbrook, 2006; Nelson et al., 2006; Spencer et al., 2007; Zhang et al., 2007). With this technique, a transparent transfer film is applied to the surface of the tissue section while the other side is bonded to a cap. The transparent transfer film is then irradiated with a pulsed infrared laser beam that is focused through the objective onto the target cells in the tissue. The special transfer film then melts and bonds to the targeted cells and consequently, the target cells and their contents become attached to the cap. The DNA, RNA, or proteins in the target cells can then be analyzed.

The PALM MicroLaser Systems manufactured by Zeiss combines laser microbeam microdissection (LMM) and laser pressure catapulting (LPC) techniques. In this system, a pulsed ultraviolet laser is focused through an objective to a beam spot size of less than 1 micrometer in diameter that is used for cutting the targeted cells out of the tissue. After the cells are cut out, the targeted cells within the confines of the cut are ejected out of the object plane and catapulted directly into the cap of a microfuge tube using a single defocused laser pulse.

LASER DOPPLER MICROSCOPE

The laser Doppler microscope is based upon the Doppler Effect. The Doppler Effect was first noticed by Johann Christian Doppler (1842) when he posited that the color of binary stars may be caused by their movement toward or away from an observer. Following the introduction of the newly-invented, rapidly-moving steam locomotive, Buijs Ballot (1845) tested Doppler's theory by placing musicians on a railroad train that traveled 40 mph past musically trained observers. The stationary observers found that notes were perceived to be a half-note sharper when the train approached and a half-note flatter when the train receded. Three years later, Russell (1848) noticed that when he was on a train moving at 50 to 60 mph, the pitch of the whistle of a stationary train was higher when the train moved toward it and lower when the train moved away. Think of a police car with its siren going. As it approaches you, the sound waves move closer together and thus the frequency (or pitch) gets higher and higher. As the police car moves away from you the sound waves move farther and farther apart, thus the frequency drops and the sound has a lower and lower pitch.

Doppler realized that if light were a wave it should also show a shift in the frequency and wavelength. Consequently, when a light emitting source approaches an observer, the waves, measured at an instant in time, appear to be closer together and thus blue-shifted to the observer. By contrast, when the light emitting source moves away from the viewer, the waves, measured at an instant of time, appear farther apart and thus red-shifted. When the waves are measured over time at a single point in space by the observer, the frequency of the waves appear to increase or decrease, depending on whether the source is approaching or receding from the observer, respectively. The temporally and spatially varying amplitude ($\Psi(x,t)$) of a wave perceived by an observer moving with a velocity (v) relative to the source is given by the following equation:

$$\Psi(x,t) = \Psi_o \cos(k_{observer} x \pm \omega_{source}[\sqrt{(c-v)}/\sqrt{(c+v)}]t)$$

where $k_{observer}$ angular wave number measured from the observer's frame of reference and ω_{source} is the angular frequency measured from the source's frame of reference. The angular wave numbers measured from the source's frame of reference and the observer's frame of reference are equal when v = 0 and the source and observer are at rest relative to each other. The above equation, which includes the relativistic Doppler Effect, implies that the speed of light (c) as measured by the properties of the propagation medium (i.e. the electric permittivity and the magnetic permeability) is not equal to the speed of light measured as the ratio of the angular frequency of the source to the angular wave

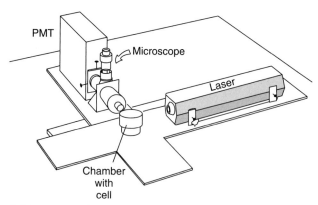

FIGURE 12-9   Diagram of a laser Doppler microscope.

number measured by the observer moving relative to the source.

The laser Doppler microscope can be used to measure the speed of moving particles within a cell or the speed of moving cells by determining how much the wavelength (or frequency) of the incident light is shifted by the moving particle. Laser Doppler velocimetry has been used to study cytoplasmic streaming (Mustacich and Ware, 1974, 1976, 1977; Langley et al., 1976; Sattelle and Buchan, 1976; Earnshaw and Steer, 1979), phloem transport in plant cells, motility of sperm (Dubois et al., 1974) and algae (Druez et al., 1989), as well as chemotaxis in bacteria (Nossal and Chen, 1973). Similar techniques are used to measure the velocity of blood flow (Stern, 1975; Tanaka et al., 1974).

In a laser Doppler microscope, light from a laser passes through a neutral density filter and strikes the cell. The light that is scattered at a given angle passes through an aperture (0.8 mm) and a half-silvered mirror and is focused on a pinhole (2 mm) that is placed immediately in front of a photomultiplier tube. The light that is reflected by the half-silvered mirror passes up through a microscope. The electrical signal from the photomultiplier tube then passes through an amplifier and a spectrum analyzer where it is added to a reference signal of 2 to 200 Hz. The two signals interfere to produce "beats." In this way the frequency of the incoming light can be determined. Figure 12-9 shows a setup for a laser Doppler microscope.

In systems like cytoplasmic streaming, where the velocity of streaming decreases with distance from the location of the motive force, an objective lens will give an average velocity over the distance equal to the depth of field. A high numerical aperture lens will let one sample the velocity in an optical section, whereas a low numerical aperture lens will give an average velocity (Staves et al., 1995).

CENTRIFUGE MICROSCOPE

The centrifuge microscope can be used to measure force and resistance (Hayashi, 1957; Hiramoto, 1967; Kaneda

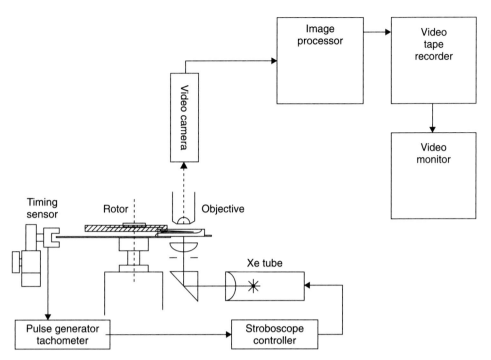

FIGURE 12-10 Diagram of a centrifuge microscope.

et al., 1987, 1990; Kamitsubo et al., 1989, 1988; Kuroda and Kamiya, 1989; Oiwa et al., 1990; Wayne et al., 1990; Takagi et al., 1991, 1992). There are simple types of centrifuge microscopes that one can build with surplus materials. These include the Harvey (1938) type and the Brown (1940) type centrifuge microscopes that use either electrical motors or air-driven motors and low magnification objectives. I will describe the centrifuge microscope of the stroboscopic type. The centrifuge microscope of the stroboscopic type is composed of a video-enhanced contrast microscope that uses bright-field optics combined with an analog image processor that creates a pseudo-relief image of the cell in real time (Kamitsubo et al., 1989). The objective lenses are long working distance objectives (Nikon 20x/0.40, 40x/0.55 LWD, Plan 60x/0.7 Ph3DL LWD CF), and the sub-stage condenser is a long working distance sub-stage condenser (NA 0.65). The optical system is completely isolated from the rotor. The rotor replaces the stage.

The rotor is 16 cm in diameter and the radius for centrifugation is adjustable in the range of 4.5 to 7.0 cm. The rotor can be rotated at 250 to 5000 rpm (4–1900 xg) while examining the specimen with a 60x objective lens. The rate of rotation can be varied by changing the voltage supply. The rate of rotation (in rpm) is recorded on the videotape and is played on the monitor.

The illumination source is a xenon bulb that flashes for 180 ns per pulse. The rate of the pulse is controlled by a trigger pulse generator. The bulb is capable of flashing at rates up to 100 Hz (6000 times per minute). The position of the specimen is detected by a photocoupler that consists of a light-emitting photodiode, a phototransistor, and an amplifier. A piece of metal attached to the rotor edge opposite the position of the specimen interrupts the light beam emitted by the photodiode once per rotation. The change in the light intensity is detected by the phototransistor and is converted into an electrical pulse, which triggers the strobe light to flash. Since the strobe flashes only when the cell is under the objective and afterwards the cell is dark, the stationary parts of the cell seem to stand still (Figure 12-10). Centrifuge microscopes can also be used to study the effect of centrifugal fields on birefringent and fluorescent specimens (Inoué et al., (2001a, 2001b).

X-RAY MICROSCOPE

X-ray microscopes, which work with electromagnetic waves with wavelengths of about 1 to 10 nm, have the potential of producing images with high spatial resolution, although currently the resolution is about 30 nm (Schmahl and Rudolph, 1984; Howells et al., 1991; Michette et al., 1992). X-rays provide image contrast by interacting with the electrons that make up the inner shells of light atoms. This provides good contrast between carbon-containing organic matter and water. Rosengren (1959) developed a method to measure the mass of small cells using an X-ray microscope. The X-rays he used had wavelengths between 0.6 and 1.2 nm because it was found that almost all biological molecules have similar absorption coefficients for wavelengths in this range and thus absorption is proportional to the concentration of matter. The x-ray microscope is similar in accuracy to an interference microscope and has a sensitivity of 1.3×10^{-16} kg.

Some X-ray microscopes use Fresnel zone plates to focus the X-rays. A zone plate consists of a silicon nitride membrane that supports thin concentric gold rings. The

X-rays that pass through the zone plate are diffracted again and brought to a focus—just as a regular glass lens diffracts again the light rays that come from an object and brings them to a focus. Some X-ray microscopes do not use lenses but record the far-field or Fraunhöfer diffraction pattern (Maio et al., 1999; Shapiro et al., 2005; Hornberger et al., 2007). The resolution attainable with lens-less microscopes is not limited by the resolution of the lens. Lens-less microscopes are also free from spherical aberration. Veit Elser (2003) has written the algorithm that converts the diffraction pattern that is produced by the object into an image to and interestingly enough, the same algorithm can be used to make and solve Sodoku puzzles (Thibault et al., 2006).

Construction of microscopes that use π-mesons for the illuminating rays is under way (Breedlove and Trammell, 1970). Perhaps there will be "antimatter microscopes" that bombard specimens with positrons and generate large numbers of gamma rays in regions of high electron density when the positrons combine with electrons. This is how positron emission tomography (PET), which is currently used in many hospitals, works.

INFRARED MICROSCOPE

Fourier transform infrared (FTIR) microscopes have been developed to determine the absorption characteristics in the infrared region of various molecules in the cell. The only unusual aspect of this microscope is that all the lenses are made of mirrors that have no chromatic aberration, so visible light can be used to focus the object while an infrared spectrum is obtained (Schiering et al., 1990).

NUCLEAR MAGNETIC RESONANCE IMAGING MICROSCOPE

I have been discussing microscopes that utilize electromagnetic waves from various regions of the spectrum, including X-rays, ultraviolet light, visible light, and infrared light, to interact with electrons in a specimen in order to derive information about the electronic properties of that specimen. Radio waves can also be used to image a specimen; however, radio waves interact with the nuclei of atoms instead of the electrons. Radio waves are utilized by nuclear magnetic resonance (NMR) imaging microscopes (Goodman et al., 1992, 1992; Callaghan 1991, 1992; Bowtell et al., 1990; Woods et al., 1989; Jelinski et al., 1989) in the same way they are used in magnetic resonance imaging (MRI). The distribution of ^1H, ^{13}C, ^{19}F, ^{23}Na, ^{31}P, or any nucleus that has an odd number of nucleons and thus a net magnetic spin can be determined with magnetic resonance imaging.

Basic NMR works by aligning the magnetic moment (M) of a nucleus in a magnetic field (B), then giving the

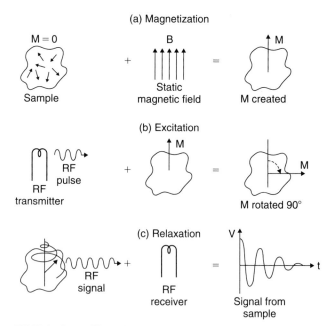

FIGURE 12-11   Illustration of how the nuclear spins respond to a radio-frequency pulse using magnetic resonance imaging.

specimen a radio frequency pulse oriented 90 degrees relative to the magnetic field vector. Then the radio frequency pulse is absorbed, and the magnetic moment reorients along the direction of the pulse. The nucleus emits the radio frequency radiation and the magnetic moment slowly returns to the direction it was facing before the pulse, which is the lowest energy state in the magnetic field. The emitted energy identifies the atom and/or the environment surrounding the atom. This energy usually is given in frequency terms or relative terms called chemical shift. The frequency, pulse width, and pulse pattern of the radio frequency radiation is chosen to selectively excite the atom of interest (e.g., ^1H, ^3C, ^{31}P). The frequency of the energy emitted depends on which element is in the magnetic field (Figure 12-11).

The magnetic fields that orient the nuclei of specific atoms move across the specimen in the x, y, and z directions. The intensity of the emitted signal at each x,y,z coordinate is reconstructed in a point-by-point manner to form an image. Nuclear magnetic resonance microscopes, which have a spatial resolution of 10^{-5} m, have been used to map the distribution of water in large single cells (*Xenopus* eggs; Aguayo et al.., 1986), as well as in whole plants and animals (Jenner et al., 1988; Goodman et al., 1992).

STEREO MICROSCOPES

Stereo microscopes, also known as dissecting microscopes, produce three-dimensional upright images enlarged up to about 600x with large and flat fields, a tremendous depth of field, and long working distances. Stereo microscopes fall into two categories: those designed according to the Greenough design principle and those designed according to the telescope principle. The Greenough design utilizes two identical, nearly side-by-side objectives. The two objectives project two separate images to two oculars. Most stereo microscopes are made according to the telescope design. These microscopes consist of two partial microscopes, which share one large objective.

Stereo microscopes now offer highly corrected, apochromatic, distortion-free, flat field optics equipped with zoom lenses. They are available with transmitted and/or reflected light bright-field optics as well as with dark-field, epi-fluorescence and polarizing optics. They come equipped with a variety of stages including rotating and glide stages. They have iris diaphragms to control the depth of field, and ports to mount video and/or digital cameras. Many fiber optic illumination sources are available for stereo microscopes that give dramatic lighting effects.

The zero perspective imaging microscope uses a 35 mm camera with a macro lens to obtain a three-dimensional image with maximal depth of field using the full resolving power of the lens (Peres and Meitchik, 1988). In this microscope, a thin sheet of light whose plane is perpendicular to the optical axis illuminates a section of the object. The line must be thinner than the depth of field of the objective lens. The camera then focuses at this depth. This specimen is then driven up or down by a motor while the camera shutter remains open. In this way every plane in the vertical axis is perfectly in focus in the image.

SCANNING PROBE MICROSCOPES

Scanning probe microscopes include the near-field scanning optical microscope, the scanning tunneling microscope, and the atomic force microscope.

Synge (1928, 1932) proposed that Abbe's limit of resolution, which is based on the formation of a Fraunhöfer or far-field diffraction pattern could be overcome by capturing the Fresnel or near-field diffraction pattern. Near-field scanning optical microscopy (NSOM) produces an image by capturing the Fresnel diffraction pattern. The near-field scanning optical microscope can be used in the transmission, reflection, or fluorescence mode. In the reflection or fluorescence mode, the specimen is illuminated by light that passes through a subwavelength aperture at the end of an optical fiber probe, and the same optical fiber probe collects the reflected light (Pohl et al., 1984; Lewis et al., 1984; Liberman et al., 1990; Betzig et al., 1991; Wiesendanger, 1998; Courjon, 2003). A feedback mechanism maintains a constant distance of a few nanometers between the specimen and the probe (Figure 12-12). The feedback voltage at each point is used to create a point-by-point image that is a function of the microtopography

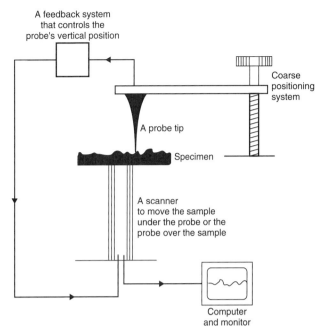

FIGURE 12-12 Diagram of a scanning probe microscope.

of the specimen. The maximum depth-of field of the near-field optical scanning microscope is about 300 nm, the minimum is 0.65 nm (Revel, 1993; Querra, 1990).

Many varieties of scanning probe microscopes have been and are being developed to probe matter at atomic dimensions (Kalinin and Gruverman, 2007; Morita, 2007). The scanning tunneling microscope (STM) differs from the near-field scanning optical microscope in that the probe acts as a conductor of electrons to or from the surface through a distance small enough (≈ 0.2 nm) to allow quantum mechanical tunneling, which is analogous to an evanescent wave, when an electrical potential is applied across the gap between the specimen and the probe. The amount of current depends on the distance between the specimen and the probe. The computer then reconstructs an image, point by point, from the position of the probe and the magnitude of current that flows through the probe at each point. The 1986 Nobel Prize in Physics went to Gerd Binnig and Heinrich Rohrer at the IBM Zurich Research Laboratory for developing the scanning tunneling microscope (Binnig et al., 1982; Amato, 1997).

Image contrast is obtained in a scanning tunneling microscope because the probe produces pressure-induced elastic deformations of the electronic orbitals, which result in molecular orbitals close to the Fermi energy and therefore, enhanced tunneling (McCormick and McCormick, 1990). Although the scanning tunneling microscope usually is used in materials science (Hawley et al., 1991), it has also been used to image cell membranes and macromolecules (Cricenti et al., 1989; Ruppersberg et al., 1989; Welland et al., 1989; Edstrom et al., 1990; Yang et al., 1990; Clemmer and Beebe, 1991).

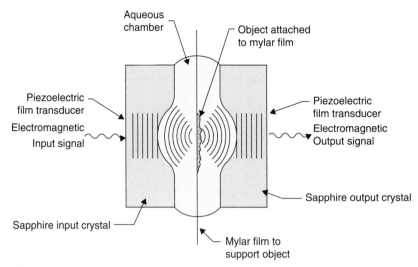

FIGURE 12-13   Diagram of an acoustic microscope.

The atomic force microscope measures one of a number of interaction forces between the tip of the probe and surface of interest (Marti and Amrein, 1993; Morris et al., 1999). The choice of probe determines the interaction force that is measured. When the probe has a tip with a small diamond, a carbon nanotube, a silicon chip, or silicon nitride chip, the atomic force microscope measures the electrostatic forces between the tip of the probe and the surface of the specimen. When the probe has a tip whose resistance changes with temperature, the atomic force microscope measures the thermal conductivity of the surface. When the probe has a tip that is susceptible to magnetic fields, then the atomic force microscope measures magnetic forces. When the probe is dragged across the surface of the specimen, then the atomic force microscope can measure frictional forces.

In the atomic force microscope, the various probes are mounted on a cantilever so that the sharp point faces the sample and it rides directly on the sample. A laser is focused on the mirrored back surface of the cantilever and the position of the reflected beam is monitored with a photocell. The specimen can be scanned across the probe with a lateral resolution of a few nanometers. A computer generates a three-dimensional image by reconstructing, point by point, the position of the probe tip in three dimensions (Robinson et al., 1991).

ACOUSTIC MICROSCOPE

Up until now I have been talking about imaging specimens with transverse electromagnetic waves from the X-ray region of the spectrum though the radio wave region of the spectrum. Specimens can also be imaged with longitudinal sound waves that have approximately the same wavelength as light waves. Sound travels through a specimen with a velocity that depends on the mechanical (viscoelastic) properties

of the specimen. For example, sound travels at a velocity of 1531 m/s in water at room temperature and at a velocity of about 300 m/s in air (Airy, 1871; Rayleigh, 1894). In general, the speed of sound is equal to the square root of the ratio of an elastic property of the medium to an inertial property of the medium. The best indicator of the elastic property is the bulk modulus (in N/m^2) and the best indicator of the inertial property is the density (in kg/m^3). In general, the more rigid the material is, the faster sound moves through it.

The wavelength of sound can be found from the dispersion relation that relates the wavelength (λ), frequency (ν), and speed (c) of a wave.

$$\lambda = c/v$$

The wavelength of a 1000 MHz sound wave traveling through water is $(1531\,\text{m/s})(1000 \times 10^6\,\text{s}^{-1})^{-1} = 1.5 \times 10^{-6}\,\text{m}$, which is close to the wavelength of visible light. Using the Rayleigh criterion, and assuming the numerical aperture of a lens in a acoustic microscope is 1.4, the limit of resolution for a 1000 MHz sound wave would be $0.61\,(1.5 \times 10^{-6}\,\text{m})/\text{NA} = 0.65 \times 10^{-6}\,\text{m}$; similar to the limit of resolution of a light microscope (Lemmons and Quate, 1974, 1975).

Creating images using the interaction of light with matter reveals information about the electrical properties of the specimen. By contrast, the interaction of sound with matter reveals information about the elastic and inertial properties of matter known as the viscoelastic properties. The viscoelastic properties of the specimen can alter both the phase and the amplitude of the sound waves traveling through the specimen. The amplitudes of the sound waves are differentially damped when they travel through media with different viscoelastic properties (density, viscosity, elasticity). For example, the greater the viscosity, the greater the absorption. There are amplitude contrast acoustic microscopes as well as phase-contrast, dark-field, and differential interference contrast acoustic microscopes (Wickramasinghe, 1989).

In an acoustic microscope, electrical energy is transformed into acoustic energy by a piezoelectric film at the surface of a sapphire crystal (Figure 12-13). The acoustic wave then propagates as a plane wave through the sapphire crystal. The other side of the sapphire crystal meets a water cell. The acoustic wave hits the sapphire-water interface and perceives it as a converging lens since the acoustical refractive index of water is greater than the acoustical refractive index of sapphire. As a result, the spherical sound waves are focused to a point. This makes up the acoustic transmitter that transmits sound waves with a wavelength of 10^{-6} m. The sound waves then pass through the specimen, which is attached to a Mylar film, and at the focus of the transmitter.

The specimen is also at the focus of the receiver. The receiver in the acoustic microscope is constructed exactly the same way as the transmitter. The sound waves pass through the specimen and through the water to the sapphire-water interface as spherical waves. The spherical sound wave experiences a converging lens that turns them into a plane wave. The plane wave then strikes a second piezoelectric film that transforms the sound waves into electrical waves.

The limit of resolution is set by the wavelength of the acoustic signal. Therefore, increasing the frequency of the sound increases the resolving power of the acoustic microscope. Unfortunately the attenuation of sound is proportional to the frequency, and this property sets a limit on the focal length of the lens. In an acoustic microscope, the specimen is translated on a mechanical stage through the acoustic beam.

The acoustic microscope has been used to contrast regions with differing viscoelastic properties in cells (Johnston et al., 1979; Israel et al., 1980; Lüers et al., 1991; Hildebrand et al., 1981; Bereiter-Hahn et al., 1995). An ultrasound imager is a macro version of an acoustic microscope.

HORIZONTAL AND TRAVELING MICROSCOPES

Microscopes can be mounted horizontally so that the stage is parallel to the vector of gravity. Although the stage is vertical, the optical system is horizontal, thus the name horizontal microscope. A horizontal microscope has been used to measure gravity sensing in plant cells (Sack and Leopold, 1985; Sack et al., 1984, 1985; Wayne et al., 1990).

A traveling microscope is a microscope that does not have a stage. It can be oriented in any direction. It is used commonly with a low-power objective as a horizontal microscope to observe an object mounted in some kind of bulky apparatus. A traveling microscope has been used in a turgor balance, an apparatus that measures noninvasively the osmotic pressure of plant cells (Tazawa, 1957). A traveling microscope also has been used to measure the density of cell components (Kamiya and Kuroda, 1957) and the tension at the surface of an endoplasmic drop (Kamiya and Kuroda, 1958).

MICROSCOPES FOR CHILDREN

The microscope can help open up the world to children of all ages. *Microscopy Today* publishes a *Microscopy Bibliography for Children* (e.g., supplement Issue #00-10, December 2000). This bibliography includes books, CD-ROMS, and videotapes on teaching microscopy, optics, the microscopic world, and specimen preparation for primary and middle school children. In order to find age-appropriate materials and learn how to buy children's microscopes, visit The Microscopy Society of America's web site at http://www.msa.microscopy.org/ProjectMicro/PMHomePage.html; the Southwest Environmental Health Sciences Center web site at http://swehsc.pharmacy.arizona.edu/exppath/micro/edu/education.html; or the microscopy.info web site at http://www.mwrn.com/microscopy/educational/books.aspx.

MICROSCOPE ACCESSORIES

Once you know how to use a microscope, it can offer you unlimited opportunities to study cells at high magnification. Many accessories are made to assist you in your goals.

There are specialized stages, including rotating stages (Abramowitz, 1990), motorized stages, temperature-controlled stages (Hartshorne, 1975, 1976, 1981; Skirius, 1984; Moran and Moran, 1987; Wilson and McGee, 1988; McCrone, 1991; Valaskoveic, 1991). If you need a stage cold enough to photograph snowflakes, you can always take your microscope outside (Nakaya, 1954; Bentley and Humphreys, 1962; LaChapelle, 1969; see references in Delly, 1998).

If you want to study dynamic processes in cells over extended times you must construct or buy special culture chambers that keep the temperature, pH, CO_2, O_2, osmolarity, nutrients, and growth factors constant (Smith, 1856; Davidson, 1975; McKenna and Wang, 1989). These chambers must also be optically transparent so that the specimen can be seen with optimal resolution and contrast. Do not forget that you can make good chambers with cover slips cut with a phonograph needle (glued with Elmer's Bonding Cement), parafilm, or VALOP (1 part vaseline, 1 part lanolin, 1 part paraffin oil).

A number of instruments, known as micromanipulators, are available that allow you to move, inject, or cut microscopic objects under the microscope. Microcapillary pipette pullers, bevellers, and microforges let you make pipettes capable of injecting large substances and organelles into cells (McNeil, 1989), measuring the electrical properties of cells and membranes, or measuring the forces that drive motile processes in cells. The only things that limit the design and applications of a microscope are the laws of physics and your own imagination! Contribute your imagination to discovering new laws of physics and to microscopy!

WEB RESOURCES

Confocal Microscopy

Marvin Minsky's Memoir on the Invention of the Confocal Microscope and Homepage: http://web.media.mit.edu/~minsky/papers/Confocal Memoir.html and http://web.media.mit.edu/~minsky/

Leica Microsystems: http://www.leica-microsystems.com/Confocal_Micro-scopes

Nikon Microscopy U: http://www.microscopyu.com/articles/confocal/confocalintrobasics.html

Molecular Expressions: http://micro.magnet.fsu.edu/primer/techniques/confocal/index.html

Zeiss: http://www.zeiss.com/4125681f004ca025/Contents-Frame/f544501 1b5a0a89f852571d200714c4d

Total Internal Reflections Fluorescence Microscopy: http://www.olympusmicro.com/primer/techniques/fluorescence/tirf/tirfhome.html

Fluorescence Resonance Energy Transfer (FRET): http://www.olympus-fluoview.com/applications/fretintro.html

Optical Tweezers

Michelle Wang's Web site: http://people.ccmr.cornell.edu/~mwang/overview.html

Laser Capture Microdissection

http://dir.nichd.nih.gov/lcm/LCM_Website_Introduction.htm)
http://www.lasercapturemicrodissection.org/
http://www.palm-microlaser.com/dasat/index.php

http://www.moleculardevices.com/pages/instruments/microgenomics.html

X-Ray Microscopes

http://xray1.physics.sunysb.edu/research/links.php

Scanning Probe Microscopes

http://www.olympusmicro.com/primer/techniques/nearfield/nearfieldintro.html
http://www.mobot.org/jwcross/spm/
http://nobelprize.org/nobel_prizes/physics/laureates/1986/index.html

Acoustic Microscopes

Sonoscan: http://www.sonoscan.com/
Sonix: http://www.sonix.com/learning/ultrasonics.php3

Accessories

World Precision Instruments: http://www.wpiinc.com/products/
Research Precision Instruments: http://www.rpico.com/company.html
Narishige: http://www.narishige.co.jp/main.htm
In vivo Scientific: http://www.invivoscientific.com/
Olympus Fluoview: http://www.olympusfluoview.com/resources/specimenchambers.html

Chapter 13

Video and Digital Microscopy

There are many ways to accomplish the goal of creating a high-resolution and high-contrast image. Once we understand how light interacts with matter and how optical components of a microscope transform the light emitted by a specimen into a high contrast image, it becomes easy to understand, by using analogous physical principles, how electronic devices can be used in conjunction with a few optical components to produce a high-contrast image. In order to take advantage of electronics in image processing, we must convert the optical signal into an electronic signal. This is the job of a video or digital camera (Allen et al., 1981a, 1981b; Allen and Allen, 1983; Allen, 1985; Cristol, 1986; Inoué, 1986; Dodge et al., 1988; Aikins et al., 1989; Spring and Lowy, 1989; Weiss et al., 1989; Varley et al., 2007).

THE VALUE OF VIDEO AND DIGITAL MICROSCOPY

Just as a diverging lens made of flint glass reverses the chromatic aberration caused by a converging lens made of crown glass, video and digital microscopy can be used to reverse the aberrations introduced by the optical system. Whereas, traditionally, the aberrations are corrected by optical means, video and digital cameras allow the aberrations to be corrected by electronic means. In addition, just as additions to the light microscope, including phase-contrast, differential interference contrast, and modulation contrast optics can be used to surpass the limitations of the bright-field microscope and produce images that are high in both resolution and contrast, video and digital microscopies can be used to overcome the same limitations.

Video and digital microscopies are also useful when doing fluorescence microscopy. The light emission from a specimen in a fluorescence microscope is relatively low and consequently, when using film cameras, one must take long exposures with high-speed film (ISO 800–1600) in order to capture an image. However, image capture is fast and easy when the image is captured with a video or digital camera since the light sensitivity of video and digital cameras can be as high as ISO 1,000,000. Today, specimens that emit only a few photons per point, be they distant galaxies or fluorescent microscopic objects, can be imaged with a photon-counting camera.

Video and digital microscopies are also useful for polarization microscopy, where resolution often is sacrificed for maximal extinction because high numerical aperture objective lenses tend to depolarize the light that strikes their surfaces. Video and digital microscopy combined with background subtraction (see Chapter 14) can be used to eliminate any background brightness without reducing the light that arises from a birefringent object, thus achieving maximal resolution and contrast simultaneously.

Moreover, video and digital microscopies let us take images of moving objects in rapid succession. Moving pictures began in 1872 when then California Governor Leland Stanford hired Eadweard Muybridge to see if all four hooves of a horse leave the ground when it gallops (Bova, 1988). Muybridge (1887, 1955) set up 12 cameras at a racetrack and took 12 consecutive photos. All four hooves do leave the ground when a horse gallops and Stanford won a $25,000 bet. Today, digital cameras can take images of moving bullets and biological objects as fast as 400,000 frames per second.

Video and digital cameras convert an optical signal into an electric signal, thus making it possible to operate on the image with inexpensive electronic means instead of expensive optical devices (de Weer and Salzber, 1986). The use of video and digital cameras for image processing has not even begun to see its full potential. Imagine putting an imaging chip in the back focal plane of an objective lens, or better yet in the aperture plane of an infinity corrected microscope. The chip in the aperture plane would collect the diffraction pattern of the specimen and convert the light intensity at each point into an electrical signal. The signal at each point could be digitized and sent to a computer where the information could be stored numerically. If the specimen were illuminated sequentially with points of light in the form of an annulus at the front focal plane of the condenser, we could obtain many high-resolution diffraction patterns of the same specimen. Then the computer could do an inverse Fourier transform on each diffraction pattern, combine the inverse Fourier transforms, and construct an image of the specimen.

Imagine that we could program the computer to remove the Fourier components that represent the diffracted light from one side or the other of the diffraction plane before it reconstructs an image. This would produce a pseudo-relief image reminiscent of an image produced by oblique illumination. Imagine illuminating the object with oblique slit illumination and then programming the computer to adjust the numerical value of Fourier components, depending on which point in the diffraction plane they represent. We would obtain a modulation contrast image if we multiplied the values of the Fourier components in the region conjugate with the slit by 0.15, and multiplied the values of the Fourier components on one side of the conjugate region by 0.01, and multiplied the values of the Fourier components on the other side by 1.

It is also possible to produce a differential interference contrast image using analog electronics by converting the light intensity of a line across the image plane into an electronic signal, then duplicate the signal, and then advance or retard the duplicated signal with respect to the original signal (see Chapter 14). When the two signals are recombined, a differential interference contrast image is obtained without the use of expensive interferometers. Use your imagination: I bet that you can think of many ways to use relatively inexpensive and versatile computers and electronics to do the job of more expensive specialized optical components.

VIDEO AND DIGITAL CAMERAS: THE OPTICAL TO ELECTRICAL SIGNAL CONVERTERS

A video or digital camera provides the interface between the optical system and the electronic system. The imaging device converts the intensity of each designated area in the optical image into an electrical signal. The video camera is attached to a microscope using an optical coupler known as a "c mount," and digital cameras can be attached to microscopes, either by c mounts, or by a variety of optical couplers that have not yet been standardized. The formation of an electrical signal depends on the interaction of a photon with an electron. Since the interaction depends on the relationship between the work function of the photosensitive surface and the energy of the photon, the response of electrical detectors is wavelength-dependent.

Semiconductors typically are used to convert light energy into electrical energy (Pankove, 1971). Semiconductors traditionally are made of silicon atoms arranged in a crystal lattice where each atom has tetrahedral bonding with four of its nearest neighbors (Figure 13-1). Semiconductors are called semiconductors because they are neither good conductors nor good insulators. In order for a material to conduct electricity, the valence electrons have to be freed from the atomic orbitals, which make up the valence band, and allowed to migrate freely through the material, in what

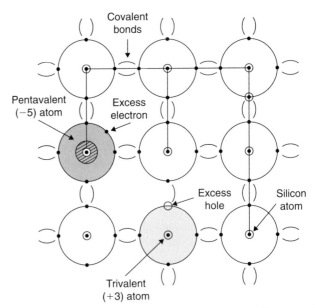

FIGURE 13-1 Diagram of a silicon wafer with excess electrons and holes.

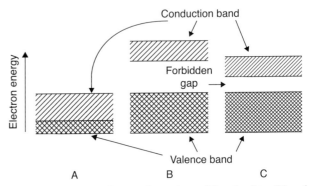

FIGURE 13-2 Energy diagram of a conductor (A), an insulator (B), and a semiconductor (C).

is known as the conduction band. A good conductor has a conduction band that is minimally separated from the valence band, and it takes only 4×10^{-21} J of energy to boost the electrons across the band gap into the conduction band. This energy is readily provided by the thermal energy available at room temperature (kT = 1.38×10^{-23} J/K × 300 K = 4×10^{-21} J). By contrast, a good insulator has a large band gap between the valence band and the conduction band, and it requires more than 10^{-18} J of energy to move an electron across the band gap from the valence band to the conduction band. An electron can be moved from the valence band to the conduction band of an insulator using the electrical energy (eV) of an applied voltage (1.6×10^{-19} C × 10 V) = 1.6×10^{-18} J). In a semiconductor, the gap between the valence band and the conduction band is intermediate between a conductor and an insulator and visible light (hc/λ = 6.63×10^{-34} Js × (2.99×10^8 m/s)/ (500×10^{-9} m) = 4×10^{-19} J) provides sufficient energy to boost an electron from the valence band to the conduction band (Figure 13-2).

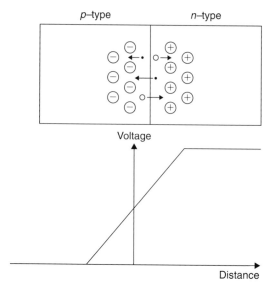

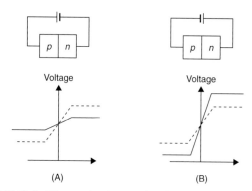

FIGURE 13-4 Diagram showing a p-n junction and the electrical potential across it when it is connected to a battery with a forward (A) or a reverse (B) bias configuration p-n junction without bias (---); p-n junction with bias (——).

FIGURE 13-3 Diagram of a p-n junction and the electrical potential across it.

The conduction of electricity through a semiconductor involves the movement of electrons through the conduction band and the movement of holes through the valence band. When an electron makes the transition from the valence band to the conduction band and becomes a free electron, it leaves behind a vacancy or hole in the valence band. Another electron in the valence band can then occupy the hole and consequently a new hole is created. When an electric field is applied to a semiconductor, the holes move to the negative pole of the battery and the electrons move to the positive pole.

The conductivity of a semiconductor can be increased by introducing impurities into the silicon, which is called doping the silicon (Shockley, 1956). When pentavalent atoms (e.g., arsenic, phosphorous, or antimony) are introduced into the tetravalent silicon, there will be one too many electrons to satisfy the quadravalent bonding system of the silicon. This extra electron can readily be removed from the valence electrons to become a free electron. When trivalent atoms (e.g., boron or gallium) are introduced as an impurity, holes are created in the quadravalent bonding system of the silicon. The semiconductors that have a surplus of negatively charged electrons are known as negative-type or n-type semiconductors, and the semiconductors that have a surplus of positively charged holes are known as positive-type or p-type semiconductors.

When a p-type semiconductor is brought in contact with an n-type semiconductor, the excess electrons in the n-type semiconductor diffuse to the p-type semiconductor and at the same time, the excess holes in the p-type semiconductor diffuse to the n-type semiconductor (Figure 13-3). Consequently, near the junction, the n-type semiconductor becomes positively charged and is an anode and the p-type semiconductor becomes negatively charged and is a cathode. The electric field set up by diffusion mitigates

the concentration-dependent diffusion of electrons to the p-type semiconductor and the concentration-dependent diffusion of holes to the n-type semiconductor and sets up an equilibrium condition. The recombination of electrons and holes right at the junction depletes the electrons from the conduction band, thus forming a nonconducting layer or a "depletion zone" at the junction. The depletion zone in the p-n junction results in a reduction in conductivity compared to the conductivities of the p-type and n-type semiconductors individually.

The depletion zone at the p-n junction can be reduced by connecting a battery in the forward-bias mode where the positive pole of the battery is connected to the p-type semiconductor and the negative pole of the battery is connected to the n-type semiconductor (Figure 13-4). In this arrangement, the electrons in the n-type semiconductor are repelled by the negative pole of the battery and move toward the junction and the holes in the p-type semiconductor are repelled by the positive pole of the battery and move toward the junction. At the junction, the recombination of electrons with holes further depletes the conduction band of electrons. The greater the forward-biased voltage, the smaller the depletion zone, and the better the p-n junction acts as a conductor.

The depletion zone in the p-n junction can be widened when a battery is connected to the p-n junction in the reverse-bias mode. In this arrangement, the positive pole of the battery is connected to the n-type semiconductor and the negative pole of the battery is connected to the p-type semiconductor. Consequently, the excess electrons in the n-type semiconductor move to the positive pole of the battery and holes in the p-type semiconductor move to the negative pole of the battery, further increasing the width of the depletion zone. The greater the reverse-bias voltage, the wider the depletion zone, and the less conducting the p-n junction becomes. The p-n junction functions as a diode because it lets current flow across it only in the forward-bias mode and not in the reverse-bias mode. When light is able to control the amount of current that flows through the p-n junction in the reverse-bias mode, that diode is known as a photodiode.

When light strikes a photodiode, held at constant voltage in the reverse-bias mode, the light excites an electron from the valence band to the conduction band, creating an electron hole pair. The electron then flows in the conduction band through the n-type semiconductor to the positive pole of the battery and the hole flows through the p-type semiconductor to the negative pole of the battery. The greater the light intensity, the greater the number of electron hole pairs created and the greater is the photocurrent. In the reverse-bias mode, the current is linear with light intensity, but this is not the case for the forward-bias mode.

As an aside, diodes, made out of aluminum gallium arsenide, aluminum arsenide phosphide, aluminum gallium phosphide, and indium gallium nitride will emit red, orange, green, and blue light, respectively, when a battery is connected in the forward bias mode. In the forward-bias mode, electrons and holes flow into the junction and recombine. Upon recombination, the electron falls from the conducting band to the valence band, converting electrical energy into radiant energy. The semiconductors used to create each color have different band gap energies. The wavelength (λ) of the radiated light depends on the energy of the band gap (E_{bg}) according to the following equation:

$$\lambda = hc/E_{bg}$$

where h is Planck's constant and c is the speed of light. Light emitting diodes (LED) and laser diodes are built on this principle. Light emitting diodes can be combined in a two-dimensional array to make high definition monitors. Some of the light emitting diodes used in these monitors are being made from semiconductors composed of organic polymers.

Millions of photodiodes can be combined into a two-dimensional photodiode array (Figure 13-5). A photodiode array captures images and converts them into electrical signals. The current emerging from each photodiode in the array is related to the light intensity that strikes that photodiode. Each photodiode acts as a picture element or pixel. There should be a one-to-one correspondence between the brightness of a point in the specimen and the current that emerges from the photodiode in the corresponding pixel. Currently, Cannon markets a digital camera (EOS 1Ds Mark

III) that has 21 million pixels packed into a 36×24 mm array. The area taken up by each pixel, known as the pixel pitch, is 6.4 μm. The arrays are made using CMOS (complementary metal-oxide semiconductor) technology. Other cameras, which are made for scientific imaging, give 21-megapixel resolution using a one-half inch, one-megapixel chip with subpixel imaging. Subpixel imaging involves the rapid shifting or wobbulation of the photodiode array within the camera using a piezoelectric device. A composite image is formed from the many images captured by a shifting array. As long as the signal from the chip can be captured quickly enough from each position, the movement of the array within the camera increases the spatial resolution, just as the movement of our eyes increases our visual acuity.

A charge-coupled device (CCD), which was invented in 1969 by Willard Boyle and George E. Smith at Bell Laboratories while trying to make a "picture phone," is similar to a photodiode array in that a CCD also converts light intensities into electrical signals. In both cases, light causes the formation of electron-hole pairs. However, a CCD differs from a photodiode array in that each element in a photodiode array acts as a diode, whereas each element in a CCD acts as a capacitor (Figure 13-6).

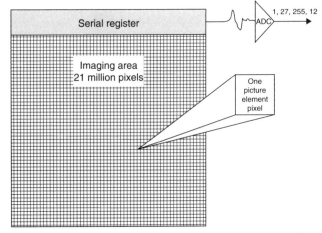

FIGURE 13-5 Millions of photodiodes in an array. Each photodiode constitutes one pixel.

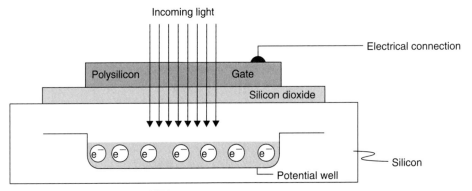

FIGURE 13-6 Diagram of a charge-coupled device (CCD).

Currently, Hasselblad produces a CCD with 39 million pixels in a 36 mm × 44 mm chip. Each pixel has a pixel pitch of about 6.8 μm. When light strikes a pixel in a CCD, it produces a free electron. An electrical potential is applied across each pixel so that the electrons freed by the photons flow from the silicon into a potential well where they are stored. The charge created by the light is stored until the well is discharged. When the well is discharged, the electrons flow from the potential well through a buried channel to the output of the CCD. The information from each pixel in a CCD chip must be transferred to a memory chip (Figure 13-7). This happens in a full-frame CCD, when the shutter closes and a sequence of voltage pulses cause the charge in each pixel of a parallel array, which is also called the parallel shift register, to be transferred one row at a time until it reaches one edge of the chip known as the serial shift register. One row of pixels at a time is transferred from the parallel register to the serial shift register. The charge travels down the serial shift register pixel by pixel, and then the charge of each pixel is converted to a voltage by an amplifier. The shutter then opens so that the next image can be acquired.

The full-frame transfer CCD is relatively slow, and consequently reliable images can be obtained only for slow or static processes. Some CCDs, with frame-transfer rates of 4 to 5 frames/second, work well under low-light conditions. The charge stored in the wells of these cameras is proportional to the product of the image intensity and the integration time. The slow cameras are also extremely precise because they use high-resolution electronics that are relatively slow; however, the precision electronics limit the temporal resolution.

Typical CCD sensors produce 30 complete images per second, which is what we perceive as real time. In order to capture images in real time, an interline transfer CCD can be used (Murphy, 2001). Interline transfer CCDs have alternating columns of imaging pixels and light-insensitive data-transfer columns. After image acquisition, the data from each imaging column is transferred to the data-transfer columns. In progressive scan cameras, the data-transfer columns are read from the top of the chip to the bottom. When each data-transfer column is read, the charges in each data-transfer column are transferred to the serial shift register. The charge travels down the serial shift register pixel-by-pixel, and then the charge of each pixel is converted to a voltage. While the data are being transferred, a new image is acquired by the pixels in the imaging columns. Image acquisition and data-transfer take less than 1/30 s.

Under low light conditions, too few electrons are stored in the CCD potential cells and the signal-to-noise ratio will be too low. The signal-to-noise ratio can be increased at the expense of spatial resolution by combining the electrons in neighboring pixels. This process is known as binning (Figure 13-8).

In a digital CCD camera, the voltage output from the serial register is digitized. That is, the voltage is converted into a series of integers that represent the spatial distribution of light intensities. These numbers then are transferred to a computer for processing and display. The bit-depth

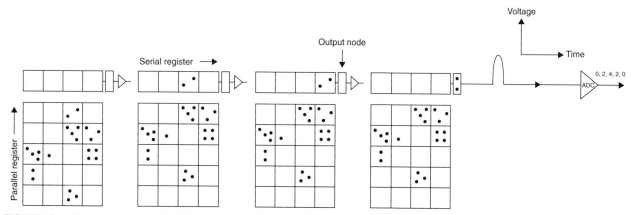

FIGURE 13-7 Producing a voltage signal by reading out a charge-coupled device.

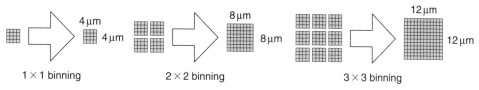

FIGURE 13-8 If the light intensity is too low, the electrons from neighboring wells can be combined in a process called binning. Binning increases the speed and decreases the spatial resolution of the imager in a way analogous to increasing the speed and decreasing the spatial resolution of film by making larger grains.

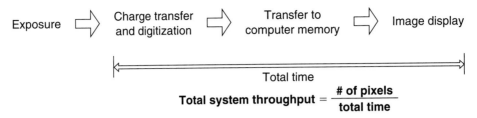

$$\text{Total system throughput} = \frac{\text{\# of pixels}}{\text{total time}}$$

FIGURE 13-9 Diagram of the processes that contribute to the total system throughput.

of the imaging chip determines the number of distinct shades of gray in which the intensity of each pixel can be expressed. The bit-depth of an n-bit analog-to-digital converter (ADC) is equal to 2^n. That is, an 8-bit, 10-bit, 12-bit, 16-bit, and 24-bit analog-to-digital converter produces 256, 1024, 4096, 65,536, and 16,777,216 shades of gray.

There are a number of ways to make a color CCD camera. One way is to spin a red, green, blue filter wheel over a single chip and capture three separate and sequential images that can be combined to give a full-colored image. Another way is to use a beam splitter that splits the incident light into three images that fall on three different chips, each one covered by a red, green, or blue filter. In a third method, each pixel is composed of three subpixels, one that is covered by a red filter, one that is covered by a green filter, and one that is covered by a blue filter. In the latter case, the voltage output of each subpixel goes through an 8-bit analog-to-digital converter to yield a 24-bit color-depth, giving $256 \times 256 \times 256 = 16,777,216$ different colors. Twenty-four bit color images can be stored as tagged image file format files (tiff).

The speed in which the signal is converted from the start of the light exposure to the final image display is known as the system throughput. The system throughput depends on the number of pixels and the number of frames per second. The system throughput is the number of pixels on the chip divided by the amount of time needed to display the image. A system with a throughput rate of 65,000 pixels/second can display an image from a 1.6 megapixel chip in about 25 seconds. One with 650,000 pixels/second can display the same image in 2.5 seconds. To display this image in real time (1/30s) would require a throughput of 48×10^6 pixels per second (Figure 13-9).

Video cameras still prove to be the best option for capturing images for extended lengths of time. The imager in a video camera is either a CCD chip, whose signal output has not been digitized, or a vacuum tube. Vacuum tubes, originally known as Crooke's tubes or cathode ray tubes, are the oldest form of technology used to convert light intensities representing an image into electronic signals. Photomultipliers are very sensitive light detectors made of vacuum tubes. Photons cause the emission of electrons from a photocathode placed in a vacuum tube. The photocathode is coated with a material whose work function is low enough to allow electrons to be released by photons in the ultraviolet and visible range. An applied voltage causes the electrons

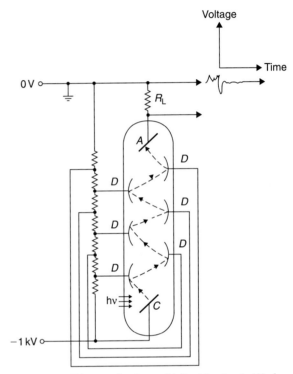

FIGURE 13-10 Diagram of a photomultiplier tube. Anode (A), dynode (D) and cathode (C).

to flow toward the anode, creating a current that is correlated with the light intensity. The photocurrent is then amplified by secondary emission by placing dynodes between the cathode and the anode.

The dynodes have potentials intermediate between the cathode and anode. When an electron is accelerated toward the first dynode, it strikes it with enough kinetic energy to release two or more secondary electrons from the dynode surface. These secondary electrons are attracted to the next dynode surface and hit with enough kinetic energy to release two or more electrons for each incoming electron. This process continues until the original electron from the photocathode causes the release of 10^6 electrons from the final dynode. Thus the original signal is amplified approximately one-million times (Figure 13-10)!

Philo T. Farnsworth and Vladamir Zworykin, among others, converted the vacuum tube into a video imaging tube by adding a light-sensitive layer (Farnsworth, 1989; Godfrey, 2001; Swartz, 2002). The light sensitive layer in video tubes can be made from a variety of light sensitive materials,

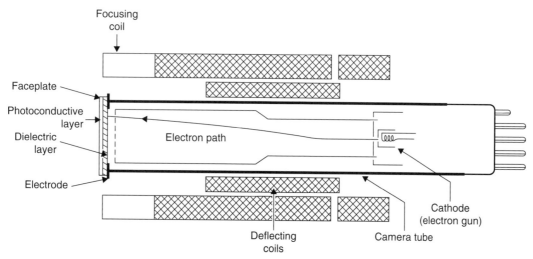

FIGURE 13-11 Diagram of a typical video tube.

including CdSe (Chalnicon), CdZnTe (Newvicon), PbO (Plumbicon), SeAsTe (Saticon), Si (Silicon or Ultricon), or Sb_2S_3 (Vidicon or Sulfide Vidicon). The light-sensitive layer is enclosed in a vacuum tube that is usually two-thirds to 1 inch in diameter and 4 to 7 inches long (Figure 13-11). The light-sensitive layer is placed at the front end of the tube. The light-sensitive layer is a sandwich that consists of a back-plate electrode, the dielectric layer, and the photoconductive layer.

The electron gun (or cathode) produces electrons that are accelerated toward the light-sensitive layer by placing a positive voltage (10–100 V) between the cathode and the back-plate electrode. The electrons are focused and guided by focusing, alignment, horizontal-, and vertical-deflecting coils. The electron beam forms a small focused spot on the target. The small focused spot defines the size of the pixel (Figure 13-12).

Farnsworth, who grew up on a farm, got the idea of scanning the beam across the tube from his day-to-day image of the back and forth motion used to plow a field. In a video camera, the horizontal- and vertical-deflecting coils sweep the beam across the target. As the beam sweeps across the target, it charges the back surface of the photoconductive layer with electrons. In the dark, the photoconductive layer acts as an insulator so that the electric charge remains on the back surface of the photoconductive layer. However, when light strikes the photoconductive layer, the resistance of the layer drops as electron-hole pairs are created. The decrease in resistance is proportional to the intensity of illumination.

Since the back-plate electrode is 10 to 100 V more positive than the cathode, the electrons begin to flow through the photoconductive layer once light causes the resistance to drop. The flow of current is then proportional to the intensity of illumination. The amount of current that flows in response to the incoming intensity of light is known

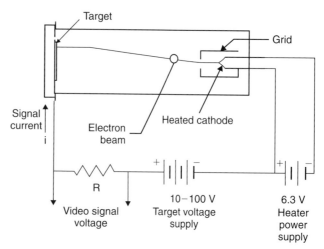

FIGURE 13-12 Diagram of how a video tube creates the voltage signal of each pixel.

as the responsivity of the tube. It is measured in μA/lm per ft^2 at 2854 K. The positive current flows from the positive terminal of the power supply (as defined by Benjamin Franklin) through a resistor, through the back-plate electrode, the dielectric layer, the photoconductive layer, the electron beam, the cathode, and then to the negative terminal of the power supply. Of course, the electrons actually move in the opposite direction. The current (I) flowing through the resistor (R) gives rise to the output voltage (V) according to Ohm's Law (V = I R).

The horizontal- and vertical-deflectors drive the raster. In general (in the 525/60 NTSC scan mode), the raster lands on a particular point on the target every 1/30 of a second. Therefore the given point or pixel accumulates light for 1/30 of a second (33.3 ms) and thus acts as an integrator of light intensity information by storing electron-hole pairs. The resistance decreases over this period, but current is not allowed to flow until the electron beam completes the circuit. When the electron beam lands on that pixel, current flows and an output voltage is created for that pixel. As the current flows, the electron beam simultaneously recharges the pixel and the resistance becomes high again.

A video camera converts a two-dimensional optical image into a sequence of electrical pulses (Figure 13-13). The magnitude of the pulse represents the intensity of the pixel and the temporal position of the pulse represents the spatial position of the pixel in the image. The image is scanned from left to right in a series of horizontal scan lines that move from top to bottom. If the scan starts at A it moves to A′. Then it is made to fly back to the beginning of the next scan line B. The fly back is much faster than the image scan and the signal is blanked out during this period. The blanking prevents the fly back trace from contributing to the signal.

The image is scanned so that every other line of the frame is converted into an electrical signal. Then the intervening horizontal lines are scanned and converted into an electrical signal. The two scans are then interlaced. This format is known as a 2:1 interlace. Thus two fields obtained at 1/60 s give rise to one frame that is made in 1/30 s.

The information about the scan rate is carried in the electrical signal that leaves the video camera. The horizontal (H) and vertical (V) sync pulses are inverted relative to the image signal so they can be distinguished from the image signal when they are combined to form the composite video signal. The composite video signal is 1.0 V peak to peak, where 0.286 V is used by the sync signal and 0.714 V is used by the image signal (Figure 13-14).

The number of horizontal scan lines in a video camera determines, in part, the vertical resolution of a camera (Figure 13-15). A typical video camera has 525 horizontal scan lines. This would be enough to distinguish 525 TV lines, which is equivalent to 263 black lines on a target separated by 262 white lines. However, the vertical resolution of video cameras in practice is not equal to the number of horizontal scan lines but is related to the number of horizontal scan lines multiplied by a factor known as the Kell factor, which is typically 0.7. Consequently, the vertical resolution of a typical video camera is about 368 TV lines.

It is easy to see why the vertical resolution of tube cameras is actually smaller than the number of scan lines. A test object, composed of three alternating horizontal bars, where each bar and each space between the bars is the height of a scan line, will be resolved by a video camera only if the bars and spaces coincide exactly with the scan lines. But if the bars fall equally across two scan lines, the three bars will be unresolved.

The 525 horizontal lines are scanned in 1/30 of a second, which means that 15,750 lines are scanned in one second and it takes 63.5 μs to scan each line. To have a horizontal resolution of 400 TV lines, which is approximately equivalent to the vertical resolution, the electronics would have to scan each vertical TV line in about 160 ns. By convention, a video camera can scan 15 TV lines accurately in 1/30 s, but the accuracy decreases as the lines get closer and closer together, because the electronic circuits in the camera are not fast enough to follow the rapid spatial changes in contrast. Consequently, while the voltage output of a video camera imaging a test pattern composed

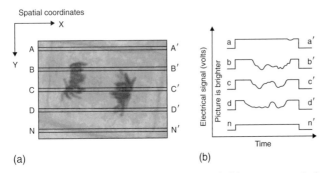

(a)   (b)

FIGURE 13-13 (a, b) Transformation of an optical image composed of points with different intensities into an electrical signal, whose amplitude at each point in time represents the intensity of a point in space.

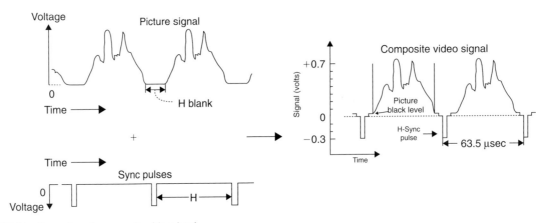

FIGURE 13-14 Generation of a composite video signal.

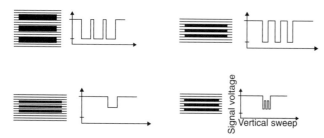

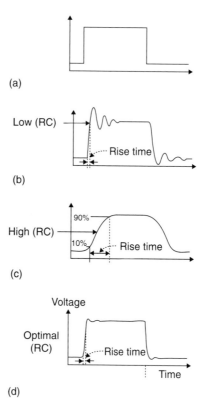

FIGURE 13-15 The vertical resolution is smaller than we would predict from the number of scan lines because it depends how the image on the imaging surface aligns with the scan lines.

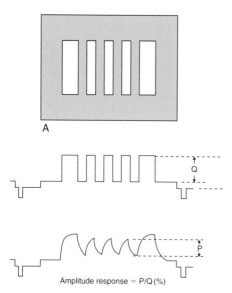

FIGURE 13-16 Mask used to determine the resolution of an imagine device (A), the ideal electrical response (Q), and the actual electrical response (P).

FIGURE 13-17 Increasing the resolution of the video camera by adjusting capacitors and resistors that influence the rise time. Square wave of object (a), under-compensated (b), over-compensated (c), optimally compensated (d).

of alternating black and white lines should be a square wave, in reality, it is a saw-tooth wave (Figure 13-16). This means that the contrast falls as the width of the black bar decreases. The resolution of the camera can be increased by adjusting the capacitance and resistance of the amplifier, which decreases the time constant of the circuit so that the rapid changes in intensity can be followed more faithfully (Figure 13-17). The downside to increasing the resolving power of the amplifier is the addition of more noise. The modulation contrast function (MTF) can be used to quantify the resolving power of a video camera, just as it was used to quantify the resolving power of film (see Chapter 5).

The Modulation Transfer Function (MTF) is a measure of how faithfully the image detail represents the object detail (Young, 1989). Here, the modulation transfer function relates the actual peak-to-peak amplitude of the video output for a target with any number of TV lines across the imaging surface (P) to the peak-to-peak output for 15 TV lines across the imaging surface (Q). Remember, the modulation transfer function is given by:

$$MTF = \frac{(H'_{max} - H'_{min})/(H_{max} - H_{min})}{(H'_{max} + H'_{min})/(H_{max} + H_{min})}$$

where H is the ideal amplitude of the output and H′ is the actual amplitude of the output. The subscripts min and max represent the minimum and maximum amplitudes, respectively. This equation usually is multiplied by 100% to give the amplitude response in percent. The resolution of a video camera is often given as the amplitude response $(P/Q \times 100\%)$ at 400 TV lines (Figure 13-18).

How good does the resolution of a video camera have to be in order to utilize the full resolving power of the microscope? Let us consider that we are using a 63x/1.4 NA objective lens and a 3.2x projection lens to view an object with 546 nm green light. The limit of resolution will be, according to the Rayleigh Criterion:

$$d_{im} = 1.22\lambda/(2NA) = 1.22(0.546\ \mu m)/((2)(1.4))$$
$$= 0.2379\ \mu m$$
$$0.2379\ \mu m = 0.0002379\ mm$$
$$= 4203\ line\ pairs\ per\ mm = 4203\ lpm$$
$$= 8406\ TV\ lines/mm$$

As a result of the magnification produced by the objective lens (63x) and the projection lens (3.2x), the limit of

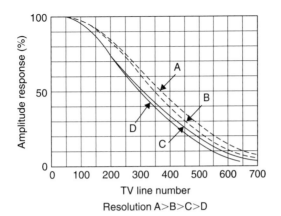

FIGURE 13-18 Graph depicting the amplitude response versus number of TV lines.

resolution at the image plane will be 8406/[(63)(3.2)] or 41.7 TV lines/mm. This is equivalent to 1059 TV lines in a camera with a 1-inch imaging surface.

Typically, tube video cameras do not have the required resolution. In order to fully utilize the resolving power of the microscope, the magnification of the projection lens has to be increased. We can also relate the pixel pitch of digital camera to the number of TV lines. A solid state CCD or CMOS imaging chip with a pixel pitch of 7 μm would be able to resolve about 428 TV lines per mm. This is more than sufficient resolving power to utilize the full diffraction-limited resolving power of the light microscope.

The limiting horizontal resolution of a video or digital camera depends on the light intensity. Normal cameras may have better resolving power than low-light intensity cameras at normal light intensities, but the low-light intensity cameras will have better resolving power than the normal camera at low-light intensities. This is because the normal camera will not be able to produce any contrast if the light intensity is too low to activate the imaging surface.

The signal output of a video camera depends on the intensity of light at the faceplate of the camera. The slope of this function, which relates the log of the signal output to the log of the signal input, is called the gamma of the video camera. The gamma is determined by the composition of the light-sensitive material and varies from 0.4 to 1.0. Many video cameras have a gamma of 0.4, which gives them a greater flexibility to capture images throughout the whole dynamic range of light intensities. Other video cameras with a gamma of 1.0 are ideal for quantitative work.

The dynamic range expresses the range of intensities on the faceplate to which the video camera responds meaningfully. Below the dynamic range, the signal is indistinguishable from the noise. Above the dynamic range, the camera is saturated and an increase in light intensity does not give rise to an increase in the signal output. The dynamic range is expressed as the ratio of maximum and minimum useful intensities and is typically 70:1 to 100:1, although it can be as high as 100,000:1 for a CCD used in astronomy.

The dynamic range curves are similar to characteristic curves (H-D curves) for film (see Chapter 5). The dynamic range of video camera can be varied with the gain control. The gain control may be manual or automatic. The automatic gain control (AGC) keeps the brightness of the image constant as the light intensity changes. Initially this is very helpful for qualitative work, but the automatic gain control must be shut off for quantitative work.

The imaging surfaces produce random signals at room temperature, because the band gap energy is not that much greater than the ambient thermal energy. A video camera can be cooled with liquid nitrogen or by thermoelectric Peltier coolers to reduce the thermal noise. This is particularly useful when doing low-light level microscopy.

Because of the finite nature of the band spectra of semiconductors, a given imaging surface does not respond equally to each and every wavelength. This must be considered when doing quantitative microscopy with more than one color of light.

The consumer electronics market has driven the development of digital imaging devices for still and video cameras and has moderated the costs. Consequently, for most applications, video and digital cameras marketed to the home electronics consumer may not only be less expensive but actually may be superior to the cameras that are developed for the scientific market, given that the cameras developed for the scientific market are often out of date before they even reach the market. This is especially true when considering cameras that will connect to rapidly evolving computers running on rapidly evolving operating systems. However, when the light microscopist is interested in specimens moving at high speeds or specimens emitting low-intensity fluorescent light, he or she should shop for a good scientific video or digital camera.

MONITORS: CONVERSION OF AN ELECTRONIC SIGNAL INTO AN OPTICAL SIGNAL

The image of the microscopic specimen taken by a video or digital camera can be viewed on a monitor or on a hardcopy printout. A monitor reverses the process that takes place in a video or digital camera by converting an electric signal into light (Figure 13-19). The vacuum tube-type video monitor is essentially the reverse of the video tube camera and has many of the same properties and features. A television can be used as a video monitor, but the video signal will first be converted into an RF signal. The RF signal has a lower bandwidth than the video signal and will thus limit the resolution.

The cathode ray tube type monitor contains a heated cathode, which is located in the neck of the picture tube. The heated cathode gives off a beam of electrons. The electron beam is accelerated by a high voltage anode. The electron beam is then focused on the phosphor screen, which is just

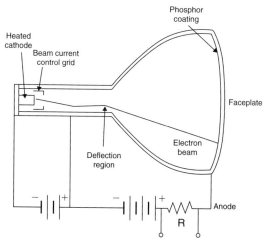

FIGURE 13.19 A monitor converts an electrical signal into an optical image.

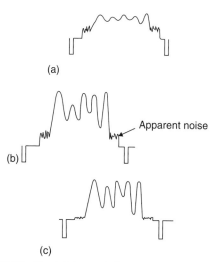

FIGURE 13-20 The brightness (offset or black level) and contrast (gain) controls on a monitor lets you process the image to get maximal resolution. Original signal (a). Signal after increasing the contrast (b) and decreasing the brightness (c).

under the faceplate of the picture tube. The accelerated electrons land on the phosphor screen and the screen gives off light. The position of the electron beam is controlled by the horizontal and vertical sync signals of the composite video signal. The brightness of the phosphor varies with the accelerating voltage and the current of the beam. The magnitude of the beam current is regulated by the amplitude of the video signal part of the composite signal. The amplitude of the video signal controls the beam current by controlling the beam current control grid.

The brightness of the monitor is not linearly proportional to the signal input. The brightness (I) is related to the signal (S) by the following formula:

$$I = S^{\gamma}$$

where γ is the gamma of a monitor and is usually between 2.2 and 2.8. The gamma of the monitor compensates the gamma of "television cameras," whose gamma is typically 0.45 and provides a camera operator with a large dynamic range. The nonlinear response of the monitor is good for a qualitative image but not for quantification. The image can be enhanced by using the brightness and contrast controls. These controls help to realize the maximal resolving power of the monitor (Figure 13-20). Cathode ray tube (CRT) technology has been shrunk to the size of a pixel. Image-viewing panels that are made from arrays of pixel-sized cathode ray tubes are known as surface-conduction electron-emitter displays (SED).

The need for displays on home computers and high definition televisions has driven the development of new and better display technologies. Now images can be viewed on liquid crystal displays (LCD) and plasma display panels (PDP). The pixels in an LCD are composed of liquid crystals sandwiched between parallel polars. The liquid crystal is a birefringent polymer that can be oriented in an electric field. In the absence of an electronic signal, the crystal is oriented so that no light passes through the crossed polars and the pixel is black. The greater the voltage, the brighter

the pixel is. A colored filter is placed after the second polar to make a colored pixel. It takes one red, one blue, and one green subpixel to make up a pixel.

The pixels in a plasma display panel are filled with a nobel gas, like neon or xenon. In the absence of an electronic signal, the gas in un-ionized and no light is emitted from the pixel and the pixel is dark. In the presence of an electronic signal, the gas becomes ionized and the ions collide with a phosphor, which then emits light. The greater the voltage, the more the gas is ionized and the more bright the pixel is. A color pixel is generated by combining a subpixel containing a red phosphor, a subpixel containing a blue phosphor, and a subpixel containing a green phosphor.

A hard copy of the image can be made with the aid of a digital printer. The resolution of the picture is determined by the number of dots per mm (typically given in dots/inch or dpi). Currently, the spatial resolution of a good ink jet printer (e.g., Canon PIXMA Pro9000) is about 4800×2400 dpi, which is equivalent to 189×94 dots per mm. Each dot is composed of a number of colors, so the resolution of the printer has to be divided between the number of inkjets used to make a single dot. In the 4800×2400 dpi printer described earlier, a single dot is composed of eight colors. Thus, the resolution is actually 24×12 dots per mm. This is equivalent to 0.04×0.08 mm between dots, which is almost at the limit of resolution of the human eye (0.07 mm between dots). The color depth of the printer is 6,144 colors. The consumer digital photography market is driving printer development, and the technology used in the printer as well as quality of the paper and the permanency of the ink are improving.

STORAGE OF VIDEO AND DIGITAL IMAGES

Hours and hours of video images can be recorded by videocassette recorders (VCR) and stored on videotapes.

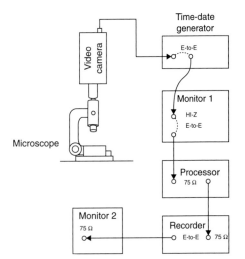

FIGURE 13-21 Diagram for connecting video components together.

Digitized images can be stored conveniently on computer hard drives, compact discs (CDs), digital video discs (DVDs), or any mobile memory stick with sufficient storage capacity. The images can be stored as is with no loss of resolution or with reduced resolution by compressing the images to save memory space.

CONNECTING A VIDEO SYSTEM

Digital imaging devices are connected to a computer through a FireWire or USB cable. Transferring analog signals require a little more care, since the last cable that passes a video signal must be set at a termination impedance of 75 Ω in order to maintain a 1.0V peak-to-peak voltage of the composite video signal (otherwise the highlights will be distorted and ghosts will occur). All other connections must be set at high impedance or Hi-Z. Video processors and videotape recorders are exceptions to this rule since they generate new video signals (Figure 13-21).

For more widespread and in-depth coverage on video and digital imaging, see Inoué (1986), Inoué and Spring (1997), and Murphy (2001).

WEB RESOURCES

Video and Digital Cameras

Low Light Intensity Cameras: (http://www.rulli.lanl.gov/) and http://www.dagemti.com/?page=product_detail&product_id=VE-1000-SIT

High Speed Cameras: (http://www.delimaging.com/products/ultrahs.htm)

High Resolution Cameras: http://www.dpreview.com, http://www.hasselblad.com/products/h-system/h3d.aspx, and http://www.manufacturingtalk.com/news/kne/kne118.html

Review of Digital Cameras: http://www.imaging-resource.com/DIGCAM01.HTM

Dage-MTI: http://www.dagemti.com/

Diagnostic Instruments: http://www.diaginc.com/cameras/

Fairchild Imaging: http://www.fairchildimaging.com/products/cameras/scientific/index.htm

Hamamatsu Photonics: http://jp.hamamatsu.com/en/product_info/index.html

Lumenera: http://www.lumenera.com/scientific/index.php

Optronics: http://www.optronics.com/

Princeton Instruments: http://www.piacton.com/

Optical Couplers for Mounting Digital Cameras on Microscopes

Edmund Optics: http://www.edmundoptics.com/onlinecatalog/DisplayProduct.cfm?productid=241

The Microscope Depot: http://www.microscope-depot.com/digadapt.asp

Microscope Vision and Image Analysis: http://www.mvia.com/Coolpix/clpxadpt.htm

Great Scopes: http://www.greatscopes.com/photo.htm

ScopeTronixs: http://www.scopetronics.com/mvp.htm

Printers

http://www.imaging-resource.com/PRINT.HTM

Chapter 14

Image Processing and Analysis

The seventeenth and eighteenth century microscopists hand-drafted the images they saw under the microscope. The images were subsequently engraved onto copper plates and printed (Espinasse; 1956; Harwood, 1989; Bennett et al., 2003; Cooper, 2003; Chapman, 2004; Jardine, 2005). The nineteenth and twentieth century microscopists, including myself, used photography to document microscopic observations and to enhance contrast. However, when taking photographic images, we had to process the film and develop the prints before we could become aware of the quality of the image and before we were able to share the image. By the time we saw the first photograph, the specimen was most likely dead, or otherwise unavailable for taking better pictures with optimal contrast and resolution.

Analog and digital processing allows instant feedback so that we may optimize the contrast and resolution in real time. In addition, measurements on the image that extract a tremendous amount of quantitative data, or image analysis, can be done on digitized images with high temporal and spatial resolution. Analog and digital image processing have been very welcome additions for the light microscopist (Walter and Berns, 1986; Bradbury, 1988, 1994; Moss, 1988).

ANALOG IMAGE PROCESSING

Above the atomic level, nature is continuous and analog, and unless we are working at the single photon limit, the intensities of light coming from the specimen are continuous and analog. Analog video cameras are used for many applications that require low light or require capturing images in rapid succession with high resolution. Even when we plan to convert information from the microscope into a digital image, we must first make sure that the analog signal is optimized to take full advantage of the dynamic range of the analog to digital converter. In fact, the introduction of digital technology has created the need for more and more varied analog circuits (Zumbahlen, 2008). We must understand how the components of those circuits can affect, both positively and negatively, the signal that is being processed.

An analog video camera transforms an optical signal into an analog electrical signal, which temporarily stores information about the light intensities of the image projected on the camera. Information of a two-dimensional image is stored in the amplitude of the electrical signal and the positional

information is stored, along with the sync signals, in the length of the electrical signal (Figure 14-1). Once the intensities of the image are encoded in an electrical signal, it is possible to use all the technology available in electrical amplifiers to enhance the image. Analog image enhancement lets us adjust the contrast in an image by manipulating the gray levels of neighboring pixels. Contrast is defined as the difference in brightness or color of two nearby pixels.

For simplicity, consider that the object is a stepped gray wedge with equal contrast between each step. The video output signal that arises from this object can be amplified with a linear amplifier so that the output signal is proportional

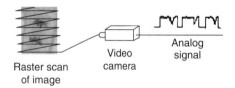

FIGURE 14-1 A video camera converts intensities into voltages.

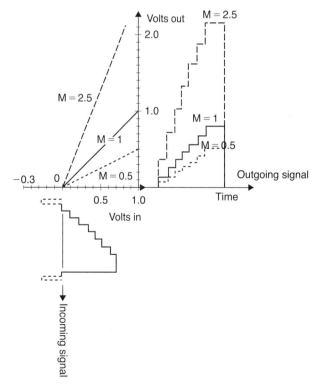

FIGURE 14-2 Linear amplification of an electrical signal that represents intensities.

to the input signal (Figure 14-2). When the amplifier has a gain of 1.0, the contrast of the video image is the same as the contrast of the optical image focused on the video camera. When the gain is greater than one, the contrast of very small details that lose contrast due to the process of video scanning is increased. When the gain is less than one, the large details, whose contrasts are not lost through the scanning process, are made more visible. Thus, a linear amplifier can selectively enhance the contrast of small or large details in the specimen. The following equation relates the output signal (in volts) of a linear amplifier to the input signal (in volts):

$$\text{output signal} = m(\text{input signal})$$

where m is the gain of the linear amplifier.

A nonlinear amplifier can selectively enhance the bright regions or the dark regions. The degree of nonlinearity of an amplifier is characterized by the gamma (γ) of the amplifier (Figure 14-3). A linear amplifier has a gamma of 1.0. When gamma is greater than one, the contrast between the brighter regions of the image is expanded. When gamma is less than one, the contrast in the darker regions of the image is enhanced.

With a nonlinear amplifier, the output signal is related to the input signal according to the following function:

$$\text{output signal} = (\text{input signal})^{\gamma}$$

We can decrease the brightness of the image by adding negative voltages to the video signal. In doing so, we redefine the black level. Normally signal voltages of 0 volts produce a black image. After adding -0.5 volts to the video signal, all input values less than 0.5 volts become black. By changing the baseline, we can increase the contrast between two neighboring bright points.

$$\text{output signal} = \text{input signal} - b$$

where b is the offset voltage.

By varying the gain, the gamma, and the offset, we can selectively increase the contrast of any given region of interest in the image and the output signal will be related to the input signal by the following equation:

$$\text{output signal} = m(\text{input signal})^{\gamma} - b$$

We can reverse the polarity of the video signal and cause the bright regions to become dark and the dark regions to become bright. The polarity control causes the input signal to be multiplied by -1 and then 0.7V is added to the product. In this way, bright regions with an input signal of $+0.7$ volts give an output signal of 0 volts, and dark regions with an input voltage of 0 volts give an output signal of 0.7 volts. The polarity control allows us to see different details in a manner analogous to how a negative phase-contrast objective brings out different details than a positive phase-contrast objective.

It is also possible to pass the signal through amplifiers with more than one stage so that the contrast of any given detail can be enhanced (Figure 14-4).

$$\text{output signal} = m_2[m_1(\text{input signal})^{\gamma 1} - b_1]^{\gamma 2} - b_2$$

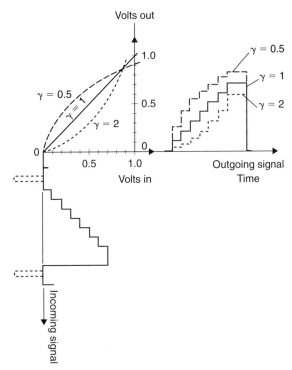

FIGURE 14-3 Nonlinear amplification of the electrical signal that represents intensities.

Although it is possible to adjust knobs and turn switches to create an image that has little relationship to reality, this is not the goal of analog image processing. Remember back to the time I discussed aberrations (in Chapters 2 and 4). An optical image produced by a single lens contains chromatic and spherical aberrations. These aberrations are mitigated by adding additional lens elements to correct the aberrations produced by the original imaging lens of the objective. Moreover, the objective lens can be considered an analog computer that diffracts the light coming from the specimen imperfectly and turns points in the specimen into Airy disks at the image plane. Therefore, if we used only a single lens to make an image, we would have a distortion of reality. The distortions of reality can be corrected either optically or electronically. Both types of corrections are not "cheating" as long as we take the advice given in Plato's Allegory of the Cave (see Chapter 1), act more like the line than the squiggle in *The Dot and the Line* (Juster, 1963), and understand the relationship between the object and the image. This is the reality of distortion. Using an analog image processor to correct an image is much like using an equalizer with a stereo to recreate the sound of the recorded music in a living room or car.

We can use an analog image processor to produce a pseudo-relief image by differentiating the input signal (Figure 14-5). Differentiating the video input signal will brighten one side and darken the other side of the image of an object with gradients in light absorption. The image will look like those produced by using oblique illumination, differential interference contrast, Hoffman modulation contrast, or single-sideband edge enhancement optics.

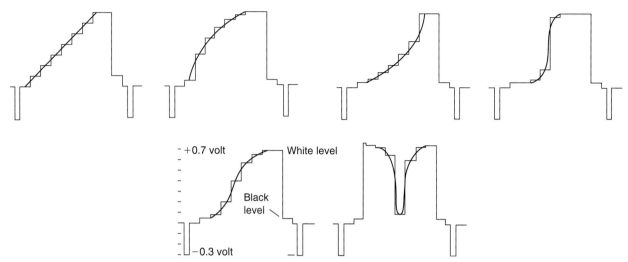

FIGURE 14-4   Examples of complex processing of the electrical signal that represents intensities.

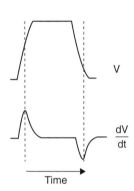

FIGURE 14-5   Taking the first derivative of the electrical signal produces a pseudo-relief image.

The fidelity of the video signal that emerges from an analog camera depends on the size of the object; the smaller the object, the less faithful is the video scan, and signals that should look like square waves appear as saw-tooth waveforms. We can use a sharpening filter to sharpen an image with an analog image processor. This works by taking the second derivative of the signal, inverting it, and adding the inverted, second derivative signal back to the original signal. This produces an image that more faithfully represents the object (Figure 14-6).

Operational amplifiers or op-amps, combined with a few electronic parts can be used to control the gain, gamma, offset, and polarity of the video signal. Operational amplifiers can also be used to differentiate and integrate the video signal. Op-amps are high gain amplifiers (Figure 14-7). They are relatively simple integrated circuits that contain approximately 24 transistors, 11 resistors, and a capacitor.

I will present a few simple cases in order to describe how op-amps can modify the video signal voltage. An op-amp has two input terminals and one output terminal.

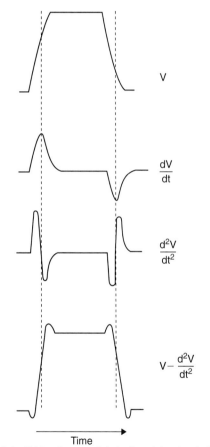

FIGURE 14-6   Taking the second derivative of the electrical signal and subtracting it from the original signal sharpens the image.

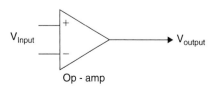

FIGURE 14-7   An operational amplifier.

An op-amp can be treated as a black box that obeys two golden rules (Horowitz and Hill, 1989):

- The output of an op-amp does whatever is necessary to make the voltage difference between the inputs zero. It does this through a feedback circuit.
- The two inputs of an op-amp draw no current.

I will describe how an op-amp can work as an inverting amplifier (Figure 14-8) using the golden rules and Ohm's Law (E = IR).

Since B is at ground, Rule I1 says that the amplifier will do everything necessary to make the input voltage A, which is connected to the feedback loop, the same voltage as ground (i.e., 0 V). This means that the voltage drop across resister R_1 (the input voltage, V_{in}) is cancelled by the voltage drop across the resistor R_2 (the output voltage, V_{out}). In order for V_A to be equal V_B, V_{out} must have the opposite sign of V_{in}.

$$V_{in}/R_1 + V_{out}/R_2 = 0$$

$$V_{out}/R_2 = -V_{in}/R_1$$

Thus the amplifier gain (V_{out}/V_{in}) is given by the following relation:

$$gain = V_{out}/V_{in} = -R_2/R_1$$

If R_2 and R_1 were equal, then the gain would be -1 and the output would be inverted. If R_2 were twice as large as R_1, the gain would be -2, if R_2 were half as large as R_1, the gain would be -0.5.

Figure 14-9 is an example of a noninverting amplifier. Again by Golden Rule I, V_{in} must equal V_A; but, what does V_A equal? V_A comes from a voltage divider as shown in Figure 14-10. According to Kirchhoff's current rule, there can be no "build-up" of current at a connection. Thus the current at V_A must equal the current coming from V_{out}.

$$I = V_A/R_1 = V_{out}/[(R_1 + R_2)]$$

We can now solve for V_{out}/V_A, which is:

$$V_{out}/V_A = (R_1 + R_2)/R_1$$

and since $V_A = V_{in}$, according to Golden Rule I,

$$gain = V_{out}/V_{in} = (R_1 + R_2)/R_1$$

$$gain = V_{out}/V_{in} = 1 + (R_2/R_1)$$

As long as $R_2 >> R_1$, the gain of the noninverting amplifier is given by R_2/R_1.

When the output of the op-amp and the negative (−) terminal is connected with a wire, then $R_2 = 0$, R_1 = infinite, and we get an amplifier with unity gain (Figure 14-11). Such an amplifier can also be used to cause a slight delay in the signal. Since an electric field travels almost at the speed of light, a current moving through a wire takes approximately 3 ns to travel 1 m. It takes more time for a signal to travel through an op-amp than through a wire the length of an op-amp. When a video signal is split into two, and one signal is subtracted from the other after one is passed through an amplifier of unity gain, a pseudo-relief image will result (O'Kane et al., 1990). Using combinations of op-amps, resistors, and capacitors hooked together in various combinations, we can get the derivative of a signal (Figure 14-12) where

$$V_{out} = -RC \, (dV_{in}/dt)$$

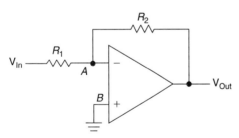

FIGURE 14-8 An operational amplifier configured to function as an inverting amplifier.

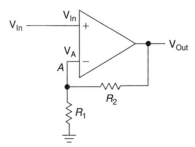

FIGURE 14-9 An operational amplifier configured to function as a noninverting amplifier.

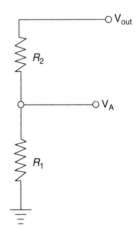

FIGURE 14-10 An equivalent circuit.

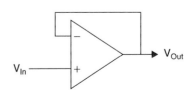

FIGURE 14-11 An operational amplifier configured as an amplifier with unity gain.

or we can get the integral of a signal

$$V_{out} = (1/RC)V_{in}dt + \text{constant}$$

using the combination shown in Figure 14-13. We can convert a logarithmic signal to a linear signal using a logarithmic converter (Figure 14-14). We can obtain the absolute value of an input signal using an active full-wave rectifier circuit (Figure 14-15). We can filter the signal with a low pass filter (Figure 14-16), a high pass filter (Figure 14-17), or a bandpass filter (Figure 14-18).

One or more video signals can be added together or subtracted from one another using adder (Figure 14-19) and subtractor (Figure 14-20) connections, respectively.

Analog image processing can be used to lower the contrast-limited limit of resolution of the light microscope proposed by Lord Rayleigh to the diffraction-limited limit of resolution proposed by Ernst Abbe. The only circuitry needed to reduce the limit of resolution to the limit imposed by diffraction is a gain (contrast) and offset (brightness) control (Figure 14-21).

We can appreciate better and deeper how the optical elements in a microscope "process an image" after we

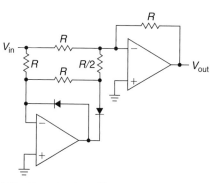

FIGURE 14-15 Operational amplifiers used to return the absolute value of the input.

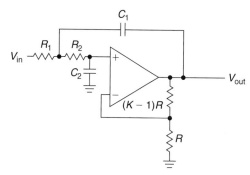

FIGURE 14-16 Operational amplifier configured as a low pass filter.

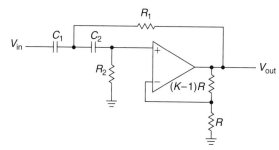

FIGURE 14-17 Operational amplifier configured as a high pass filter.

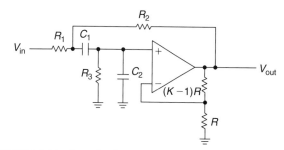

FIGURE 14-18 Operational amplifier configured as a band pass filter.

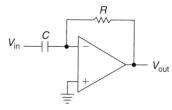

FIGURE 14-12 An operational amplifier configured as a differentiator.

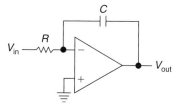

FIGURE 14-13 An operational amplifier configured as an integrator.

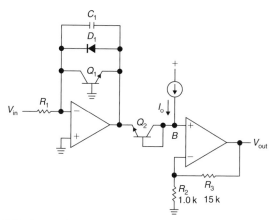

FIGURE 14-14 Operational amplifiers configured as a logarithmic converter.

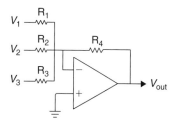

FIGURE 14-19 An operational amplifier configured as an adder connection.

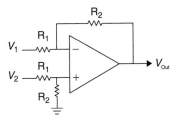

FIGURE 14-20 An operational amplifier configured as a subtractor connection.

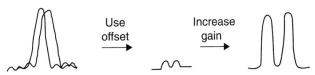

FIGURE 14-21 Beating the Rayleigh criterion using the offset and gain controls of an analog image processor.

understand the analogous process of signal conditioning by analog components.

DIGITAL IMAGE PROCESSING

Just as analog image processing put the power and versatility of electrical amplifiers at our disposal for image enhancement, digital image processing puts the power and versatility of computers at our disposal for image enhancement and image analysis (Fleagle and Skorton, 1991). A computer is an extremely powerful instrument and its image processing capabilities are almost limitless, especially if the user is able to program. In order to use a computer for image processing, we must use a digital camera or digitize the analog output from a video or analog CCD camera.

I will describe how to do digital image processing using two widely available digital image-processing programs—Image Pro Plus and NIH Image J. Image Pro Plus is a commercial digital image processor produced by Media Cybernetics Inc. (www.mediacy.com), and Image J is available from the National Institutes of Health as freeware (http://rsb.info.nih.gov/ij/). The National Institutes of Health also produces a program known as NIH Image, which can be used directly with a Macintosh computer (www.rsb.info.nih.gov/nih-image/). A PC-compatible version that can be used with a frame grabber board from Scion Corp is available free from Scion Corp (http://scioncorp.com/index.html). Many other commercial programs are available for image processing and for performing specialized techniques, including fluorescence resonance energy transfer (FRET), fluorescence recovery after photobleaching (FRAP), fluorescence in-situ hybridization (FISH), ratio-imaging, three-dimensional reconstruction, tracking movement, deconvolution, and more. The processes described in this chapter will be found in most commercial image software packages. The commands will have different, although similar names.

A digitizer in general converts an analog signal into an array of numbers, where the array of numerical values represents the spatial distribution of intensities in the optical image (Figure 14-22). The digitizer produces a discrete value for each pixel that represents the light intensity at each point in the optical image. The discrete number is known as the gray value of the pixel. Digitizers produce binary–coded gray values. The number of shades of gray is known as the bit-depth. The bit-depth produced by a digitizer equals 2^n, where n is the number of bits in the analog-to-digital converter. The greater the number of bits in the analog-to-digital converter, the more faithful is the image in terms of shades of gray, but the longer it takes to digitize each pixel. The number of colors that can be produced by a color digitizer is known as the color-depth and is given by 2^n, where n is the number of bits in the digitizer.

The numerical value of the brightness of each pixel can be viewed using the BITMAP ANALYSIS command in the Image Pro Plus processor. The length of the analog video signal that must be sampled before the digitizer is able to convert the information stored in the amplitude of the video signal into an integer defines the size of a pixel. If the digitizer is fast, the length of the video signal that must be sampled is reduced and the spatial resolution of the digitizer is improved. If the digitizer is slow, the length of the video signal that must be sampled is increased and the spatial resolution is worsened. The greater the bit-depth or color-depth of a digitizer, the slower the digitizer. It is currently possible to have great spatial resolution and color-depth, but only if the images are produced at a rate of less than 30 frames/second, which we perceive as real time.

According to sampling theory, the sampling interval must be at least twice as fast as the smallest variation in the analog signal that we want to resolve. This is known as the Nyquist criterion (Shotton, 1993; Castleman, 1993). It will be nice when 20 million pixels can be sampled in 1/30 second with 24-bit color-depth. For this to occur, the digitizer must be fast enough to digitize each pixel in 1.7 nanoseconds. This is equivalent to a throughput of 6×10^8 Hz or 600 MHz.

Once the image is digitized it must be stored in an active digital image memory known as a frame buffer. Image processing boards, known as frame grabbers, typically have more than one frame buffer so that images can be added to or subtracted from each other in real time. Before an image is digitized, it is imperative to make sure that the video signal that enters the digitizer has not saturated. This can be tested with the SIGNAL command in the Image Pro Plus digital processor.

The SNAP command digitizes a single video frame and places it in a frame buffer. The signal-to-noise ratio in the digital image can be increased with the INTEGRATE command, which gives an average value for each pixel in the frames that it sums. The INTEGRATE command also allows us to sum a given number of frames, but divide the

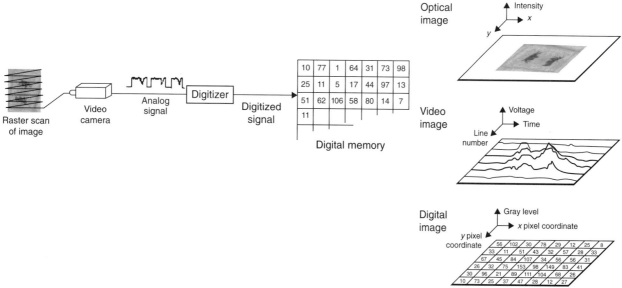

FIGURE 14-22   The formation of a digital image.

summed pixel values by any smaller number of frames, so that under low light conditions, the digital image processor acts as an integrator. There is a limit to the number of frames that can be integrated when imaging a moving specimen. A moving specimen will appear more and more blurry as more and more frames get summed.

It is still possible to increase the signal-to-noise ratio in a moving object by using the AVERAGE SEQUENCE function. This function averages a number of frames, stores the average as an image, then averages the same number of frames again to make the next image. The AVERAGE SEQUENCE is a good way to visualize movement. Digital image processors are also capable of performing RUNNING AVERAGES. With this function every time a new frame is added to the average image of N frames, the oldest stored image is lost. Although the RUNNING AVERAGEd image may have the same number of frames as the AVERAGE SEQUENCEd image, the RUNNING AVERAGE image contains information from the last video frames and the movement of the object will appear smooth, whereas the AVERAGE SEQUENCE will show "jumping movement" because n frames will be averaged, then the next n, then the next n, and so on, and the object appears to jump from average position to average position. The running average is useful when smooth path data of a moving object is desired.

ENHANCEMENT FUNCTIONS OF DIGITAL IMAGE PROCESSORS

I will describe a few functions of the Image J and the Image Pro Plus software. In each description, the first command is from Image J and the second command is from Image Pro Plus. Other image processing programs will have similar functions and command names. If you download the Image

J software and the accompanying sample images, you can perform the image processing operations described in this chapter while reading.

The function of a digital image processor is, in part, to reverse the degradation process that takes place in the optical system of the microscope. However, the reversal of the degradation process is made difficult because, as a consequence of diffraction and unintentional spatial filtering, identical images can be formed by different objects. Thus we must really understand the object and the imaging system in order to use digital image processing to give a more faithful representation of the object. Similar signal processing techniques can be used to our advantage when using chemical techniques such as optical spectroscopy. In this case, digital processing of the data is superior to using a chart recorder in that the digital processing can compensate for instrument artifacts, increase the signal-to-noise ratio through averaging, decompose complex signals into their component parts, and increase the ability to resolve overlapping peaks.

The signal-to-noise ratio in the image can be improved by reducing the glare in the microscope, which can come from many sources. The glare that cannot be removed optically can be removed digitally using background subtraction. Background subtraction is handy when doing fluore-scence microscopy—we can sharpen the wanted information by eliminating the out-of-focus fluorescence. In order to remove the out-of-focus light, we can slightly defocus the specimen and capture an image using the SUBTRACT BACKGROUND/BACKGROUND CORRECTION command. Then refocus the specimen, capture an image, and subtract the background image from the specimen plus the background to get a clean defect-free image. This function is also valuable when doing polarization light microscopy to eliminate the background light that passes through the analyzer because of the depolarization of light by the round surfaces of high numerical aperture objective lenses. First, we capture an image in the absence of a specimen and then subtract this image from the one that includes the specimen.

Typically, the contrast of biological specimens is extremely poor and the variation in gray levels in neighboring pixels is limited. The IMAGE ADJUST/CONTRAST ENHANCEMENT commands vary the contrast and brightness of the image manually. The pixel value of the enhanced image is related to the pixel value of the original image in the following manner:

$$\text{enhanced pixel value} = m(\text{original pixel value})^{\gamma} - b$$

where m is the linear change in contrast, γ is the nonlinear change in contrast, and b is the brightness. Changing these values digitally accomplishes the same goals as changing these values with an analog image processor as described earlier. The brightness of each of the colors that make up a color image can be adjusted individually using the ADJUST COLOR BALANCE/CONTRAST ENHANCEMENT commands. The mathematical equation that transforms the input values of a frame to the output values is known as a look-up table (LUT). Look-up tables can be used irreversibly before or reversibly after an image has been saved.

The image should be adjusted so that the full bit-depth is utilized and gray level values between black (0) and white (2^n) are used to make up the image. The ENHANCE CONTRAST/BEST FIT commands allow us to stretch out the gray scale in the image so that the darkest object detail becomes black and the brightest object detail becomes white. This increases the variations in brightness of the image. When only a limited number of gray levels are used, there is a high probability that a point in the image will have the same brightness as the neighboring points, and the image contrast will be low. Consequently, we may not be able to resolve two neighboring points. This is how increasing the range of intensities in the image can increase the contrast. The distribution of gray values can be plotted using the HISTOGRAM/DISPLAY RANGE commands. The bit-depth of the image can be reduced to 1-bit to produce a binary image using the THRESHOLD/THRESHOLD commands. A white-on-black image can bring out different details than the same specimen presented as a black-on-white image. We can produce either kind of image with the INVERT/INVERT IMAGE commands. The human eye is more sensitive to variations in color than it is to variations in gray levels. Therefore, the LOOKUP TABLES/PSEUDOCOLOR commands are used to change the gray scale to a pseudo-color display.

The Image Pro Plus system has a TEST STRIPS command, which allows one to obtain an optimum image relatively quickly, much like a photographer used to do in a darkroom. The TEST STRIPS command automatically produces a series of images with differing brightness, contrast, or gamma as well as images where any of the two functions are varied simultaneously.

Up to this point, I have discussed point operations where the gray value of each output pixel is directly related to the gray level value of the input pixel. Point operations do not change the spatial relations of the image—they affect only the contrast and brightness. On the other hand, spatial filtering modifies the value of a pixel based on the values of the neighboring pixels. Spatial filtering allows us to eliminate the high spatial angular wave numbers, which might represent noise, or the low spatial angular wave numbers that might represent out-of-focus light that could obscure the details of interest (Shaw, 1993). In fact, if we do spatial filtering on an image of a diffraction grating, we can repeat Abbe's experiment in which he changed the image of the grating by masking spots in the diffraction pattern (see Chapter 3). The Image J and Image Pro Plus systems perform spatial filtering in two ways: by operating on the diffraction plane, using Fourier mathematics and the FFT/FFT commands, and by operating on the image plane using spatial convolutions and the FILTERS/FILTERS commands.

The Fast Fourier Transform (FFT) command performs spatial filtering by transforming distance between points in an image into a series of sinusoidal terms with increasingly larger spatial angular wave numbers and smaller amplitudes. The image processing programs represent the Fourier transform as a diffraction pattern, with the smallest spatial angular wave numbers closest to the origin and the largest spatial angular wave numbers farthest from the origin. In the Image J system, we eliminate regions of interest in the Fourier Transform using an area selection tool. Once we select a spot that represents the spatial angular wave number that we wish to eliminate, we mask the spot using the FILL command. Once all the unwanted spots are removed from the Fourier Transform, we use the INVERSE FFT command to create the spatially filtered image. In the Image Pro Plus system, LOW PASS, HI PASS, SPIKE CUT, and SPIKE BOOST commands eliminate or enhance any desired spatial angular wave number. We then perform an INVERSE FFT on the Fourier transform to produce the spatially filtered image (Figure 14-23).

A pseudo-relief image, similar to those obtained using oblique illumination, differential interference contrast optics, or Hoffman modulation contrast optics can be rendered from a relatively transparent object using spatial filtering. First, we perform a FFT on the image. We then mask the left half, the right half, the top half, or the bottom half of the Fourier transform, and perform an INVERSE FFT to get a pseudo-relief image. We can produce a dark-field-like image by masking the central spot of the Fourier transform and then performing an INVERSE FFT. Spatial filtering offers unlimited opportunities to mimic optical processes that introduce the desired contrast into an image.

Spatial filtering can be done with the Image J and Image Pro Plus software on the image plane using convolution filters. Convolution filters are arrays of numbers. For example a 3 × 3 smoothing or low pass filter, which blurs the image

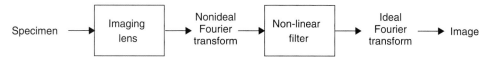

FIGURE 14-23 Use Fourier transforms to correct for the degradation of the image produced by the optical system itself.

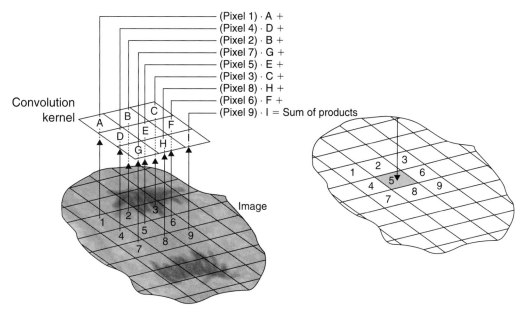

FIGURE 14-24 The convolution filtering process.

by making a target pixel more like its neighbors, looks like so:

1/9	1/9	1/9
1/9	1/9	1/9
1/9	1/9	1/9

The SMOOTHING/LO-PASS filter is overlaid on the image to be filtered. The values in each of the nine pixels in the image that coalign with the nine cells of the convolution kernel are multiplied by the corresponding value in the kernel. Then the nine products are summed and this sum is divided by the sum of the nine numbers in the kernel, which in the preceding kernel equals unity. The original value of the pixel in the center is replaced with the value of the sum of the products divided by the sum of the numbers in the kernel (Figure 14-24). The kernel is moved over the entire image, so that each pixel gets to be the target pixel. The result of spatial filtering with a low pass filter is the loss of high spatial angular wave number information. The subtraction of an image, which has been spatially filtered with a low pass filter, from itself, brings out the highlights in an image. It is one way to rid an image taken in a wide-field fluorescence microscope of the annoying out-of-focus fluorescence.

The SHARPENING/HI-PASS filter commands enhance the high spatial angular wave numbers in an image. A

3 × 3 sharpening or high pass filter, which sharpens an image by accentuating the differences between a target pixel and its neighbors looks like so:

−1	−1	−1
−1	8	−1
−1	−1	−1

Pseudo-relief images, reminiscent of those produced by oblique illumination, differential interference contrast microscopy, or Hoffman modulation contrast microscopy, can be produced from bright-field images by using the SHADOWS/HORIZONTAL EDGE commands. These commands produce a pseudo-relief image, using convolution kernels that perform the first spatial derivative of the target pixel with respect to the x-axis. The convolution kernels look like so:

−1	0	1
−1	0	1
−1	0	1

1	0	−1
1	0	−1
1	0	−1

or convolution kernels that perform the first spatial derivative of the target pixel with respect to the y-axis with convolution kernels that look like so:

−1	−1	−1
0	0	0
1	1	1

1	1	1
0	0	0
−1	−1	−1

Image J and Image Pro Plus have many different convolution kernels that let us selectively enhance any given detail. Both programs let us create original convolution kernels. An interactive Java-based tutorial in the use of convolution kernels can be found at the Molecular Expressions web site (http://micro.magnet.fsu.edu/primer/java/digitalimaging/processing/kernelmaskoperation/).

Many image enhancement programs use the "point spread function" to estimate how the optical system converts a point of light in the object to an Airy disk on the image plane. The image that we observe in the microscope is a composite image composed of the geometric image and an inflated image due to diffraction. Once the point spread function is known, all the light that is diffracted (i.e., convoluted) into the diffraction rings above, below, and in the plane of the point can be eliminated (i.e., deconvoluted) from the regions outside the point and put back into the point. In this way, the point becomes brighter, the surround becomes darker, and the resolution and contrast are maximized and the depth of field is minimized.

The point spread function is determined by considering what a point should look like in the image plane as a result of geometrical optics and in the absence of diffraction. The point spread function may be different in regions close to the optical axis and farther away. It is possible to image a test pattern composed of an array of points, and then have the computer minimize the difference between the brightness of the pixels in the real image and the predicted image. The deconvolution function minimizes this difference. The computer can store these functions for each point on the image plane and the deconvolution can then be used to eliminate the diffraction artifacts introduced into the image by the limitations of the optical system. Many commercially available software packages are capable of doing deconvolutions.

ANALYSIS FUNCTIONS OF DIGITAL IMAGE PROCESSORS

Digital image processors are very good at making quantitative measurements of counts, length, duration, and intensity, and thus can be used to analyze an image. In order to get real values for spatial measurements, the image processor must be calibrated using a stage micrometer and the CALIBRATE/SPATIAL CALIBRATION commands. Once the analysis system is calibrated in two dimensions, we can use a number of tools to measure the lengths and areas of regular or irregular objects. In the Image Pro Plus system, intensities at a point, along a line, or in various regions of interest (ROI) can be determined with the LINE PROFILE command. The INTENSITY CALIBRATION command allows us to convert the intensities into absolute units. Plug-ins for Image J that allow us to measure intensities along a line, intensities in various regions of interest (ROI), and the change in intensity over time are available at http://rsb.info.nih.gov/ij/plugins/index.html.

Plug-ins for Image J and macros in Image Pro Plus can be programmed to automatically count, measure, and analyze objects and detect how the positions of the objects change from frame to frame.

Digital image processing can be used to determine a multitude of parameters in a living cell simultaneously (Waggoner et al., 1989; Conrad et al., 1989). For example, one can correlate the distribution of proteins labeled with different fluorescent probes (e.g., for actin, myosin, and tubulin), or visualize fluorescently labeled actin at the same time one visualizes, with other fluorescent probes, the local $[Ca^{2+}]$ and pH, two factors that affect the polymerization of actin. To do so, one must capture a number of separate images taken with different excitation or emission wavelengths. A small amount of chromatic aberration in the objective lens would cause the excitation of different wavelengths to focus at slightly different depths in the sample and the different emission wavelengths to focus at slightly different image planes.

A digital image processor can be used to bring the images produced by different colors into register in the image plane by adding multipally-stained stained beads to the sample so that all the captured images contain a few beads (Figure 14-25). The center of mass of the beads is

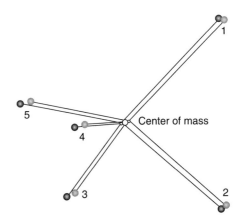

FIGURE 14-25 Using multipaly-stained beads to put in register images from multiply stained specimens.

determined by the image-processing program and used to align the different images so that they are in the same register with the same magnification. Multiple spectral parameter imaging is a powerful tool for studying dynamic processes that involve many things in live cells.

THE ETHICS OF DIGITAL IMAGE PROCESSING

As a final project, one of my students presented what I thought was his project on mitosis. He showed the rest of the class and me digital images of cells going through mitosis and described the process of mitosis and the mechanisms that brought them about. The images and the presentation on mitosis were beautiful. As he ended his presentation, he stunned us when he told us that the presentation was not about mitosis at all, but about the ethics of digital image processing. He told us that the cells he showed us were mature cells that no longer divided. He had "cut" chromosomes and mitotic figures from images of other dividing cells and flawlessly "pasted" them into the nondividing cells using Adobe Photoshop.

Throughout this book, I have discussed the relationship between the image and reality using examples from sensory psychology, Greek philosophy, geometric optics, physical optics, Fourier optics, all kinds of microscopy, as well as examples from analog and digital image processing. When studying microscopy, I have discussed the reality of distortion and the distortion of reality throughout each step in the process of forming an image (Figure 14-26). As presented in the last two chapters, the invention of transistors used to build analog and digital image processors by William Shockley, John Bardeen, and Walter Brattain has been an enormous benefit to microscopists (Riordan and Hoddeson, 1997; Shurkin, 2006). Nevertheless, like any new technology, digital image processing presents many ethical challenges. There are digital image processing techniques that are appropriate, inappropriate, or questionable, and digital imaging techniques should not be used as a substitute for correctly aligning the microscope. To ensure honesty and to protect your reputation, always archive the original unprocessed image and state any techniques used that may be questionable (Rossner and Yamada, 2004; Pearson, 2005; Couzin, 2006; Editorial, 2006a, 2006b, 2006c; MacKenzie et al., 2006). The rapidity in which an image can be obtained using digital techniques should not inadvertently lead you to think that a single image is representative of the specimen. As Simon Henry Gage (1917, 1941) wrote nearly a century ago, the "image, whether it is made with or without

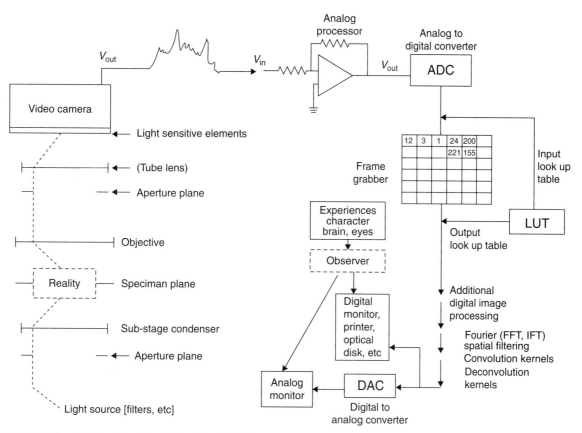

FIGURE 14-26   The distortion of reality and the reality of distortion.

the aid of the microscope, must always depend upon the character and training of the seeing and appreciating brain behind the eye." In the end, character and training is important in relating the image to reality.

It has been my pleasure to share my class with you. I will end this book with the words Francesco Maurolico (1611) used to end *Photismi de Lumine*: "Farewell most penetrating Reader! If you find the time, push these investigations further: or if, fortunate man [and woman], you stumble upon something better, generously share it with us."

WEB RESOURCES

Animation of *The Dot and the Line* made by Chuck Jones: *http://www. youtube.com/watch?v=OmSbdvzbOzY*

Commercial Digital Image Processors

BD Biosciences: http://www.scanalytics.com/
Media Cybernetics: http://www.mediacy.com/
Microscope Vision and Image Analysis: http://www.mvia.com/IASoftware/ia_software.html
Molecular Devices: http://www.moleculardevices.com/home.html
Volocity: http://www.improvision.com/products/velocity/

Helpful Web Sites on Digital Image Processing

Fred's ImageMagick Scripts: http://www.fmwconcepts.com/imagemagick/index.php
McMaster Biophotonics Facility: http://www.macbiophotonics.ca/imagej/index.html

Ethics of Digital Image Processing

www.swehsc.pharmacy.arizona.edu/exppath/micro/digimage_ethics.html

Free Publications

The field of digital imaging is changing rapidly. Links to free publications that report on digital microscopy are given here:
Microscopy Today: http://www.microscopy-today.com/
Microscopy & Analysis: http://www.microscopy-analysis.com/
Advanced Imaging: http://www.advancedimagingpro.com/
Biophotonics International: http://www.photonics.com/bioPhotonicsHome.aspx
BioTechniques: http://www.biotechniques.com/
The Spectrum: http://www.bgsu.edu/departments/photochem/research/spectrum.html

Chapter 15

Laboratory Exercises

LABORATORY 1: THE NATURE OF LIGHT AND GEOMETRIC OPTICS

The Spectral Composition of Light: The Decomposition and Recombination of White Light

Observe the sky light through red, yellow, green, and blue plastic filters. Describe and explain your results.

Observe a narrow beam of sunlight through a prism. How does the prism split the light? Why does the round sun appear as an elongated ellipse?

Observe light from a tungsten lamp, the fluorescent room light, a hydrogen lamp, and a sodium lamp with a spectroscope. Describe and explain your observations. Gas lamps are available from Arbor Scientific (http://www.arborsci.com/detail.aspx?ID=927).

Observe white stripes on black cards and black stripes on white cards using the water-filled prism. How does the image of the line depend on the thickness of the line and the background color? The water-filled prism is available from Carolina Biological Supply (http://www.carolina.com/product/physical+science/physics/light/water+prism.do?sortby=bestMatches).

Using three light sources, one with a red filter, one with a green filter, and one with a blue filter, shine the light on a white screen. Vary the intensity of each color to create magenta, cyan, yellow, and white. The color addition set is available from Arbor Scientific http://www.arborsci.com/detail.aspx?ID=419).

Light Travels in Straight Lines

1. Use the laser to prove to yourself that light travels in a straight line through air. Fill a Plexiglas tank with water and a scattering agent. Allow the laser beam to strike the tank perpendicular to the surface. Does light still travel in a straight line as it goes from air to water?

2. Can you find conditions in which light does not travel in a straight line as it propagates from air to water? The scattering tank and solution are available from Industrial Fiber Optics (http://www.i-fiberoptics.com/laser-kits-projects-detail.php?id=2450).

Demonstration of the Inverse Square Law

1. Use a quantum radiometer to measure the intensity of a candle, a tungsten light bulb, a fluorescent lamp, and a laser at various distances from the source.

2. Graph and describe your results. Do your data support the inverse square law? Why or Why not? The quantum radiometer is available from Li-cor Biosciences (http://www.licor.com/env/Products/Sensors/rad.jsp).

3. How do you think Benjamin Thompson (1794) determined the inverse square law before the invention of the quantum radiometer?

Geometrical Optics: Reflection

1. Demonstrate that the angle of reflection equals the angle of incidence by placing the edge of a front-surfaced mirror on the floor and the back against a wall. Shine a laser on the center of a mirror so that the incident and reflected beams are superimposed. Using a felt-tip pen, mark the point where the laser beam strikes the mirror.

2. Mark the position of the aperture of the laser on the floor with a piece of tape. Measure the distance (y, in m) from the aperture to the mirror with a tape measure.

3. Move the laser as far as possible (≈ 3 m) to the right of the original position in 0.05 m steps. At each position, called distance x, aim the laser at the spot on the mirror and measure how far the reflected beam is from the original laser position (distance x′).

4. Calculate the angle of incidence using the relation $\tan \theta = x/y$. To get the angle of incidence in degrees, make sure your calculator is set for degrees, then input x, divide it by y, and press $\tan^{-1}$. To calculate the angle of reflection use the relation $\tan \theta = x'/y$. Graph your results.

5. Do your data support the Law of Reflection? The laser is available from Industrial Fiber Optics (www.i-fiberoptics.com).

6. Using the laser and the setup just used, distinguish between specular and diffuse reflection. Aim the laser at a back surface mirror, paper, wood, and metal placed where the mirror was placed. Observe the light that is reflected from these surfaces by holding a piece of

glossy white paper at the spot where you expect to find the most reflected light. Describe the appearance of the reflected light.

7. Make a beam splitter based on the phenomenon of partial reflection. Pass the laser beam through a clean microscope slide that is positioned perpendicular (90°) to the beam. Then orient the microscope slide so that it is 45° relative to the beam. Notice that some light is transmitted and some is reflected. Observe the relative intensity of the transmitted and reflected light.

Geometrical Optics: Refraction

1. Add water to the refraction tank until it is one-half full (Figure 15-2). Start with air ($n_i = 1$) in the incident medium and water ($n_t = 1.3330$) in the medium of transmission. Vary the angle of incidence and measure the angle of refraction.
2. Do your results confirm the Snell-Descartes Law ($n_i \sin \theta_i = n_t \sin \theta_t$)? The refraction tank is available from Arbor Scientific (http://www.arborsci.com/detail.aspx?ID=934).
3. When the incident medium is air, the Snell-Descartes Law reduces to $n_t \sin \theta_t = \sin \theta_i$ or $n_t = \sin \theta_i / \sin \theta_t$.

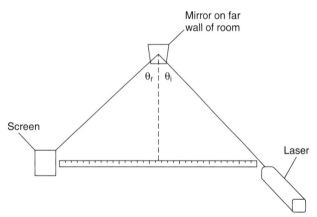

FIGURE 15-1   Setup for proving the law of reflection.

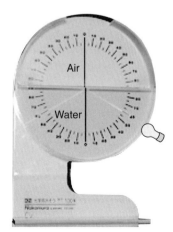

FIGURE 15-2   Setup for demonstrating the Snell-Descartes Law.

Assuming the validity of the Snell-Descartes Law, find the refractive index of the water.

Measure the Critical Angle of Reflection

1. Using the refraction tank, lower the light so that the incident medium is composed of water and the medium of transmission is composed of air. Do your results still confirm the Snell-Descartes Law ($n_i \sin \theta_i = n_t \sin \theta_t$)?
2. When the angle of transmission is greater than 90 degrees, the light is reflected back into the water instead of entering the air. Determine the critical angle that results in total internal reflection at a water/air interface. How does this angle compare with the calculated critical angle $1/n_i = \sin \theta_i$?
3. Orient a glass prism so that the incident light is totally reflected. Let your finger approach the air/glass interface where the light is totally reflected. You can see the evanescent wave jump from the glass to your finger.
4. Optical light guides (fiber optics) use total internal reflection to transmit light long distances along a cable with very little loss to the outside. Illuminate one end of a light guide with the laser and slowly bend the light guide until it is no longer able to confine the laser light. Why does the light exit the light guide?
5. Read a newspaper through a light guide that scrambles the image and one that transmits the image faithfully. How do you think the two types of light guides are constructed?

Refraction through Lenses

1. Put a double convex lens in a black bottom Plexiglas scattering tank that contains water and a scattering medium. Let the laser light propagate along the principal axis.
2. Raise and lower the laser so that the laser light is parallel to the principal axis. Can you find the focus of the lens? Do you see spherical aberration?
3. Repeat with a double concave lens. How would your results differ if the refractive index of the medium was greater than the refractive index of the lens?
4. The scattering tank and scattering solution are available from Web-tronics (http://www.web-tronics.com/opel.html\).

Measure the Refractive Index of a Liquid

1. Use a hand-held refractometer to measure the refractive index of distilled water (0% sucrose), a 2.5% (w/v) sucrose solution, a 5.0% (w/v) sucrose solution, a 7.5% (w/v) sucrose solution and a 10.0% (w/v) sucrose solution. Graph and explain your results.

2. Determine the refractive index of a 2.5% (w/v) aqueous solution of bovine serum albumin (BSA). How does the value of the protein solution compare with the values of the four sucrose solutions? Explain. In your own words, describe what the index of refraction of a material is.

LABORATORY 2: PHYSICAL OPTICS

In Laboratory 1, we experimented with geometrical optics, treating light as if it traveled as corpuscles in straight lines. However, we noticed that the inverse square law did not hold for all the light sources because some sources produce coherent light. Coherent light is defined as light composed of waves that maintain a constant phase difference between each other. These waves can interfere with one another. In this laboratory, we will see that light does not necessarily travel in straight lines, and that its behavior is consistent with the wave theory of light.

Observation of the Diffraction Patterns of Opaque Rectangles

1. Set up a laser and a lens or two on an optical bench so that a slip of card approximately 1 cm in width is illuminated with plane waves (Figure 15-3). View the far-field or Fraunhöfer diffraction pattern of the slip of card. Then insert a second lens between the object and the screen so that an image of the object is in focus on the screen (Figure 15-4). Describe the Fraunhöfer diffraction pattern and the image.
2. Repeat using slips of cards of various widths from 1 cm to 1 mm. Describe the Fraunhöfer diffraction patterns and the images.
3. How do they change as the width of the card decreases? Did you see what Newton missed?

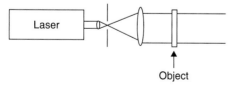

FIGURE 15-3 Setup for studying far-field diffraction.

Observation of Diffraction Patterns of Objects of Various Shapes

1. Set up a laser and a lens or two on an optical bench so that an object is illuminated with plane waves (Figure 15-3). View the far-field or Fraunhöfer diffraction pattern of the tip of a needle and the head of a pin positioned side by side on a screen. Then insert a second lens between the object and the screen so that an image of the object is in focus on the screen (Figure 15-4). In the presence of the second lens, the Fraunhöfer diffraction pattern will be in focus at the back focal plane of this lens, and can be viewed with a piece of ground glass. The focal lengths of the lenses will be determined by the geometry of the setup and the size of the room. The test objects are available in the physical optic lab sold by Industrial Fiber Optics, Inc. (www.i-fiberoptics.com).
2. After removing the second lens from the optical path, view the near-field or Fresnel diffraction patterns, of the tip of the needle and the head of the pin, 5 cm and 10 cm behind the objects. Draw the diffraction patterns. Draw the far-field or Fraunhöfer diffraction patterns of the tip of the needle and the head of a pin that are projected on the screen. Describe the relationship between the objects and their diffraction patterns.
3. Insert the second lens in the optical path. Describe the image of the objects on the screen. Use a ground glass to view the diffraction pattern at the back focal plane of the second lens. Describe this diffraction pattern.
4. Repeat this experiment, replacing the needle and pin with the edge of a razor blade, a square aperture, and a circular aperture. Describe the object, image, Fresnel, and Fraunhöfer diffraction pattern of each of the objects.

The Effect of Slit Width on the Diffraction Pattern

1. Observe the Fraunhöfer diffraction pattern produced by a slit made from two parallel razor blade edges 1 cm apart. Gradually move the edges closer together and observe what happens to the diffraction pattern. Describe your observations.

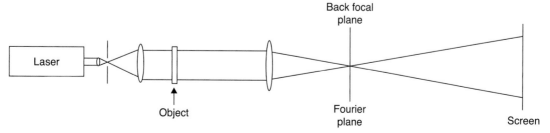

FIGURE 15-4 Setup for observing and operating on the diffraction pattern in the back focal plane.

2. Observe the diffraction pattern formed by two slits that are separated by d = 5.8×10^{-5} m. How many diffraction orders can you see? Do you see the same number when the room lights are on? What can you say about resolution and contrast?

3. Measure the distance between the slits and the screen (D, in m). Measure the distance between two distant diffraction spots and divide by the number of spaces between the spots (x, in meters). If the wavelength (λ) of the laser is 632.8×10^{-9} m, then the distance between the slits is given by:

$$d = (\lambda)(D/x)$$

4. Repeat this experiment for the second pair of slits that are 4.5×10^{-5} m apart, the third pair of slits that are 7.5×10^{-5} m apart, and the fourth pair of slits that are 10×10^{-5} m apart. Can you confirm that the distances between the two slits are correct? If you did not know the wavelength of the laser, but did know the distance between two slits, you could determine the wavelength of the laser from the formula.

Observation of Images and Fraunhöfer Diffraction Patterns of Slits and Grids

1. Observe the images and Fraunhöfer diffraction patterns of various objects, including single slits with various widths and coarse and fine grids. Vary the orientation of the various objects.

2. Describe the relationships between the object, Fraunhöfer diffraction pattern, and the image.

Observation of the Fourier Transform of an Object

1. For this section, note the distinction between spatial angular wave number and the angular wave number of the illuminating light waves.

2. Use a slide of a person's head as an object. View the image. View the diffraction pattern formed at the back focal plane of the second lens.

Spatial Filtering

1. Use a slide with a circular aperture as a mask. Using the slide of a person's head as an object, place the mask at the back focal plane of the second lens. Block out the higher diffraction orders with the mask. How does this affect the image?

2. In the back focal plane of the second lens, vary the position of the mask and describe the influence of the mask's position on the image.

3. Repeat this experiment using a grid as an object. Describe the diffraction pattern formed at the back focal plane of the second lens. Use either a circular aperture

or a slit as a mask. Draw and describe the image that is formed on the screen when:
- Only the central dot of the diffraction pattern is allowed to pass.
- Only one noncentral dot of the diffraction pattern is allowed to pass.
- Nine central spots of the diffraction pattern are allowed to pass.
- The central vertical row of spots of the diffraction pattern is allowed to pass.
- The central horizontal row of spots of the diffraction pattern is allowed to pass.
- A diagonal row of spots of the diffraction pattern is allowed to pass.

4. What generalizations can you make about the relationship between the diffraction pattern and the image?

LABORATORY 3: THE BRIGHT-FIELD MICROSCOPE AND IMAGE FORMATION

Illuminate a Specimen with Köhler Illumination Using a Microscope with a Mirror and a Separate Light Source

Establish Köhler Illumination:

1. Put a stained prepared slide on the stage.

2. Position a lamp with a coil filament so that the field diaphragm is about 9 inches from the microscope mirror. Tilt the lamp housing until an enlarged image of the filament is centered on the mirror.

3. Using the 10x objective and a low magnification ocular, move the concave mirror until the light is seen passing through the sub-stage condenser to the slide. Focus perfectly on the object. Adjust the mirror slightly to make the light central.

4. Rack up the sub-stage condenser until it is nearly touching the slide. Close the aperture diaphragm. Then focus the collecting lens of the lamp until a sharp, enlarged image of the filament is seen on the closed aperture diaphragm.

5. Close down the field diaphragm and focus the sub-stage condenser up or down until a sharp image of the field diaphragm on the lamp appears in the plane of the specimen. You may have to move the mirror slightly to keep the image of the field diaphragm in the center of the field.

6. Take out an ocular and observe the light in the microscope tube. Open up the aperture diaphragm until the light fills approximately 80 percent of the back focal plane of the objective lens.

7. Open the field diaphragm on the lamp until the light just fills the field.

8. Observe the prepared specimen.

Illuminate a Specimen with Critical Illumination Using a Microscope with a Mirror and a Separate Light Source

1. Put a stained prepared slide on the stage.
2. Remove the sub-stage condenser.
3. Position a lamp with a ribbon filament bulb so that the field diaphragm of the lamp is about 9 inches from the microscope mirror. Tilt the light source and the microscope vertically so that the rays from the light source strike the concave mirror surface at a 90-degree angle relative to the optical axis.
4. Using the 10x objective and a low magnification ocular, move the mirror until the ribbon filament is focused on the specimen. Adjust the mirror slightly to make the light central.
5. Observe the stained prepared specimen.

Establish Köhler Illumination on the Olympus BH-2

1. Make a thin hand section of a piece of cork, as Robert Hooke did in 1665. Put it in a drop of lens cleaner on a microscope slide and cover it with a #1½ cover glass. The lens cleaner helps eliminate air bubbles. Focus on the specimen with the coarse and fine focus knobs.
2. Close the field diaphragm and focus an image of the field diaphragm on the specimen plane by raising or lowering the sub-stage condenser.
3. Open the field diaphragm so that it almost fills the field and center it with the centering screws on the sub-stage condenser.
4. Open and close the aperture diaphragm so that you optimize resolution and contrast. Remove the ocular and look down the microscope tube.
5. At the position where the resolution and contrast are optimal, the light will fill about 80 percent of the width of the optical tube (Abramowitz, 1985, 1989). All microscope work from here on will be done with the room lights off. You can use a flashlight to find things, prepare specimens, and write in your notebook.

Observe the Diffraction Pattern of a Ruled Grating with the Olympus BH-2

1. Use a piece of exposed and processed Polaroid Instant 35 mm color slide film as a diffraction grating. Observe the film with the microscope and notice that it is made out of thousands of thin red, green, and blue lines. We will examine the diffraction patterns of this specimen in the back focal plane of the objective.
2. Insert a green interference filter on top of the field diaphragm. This filter blocks out the image of most of the red and blue lines of the grating. Set up Köhler illumination using a 10x objective and focus the specimen.
3. Close the aperture diaphragm as far as possible to produce axial illumination. You may also need to lower the sub-stage condenser to produce axial illumination.
4. Remove an ocular and insert the centering telescope. Focus the centering telescope on the diffraction pattern. Do not change the specimen focus.
5. Make a drawing of the observed diffraction pattern and identify the zeroth, first, and second orders. Turn the specimen.
6. What is the relationship of the orientation of the diffraction pattern with respect to the long axis of the grating? How would you expect the diffraction pattern to change if the lines of the grating were closer together?
7. Compare the diffraction pattern of the same grating produced by a 10x objective lenses with numerical apertures of 0.25 and 0.30.

Observe the Effect of the Numerical Aperture of the Objective on Resolution

1. Get a slide of the diatom *Pleurosigma*. Set up Köhler illumination. Focus on the silica cell wall of *Pleurosigma* with the 100x SPlan Apochromatic oil immersion objective lens equipped with an iris diaphragm. Put immersion oil on the slide and the top of the substage condenser.
2. Insert the green interference filter in the light path. Observe the diffraction pattern in the back focal plane of the objective using the centering telescope. Describe the diffraction pattern and compare it with the image observed with the ocular. Remove the green interference filter. Describe the diffraction pattern and image produced by white light.
3. Reduce the numerical aperture of the objective by closing the iris diaphragm in the objective. Close the iris diaphragm until the first- and higher-orders diffracted light are eliminated. Compare the diffraction pattern to the image of the specimen. Increase the aperture of the iris diaphragm in the objective and observe its influence on the diffraction pattern and the image.
4. Vary the opening of the field diaphragm. How does this affect the diffraction pattern and the image? What effect does the size of the opening of the field diaphragm have on glare?

Lens Aberrations

There are a number of possible aberrations of objective lens and these all can be corrected, but at a price. In this exercise, we will test a number of lenses in a series for various aberrations to see whether the correction is worth the cost.

1. To test objective lenses for chromatic aberration, place a slide of carbon dust on the microscope stage. Using white light, focus on the carbon particles and note the color fringes above and below the plane of perfect focus. The following effects will occur, depending on the type of objective:

   Achromat: Pronounced color fringes. Above the focus, yellow-green, below the focus, red or purple. The wider the fringes, the poorer the lens is corrected for color.

   Fluorite or semi-apochromat: Narrow color fringes of green and red. The narrower the width of the color fringes, the better the lens.

   Apochromat: No color fringes above and below the focus.

2. To test objective lenses for spherical aberration, place an Abbe Test Plate on the microscope stage (Slayter, 1957; Fletcher 1988).

3. Find the cover glass thickness marked on the test plate that is optimal for the objective lens being tested (0.17). Search for a minute, circular, brilliant spot of light in the film that shows several diffraction rings when in focus.

4. To perform the star test, focus above and below the plane of perfect focus with the fine focus knob. With an ideal objective that is free from spherical aberration, the concentric rings should be perfect concentric circles with equal spacing and brilliance and the rings should be identical at equal distances above and below the focus. If the downward rings are plainer than the upward rings, or the upward rings are plainer than the downward downward, the lens has spherical aberration. The greater the symmetry, the better is the correction for spherical aberration.

5. To test the effect of cover glass thickness on spherical aberration of the objective lenses, move the Abbe test plate to regions that have a cover glass thickness greater than optimal and less than optimal.

6. At each cover glass thickness, perform the star test by checking the symmetry of the diffraction rings as you over-focus and under-focus.

7. To test the objective lenses for flatness of field, place a stained prepared slide on the stage. Is the center of the image and the edge of the image in focus at the same time?

8. To test the objective lenses for distortion, place a specimen of a very fine grid, like the kind used for electron microscopy, on the microscope stage. Are all the lines parallel? Is the image of the grid distorted? Is there pincushion or barrel distortion?

Measurements with a Microscope

1. To measure the length of an object, insert an ocular micrometer in the ocular and place the stage micrometer on the stage. Carefully focus the stage micrometer. Both the stage micrometer and the ocular micrometer will be in focus and sharply defined. Turn the ocular so that the lines of the eyepiece are parallel to each other.

2. Determine how many intervals on the ocular micrometer correspond to a certain distance on the stage micrometer and then calculate the length that corresponds to one interval of the ocular micrometer. For example, if 35 intervals correspond to $200\,\mu m$ (0.2 mm), then one interval equals $(200/35) = 5.7\,\mu m$. This value is specific for each objective. Calibrate the ocular micrometer for each objective on your microscope. Put the calibration in a convenient place in your laboratory notebook.

3. Find the field of view number on the ocular. Divide the field of view number by the magnification of the objective to get the diameter (in mm) of the field of view. Remember to divide this distance again by the magnification introduced by any additional intermediate pieces. List the diameter of the field for each objective next to their calibration.

4. Measure the length of *Pleurosigma*. *Pleurosigma* is _____ µm long.

Many simple experiments that demonstrate the optical properties of the light microscope can be found in Quekett (1848, 1852), Wright (1907), Gage (1891, 1894, 1896, 1917, 1925, 1941), Belling (1930), and Oldfield (1994). Numerous suggestions of good microscopic objects can be found in Hogg (1898), Clark (1925), Beavis (1931), Popular Science Staff (1934), and Ealand (no date).

LABORATORY 4: PHASE-CONTRAST MICROSCOPY, DARK-FIELD MICROSCOPY, RHEINBERG ILLUMINATION, AND OBLIQUE ILLUMINATION

In this lab, we will observe transparent, nearly-invisible specimens under the microscope. We will use Gold Seal microscope slides and Gold Seal #1½ cover glasses, which are available from Ted Pella (http://www.tedpella.com/). Keep the lids closed so the slides and cover glasses remain dry and dust-free.

Phase-Contrast Microscopy

1. Before you obtain cells to view with qualitative phase-contrast microscopy, put a drop of water on a microscope slide so that the cells you will obtain will remain hydrated.

2. Make a peel of the epidermis from the convex side of the bulb scale by cutting out a 2 cm × 2 cm piece of a bulb scale that is four or five bulb scales deep into the onion. Snap the bulb scale so that most of it breaks in

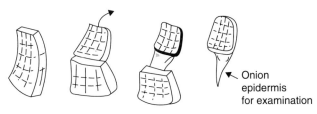

Onion
epidermis
for examination

FIGURE 15-5 How to make an epidermal peel form the convex side of an onion scale.

two (Figure 15-5). The epidermal layer will not break clean. Pull this layer back with forceps and place it on a drop of water on a microscope slide.

3. To make a peel of the epidermis on the concave side of the bulb scale, remove a bulb scale that is four or five layers deep within the onion. Make a checkerboard pattern of cuts with a razor blade on the concave side of the bulb scale. Pick up several 3 mm × 3 mm squares of epidermal tissue with forceps, and place the epidermal tissue sections on the drop of water.

4. Set up Köhler illumination in the Olympus BH-2 microscope. Observe an onion epidermal cell with bright-field optics using the 10x phase-contrast objective lens. Make sure that the condenser ring is in the 0 position. The cells are virtually invisible and the contrast is best when the specimen is slightly defocused.

5. Turn the sub-stage condenser turret to the "10" position to observe the cells with phase-contrast microscopy. In general, the number on the turret matches the magnification of the objective lens. Center the phase ring in the sub-stage condenser turret by removing one ocular and inserting the centering telescope. Focus the centering telescope so that the phase ring and the phase plate are in focus. Use the phase annulus centering screws to center the phase ring and align it with the phase plate. Remove the centering telescope and replace the ocular. Repeat for the 40x phase objective (Abramowitz, 1987).

6. Observe the onion epidermal cells with the 40x PL phase-contrast objective. Are the epidermal cells on the convex side different from the epidermal cells from the concave side? Can you see the structure in the nucleus? Can you see mitochondria moving, even dividing? Can you see the peripheral endoplasmic reticulum?

7. Document your observations with photographs. When you take photomicrographs, make a note of the specimen identity, the type of objective you use, its numerical aperture, its magnification, the total magnification, and the type of microscopy (e.g., phase-contrast). Also, note the camera and/or film type and the exposure adjustment. How do you set the exposure adjustment?

8. If the specimen is completely distributed throughout the bright background, set the exposure adjustment knob to 1x. If the specimen covers approximately 25 percent of the bright background, set the exposure adjust to 0.5x, to make the exposure longer. If the specimen covers less than 25 percent of the bright background, set the exposure adjust to 0.25x to make the exposure even longer.

9. Look at the pollen tubes of periwinkle (*Catharanthus roseus*) that have been growing on modified Brewbaker-Kwack medium for about two hours. Compare the images obtainable with the 40x positive and the 40x negative phase-contrast objectives. Which structures and processes show up best with the negative phase-contrast objective lens? Which structures and processes show up best with the positive phase-contrast lens? Do you see any halos?

10. Document your observations with photographs.

11. To make 10 ml of modified Brewbaker-Kwack Medium, mix together 3 g sucrose, 1 mg boric acid, 2 mg $MgSO_4 \cdot 7H_2O$, 1 mg KNO_3, 3 mg $Ca(NO_3)_2 \cdot 4H_2O$ and 58 mg 2-(N-morpholino) ethanesulfonic acid (MES). Add 7 ml distilled water and titrate to pH 6.5 with 1 N NaOH. Bring up to 10 ml with distilled water.

12. To practice quantitative phase-contrast microscopy, we will use the Olympus BH-2 microscope to measure the refractive index of the cytoplasm in endoplasmic drops made from *Chara*.

13. To make a 100 ml aqueous solution of artificial cell sap (ACS), add 0.596 g KCl, 0.147 g $CaCl_2$, 0.175 g NaCl, and 0.203 g $MgCl_2$. Take three 1 ml aliquots of ACS and make a 2% (w/v in ACS) bovine serum albumin (BSA) solution, a 4% (w/v ACS) bovine serum albumin solution, and a 6% (w/v ACS) bovine serum albumin solution. Measure the refractive indices of these solutions with a hand-held refractometer, and graph the refractive index vs. the concentration of bovine serum albumin.

14. Gently dry an internodal cell of the alga, *Chara corallina* with toilet paper. As soon as the cell gets a matte finish, cut one end off the cell and squeeze the cell contents into a drop of artificial cell sap (ACS), a drop of 2% BSA in ACS, a drop of 4% BSA in ACS, and a drop of 6% BSA in ACS.

15. Using the 40x PL objective lens, and a double counter, count the number of endoplasmic drops that are darker than the surround or lighter than the surround and plot the percentage of cells that are darker than the background vs. the refractive index of the medium.

16. Find the refractive index where 50 percent of the endoplasmic drops would be dark. This gives the percent bovine serum albumin that will make the average droplet invisible. Find the refractive

index from your graph that is equivalent to this concentration of bovine serum albumin. This is the refractive index of the cytoplasm in the endoplasmic drops.

17. The refractive index of the cytoplasm in the endoplasmic drops is: _____.

18. Your own cheek cells provide a readily available specimen that can be used for aligning a phase-contrast microscope at a moment's notice. You may want to observe your cheek cells.

Dark-Field Microscopy

1. Focus on the diatoms on a diatom exhibition slide or a diatom test plate. These slides are available from Klaus D. Kemp, Microlife Services (http://www.diatoms.co.uk/pg.htm). Raise the dark-field condenser on the Olympus BH-2 microscope so that it almost touches the bottom of the slide.

2. Turn the sub-stage condenser turret to DF. Open the field diaphragm only until all the specimens in the field are evenly illuminated, and observe the diatoms (Abramowitz, 1991). Do some of the diatoms appear colored? Why? How is the image influenced by putting water or immersion oil between the top lens of the condenser and the bottom of the slide? Document your observations with photographs.

3. If the specimen is completely distributed throughout the black background, set the exposure adjustment knob to 1X. If the black background is half-filled with the specimen, set the exposure adjustment knob to 2X, so that the exposure is twice as fast. If the black background is only 25 percent filled with the specimen, set the exposure adjustment knob to 4X, so that the exposure is four times as fast. If the black background is sparsely dotted (<25%) with specimens, set the exposure adjustment knob at 4X and turn the ISO dial to a higher number, to get an even shorter exposure.

4. Put a drop of pond water, a drop from a soil-water mixture, a drop of pepper-water or a drop of hay infusion on a microscope slide and cover with a cover glass. Observe the animalcules that Leeuwenhoek saw 300 years ago. If you have a slowly moving organism in the preparation, document your observations with photographs.

Rheinberg Illumination

1. You can make Rheinberg filters using colored theatre gels manufactured by Rosco International (http://www.rosco.com/). Choose a color for the outer part of the filter and cut an 18 mm × 18 mm square of this color. Punch a 3 mm hole in the center of the colored filter with a cork borer, and insert two or three layers of the color you want for the central stop in this hole.

2. The central color will give the color to the background. The color of the outer region will give the color to the specimen. Tape the colored filters to a microscope slide.

3. Hold the microscope slide against the sub-stage condenser of the Olympus BH-2 and translate the slide around until the background color at the image plane is uniform.

4. Tape the microscope slide in place to the sub-stage condenser. You are ready to view microscopic specimens on the stage using this quick and dirty form of Rheinberg illumination.

5. Establish Köhler illumination on your microscope and observe a drop of pond water. Document your observations with photography if the specimens are moving slowly enough.

6. If you would like to make better Rheinberg filters, make the filters so that the outer diameter of the filter is the same size as the filter holder at the front focal plane of the sub-stage condenser on your microscope. Make the hole for the central stop 1 mm larger than is necessary to fill the image plane. Place a black ring around the edge of the central colored spot.

7. Mount the colored filters between two pieces of round cover glass, using clear mounting medium. The cover glass should be the same size as the inner diameter of the filter holder. You can also make tricolor Rheinberg filters.

Oblique Illumination

1. Put an epidermal peel from the top of a vanilla orchid on a drop of water on a microscope slide and observe it with the Olympus BH-2 microscope set up for Köhler illumination. Then slightly rotate the sub-stage condenser turret until a pseudo-relief image appears. The nucleus in each cell will appear as prominent as they did to Robert Brown (1831), who discovered the nucleus in the epidermal cells of orchids.

2. Do you remember how difficult it was to see the nucleus in high school and freshmen biology, before you knew about oblique illumination? Document your observations with photographs.

Use of *camera lucida* (optional)

1. Put a prepared slide in the single ocular microscope equipped with a *camera lucida*.

2. Put a piece of blank paper under the mirror, and hold a pencil over the paper.

3. Look in the ocular; you should be able to see the image of the specimen and image of the tip of the pencil simultaneously. If you do not see the pencil tip, put a piece of white paper on top of the cover glass, and then translate and rotate the mirror so that you see the pencil

tip while looking through the *camera lucida*. Remove the paper from on top of the cover glass.

4. To get the correct contrast between the specimen and the pencil tip, rotate the wheel with neutral density filters in the *camera lucida* until the specimen and the pencil tip can be seen clearly . You may also need to vary the intensity of the microscope illumination and the intensity of the light used to illuminate the paper.

5. Trace the image of the specimen on the piece of paper.

LABORATORY 5: FLUORESCENCE MICROSCOPY

Setting up Köhler Illumination with Incident Light

Before you begin, set up Köhler illumination for both the transmitted light and the incident light on the Olympus BH-2. To set up Köhler illumination for the incident mercury light (Abramowitz, 1993):

1. Turn on the Hg Lamp
2. Place a slide on the stage.
3. Open the shutter slider all the way.
4. Rotate the iris diaphragm (A) and the field diaphragm (F) counterclockwise so they are open maximally.
5. Close the field diaphragm so that you can just see the edges. Center the field diaphragm by adjusting the centering screws. Open the field diaphragm.
6. Center the lamp carefully and gently with the two lamp centering screws, until the center of the field is maximally bright.
7. Adjust the collector lens with the focusing handle until the field is maximally bright and evenly illuminated.
8. You may find that you get better contrast with some specimens and even-enough illumination by focusing the lamp on the image plane.

Visualizing Organelles with Fluorescent Organelle-Selective Stains

1. In order to observe the endoplasmic reticulum of onion epidermal cells, make a stock solution of $DiOC_6(3)$ by dissolving 1 mg of $DiOC_6(3)$ in ethanol, and dilute the stock solution with water (1:1000) to make the working solution.
2. Prepare several pieces of onion epidermis and mount them on a drop of staining solution. Wait 5 to 10 minutes. Remove the staining solution with a pipette and immediately replace it with 0.05% n-propylgallate.
3. Observe the endoplasmic reticulum using the blue excitation cube. Document your observations with photographs.
4. To take micrographs of bright fluorescent specimens scattered over a dark field, typically you will have to

set the exposure meter so that the camera takes shorter exposures. If the staining is too intense and there is too much background staining, dilute to working solution down to 1:50 with distilled water. As a rule, as you get to know the specimen better, you require less stain and consequently achieve better selectivity and contrast.

5. To observe actin microfilaments in onion epidermal cells, prepare 2 ml of the staining solution by mixing 1.8 ml of Part A with 0.2 ml of Part B. To make 10 ml of Part A, add 5.5 ml of a 100 mM stock solution of Piperazine-1,4-bis(2-ethanesulfonic acid) (PIPES) buffer (pH 7.0), 0.055 ml of a 10 percent stock solution of Triton X-100 (to permeabilize the cells), 0.55 ml of a 100 mM stock solution of $MgCl_2$, 0.275 ml of a stock solution of ethylene glycol tetraacetic acid (EGTA, pH 7), 0.165 ml of a 100 mM stock solution of dithiothreitol (DTT), 0.165 ml of a 100 mM stock solution of phenylmethylsulphonyl fluoride (PMSF), 0.275 ml 200 mM Na^+ phosphate buffer (pH 7.3), and 0.44 g NaCl. To make 10 ml of 200 mM Na^+ phosphate buffer (pH 7.3), mix together 2.3 ml of 200 mM monobasic sodium phosphate and 7.7 ml of 200 mM dibasic sodium phosphate. Part B consists of a 3.3 μM stock solution of rhodamine labeled phalloidin dissolved in methanol.

6. Place several peels of the onion epidermis in the staining solution for 10 minutes in a warm place (35° C). Mount the epidermal peels in phosphate-buffered saline (PBS) that contains 0.05% (w/v) n-propylgallate. To make PBS, add 0.85 g NaCl, 0.039 g $NaH_2PO_4·H_2O$, 0.0193 g $Na_2HPO_4·7H_2O$ and enough distilled water to bring the solution up to 100 ml.

7. Observe the actin microfilaments using the green excitation cube. Document your observations with photographs.

Observe Organelles (e.g., Mitochondria and/or Peroxisomes) in Tobacco Cells Transformed with Organelle-Targeted Green Fluorescent Protein (GFP)

1. Make several hand sections of a piece of tobacco leaf in water or make epidermal peels of the leaf. Place the sections or peels in a drop of water on a microscope slide.

2. Using the blue excitation cube, look for the organelles in the epidermal hairs. Document your observations with photographs.

3. It is important to shut off the mercury lamp when you finish with your observations, since the lamp has a limited lifetime, and may explode if left on for extended periods when you leave the laboratory. Never turn a hot mercury lamp back on until it has cooled.

Determine the Resolution of Journal Plates

1. Place the 7x Bausch and Lomb measuring magnifier on a piece of white paper and rotate the lens until the scale appears in sharp focus.
2. Place the focused measuring magnifier over micrographs in a variety of scientific journals and determine the number of dots per mm or dots per inch (dpi) used to print the micrographs in those journals.

LABORATORY 6: POLARIZED LIGHT

Observation of CuSO$_4$ and Urea

1. Install the polarizing attachments to the Olympus BH-2 microscope. Make sure the polarizer and analyzer are in the crossed position.
2. Place drops of water on two slides and then add a few crystals of copper sulfate to one slide and urea to another. Watch what happens as the crystals dissolve under the polarizing light microscope with and without the first order red plate.
3. Describe what happens to the dissolving crystals in terms of the Michel-Lévy color chart (Delly, 2003).

Observation of Bordered Pits

1. Obtain a lightly stained prepared slide of bordered pits or make your own slide out of pine.
2. Observe the bordered pits under crossed polars with the polarized light microscope with and without the first-order wave plate. Document your observations with photographs, making sure to document the orientation of the polarizer, the analyzer and the slow axis (z′) of the first-order wave plate.
3. The cellulose microfibrils that make up the bordered pit are positively birefringent. How are they oriented in the bordered pit? Using the Michel-Lévy color chart, estimate the retardation of the bordered pit.

Observation of the Starch Grains of Potato

1. Grind a chunk of potato in water with a mortar and pestle. Pipette a drop of the starch grain solution on a microscope slide and cover with a cover glass.
2. Observe the starch grains under crossed polars with the polarized light microscope with and without the first-order wave plate. Document your observations with photographs, making sure to document the orientation of the polarizer, the analyzer, and the slow axis (z′) of the first-order wave plate.
3. The starch molecules that make up the starch grains are positively birefringent. How are the starch molecules oriented in the grain?

4. Using the Michel-Lévy color chart, estimate the retardation of the starch grain. If you like, prepare a thin section of the potato to see the starch grains *in situ*.

Observation of DNA

1. Using forceps, mount thin strands of DNA in a drop of lens cleaner on a microscope slide.
2. Observe with a polarized light microscope under crossed polars with and without the first-order wave plate. Document your observations with photographs, making sure to document the orientation of the polarizer, the analyzer, and the slow axis (z′) of the first-order wave plate.
3. The strands of DNA are negatively birefringent. Do the stands, whose physical axes are parallel to the slow axis of the first-order wave plate, show additive colors (blue) or subtraction colors (yellow-orangish)?
4. Put a lot of DNA on the slide and observe retardation-dependent colors with and without the first-order wave plate. Herring sperm deoxyribonucleic acid is available from Sigma Aldrich (http://www.sigmaaldrich.com/).

Observations of the Orientation of Microfibrils in the Cell Walls

1. Make a thin transverse hand-section of an *Asparagus* root using a razor blade. Mount it in distilled water.
2. Observe it with a polarized light microscope with and without the first-order wave plate. Document your observations with photographs. What can you say about the orientation of the positively birefringent cellulose microfibrils?
3. Using the Michel-Lévy color chart, estimate the retardation of the walls.
4. Make an epidermal peel from the bottom of a vanilla leaf. Mount it in distilled water.
5. Observe the stomata with a polarized light microscope with and without the first-order wave plate. Document your observations with photographs. What can you say about the orientation of the positively birefringent cellulose microfibrils in the guard cells?
6. Using the Michel-Lévy color chart, estimate the retardation of the walls.

Art with Polarized Light (optional)

1. Draw a mosaic-like picture on a piece of paper and cover it with a piece of glass.
2. Following the pattern drawn on the paper, cover the glass over each of the mosaic pieces with pieces of cellulose sheet.

3. Place more than one layer in some places and vary the orientation over other places. Cover the cellulose layers with another piece of glass and secure the two pieces of glass together with tape.
4. Observe your art between crossed polars. What happens when you turn one of the polarizers?

LABORATORY 7: POLARIZING LIGHT MICROSCOPY

Measuring the Retardation of Cell Walls Using a de Sénarmont Compensator (optional)

1. Insert the λ/4 plate into the Zeiss photomicroscope to make a de Sénarmont compensator.
2. Make an epidermal peel of the leaf of a vanilla orchid. Mount it in distilled water and observe it with crossed polars under monochromatic green light.
3. Rotate the analyzer until the walls that are diagonal to the slow axis of the λ/4 plate are brought to extinction.
4. The retardation of the cell walls is _____ nm.

Measuring the Retardation of Stress Fibers Using a Brace-Köhler Compensator (optional)

1. Make a slide of your cheek cells and observe them in the Zeiss photomicroscope with crossed polars.
2. Illuminate the cells with monochromatic green light. Find cells whose birefringent stress fibers are perpendicular to the slow axis of the λ/30 Brace-Köhler compensator.
3. Determine the retardation of the spindle using the traditional method and the Bear-Schmitt method.
4. The retardation of the stress fibers is _____ nm when using the traditional method and _____ nm when using the Bear-Schmitt method.

Procedure for Testing the Amount of Strain Birefringence in an Objective (optional)

1. Using the Zeiss photomicroscope, with monochromatic green light, put a clean slide on the microscope and set up Köhler illumination for the objective of interest.
2. Close the field diaphragm and set the analyzer to 0. Turn the polarizer until you get maximal extinction.
3. Replace one of the oculars with a centering telescope and observe the back focal plane of the objective lens. You should see a dark, wide, and symmetrical Maltese cross in the back focal plane of all strain-free objectives. If you do not see a dark, wide, and symmetrical cross, then your lens has strain birefringence.

4. Rotate the analyzer. The Maltese cross should open symmetrically into hyperbolas (in one dimension at a time). The arms of the hyperbolas should remain dark until they disappear beyond the field of view.
5. The objective lens is free from lateral and local strain birefringence if the hyperbola arms remain undistorted when the objective lens is rotated. The objective lens is free from radially symmetrical strain birefringence if the Maltese cross loses contrast and fades away with minimal change in its shape or position when a Brace-Köhler compensator is rotated. If the objective has radially symmetrical strain birefringence, the cross will open symmetrically into hyperbolas as it does in strain-free lenses when the analyzer is turned.

LABORATORY 8: INTERFERENCE MICROSCOPY

Qualitative Image Duplication Interference Microscopy Using the AO-Baker Interference Microscope

1. Attach the 40x shearing objective and sub-stage condenser pair to the microscope. Make a slide of your cheek cells.
2. Set up bright-field illumination by turning the polarizer to the OFF position and sliding the λ/4 plate so that you can see the λ/4 markings.
3. Rotate the analyzer to the unnumbered line position (where the analyzer is out). Focus the check cells and set up Köhler illumination.
4. To set up the microscope for interference microscopy, turn the polarizer to the INT position. Slide the λ/4 plate so that the λ/4 does not show. Rotate the analyzer to the 90-degree setting when using the 40x SH objective.
5. Find the astigmatic image of the cheek cell. This is the region through which the reference wave for the cell of interest passes. Make sure that there are no cells or schmootz in this position. The reference wave is 160 μm away from the specimen wave.
6. Remove an ocular and put in the centering telescope. Examine the back focal plane of the objective. When using diffuse white light you will see an interference pattern. This interference pattern will be sharp and clear if you are focused on a clear portion of the slide.
7. Broaden the first-order red fringe so that it fills the back focal plane of the objective as uniformly as possible. You can accomplish this by adjusting the two screws on the substage condenser screws. The more carefully you broaden the first-order red fringe and the more uniformly you illuminate the back focal plane of the objective, the more dependable your measurements will be. If the fringe does not cover the entire back

focal plane of the objective, close down the aperture diaphragm until the color is uniform. The better the initial broadening, the better the image. These adjustments must be made every time you change the slide.

8. Replace the ocular and view the specimen. It should be brilliantly colored. Vary the position of the analyzer and the colors will change. The image duplication interference microscope converts differences in phase in transparent specimens into visible differences in color.

Using the AO-Baker Interference Microscope to Weigh a Live Nucleus

1. After setting up the microscope as just described, insert a green (550×10^{-7} cm) filter in front of the light source. Mount some of your cheek cells in a drop of distilled water.

2. Select a cell. Measure the lengths (in cm) of the major axis (a) and the minor axis (b) of a nucleus using your optical micrometer and a stage micrometer. Calculate the cross-sectional area (A, in cm^2) of the nucleus with the following formula:

$$A = (\pi/4) \, ab$$

3. Rotate the analyzer until the background is maximally dark and note the reading of the calibration scale through the small lens. If the maximally dark area occurs in the blank area, turn the analyzer in the other direction until the background is maximally dark again. This reading is θ_1.

4. Turn the analyzer in the counterclockwise direction (so that the numbers increase) until the nucleus is brought to extinction. Note this reading on the calibrated scale. This reading is θ_2.

5. Repeat the preceding two steps until you have three readings for each in order to minimize subjective errors of judgment in determining where the darkest settings truly are.

6. Average the three pairs of readings and find an average θ_1 and θ_2.

7. To calculate the optical path difference (OPD, in cm), use the following formula:

$$OPD = [(\theta_2 - \theta_1)180°] \, 550 \times 10^{-7} \text{cm}$$

8. Calculate the mass (m, in g) of the nucleus using the specific refractive increment (0.0018 (100 ml/g), which equals 0.18 cm^3/g, and the following formula:

$$m = [(OPD)A]/0.18 \text{ cm}^3/\text{g}$$

9. The average mass of a nucleus is: _____ g. Note that DNA accounts for approximately half of the mass and protein accounts for the other half.

LABORATORY 9: DIFFERENTIAL INTERFERENCE CONTRAST MICROSCOPY AND HOFFMAN MODULATION CONTRAST MICROSCOPY

Differential Interference Contrast Microscopy

1. Mount cheek cells on a slide and focus on the slide using bright-field optics and Köhler illumination. Use the SPLAN objectives.

2. Slide in the polarizer that is attached to the sub-stage condenser. Slide in the differential interference contrast beam-recombining prism with the built in analyzer and first-order wave plate.

3. Rotate the sub-stage condenser turret to the red number that matches the magnification of the objective lens. Vary the contrast by turning the knob on the beam-recombining prism. Which color gives the best contrast? How does the direction of shear influence the image?

4. Optically section the cells. Document your observations with photographs.

5. Place the Diatom Test Plate with 8 forms on the microscope. The longest diatom is *Gyrosigma balticum*, the one next to that is *Navicula lyra*, the next one is *Stauroneis phoenocenteron*, then *Nitzschia sigma*, *Surirella gemma*, *Pleurosigma angulatum*, *Frustulia rhomboides*, and *Amphipleura pellucida*. Observe with the SPLAN 40 objective.

6. Rotate the stage so that the long axes of the diatoms are either parallel or perpendicular to the direction of shear. How does the orientation of the diatom affect the image? Pay particular attention to the dots and striations in *Stauroneis phoenocenteron* and *Nitzschia sigma*.

Hoffman Modulation Contrast Microscopy

1. To use Hoffman modulation contrast optics, remove the regular sub-stage condenser and replace it with the Hoffman modulation contrast substage condenser, and remove the regular objectives and replace them with the Hoffman modulation contrast objectives. Make a slide of cheek cells and place the microscope slide on the stage.

2. Rotate the Hoffman modulation contrast substage condenser turret to the bright-field position (no colored dot). Set up Köhler illumination using the 10x objective lens.

3. Remove an ocular and replace it with a centering telescope. Focus the centering telescope to get a sharp image of the modulator plate in the back focal plane of the objective.

4. Move the specimen out of the way so that a clear area of the slide is visible. Set the Turret to the 10x position (yellow dot).

5. Put the polarizer (called the contrast control filter) above the field diaphragm. Note that rotating the polarizer causes a change in brightness of half of the slit. The region whose brightness changes is called region 2.

6. Move the slit at the front focal plane of the substage condenser, using the knurled knob until region 1 of the slit is superimposed on the gray region of the modulator plate and region 2 of the slit is in the clear region of the modulator plate. When the slit is correctly aligned, region 2 of the slit will be completely black when the polarizer is rotated to the crossed position.

7. Repeat the slit alignment step for each objective making sure to match the correct turret position with each objective. Use the green dot position for the 20x objective, the blue dot position for the 40x objective, and the white dot position for the 100x objective. Remove the centering telescope and replace it with the ocular.

8. Reestablish Köhler illumination. As you view the specimen, adjust the contrast control filter to get optimal contrast. The graininess of the image increases as the contrast increases, so find a good compromise. Document your observations with photographs.

Seeing "Animalicules" or "Living Atoms" in Pseudo-Relief

1. Make a slide with a drop of pond water and/or make a variety of slides, each containing a different organism. View your slides with differential interference microscopy and Hoffman modulation contrast microscopy. Compare and contrast the images obtained with each method.

2. Take photographs of the organisms that move slowly. Increasing the viscosity of the sample with Protoslow or methylcellulose will slow down the organisms. Samples of the amoeba, *Chaos carolinensis*; the ciliates, *Blepharisma*, *Paramecium bursaria*, *Stentor* and *Vorticella*; the coelenterate, *Hydra*; the nematode, *Turbatrix aceti*; the rotifer, *Brachionus*; the tardigrade, either *Milnesium* or *Hypsibius*; and the crustaceans, *Gammarus*, *Cyclops*, *Daphnia pulex* and *Daphnia magna* are available. They can be obtained from Carolina Biological Supply (http://www.carolina.com/).

LABORATORY 10: VIDEO AND DIGITAL MICROSCOPY AND ANALOG AND DIGITAL IMAGE PROCESSING

Making Videos of Moving Microscopic Objects

1. Remove the semiautomatic camera from the microscope and replace it with the c-mount adapter.

Attach a video camera to the c-mount adapter. Attach a coaxial cable with BNC connectors from the VIDEO OUT terminal of the video camera to the VIDEO IN terminal in the videocassette recorder (VCR). Connect the VIDEO OUT on the back of the VCR to the VIDEO IN terminal of the monitor. Make sure that you set the IMPEDANCE SELECT switch on the monitor to 75 ohms.

2. Turn on the microscope, the video camera, the VCR, and the monitor. Set the beam splitter so that either 80 or 100 percent of the light goes to the video camera and 20 or 0 percent goes to the oculars. You may also use the digital camera to capture images by using the continuous mode. Plug the digital camera (e.g., Nikon 4500 attached to an optical coupler) into a video capture card (http://www.winnov.com/Home.aspx).

3. Choose one of the specimens that show motility. For example, mitosis in *Tradescantia* stamen hairs, pollen tube growth in *Catharanthus*, cytoplasmic streaming in *Tradescantia* stamen hairs, onion epidermal cells or *Chara* rhizoids, light-induced chloroplast movements in *Elodea* or *Mougeotia*, and the various swimming movements that occur in the protozoa that live in pond water.

4. Choose the type of microscopy that best brings out the process that you want to study (e.g., bright-field, dark-field, phase-contrast, polarization, fluorescence, differential interference microscopy, or Hoffman modulation contrast microscopy.

5. Start taping the sequence when you are ready. You must make a five-minute video for this course on any specimen or process you like that takes advantage of the strengths of video microscopy in studying the movements that occur in microscopic objects. We will show the videos after we eat dinner at my house, so they must also be entertaining. You can dub music over them and add titles. You must write a one- or two- page paper that describes what organism or process you are studying and the type of optics and video equipment you are using.

6. Remember the x-y rule: Take a sequence that is long enough before moving the specimen. It is annoying to the viewer if you take too many short sequences trying to find a better position. Remember the z rule: It is also annoying to the viewer if you try to incessantly improve the focus while recording.

Analog Image Processing

1. Remove the semiautomatic camera from the microscope and replace it with the c-mount adapter. Attach an analog CCD camera (e.g., Dage-MTI CCD 72) to the c-mount adapter. Notice the specifications of the camera. What do the specifications mean?

2. Display the gray level test pattern generated by the camera. Adjust the brightness and contrast of the monitor so that all the gray levels are clear.

3. Turn on the Olympus BH-2 microscope, the camera control box, the computer, and the monitor. Set the beam splitter to the 80/20 position so that 80 percent of the light goes to the video camera and 20 percent goes to the oculars.

4. Set up the bright-field microscope with Köhler illumination. Focus with the SPLAN 40 lens. The cells should be visible when the iris diaphragm is closed and invisible when the iris diaphragm is open and the resolution is maximal. Keep the iris diaphragm most of the way open.

5. Adjust the manual gain (contrast) and manual black level (brightness) so that the cheek cells, which are invisible when looking through the oculars, become visible on the video monitor. Vary the gain and black level to maximize the resolving power of the bright-field microscope. The camera box, which is a sophisticated analog image processor, has the following controls.

6. The GAIN control affects the amplitude of the video signal by acting like a linear amplifier. When the GAIN SWITCH is in the AUTOMATIC position, the camera automatically tracks and adjusts the whitest portion of the image to an internally preset value. When the switch is in the MANUAL position, the camera maintains a fixed gain that is set by adjusting the GAIN KNOB.

7. The BLACK LEVEL control affects the amplitude of the video signal by changing the voltage that produces a black image. When the BLACK LEVEL is on PRESET AUTOMATIC, the camera automatically sets the darkest level to an internally preset value. The VARIABLE AUTOMATIC setting automatically adjusts the image to the black level set by the BLACK LEVEL KNOB. When this switch is on VARIABLE MANUAL, the camera maintains a fixed black reference that' is set by adjusting the BLACK LEVEL knob.

8. The GAMMA control affects the amplitude of the video signal by acting like a nonlinear amplifier. Turning the gamma from 1.0 to 0.45 increases the contrast of the low light details in the specimen and decreases the contrast of the brighter details. The POLARITY switch changes black to white and white to black by passing the signal through an inverting amplifier.

9. The camera box has other analog image processing capabilities. The STRETCH switch gives additional gain to low light level signals by turning on a nonlinear amplifier. The BANDWIDTH control increases the speed of the amplifier so you get better resolution at the cost of more noise. The ENHANCE knob is used to sharpen the edges of the specimen.

10. What happens to the image of the cheek cells as you vary the black level and gain? What happens to the image of the cheek cells when you change the polarity? What happens to the image of the cheek cells as you vary the gamma? What effect does varying the bandwidth have on the image?

Digital Image Processing

1. Attach a digital camera (e.g., Hitachi KP-D50 color digital camera) to the Integral flashpoint 128 frame grabber board in the computer. Start the Image Pro Plus software.

2. Go to the Acquire menu and explore the various options. Then capture an image with the SNAP command.

3. Close the Acquire menu and open the File menu. Save the unenhanced image using the SAVE command.

4. Open the Enhance menu and click DISPLAY RANGE to see the distribution of pixel brightness. Close the DISPLAY RANGE window and click CONTRAST ENHANCEMENT. Experiment with the brightness, contrast, gamma, and invert controls. View the image before and after the enhancement and look at the graph of the lookup table, which relates the output intensity to the original intensity. Keeping the CONTRAST ENHANCEMENT window open, go to the EQUALIZE command in the Enhance menu, and experiment with the BEST FIT command. Notice how the image and the lookup table changes.

5. Acquire another image and save it. Defocus the specimen or move it out of the way, and capture another image and save it.

6. Go to the Process menu and click the BACKGROUND CORRECTION command. Subtract the background image from the image of the specimen.

7. Experiment with spatial filtering using Fourier transforms and convolution kernels. First go to the Edit menu and convert any color image to a black-and-white image, using the CONVERT TO command.

8. Then go to the Process menu and click the FFT command. Click the Forward FFT command to do a Fourier transform on the image. Observe the Fourier transform. Using the SPIKE CUT command and the ASYMMETRICAL option, remove selected Fourier spots, and then click INVERSE FFT. How does the image produced by the inverse Fourier transform compare with the original image?

9. Also experiment with the LO-PASS, HI-PASS, and SPIKE-BOOST commands. Close the Process menu.

10. After experimenting with Fourier transforms, try spatial filtering with convolution filters. In the Process

menu, click FILTERS, ENHANCE, and LO-PASS. How does the image look?

11. Click UNDO and then click HI-PASS. How does the image look now? Experiment with various convolution kernels, varying the size, strength, and number of passes.

12. Focus the microscope on a stage micrometer. Then capture and save an image of the stage micrometer.

13. Pull down the Measure menu. Click CALIBRATION, SPATIAL, and IMAGE. Drag a line between two distant bars on the stage micrometer and type the actual distance. The program will respond with the number of pixels/micrometer for the magnification used. Then click the MEASUREMENT command, select the line tool, and begin measuring lengths in one of your images. Experiment with the various MEASUREMENT and CALIPER tools.

14. Pull down the Macros menu and follow the commands given to get a feeling for what the Image Pro Plus digital image processor can do.

15. Copy all your saved images onto a memory stick. You will have to print them yourself.

COMMERCIAL SOURCES FOR LABORATORY EQUIPMENT AND SPECIMENS

Arbor Scientific: http://www.arborsci.com/detail.aspx? ID=934
Carolina Biological Supply: http://www.carolina.com/
Edmund Scientific: http://scientificsonline.com/
Industrial Fiber Optics, Inc: http://www.i-fiberoptics.com/
Li-cor Biosciences: http://www.licor.com/
Pasco: http://www.pasco.com/
Ted Pella: http://www.tedpella.com/
Web-tronics: http://www.web-tronics.com/

References

Abbe, E. [1876–1878]. (1921–1923). Some Abbe letters. Read by Professor Cheshire. *J. Roy. Microscop. Soc.*, 163–170, 221–236.

Abbe, E. (1881). On the estimation of aperture in the microscope. *J. Roy. Microscop. Soc. Ser. II* **1**: 349–388.

Abbe, E. (1887). On improvements of the microscope with the aid of new kinds of optical glass. *J. Roy. Microscop. Soc.* February: 20–34.

Abbe, E. (1889). On the effect of illumination by means of wide-angled cones of light. *J. Roy. Microscop. Soc.* **9**: 721–724.

Abbe, E. (1906). *Statut der von Ernst Abbe errichteten Carl Zeiss-Stiftungzu Jena.* G. Fischer, Jena, Germany.

Abbe, E. (1921). *Sozialpolitische Schriften.* G. Fischer, Jena.

Abercrombie, M. and G. A. Dunn. (1975). Adhesion of fibroblasts to substations during contact inhibition observed by interference reflection microscopy. *Exp. Cell Res.* **92**: 57–62.

Abraham, M. and R. Becker. (1932). *The Classical Theory of Electricity and Magnetism.* Blackie & Son, London, UK.

Abramowitz, M. (1985). *Microscope Basics and Beyond.* Olympus Corporation, Lake Success, New York.

Abramowitz, M. (1987). *Contrast Methods in Microscopy. Transmitted Light.* Olympus Corporation. Lake Success, New York.

Abramowitz, M. (1989). Köhler illumination. *American Laboratory* **21**(4): 106.

Abramowitz, M. (1990). Centering a centerable stage or centerable objectives. *American Laboratory* July: 44.

Abramowitz, M. (1991). Darkfield illumination. *American Laboratory* November: 60–61.

Abramowitz, M. (1993). *Fluorescence Miscroscopy. The Essentials.* Vol. 4, Basics and Beyond Series. Olympus Corporation, Lake Success, New York.

Ackerman, C. W. (1930). *George Eastman.* Houghton Mifflin, Boston, MA.

Adams, A. E., W. S. MacKenzie, and C. Guilford. (1984). *Atlas of Sedimentary Rocks Under the Microscope.* Longman, London.

Adams, G. (the elder). (1746). *Micrographia illustrata*, or, The knowledge of the microscope explain'd. Printed for, and sold by the author. London.

Adams, G. (the elder). (1747). *Micrographia illustrata*, or, The knowledge of the microscope explain'd, second edition. Printed for, and sold by the author. London.

Adams, G. (the elder). (1771). *Micrographia illustrata*, or, The knowledge of the microscope explain'd, fourth edition. Printed for, and sold by the author. London.

Adams, G. (1787). *Essays on the Microscope.* Dillon and Keating, London.

Adams, G. (1798). *Essays on the Microscope*, second edition. Dillon and Keating, London.

Aguayo, J. B., S. J. Blackband, J. Schoeniger, M. A. Mattingly, and M. Hintermann. (1986). Nuclear magnetic resonance imaging of a single cell. *Nature* **322**: 190–191.

Aikins, R. S., D. A. Agard, and J. W. Sedat. (1989). Solid-state imagers for microscopy. *Methods in Cell Biology* **29**: 291–313.

Airy, G. B. (1866). *On the Undulatory Theory of Optics.* Macmillan, London.

Airy, G. B. (1871). *On Sound and Atmospheric Vibrations with the Mathematical Elements of Music.* Macmillan, London.

Aist, J. R. (1995). Independent nuclear motility and hyphal tip growth. *Can. J. Bot.* **73**: S122–S125.

Aist, J. R., C. J. Bayles, W. Tao, and M. W. Berns. (1991). Direct experimental evidence for the existence, structural basis and function of astral forces during anaphase B *in vivo*. *J. Cell Sci.* **100**: 279–288.

Aist, J. R. and M. W. Berns. (1981). Mechanics of chromosome separation during mitosis in *Fusarium* (Fungi imperfecti): New evidence from ultrastructural and laser microbeam experiments. *J. Cell Biol.* **91**: 446–458.

Aist, J. R. and N. R. Morris. (1999). Mitosis in filamentous fungi: How we got where we are. *Fungal Gen. Biol.* **27**: 1–25.

Allen, R. D., G. B. David, and G. Nomarski. (1969). The Zeiss-Nomarski differential interference equipment for transmitted-light microscopy. Z.F. Wissensch. *Mikro und mickro. Tech.* **69**: 193–221.

Allen, R. D., N. S. Allen, and J. L. Travis. (1981a). Video-enhanced contrast, differential interference contrast (AVEC-DIC) microscopy: A new method capable of analyzing microtubule-related motility in the reticulopodial network of *Allogromia laticollaris*. *Cell Motility* **1**: 291–300.

Allen, R. D., J. L. Travis, N. S. Allen, and H. Yilmaz. (1981b). Video-enhanced contrast polarization (AVEC-POL) microscopy: A new method applied to the detection of birefringence in the motile reticulopodial network of *Allogromia laticollaris*. *Cell Motility* **1**: 275–289.

Allen, R. D. and N. S. Allenm. (1983). Video-enhanced microscopy with a computer frame memory. *J. Microscopy* **129**: 3–17.

Allen, R. D. (1985). New observations on cell architecture and dynamics by video-enhanced contrast optical microscopy. *Ann. Rev. Biophys. Biophys. Chem.* **14**: 265–290.

Amato, I. (1997). Atomic imaging: Candid cameras for the nanoworld. *Science* **276**: 1982–1985.

Amici, G. B. (1818). Osservazioni sulla circolazione del succhio nella Chara. *Memorie della Società Italiana delle Scienze* **18**: 183–198.

Anderson, C. D. (1932). The apparent existence of easily deflectable positives. *Science* **76**: 238.

Anderson, C. D. (1933). The positive electron. *Phys. Rev.* **43**: 491.

Anderson, C. D. (1999). *The Discovery of Antimatter*. World Scientific, Singapore.

Anon. (1803). Bakerian lecture on light and colours. *Edinburgh Review* **1**: 450–460.

Anon. (1804). Dr Young's Bakerian lecture. *Edinburgh Review* **5**: 97–103.

Arago, F. (1857). *Biographies of Distinguished Scientific Men*. Longman, Brown, Green, Longmans & Roberts, London.

Arago, F. (1859a). Malus. 117–170. *Biographies of Distinguished Scientific Men, second series*. Tichnor and Fields, Boston.

Arago, F. (1859b). Fresnel. 171–279. *Biographies of Distinguished Scientific Men, second series*. Tichnor and Fields, Boston, MA.

Arago, F. (1859c). Thomas Young. 280–349. *Biographies of Distinguished Scientific Men, second series*. Tichnor and Fields, Boston, MA.

Aristophanes. 423 BC. The Clouds. Translated by B. B. Rogers. In: *Great Books of the Western World*, Volume 5. Encyclopedia Britannica, Chicago.

Arnold, H. J. P. (1977). William Henry Fox Talbot. *Pioneer of Photography and Man of Science*. Hutchinson Benham, London.

Ash, E. A. and G. Nicholls. (1972). Super-resolution aperture scanning microscope. *Nature* **237**: 510.

Ashkin, A. (1970a). Acceleration and trapping of particles by radiation pressure. *Phys. Rev. Lett.* **24**: 156–159.

Ashkin, A. (1970b). Atomic-beam deflection by resonance-radiation pressure. *Phys. Rev. Lett.* **25**: 1321–1324.

Ashkin, A. (1978). Trapping of atoms by radiation pressure. *Phys. Rev. Lett.* **40**: 729–732.

Ashkin, A. and J. M. Dziedzic. (1987). Optical trapping and manipulation of viruses and bacteria. *Science* **235**: 1517–1520.

Ashkin, A. and J. M. Dziedzic. (1989). Internal cell manipulation using infrared laser traps. *Proc. Natl. Acad. Sci. USA,* **86**: 7914–7918.

Ashkin, A., J. M. Dziedzic, and T. Yamane. (1987). Optical trapping and manipulation of single cells using infrared laser beams. *Nature* **330**: 769–771.

Ashkin, A., K. Schutze, J. M. Dziedzic, U. Enteneuer, and M. Schliwa. (1990). Force generation of organelle transport measured in vivo by and infrared laser trap. *Nature* **348**: 346–348.

Asimov, I. (1966). *Understanding Physics: Light, Magnetism, and Electricity*. New American Library, Inc, New York.

Auerbach, F. (1904). The Zeiss Works and The Carl-Zeiss Stiftung in Jena. *Their Scientific Technical and Sociological Development and Importance Described*. Marshall, Brookes and Chalkley, Ltd, London.

Axelrod, D. (1984). Total internal reflection fluorescence in biological systems. *J. Luminescence* **31–32**: 881–884.

Axelrod, D. (1990). Total internal reflection fluorescence at biological surfaces. In: *Noninvasive Techniques in Cell Biology*. Wiley-Liss, New York.

Axelrod, D., D. E. Koppel, J. Schlessinger, E. Elson, and W. W. Webb. (1976). Mobility measurement by analysis of fluorescence photobleaching recovery kinetics. *Biophys. J.* **16**: 1055–1069.

Axelrod, D., N. L. Thompson, and T. P. Burghart. (1982). Total internal reflection fluorescence microscopy. *J. Micros.* **129**: 19–28.

Babbage, C. (1830). *Reflections on the Decline of Science in England, and on Some of Its Causes*. Printed for B. Fellowes, London.

Baden-Powell. (1833). *A Short Elementary Treatise on Experimental and Mathematical Optics*. D. A. Talboys, Oxford.

Baierlein, R. (1992). Newton to Einstein. *The Trail of Light*. Cambridge University Press, Cambridge.

Baker, B. B. and E. T. Copson. (1987). *The Mathematical Theory of Huygen's Principle*. Chelsea Publishing Co, New York.

Baker, C. (1985). *Through the Kaleidoscope*. Beechcliff Books, Annapolis, MD.

Baker, C. (1987). *Through the Kaleidoscope...and Beyond*. Beechcliff Books, Annapolis, MD.

Baker, C. (1990). *Kaleidorama*. Beechcliff Books, Annapolis, MD.

Baker, C. (1993). *Kaleidoscope Renaissance*. Beechcliff Books, Annapolis, MD.

Baker, C. (1999). *Kaleidoscopes: Wonders of Wonder*. C&T Publishing, Concord, CA.

Baker, H. (1742). *The Microscope Made Easy or, I. The Nature, Uses, and Magnifying Powers of the Best Kinds of Microscopes Described, Calculated, and Explained: ... II. An Account of What Surprising Discoveries Have Been Already Made by the Microscope*. Printed for R. Dodsley, London.

Baker, H. (1743). *The Microscope Made Easy or, I. The Nature, Uses, and Magnifying Powers of the Best Kinds of Microscopes Described, Calculated, and Explained: ... II. An Account of What Surprising Discoveries Have Been Already Made by the Microscope*, second edition. Printed for R. Dodsley, London, UK.

Baker, H. (1769). *The Microscope Made Easy or, I. The Nature, Uses, and Magnifying Powers of the Best Kinds of Microscopes Described, Calculated, and Explained: ... II. An Account of What Surprising Discoveries Have Been Already Made by the Microscope*, fifth edition. Printed for R. Dodsley, London, UK.

Barakat, N. and A. A. Hamza. (1990). *Interferometry of Fibrous Materials*. Adam Hilger, Bristol.

Barbaro, D. (1568). *La Pratica della Perspettiva: opera molto vtile a pittori, a scvltori, & ad architetti con due tauole, una de' capitoli principali, l'altra delle cose piu notabili contenue nella presente opera*. Camillo & Rutilio Borgominieri fratelli, Venice.

Barenboim, G. M., A. N. Domanskii, and K. K. Turoverov. (1969). *Luminescence of Biopolymers and Cells*. Plenum Press, New York.

Barer, R. (1952a). A vector theory of phase contrast and interference contrast. I. Positive phase contrast. *Journal of the Royal Microscopical Society* **72**: 10–38.

Barer, R. (1952b). A vector theory of phase contrast and interference contrast. II. Positive phase contrast (continued). *Journal of the Royal Micro-scopical Society* **72**: 81–88.

Barer, R. (1952c). Interference microscopy and mass determination. *Nature* **169**: 366–367.

Barer, R. (1953a). A vector theory of phase contrast and interference contrast. III. Negative phase contrast. *Journal of the Royal Microscopical Society* **73**: 30–39.

Barer, R. (1953b). A vector theory of phase contrast and interference contrast. IV. Type B phase contrast. *Journal of the Royal Microscopical Society* **73**: 206–215.

Barer, R. (1953c). Determination of dry mass, thickness, solid and water concentration in living cells. *Nature* **172**: 1097–1098.

Barer, R. (1955). Phase-contrast, interference contrast, and polarizing microscopy. In: *Analytical Cytology*. R. C. Mellors, ed. McGraw-Hill, New York.

Barer, R. (1956). *Lecture Notes on the Use of the Microscope*, second edition. Charles C. Thomas, Springfield, IL.

Barer, R. (1966). Phase contrast and interference microscopy in cytology. In: *Physical Techniques in Biological Research*. A. W. Pollister, ed. Volume III, Part A. Cells and Tissues. Academic Press, New York.

Barer, R., K. R. A. Ross, and S. Tkaczyk. (1953). Refractometry of living cells. *Nature* **171**: 720–724.

Barer, R. and D. A. T. Dick. (1957). Interferometry and refractometry of cells in tissue culture. *Exp. Cell Res.* **4**(Suppl.): 103–135.

Barer, R. and S. Joseph. (1954). Refractometry of living cells. *Quart. J. Microscop. Sci.* **95**: 399–423.

Barer, R. and S. Joseph. (1955). Refractometry of living cells. II. The immersion method. *Quart. J. Microscop. Sci.* **96**: 1–26.

Barer, R. and S. Joseph. (1958). Concentration and mass measurements in microbiology. *J. Appl. Bact.* **21**: 146–159.

Barnes, J., ed. (1984). *The Complete Works of Aristotle*. Princeton University Press, Princeton, NJ.

Baron-Epel, O., D. Hernandez, L.-W. Jiang, S. Meiners, and M. Shindler. (1988). Dynamic continuity of cytoplasmic and membrane compartments between plant cells. *J. Cell Biol.* **106**: 715–721.

Bartels, P. H. (1966). Sensitivity and evaluation of microspectrophotometric and microinterferometric measurements. 93–105. In: *Introduction to Quantitative Cytochemistry*. G. L. Wied, ed. Academic Press, New York.

Bartels, P. H. (1966). Principles of polarized light. 519–538. In: *Introduction to Quantitative Cytochemistry*. G. L. Wied, ed. Academic Press, New York.

Bartholinus, E. (1669). *Versochemit dem Doppeltbrechenden Isländischen Kristall*. Published by Academische Verlag. Leipzig in 1922, and Experiments with the double refracting island crystal. Translated by Werner Brandt. Sept. 1959 (Published privately).

Baskin, T. I., M. S. Bret-Harte, and P. B. Green. (1987). Rates of cell wall accumulation in higher plants; measurement with interference microscopy. 258–261. In: *Physiology of Cell Expansion during Plant Growth*. D. J. Cosgrove and D. P. Knievel, eds. Amer. Soc. Plant Physiol, Rockville, MD.

Bayles, C. J., J. R. Aist, and M. W. Berns. (1993). The mechanics of anaphase B in a basidiomycete as revealed by laser microbeam microsurgery. *Exptl. Mycol.* **17**: 191–199.

Beale, L. S. (1880). *How to Work with the Microscope*, fifth edition. Harrison, Pall Mall, London.

Beavis, G. (1931). *The Book of the Microscope*. J. B. Lippincott, Philadelphia, PA.

Beck, C. (1924). *The Microscope. Part II. An Advanced Handbook*. R. & J. Beck, London.

Beck, R. (1865). *A Treatise on the Construction, Proper Use, and Capabilities of Smith, Beck, and Beck's Achromatic Microscopes*. Printed for Smith, Beck and Beck, London.

Beller, M. (1999). *Quantum Dialog. The Making of a Revolution*. University of Chicago Press, Chicago. IL.

Belling, J. (1930). *The Use of the Microscope*. McGraw-Hill, New York.

Beneke, G. (1966). Application of interference microscopy to biological material. 63–92. In: *Introduction to Quantitative Cytochemistry*. G. L. Wied, ed. Academic Press, New York.

Bennett, A. H., H. Osterberg, H. Jupnik, and O. W. Richards. (1951). *Phase Microscopy. Principles and Applications*. John Wiley & Sons, New York.

Bennett, J., M. Cooper, M. Hunter, and L. Jardine. (2003). *London's Leonardo: The Life and Work of Robert Hooke*. Oxford University Press, Oxford.

Bennett, H. S. (1950). The microscopical investigation of biological materials with polarized light. 591–677. In: *McClung's Handbook of Microscopical Technique*, third edition. R. McClung Jones, ed. Paul B. Hoeber, New York.

Bentley, W. A. and W. J. Humphreys. (1962). *Snow Crystals*. Dover Publications, New York.

Bereiter-Hahn, J., I. Karl, H. Lüers, and M. Vöth. (1995). Mechanical basis of cell shape: investigations with the scanning acoustic microscope. *Biochem. Cell Biol.* **73**: 337–348.

Bergmann, L. and C. Schaefer. (1999). *Optics of Waves and Particles*. Water de Gruyter, Berlin.

Berkeley, G. [1709]. (1910). *A New Theory of Vision and Other Writings*. J. M. Dent & Sons, London.

Berlyn, G. P. and J. P. Miksche. (1976). *Botanical Microtechnique and Cytochemistry*. Iowa State University Press, Ames, IA.

Berns, M. W., J. Aist, J. Edwards, K. Strahs, J. Girton, P. McNeill, J. B. Rattner, M. Kitzes, M. Hammer-Wilson, L.-H. Liaw, A. Siemens, M. Koonce, S. Peterson, S. Brenner, J. Burt, R. Walter, P. J. Bryant, D. van Dyk, J. Coulombe, T. Cahill, and G. S. Berns. (1981). Laser microsurgery in cell and developmental biology. *Science* **213**: 505–513.

Berns, M. W. (1991). Laser surgery. *Scientific American* (June) **265**: 84–90.

Berns, M. W., W. T. Wright, and R. Wiegand Stenbing. (1991). Laser microbeam as a tool in biology and medicine. *Int. Rev. Cytol.* **129**: 1–44.

Beth, R. A. (1936). Mechanical detection and measurement of the angular momentum of light. *Phys. Rev.* **50**: 115–125.

Bethe, H. 1944. Theory of diffraction by small holes. *Phys. Rev.* **66**: 163–182.

Betzig, E., J. K. Trautman, T. D. Harris, J. S. Weiner, and R. L. Kostelak. (1991). Breaking the diffraction barrier: Optical microscopy on a nanometric scale. *Science* **251**: 1468–1470.

Beyer, H. (1965). *Theorie und Praxis des Phasenkontrastverfahrens*. Akademische Verlagsgesellschaft, Geest & Portig, K.-G, Leipzig.

Beyer, H. (1971). Epival interphako incident light interference microscope. *Jena Review* **16**: 82–88.

Bialynicki-Birula, I., M. Cieplak, and J. Kaminski. (1992). *Theory of Quanta*. Oxford University Press, New York.

Binnig, G., H. Rohrer, Ch. Gerber, and E. Weibel. (1982). Surface studies by scanning tunneling microscopy. *Phys. Rev.* **49**: 57–61.

Birch, T. [1756–1757]. (1968). *The History of the Royal Society of London*. Johnson Reprint Corp. New York.

Bitter, F. and H. A. Medicus. (1973). *Fields and Particles. An Introduction to Electromagnetic Wave Phenomena and Quantum Physics*. American Elsevier, New York.

Bjorkholm, J. E., A. Ashkin, and D. B. Pearson. (1975). Observation of resonance radiation pressure on an atomic vapor. *Appl. Physics Lett.* **27**: 534–537.

Blackett, P. M. S. (1948). Cloud chamber researches in nuclear physics and cosmic radiation. *Nobel Lecture*. December 13, 1948.

Blackett, P. M. S. and G. P. S. Occhialini. (1933). Some photographs of the tracks of penetrating radiation. *Proceedings of the Royal Society* **139A**: 699–727.

Blanchard, D. C. (1998). *The Snowflake Man. A Biography of Wilson A. Bentley*. McDonald & Woodward, Blacksburg, VA.

Bock, J. H., M. A. Lane, and D. O. Norris. (1988). *Identifying Plant Food Cells in Gastric Contents for Use in Forensic Investigations A Laboratory Manual*. U.S. Department of Justice, National Institute of Justice, Washington, D.C.

Block, S. M., L. S. B. Goldstein, and B. J. Schnapp. (1990). Bead movement by single kinesin molecules studied with optical tweezers. *Nature* **348**: 348–352.

Bohm, D. (1979). *Quantum Theory*. Dover, New York.

Bohr, N. (1913). On the constitution of atoms and molecules. *Phil. Mag.* **26**: 1–25. 476–502, 857–875.

Bohr, N. (1922). The structure of the atom. *Nobel Lecture* December 11, 1922.

Bohr, N. (1934). *Atomic Theory and the Description of Nature*. Cambridge University Press, Cambridge.

Bohr, N. (1958). *Atomic Physics and Human Knowledge*. John Wiley & Sons, New York.

Bohr, N., H. A. Kramers, and J. C. Slater. (1924). The quantum theory of radiation. *Phil. Mag.* **47**: 785–802.

Bolles, E. B. (2004). *Einstein Defiant*. Joseph Henry Press, Washington, DC.

Bondi, H. (1957). Negative mass in general relativity. *Rev. Mod. Phys.* **29**: 423–428.

Bonner, R. F., M. R. Emmert-Buck, K. Cole et al. (1997). Laser capture microdissection. Molecular analysis of tissue. *Science* **278**: 1481–1483.

Boorse, H. A. and L. Motz. (1966). *The World of the Atom*. Two Volumes. Basic Books, New York.

Boorstin, D. (1961). *The Image: A Guide to Pseudo-Events in America*. Harper Colophon Books, New York.

Born, M. (2005). *The Born-Einstein Letters 1916–1955. Friendship, Politics and Physics in Uncertain Times*. Macmillan, New York.

Born, M. and N. S. Nagendra Nath. (1936a). The neutrino theory of light. *Proc. Indian Acad. Sci.* **3**: 318–337.

Born, M. and N. S. Nagendra Nath. (1936b). The neutrino theory of light. II. *Proc. Indian Acad. Sci.* **4**: 611–620.

Born, M. and E. Wolf. (1980). *Principles of Optics*. Pergamon Press, Oxford.

Bose, J. C. (1927). *Collected Physical Papers*. Longmans, Green, London.

Bose, S. N. (1924). Planck's law and the light quantum hypothesis. In: *Amer. J. Phys.* 1976, **44**: 1056–1057 and *J. Astrophys. Astr.* 1994. **15**: 3–11.

Bova, B. (1988). *The Beauty of Light*. John Wiley & Sons, New York.

Bown, W. (1992). Brownian motion sparks renewed debate. *New Scientist* **133**: (February 15): 25.

Bowtell, R. W., G. D. Brown, P. M. Glover, M. McJury, and P. Mansfield. (1990). Resolution of cellular structures by NMR microscopy at 11.7T. *Philos. Trans. R. Soc. London,* [A] **33**: 457–467.

Boyle, R. (1664). *Experiments and Considerations Touching Colours*. Printed for Henry Herringman, London.

Boyle, R. (1672). *Essay about the Origine and Virtues of Gems*. Printed by William Godbid, London.

Bracegirdle, B. (1993). Microscopical illumination in the 1890's. *Proc. RMS* **28**: 196–201.

Bradbury, S. (1988). Processing and analysis of the microscope image. *Microscopy* **36**: 23–39.

Bradbury, S. (1988). It's all done by mirrors! Some aspects of the development of the reflection microscope. *Microscopy* **36**: 14–22.

Bradbury, S. (1994). Recording the image-past and present. *Queckett Journal of Microscopy* **37**: 281–295.

Braddick, H. J. J. (1665). *Vibrations, Waves and Diffraction*. McGraw-Hill, New York.

Bradbury, S., P. T. Evennett, H. Haselmann, and H. Piller. (1989). *Dictionary of Light Microscopy*. Oxford University Press, Oxford.

Bragg, W. H. (1907a). A comparison of some forms of electric radiation. *Trans. and Proc and Rep. of the Royal Soc. of South Australia* **31**: 79.

Bragg, W. H. (1907b). The nature of Röntgen rays. *Trans. and Proc and Rep. of the Royal Soc. of South Australia,* **31**: 94.

Bragg, W. H. (1907c). On the properties and nature of various electric radiations. *Phil. Mag. Series 6* **14**: 429.

Bragg, W. H. (1911). Corpuscular radiation. *Rep. Brit. Assoc. Adv. Sci.* 340.

Bragg, W. H. and J. P. V. Madsen. (1908). An experimental investigation of the nature of γ rays. *Proc. Physical Soc. London,* **21**: 261.

Brakenhoff, G. J., E. A. van Spronsen, H. T. M. van der Voort, and N. Nanninga. (1989). Three-dimensional confocal fluorescence microscopy. *Methods in Cell Biology* **30**: 379–398.

Brayer, E. (1996). *George Eastman: A Biography*. Johns Hopkins University Press, Baltimore, MD.

Breedlove, J. R. Jr. and G. T. Trammell. (1970). Molecular microscopy: Fundamental limitations. *Science* **170**: 1310–1313.

Bret-Harte, M. S., T. I. Baskin, and P. B. Green. (1991). Auxin stimulates both deposition and breakdown of material in the pea outer epidermal cell wall, as measured interferometrically. *Planta* **185**: 462–471.

Brewster. (1815a). On the laws which regulate the polarisation of light by reflexion from transparent bodies. *Phil. Trans. Roy. Soc. Lond.* **105**: 125–159.

Brewster. (1815b). Additional observations on the optical properties and structure of heated glass and unannealed glass drops. *Phil. Trans. Roy. Soc. Lond.* **105**: 1–8.

Brewster. (1815c). On the effects of simple pressure in producing that species of crystallization which forms two oppositely polarised images, and exhibits the complementary colours by polarised light. *Phil. Trans. Roy. Soc. Lond.* **105**: 60–64.

Brewster, D. (1818). *The Kaleidoscope. Its History, Theory, and Construction*. John Murray, London.

Brewster, D. (1833a). On the colours of natural bodies. *Roy. Soc. Edinb. Trans.* **12**: 538–545.

Brewster, D. (1833b). *Treatise on Optics*, First American Edition. Carey, Lea & Blanchard, Philadelphia, PA.

Brewster, D. (1835). *Letters on Natural Magic*. Harper & Bros., New York.

Brewster, D. (1837). *A Treatise on the Microscope*. A. and C. Black, Edinburgh.

Brewster, D. (1858). *The Kaleidoscope, Its History, Theory, and Construction*, second edition. John Murray, London.

Brode, W. R. (1943). *Chemical Spectroscopy*, second edition. John Wiley & Sons, New York.

Brouillette, J. (1990). Microscopical analysis of Whistler's "Rose et Vert L'Iris: Portrait of Miss Kinsella". *Microscopy* **38**: 271–279.

Brown, R. (1828). A brief account of microscopical observations made in the months of June, July, and August, 1927, on the particles contained in the pollen of plants; and on the general existence of active molecules in organic and inorganic bodies. Richard Taylor, London. 1–16. Republished by the Philosophical Magazine and Annals of Philosophy. *New Series* **4**: 161–173.

Brown, R. (1829). *Additional Remarks on Active Molecules*. Richard Taylor, London. 17–20. Republished by the Ray Society 1866. 479–486.

Brown, R. (1831). On the organs and mode of fecundation in Orchideae and Asclediadeae. *Trans. Linn. Soc. Lon.* **16**: 685–745.

Brown, R. H. J. (1940). The protoplasmic viscosity of *Paramecium*. *J. Exp. Biol.* **17**: 21–324.

Buchwald, J. Z. (1983). Fresnel and diffraction theory. *Arch. Internat. Hist. Sci.* **33**: 36–111.

Buchwald, J. Z. (1989). *The Rise of the Wave Theory of Light*. University of Chicago Press, Chicago, IL.

Buder, J. (1918). Die Inversion des Phototropismus bei Phycomyces. *Bet. Dtsch. Bot. Ges.* **36**: 104–105.

Buder, J. (1920). Neue phototropische Fundamenalversuche. *Bet. Dtsch. Ges,* **38**: 10–19.

Buijs Ballot, C. H. D. (1845). Alustische Versuche auf der Niederl Eisenbahn, nebst gelegentliche Bemerkungen zur Theorie des Herrn Prof. Doppler. *Ann d Phys. Ser. 2.* **66**: 321–351.

Cachon, J., H. Sato, M. Cachon, and Y. Sato. (1989). Analysis by polarizing microscopy of chromosomal structure among dinoflagellates and its phylogenetic involvement. *Biology of the Cell* **65**: 51–60.

Cahan, D. (1996). The Zeiss Werke and the ultramicroscope: The creation of a scientific instrument in context. 67–115. In: *Scientific Credibility and Technical Standards in 19th and early 20th Century Germany and Britain*. J. Z. Buchwald, ed. Kluwer, London.

Cai, S. and C. C. Lashbrook. (2006). Laser capture microdissection of plant cells from tape transferred paraffin sections promotes recovery of structurally intact RNA for global gene profiling. *Plant J* **48**: 628–637.

Cajori, F. (1916). *A History of Physics*. Macmillan, New York.

Callaghan, P. T. (1991). *Principles of Nuclear Magnetic Resonance Microscopy*. Academic Press, Oxford.

Callaghan, P. T. (1992). Nuclear magnetic resonance microscopy. *Proc. RMS* **27/2**: 67–70.

Camilleri, K. (2006). Heisenberg and the wave-particle duality. Studies in History and Philosophy of Science. Part B. *Studies in History and Philosophy of Modern Physics* **37**: 298–315.

Campbell, L. and W. Garnett. (1884). *The Life of James Clerk Maxwell*. Macmillan, London.

Campos, R. A. (2004). Still shrouded in mystery: The photon in 1925. http://arxiv.org/ftp/physics/papers/0401/0401044.pdf

Cannizzaro, S. [1858]. (1961). *Sketch of a Course of Chemical Philosophy*. *Alembic Club Reprints*, Re-issue Edition. E. & S. Livingstone, Edinburgh.

Carboni, G. (2003). High-magnification stereoscopy. http://www.funsci.com/fun3_en/hmster/hmster_en.htm#11

Carpenter, W. B. [1856]. (1883). *The Microscope and Its Revelations*, sixth edition. William Wood & Co., New York.

Carson, T. R. (2004). Steps to the Planck Function: A centenary reflection. ArXiv:astro-ph/0011219v1. 10 Nov 2000.

Castle, E. S. (1930). Phototropic "indifference" and the light-sensitive system of Phycomyces. *J. Gen. Physiol.* **13**: 421–435.

Castle, E. S. (1932). On "reversal" of phototropism in Phycomyces. *J. Gen. Physiol.* **15**: 487–489.

Castle, E. S. (1938). Orientation of structure in the cell wall of Phycomyces. *Protoplasm* **31**: 331–345.

Castle, E. S. (1961). Phototropic inversion in Phycomyces. *Science* **133**: 1424–1425.

Castle, E. S. (1966). Light responses of Phycomyces. *Science* **154**: 1416–1420.

Castleman, K. R. (1993). Resolution and sampling requirements for digital image processing, analysis and display. 71–93. In: *Electronic Light Microscopy*. D. Shotton, ed. Wiley-Liss, Inc, New York.

Chalfie, M., Y. Tu, G. Euskirchen, W. Ward, and D. Prasher. (1994). Green fluorescent protein as a marker for gene expression. *Science* **263**: 802–805.

Chamberlain, C. J. (1924). *Methods in Plant Histology*, fourth revised edition. University of Chicago Press, Chicago, IL.

Chamot, E. M. (1921). *Elementary Chemical Microscopy*. John Wiley & Sons, New York.

Chamot, E. M. and C. W. Mason. (1958). *Handbook of Chemical Microscopy*, third edition. John Wiley & Sons, New York.

Chapman, A. (2004). *England's Leonardo: Robert Hooke and the Seventeenth-century Scientific Revolution*. Institute of Physics, Bristol.

Chayen, J. and E. F. Denby. (1968). *Biophysical Technique as Applied to Cell Biology*. Methuen & Co, London.

Chen, L. B. (1989). Fluorescent labeling of mitochondria. *Methods in Cell Biology* **29**: 103–123.

Chen, X. (1997). The debate on the "Polarity of Light" during the optical revolution. *Archive for History of Exact Sciences* **50**: 359–393.

Chen, X. and P. Barker. (1992). Cognitive appraisal and power: David Brewster, Henry Brougham, and the tactics of the emission-undulatory controversy during the early 1850s. *Studies in the History and Philosophy of Science* **23**: 75–101.

Cheshire, F. J. (1905). Biographical chart of some optical workers. *Proceedings of the Optical Convention, 1905*, 8–9. London: Norgate & Williams.

Chu, S. (1997). The manipulation of neutral particles. *Nobel Lecture* December 8, 1997.

Clark, C. H. (1925). *Practical Methods in Microscopy*, fifth edition. D. C. Heath and Co., Boston.

Clay, R. S. and T. H. Court. (1932). *The History of the Microscope*. Charles Griffin and Co., London.

Clayton, R. K. (1970). *Light and Living Matter, Volume 1: The Physical Part*. McGraw-Hill, New York.

Clemmer, C. R. and T. P. Beebe Jr. (1991). Graphite: A mimic for DNA and other biomolecules in scanning tunneling microscopic studies. *Science* **251**: 640–642.

Cohen, I. B. (1952). Maxwell's poetry. *Sci. Amer.* **187**(3): 62–63.

Cohen-Tannoudji, C. N. (1997). Manipulating atoms with photons. *Nobel Lecture* December 8, 1997.

Colby, R. H. (1971). Intrinsic birefringence of glycerinated myofibrils. *J. Cell Biol.* **51**: 763–771.

Collett, E. (1993). *Polarized Light. Fundamentals and Applications*. Marcel Dekker, New York.

Commins, S. and R. N. Linscott. (1947). *The World's Great Thinkers. Man and the Universe. The Philosophers of Science*. Random House, New York.

Compton, A. H. (1923). A quantum theory of the scattering of x-rays by light elements. *Phys. Rev.* **21**: 483–502.

Compton, A. (1929a). What things are made of—I. *Sci. Amer.* **140**(2): 110–113.

Compton, A. (1929b). What things are made of—II. *Sci. Amer.* **140**(3): 234–236.

Conn, H. J. (1933). *The History of Staining*. Biological Stain Commission, Geneva, NY.

Conrad, P. A., M. A. Nederlof, I. M. Herman, and D. L. Taylor. (1989). Correlated distribution of actin, myosin, and micrtotubules at the leading edge of migrating swiss 3T3 fibroblasts. *Cell Motility and Cytoskeleton* **14**: 527–543.

Cooper, M. (2003). *A More Beautiful City: Robert Hooke and the Rebuilding of London after the Great Fire*. Sutton, Stroud.

Cornford, F. M. (1945). *The Republic of Plato*. Oxford University Press, New York.

Cornford, F. M. (1966). *Microcosmographia Academica. Being a Guide for the Young Academic Politician*. Reprinted from the 1908 edition published by Bowes & Bowes, Cambridge. Barnes & Noble, New York, NY.

Corson, D. R. and P. Lorrain. (1962). *Introduction to Electromagnetic Fields and Waves*. W. H. Freeman and Co., San Francisco, CA.

Courjon, D. (2003). *Near-Field Microscopy and Near-Field Optics*. Imperial College Press, London.

Couzin, J. (2006). Don't pretty up that picture just yet. *Science* **314**: 1866–1868.

Crawford, F. S. Jr. (1968). *Waves*. Berkeley Physics Course. Vol. 3. McGraw-Hill, New York.

Crew, H. (1981). *The Wave Theory. Light and Spectra*. Arno Press, New York.

Cricenti, A., S. Selci, A. C. Felici, R. Generosi, E. Gori, W. Djaczenko, and G. Chiarotti. (1989). Molecular structure of DNA by scanning tunneling microscopy. *Science* **245**: 1226–1227.

Cristol, Y. (1986). *Solid State Video Cameras*. Pergamon, Oxford.

Curtis, A. S. G. (1964). The mechanism of adhesion of cells to glass, a study by interference reflection microscopy. *J. Cell Biol.* **20**: 199–215.

Dake, H. C. and J. de Ment. (1941). *Fluorescent Light and Its Applications*. CRC Publishing Co., Brooklyn, NY.

D'Alembert. (1747). Investigation of the curve formed by a vibrating string. 119–123. In: *Acoustics: Historical and Philosophical Development*. R. B. Lindsay, ed. Dowden, Huthinson & Ross, Stroudsburg, PA.

Darrigol, O. (2000). *Electrodynamics from Ampère to Einstein*. Oxford University Press, Oxford.

Darwin, C. (1889). *The Origin of Species by Means of Natural Selection or the Preservation of the Favored Races in the Struggle for Life*. Volume II. D. Appleton & Co, New York.

Darwin, F., ed. (1887). *The Life and Letters of Charles Darwin Including an Autobiographical Chapter, second edition*. Vol. II. John Murray, London.

Davidson, B. M. (1990). Sources of illumination for the microscope 1650–1950. *Microscopy* **36**: 369–386.

Davidson, J. A. (1975). Pressure cells for optical microscopy. *The Microscope* **23**: 61–71.

Davidson, M. W. (1990). Fabrication of unusual art forms with multiple exposure color photography. *The Microscope* **38**: 357–365. plus color plates.

Davidson, M. W. (1998). *Magical Display. The Art of Photomicrography*. Amber Lotus, Portland, OR.

Davies, H. G. and M. H. F. Wilkins. (1952). Interference microscopy and mass determination. *Nature* **169**: 541.

Davies, H. G., E. M. Deeley, and E. F. Denby. (1954). Attempts at measurement of lipid, nucleic acid and protein content of cell nuclei by microscopeinterferometry. *Exp. Cell Res.* **4**(Suppl.): 136–149.

Davies, H. G., M. H. F. Wilkins, J. Chayen, and L. F. La Cour. (1954). The use of the interference microscope to determine dry mass in living cells and as a quantitative cytochemical method. *Quart. J. Microscopical Sci.* **95**: 271–304.

Da Vinci, L. (1970). *The Notebooks of Leonardo Da Vinci*. Compiled and edited from the original manuscripts by J. P. Richter. Dover Publications, New York.

de Broglie, L. no date. *An Introduction to the Study of Wave Mechanics*. E. P. Dutton and Co., New York.

de Broglie, L. (1922). Black radiation and light quanta. 1–7. In: *Selected Papers on Wave Mechanics by Louis de Broglie and Léon Brillouin*. (1929). Blackie & Son, London.

de Broglie, L. (1924). A tentative theory of light quanta. *Phil. Mag.* **47**: 446–458.

de Broglie, L. (1925). Theory of quanta. Doctoral Thesis. http://www.ensmp.fr/aflb/LDB-oeuvres/De_Broglie_ Kracklauer.pdf.

de Broglie, L. (1932a). Sur une analogie entre l'électron de Dirac et l'onde électromagnétique. *Comptes rendus.* **195**: 536–537.

de Broglie, L. (1932b). Remarques sur le moment magnétique et le moment de rotation de l'électron. *Comptes rendus.* **195**: 577–588.

de Broglie, L. (1932c). Sur le champ électromagnétique de l'onde lumineuse. *Comptes rendus.* **195**: 862–864.

de Broglie, L. (1933). Sur la densité de l'énergie dans la théorie de la lumiére. *Comptes rendus.* **197**: 1377–1380.

de Broglie, L. (1934a). Sur la nature du photon. *Comptes rendus.* **198**: 135–138.

de Broglie, L. (1934b). L'équation d'ondes du photon. *Comptes rendus.* **199**: 445–448.

de Broglie, L. (1934c). Sur l'expression de la densité dans la nouvelle théorie du photon. L'équation d'ondes du photon. *Comptes rendus.* **199**: 1165–1168.

de Broglie, L. [1937]. (1946). *Matter and Light. The New Physics*. Dover, New York.

de Broglie, L. (1953). *The Revolution in Physics*. Farrar, Straus & Cudahy, New York.

de Broglie, L. (1955). *Physics and Microphysics*. Pantheon Books, New York.

de Broglie, L. (1962). *New Perspectives in Physics*. Oliver & Boyd, Edinburgh.

de Broglie, L. and J. Winter. (1934). Sur le spin du photon. *Comptes rendus.* **199**: 813–816.

de. L. Kronig, R. (1935a). Zur Neutrinotheorie des Lichtes. *Physica* **2**: 491–498.

de. L. Kronig, R. (1935b). Zur Neutrinotheorie des Lichtes II. *Physica* **2**: 854–860.

de. L. Kronig, R. (1935c). Zur Neutrinotheorie des Lichtes III. *Physica* **2**: 968–980.

de Weer, P. and B. M. Salzberg. (1986). *Optical Methods in Cell Physiology*. Society of General Physiologists and Wiley-Interscience, New York.

Day, P. (1999). *The Philosopher's Tree, Michael Faraday's Life and Work in His Own Words*. Institute of Physics, Bristol.

Delly, J. G. (1980). *Photography through the Microscope*. Eastman Kodak Corp. Rochester, New York.

Delly, J. G. (1988). *Photography through the Microscope*, second edition. Eastman Kodak Co., Rochester, New York.

Delly, J. G. (1998). Stars of Snow. *Microscopy Today* November(Issue 98–9): 8–10.

Delly, J. G. (2003). The Michel-Lévy interference chart—Microscopy's magical color key. www.modernmicroscopy.com; http://www.modernmicroscopy.com/main.asp?article=15&page=1.

De Morgan, A. (1872). *A Budget of Paradoxes*. Longmans, Green, London.

Denk, W., J. H. Strickler, and W. W. Webb. (1990). Two-photon laser scanning fluorescence microscopy. *Science* **248**: 73–76.

Dennison, D. S. (1959). Gallic acid in *Phycomyces* sporangiophores. *Nature* **184**: 2036.

Dennison, D. S. and T. C. Vogelmann. (1989). The Phycomyces lens: Measurement of the sporangiophore intensity profile using a fiber optic microprobe. *Planta* **179**: 1–10.

Descartes, R. [1637]. (1965). *Discourse on Method, Optics, Geometry, and Meteorology*. Bobbs-Merrill Co, Indianapolis, IN.

Deutsch, D. H. (1991). Did Robert Brown observe Brownian Motion: Probably not. *Bulletin of the American Physical Society* **36**: 1374.

Dickinson, E. (1924). *The Complete Poems of Emily Dickinson*. Little Brown, Boston, MA.

Dirac, P. A. M. (1924). Note on the Doppler principle and Bohr's frequency condition. *Proc. Cam. Phil. Soc.* **22**: 432–433.

Dirac, P. A. M. (1927). The quantum theory of the emission and absorption of radiation. *Proc. Roy. Soc.* **A114**: 243–265.

Dirac, P. A. M. (1930). A theory of electrons and protons. *Proc. Roy. Soc.* **A126**: 360–365.

Dirac, P. A. M. (1933). Theory of electrons and positrons. *Nobel Lecture,* December 12, 1993.

Dirac, P. A. M. (1958). *The Principles of Quantum Mechanics*, fourth edition. (revised). Oxford University Press, Oxford.

Disney, A. N., C. F. Hill, and W. E. Watson Baker. (1928). *Origin and Development of the Microscope*. The Royal Microscopical Society, London.

Djerassi, C. (1960). *Optical Rotary Dispersion*. McGraw-Hill, New York.

Dobell, C. [1932]. (1960). *Antony van Leeuwenhoek and his "Little Animals"*. Dover, New York.

Dobell, C. and F. W. O'Connor. (1921). *The Intestinal Protozoa of Man*. John Bale, Sons & Danielsson, Ltd, London.

Dodge, A. V., S. V. Dodge, and K. Jones. (1988). An introduction to video recording at the microscope. *Microscopy* **36**: 43–53.

Dollond, J. (1758). Account of some experiments concerning the different refrangibility of light. *Phil. Trans. Roy. Soc. Lond.* 733–743.

Doppler, C. (1842). On the coloured light of the double stars and certain other stars of the heavens. Translated by A. Eden. 1992. In: *The Search for Christian Doppler*. Eden, A. ed. Springer-Verlag, Wien.

Draper, J. W. (1878). *Scientific Memoirs, Being Experimental Contributions to a Knowledge of Radiant Energy*. Harper, New York.

Drude, P. (1939). *The Theory of Optics*. Longmans, Green, London.

Druez, D., F. Marano, R. Calvayrac, B. Volochine, and J. C. Soufir. (1989). Effect of gossypol on the morphology, motility and metabolism of a flagellated protest, *Dunaliella biciliata*. *J. Submicrosc. Cytol. Pathol.* **21**: 367–374.

Drummond, T. (1826). On the means of facilitating the observation of distant stations in geodaetical operations. *Phil. Trans. Roy. Soc. Lond.* **116**: 324–337.

Dubois, M., P. Jouannet, P. Bergé, and G. David. (1974). Study of spermatozoa motility in human cervical mucus by inelastic light scattering. *Nature* **252**: 711–713.

Dunitz, J. (1989). *Pacific Light*. Light Press, Hillsboro, Oregon.

Dunn, G. A. (1991). Quantitative interference microscopy. 91–118. In: *New Techniques in Optical Microscopy and Microspectroscopy*. R. J. Cherry, ed. CRC Press.

Dutrochet, R. (1924). *Recherches anatomiques et physiologiques, sur la structure intime des animaux et des végétaux, et sur leur motilité*. Chez J. B. Baillière, Paris.

Dvoeglazov, V. V. (1999). Speculations on the neutrino theory of light. *Annales de la Fondation Louis de Broglie* **24**: 111.

Dyson, F. W., A. S. Eddington, and C. Davidson. (1920). A determination of the deflection of light by the sun's gravitational field, from observations made at the total eclipse of May 29, 1919. *Phil. Trans. Roy. Soc. Lond.* Series A. **220**: 291.

Dyson, J. (1961). Interference microscopy instrument classification and applications. 412–420. In: *Encyclopedia of Microscopy*. G. L. Clark, ed. Reinhold, New York.

Ealand, C. A. no date. *The Romance of the Microscope*. Seeley, Service & Co., London.

Earnshaw, J. and M. Steer. (1979). Laser doppler microscopy. *Proc. Roy. Micro. Soc.* **14**: 108–110.

Eddington, A. S. (1920). *Space, Time and Gravitation*. Harper & Brothers, New York.

Eder, J. M. (1945). *History of Photography*. Columbia University Press, New York.

Edidin, M., Y. Zagyansky, and T. J. Lardner. (1976). Measurement of membrane protein lateral diffusion in single cells. *Science* **191**: 466–468.

Editorial. (2006a). Beautification and fraud. *Nature Cell Biol.* **8**: 101–102.

Editorial. (2006b). Appreciating data: Warts, wrinkles and all. *Nature Cell Biol.* **8**: 203.

Editorial. (2006c). Not picture perfect. *Nature* **439**: 891–892.

Edstrom, R. D., X. Yang, G. Lee, and D. F. Evans. (1990). Viewing molecules with scanning tunneling microscopy and atomic force microscopy. *FASEB J.* **4**: 3144–3151.

Einstein, A. (1905a). Concerning an heuristic point of view toward the emission and transformation of light. Translated by Arons, A. B. and M. B. Peppard. (1965). *Amer. J. Phys.* **33**: 367–374.

Einstein, A. (1905b). On the electrodynamics of moving bodies. In: *The Principle of Relativity*. Dover, New York.

Einstein, A. (1909a). On the development of our views concerning the nature and constitution of radiation. Doc. 60. *The Collected Papers of Albert Einstein*. Volume 2. Princeton University Press, Princeton, NJ.

Einstein, A. (1909b). On the present status of the radiation problem. Translated by Anna Beck. *The Collected Papers of Albert Einstein*. Volume 2. Princeton University Press, Princeton, NJ.

Einstein, A. (1917). The quantum theory of radiation. 888–901. In: *The World of the Atom*. H. A. Boorse and L. Motz, eds. Basic Books, New York.

Einstein, A. (1926). *Investigations on the Theory of the Brownian Movement*. Methuen & Co. Ltd, London.

Ellis, G. W. (1978). Advances in visualization of mitosis *in vivo*. 465–476. In: *Cell Reproduction: In honor of Daniel Mazia*.

E. R. Dirksen, D. M. Presco, and C. F. Fox, eds. Academic Press, New York.

Elmore, W. C. and M. A. Heald. (1969). *Physics of Waves*. Dover Publications, New York.

Elser, V. (2003). Random projections and the optimization of an algorithm for phase retrieval. *J. Phys. A: Math. Gen.* **36**: 2995–3007.

Emmert–Buck, M. R., R. F. Bonner, P. D. Smith, R. F. Chaqui, Z. Zhuang, S. R. Goldstein, R. A. Weiss, and L. A. Liotta. (1996). Laser capture microdissection. *Science* **274**: 998–1001.

Ermert, L., M. Ermert, M. Goppelt-Struebe, D. Walmrath, F. Grimminger, W. Steudel, H. A. Ghofrani, C. Homberger, H. R. Duncker, and W. Seege. (1998). Cyclooxygenase isoenzyme localization and mRNA expression in rat lungs. *Am. J. Respir. Cell Mol. Biol.* **18**: 479–488.

Ermert, L., M. Ermert, M. Goppelt-Struebe, M. Merkle, H. R. Duncker, F. Grimminger, and W. Seeger. (2000). Rat pulmonary cyclooxygenase-2 expression in response to endotoxin challenge: differential regulation in the various types of cells in the lung. *Am. J. Pathol.* **156**: 1275–1287.

Ermert, L., A. C. Hocke, H.-R. Duncker, W. Seeger, and M. Ermert. (2001). Comparison of different detection methods in quantitative microdensitometry. *Am. J. Pathol.* **158**: 407–417.

'Espinasse, M. (1956). *Robert Hooke*. William Heinemann, London.

Evennett, P. (1993). Köhler illumination: A simple interpretation. *Proc. RMS* **28**: 189–192.

Everitt, C. W. F. (1975). *James Clerk Maxwell. Physicist and Natural Philosopher*. Charles Scribner's Sons, New York.

Faraday, M. [1845]. (1952). Experimental Researches in Electricity. In: *Great Books of the Western World, R. M. Hutchins, ed., Encyclopedia Britannica, Chicago, IL.

Farnsworth, Elma Gardner. (1989). *Distant Vision: Romance & Discovery on an Invisible Frontier*. Pemberley Kent, Salt Lake City, UT.

Feffer, S. M. (1996). Ernst Abbe, Carl Zeiss, and the transformation of microscopical optics. 23–66. In: *Scientific Credibility and Technical Standards in 19th and early 20th Century Germany and Britain*. J. Z. Buchwald, ed. Kluwer, London.

Fermi, E. (1932). Quantum theory of radiation. *Rev. Mod. Phys.* **4**: 87–132.

Fermi, E. and C. N. Yang. (1949). Are mesons elementary particles? *Phys. Rev.* **76**: 1739–1743.

Feynman, R. P., R. B. Leighton, and M. Sands. (1963). *The Feynman Lectures in Physics*. Addison-Wesley Publishing Co., Reading, MA.

Feynman, R. P. (1965). *The Character of Physical Law*. MIT Press, Cambridge, MA.

Feynman, R. P. (1988). *QED: The Strange Theory of Light and Matter*. Princeton University Press, Princeton, NJ.

Fine, A. (1986). *The Shaky Game. Einstein Realism and the Quantum Theory*. University of Chicago Press, Chicago, IL.

Fineman, M. (1981). *The Nature of Visual Illusion*. Dover, New York.

Finkelstein, D. (2003). What is a photon? *Optics & Photonics News*. October: S12–S17.

FitzGerald, G. F. (1896). On the longitudinal component of light. *Phil. Mag.* **42**: 260–271.

Fizeau, H. (1849a). Sur une expérience relative a la vitesse de propagation de la lumiére. *Comptes rendus des séances de l'Académie des Science* **29**: 90–92.

Fizeau, H. (1849b). Sur une expérience relative a la vitesse de propagation de la lumiére. *Revue scientifique et industrielle* **36**: 393–397.

Fizeau, H. and L. Bréguet. (1850). Note sur l'expérience relative a la vitesse comparative de la lumière dans l'air et dans l'eau. *Comptes rendus des séances de l'Académie des Sciences* **30**: 562–563.

Fleagle, S. R. and D. J. Skorton. (1991). Quantitative methods in cardiac imaging: An introduction to digital image processing. 72–86. In: *Cardiac Imaging*. M. L. Marcus, H. R. Schelbert, D. J. Skorton and G. L. Wolf, eds. W. B. Saunders Co., Philadelphia.

Fletcher, J. R. (1988). The star test for microscope optics. *Microscopy* **36**: 154–159.

Flemming, W. [1880]. (1965). Contributions to the knowledge of the cell and its vital processes. *J. Cell Biol.* **25** (Supplement on Mitosis): 1–69.

Flint, O. (1994). *Food Microscopy: A Manual of Practical Methods, Using Optical Microscopy*. BIOS Scientific Publishers, Oxford.

Ford, B. J. (1983). What were the missing Leeuwenhoek microscopes really like? *Proc. Royal Micro. Soc.* **18**: 118–124.

Ford, B. J. (1985). *Single Lens: The Story of the Simple Microscope*. William Heinemann, London.

Ford, B. J. (1991). Robert Brown, Brownian movement and teethmarks on the hat brim. *The Microscope* **39**: 161–171.

Ford, B. J. (1992a). Proof of the Brownian movement. (Abstract) *The Microscope* **40**: 181–182.

Ford, B. J. (1992b). Brownian Movement in *Clarkia* pollen: A reprise of the first observations. *The Microscope* **40**: 235–241.

Ford, B. J. (1992c). The controversy of Robert Brown and Brownian movement. *Biologist* **39**: 82–93.

Forer, A. (1965). Local reduction of spindle fiber birefringence in living *Nephrotoma suturalis* (Loew) spermatocytes induced by ultraviolet microbeam irradiation. *J. Cell Biol.* **25**: 95–117.

Forer, A. (1991). In ultraviolet microbeam irradiations, characteristics of the monochromater and lamp affect the spectral composition of the ultraviolet light and probably the biological results. *J. Cell Sci.* **98**: 415–422.

Forer, A. and R. D. Goldman. (1972). The concentrations of dry matter in mitotic apparatuses *in vivo* and after isolation from sea-urchin zygotes. *J. Cell Sci.* **10**: 387–418.

Forer, A., T. Spurck, and J. D. Pickett-Heaps. (1997). Ultraviolet microbeam irradiations of spindle fibres in crane-fly spermatocytes and newt epithelial cells: Resolution of previously conflicting observations. *Protoplasma* **197**: 230–240.

Foucault, L. (1850). Méthode générale pour mesurer la vitesse de la lumiére dans l'air et les milieux transparents. Vitesses relatives de la lumiére dans l'air et dans l'eau. Projet d'expérience sur la vitesse de propagation du calorique rayonnant. *Comptes rendus des séances de l'Académie des Science* **30**: 551–560.

Foucault, L. (1862). Détermination expérimentale de la vitesse de la lumiére; parallaxe du Soleil. *Comptes rendus des séances de l'Académie des Science* **55**: 501–503.

Fourier, J. B. (1822). *Théorie Analytique de la Chaleur*. Chez Firmin Didot, Paris.

Fox, D. L. (1936). Structural and chemical aspects of animal coloration. *Amer. Nat.* **70**: 477–493.

Fox, D. L. and J. R. Wells. (1971). Schemochromic blue leaf-surfaces of *Selaginella*. *Amer. Fern. J.* **61**: 137–139.

Francon, M. (1961). *Progress in Microscopy. Modern Trends in Physiological Sciences*. Vol. 9. General editors, P. Alexander and Z.M. Bacq. Row, Peterson and Co, Evanston, IL.

Frank, N. H. (1940). *Introduction to Electricity and Optics*. McGraw-Hill, New York.

French, A. P. (1971). *Vibrations and Waves*. W. W. Norton & Co, New York.

Frercks, J. (2000). Creativity and technology in experimentation: Fizeau's terrestrial determination of the speed of light. *Centaurus* **42**: 249–287.

Fresnel, A. (1819). Memoir on the diffraction of light. In: *The Wave Theory. Light and Spectra*. H. Crew, ed. Arno Press, New York. 1981.

Fresnel, A. (1827–1829). Elementary view of the undulatory theory of light. *The Quarterly Journal of Science Literature and Art* 127, 159, 441, 113, 168, 198, 431.

Freud, S. (1989). *Civilization and Its Discontents*. Re-issue edition. W. W. Norton & Company, New York.

Freund, I. (1906). *The Study of Chemical Composition*. Cambridge University Press, Cambridge.

Frey-Wyssling, A. (1953). *Submicroscopic Morphology of Protoplasm*. Elsevier, Amsterdam.

Frey-Wyssling, A. (1957). *Macromolecules in Cell Structure*. Harvard University Press, Cambridge.

Frey-Wyssling, A. (1959). *Die Pflanzliche Zellwand*. Springer-Verlag, Berlin.

Frey-Wyssling, A. (1976). *The Plant Cell Wall*, third revised edition. Barntrceger, Berlin.

Friedberg, R. (1994). Einstein and stimulated emission: A completely corpuscular treatment of momentum balance. *Am. J. Phys.* **62**: 26–32.

Frisch, O. R. (1933). Experimenteller Nachweis des Einsteinschen Strahlungsrückstosses. *Z. Phys.* **86**: 42.

Gage, S. H. (1891). *The Microscope and Histology. Part I. The Microscope and Microscopical Methods,* third edition. Printed by Andrus & Church, Ithaca, NY

Gage, S. H. (1894). *The Microscope and Microscopical Methods*, fifth edition. Comstock Publishing Co., Ithaca, NY.

Gage, S. H. (1896). *The Microscope and Microscopical Methods*, sixth edition. Comstock Publishing Co, Ithaca, NY.

Gage, S. H. (1908). *The Microscope. An Introduction to Microscopic Methods and to Histology*, tenth edition. Comstock Publishing Co., Ithaca, NY.

Gage, S. H. (1917). *The Microscope. An Introduction to Microscopic Methods and to Histology*, twelfth edition. Comstock Publishing Co., Ithaca, NY.

Gage, S. H. (1920). Modern dark-field microscopy and the history of its development. *Trans. Amer. Micros. Soc.* **39**: 95–141.

Gage, S. H. (1925). *The Microscope. Dark-Field Edition*. Comstock Publishing Co., Ithaca, NY.

Gage, S. H. (1941). *The Microscope*, seventeenth edition. Comstock Publishing Co, Ithaca, NY. revised.

Gage, S. H. and H. P. Gage. (1914). *Optic Projection*. Comstock Publishing Co., Ithaca, NY.

Galilei, Galileo. (1653). *Sidereus nuncius*. Typis Jacobi Flesher, prostat apud Cornelium Bee, Londini.

Galjaard, H. and J. A. Szirmai. (1964). Determination of the dry mass of tissue sections by interference microscopy. *J. Royal Microscopical Soc.* **84**: 27–42.

Gamow, G. (1988). *The Great Physicists from Galileo to Einstein*. Dover, New York.

Gao, K., R. E. Morris, and C. Wie. (1995). Epipolarization microscopic immunogold assay: A combination of immunogold silver staining, enzyme-linked-immunosorbent assay and epipolarization microscopy. *Biotech. Histochem* **70**: 211–216.

Gao, K. and E. L. Cardell. (1994). Pseudocolor image processing of PEPCK subcellular distribution in rat hepatocytes shown with IGSS and epipolarization microscopy. *J. Histochem. Cytochem.* **42**: 1651–1653.

Georgi, H. (1993). *The Physics of Waves*. Prentice Hall, Englewood Cliffs, NJ.

Gerlach, W. and A. Golsen. (1923). Untersuchung an Radiometern. II. Eine neue Messung des Strahlungsdruckes. *Z. Phys.* **15**: 1–7.

Gernsheim, H. (1982). *The Origins of Photography*. Thames and Hudson, London.

Giepmans, B. N. G., S. R. Adams, M. H. Ellisman, and R. Y. Tsien. (2006). The fluorescent toolbox for assessing protein location and function. *Science* **312**: 217–224.

Gill, R. W. (1974). *Basic Perspective*. Thames and Hudson, London.

Gillespie, C. C. (1960). *The Edge of Objectivity*. Priceton University Press, Princeton, NJ.

Ginzburg, M., B. Z. Ginzburg, and R. Wayne. (1999). Ultrarapid endocytotic uptake of large molecules in *Dunaliella* species. *Protoplasma* **206**: 73–86.

Glazebrook, R. T. (1896). *James Clerk Maxwell and Modern Physics*. Cassell and Co., London.

Godfrey, D. G. (2001). *Philo T. Farnsworth: The Father of Television*. University of Utah Press, Salt Lake City, UT.

Goethe, W. (1840). *Goethe's Theory of Colours*. J. Murray, London.

Goldstein, D. J. (1964). Relation of effective thickness and refractive index to permeability of tissue components in fixed sections. *J. Royal Microscop. Soc.* **84**: 43–54.

Goldstein, D. J. (1969). Detection of dichroism with the microscope. *J. Microscopy* **89**: 19–36.

Goldstein, D. J. (1990). Quantitative theory of ideal phase-contrast microscopy, taking object width into account. *J. Microscopy* **164**: 127–142.

Goldstein, D. H. (2003). *Polarized Light*. second edition. Optical Science and Engineering. Vol. 83. CRC Press.

Goodman, B. A., B. Williamson, and J. A. Chudek. (1992). Non-invasive observation of the development of fungal infection in fruit. *Protoplasma* **166**: 107–109.

Gorbsky, G. J., P. J. Sammak, and G. G. Borisy. (1988). Microtubule dynamics and chromosome motion visualized in living anaphase cells. *J. Cell Biol.* **106**: 1185–1192.

Gordon, J. W. (1907). The use of a top stop for developing latent powers of the microscope. *J. Roy. Microscop. Soc.*: 1–13.

Goring, C. R. and A. Pritchard. (1837). *Micrographia, containing practical essays on reflecting, solar, oxy-hydrogen gas microscopes, micrometers, eye-pieces, etc*. Whittaker and Co, London.

Gorman, J., A. Chowdhury, J. A. Surtees, J. Shimada, D. R. Reichman, E. Alani, and E. C. Greene. (2007). Dynamic basis for one-dimensional DNA scanning by the mismatch repair complex Msh2-Msh6. *Molecular Cell* **28**: 359–370.

Grabau, M. (1938). Polarized light enters the world of everyday life. *J. Appl. Phys.* **9**: 215–225.

Gregory, R. L. (1973). *Eye and Brain. The Psychology of Seeing.* McGraw Hill, New York.

Green, P. B. (1958). Structural characteristics of developing *Nitella* internodal cell walls. *J. Biophys. and Biochem. Cytol.* **4**: 505–519.

Green, P. B. (1960). Multinet growth in the cell wall of *Nitella*. *J. Biophys. and Biochem. Cytol.* **7**: 289–297.

Gregory, D. (1715). *Dr. Gregory's Elements of Catoptrics and Dioptrics.* Printed for E. Curll, J. Pemberton, and W. Taylor, London.

Gregory, D. (1735). *Dr. Gregory's Elements of Catoptrics and Dioptrics,* second edition. Printed for E. Curll, London.

Greulick, K. O. (1992). Moving particles by light: no longer science fiction. *Proc. Royal Microscop. Soc.* **27**: 3–8.

Grew, N. (1672). *The Anatomy of Vegetables Begun: With a General Account of Vegetation Founded Thereon.* Printed for Spencer Hickman, London.

Grew, N. (1682). *The Anatomy of Plants: With An idea of a Philosophical History of Plants, and Several Other Lectures* Printed by W. Rawlins, for the author, London.

Griffiths, D. J. (1989). *Introduction to Electrodynamics.* Prentice Hall, Englewood Cliffs, NJ.

Griffiths, D. J. (2005). *Introduction to Quantum Mechanics.* Pearson Prentice Hall, Upper Saddle River, NJ.

Grimaldi, F. M. (1665). *Physico-mathesis de Lumine.* H. Berniae, Bononiae.

Grotha, R. (1983). Chlorotetracycline-binding surface regions in germlings of *Riella heliocophylla* (Bory et Mont). *Mont. Planta* **158**: 473–481.

Gundlach, H. (1993). Köhler illumination in modern light microscopy. *Proc. RMS* **28**: 194–196.

Haas, A. (1925). *Introduction to Theoretical Physics.* D. van Nostrand, New York.

Haigler, C. H., R. M. Brown Jr., and M. Benziman. (1980). Calcofluor white ST alters the *in vivo* assembly of cellulose microfibrils. *Science* **210**: 903–906.

Hakfoort, C. (1995). *Optics in the Age of Euler.* Cambridge University Press, Cambridge.

Hale, A. J. (1958). *The Interference Microscope in Biological Research.* E. & S. Livingstone, Edinburgh.

Hale, A. J. (1960). The interference microscope as a cell balance. 173–186. In: *New Approaches in Cell Biology.* P. M. B. Walker, ed. Academic Press, London.

Hall, A., M. Browne, and V. Howard. (1991). Confocal microscopy—The basics explained. *Proc. Royal Microscopical Soc.* **26**: 63–70.

Hallimond, A. F. (1947). Production of contrast in the microscope image by means of opaque diaphragms. *Nature* **159**: 851–852.

Hamilton, W. R. (1833). On a general method of expressing the paths of light, and of the planets, by the coefficients of a characteristic function. *Dublin University Review*: 795–826.

Hammond, J. H. (1981). *The Camera Obscura. A Chronicle.* Adam Hilger Ltd., Bristol.

Harburn, G., C. A. Taylor, and T. R. Welberry. (1975). *Atlas of Optical Transforms.* Cornell University Press, Ithaca, NY.

Hardy, A. C. and F. H. Perrin. (1932). *The Principles of Optics.* McGraw-Hill, New York.

Harnwell, G. P. (1949). *Principles of Electricity and Electromagnetism,* second edition. McGraw-Hill, New York.

Harris, H. (1999). *The Birth of the Cell.* Yale University Press, New Haven, CT.

Hartley, W. G. (1980). The usefulness of oblique illumination. *Proc. RMS* **15**: 422–424.

Hartshorne, N. H. (1975). A hot-wire stage and its implications. *The Microscope* **23**: 177–190.

Hartshorne, N. H. (1976). The hot-wire stage II. The temperature gradient and determination of temperatures. *The Microscope* **24**: 217–226.

Hartshorne, N. H. (1981). The hot-wire stage III. New Design avoiding the use of materials containing asbestos. *The Microscope* **29**: 117–120.

Harvey, E. N. (1920). *The Nature of Animal Light.* J. B. Lippincott Co, Philadelphia, PA.

Harvey, E. N. (1938). Some physical properties of protoplasm. *J. Appl. Phys.* **9**: 68–80.

Harvey, E. N. (1940). *Living Light.* Princeton University Press, Princeton, NJ.

Harvey, E. N. (1957). *A History of Luminescence from the Earliest Times until 1900.* American Philosophical Society, Philadelphia, PA.

Harwood, J. T. (1989). Rhetoric and graphics in Micrographia. 119–147. In: *Robert Hooke: New Studies.* M. Hunter and S. Schaffer, eds. Boydell Press, London.

Haselmann, H. (1986). One hundredth anniversary of Abbe's Apochomats. *Proc. Royal Microscop. Soc.* **21**: 197–202.

Haselmann, H. (1993). One hundredth anniversary of Köhler illumination. *Proc. RMS,* **28**: 186–188.

Hatschek, E. (1919). *An Introduction to the Physics and Chemistry of Colloids.* P. Blakiston's Son & Co, Philadelphia, PA.

Haugland, R. P. (1989). *Handbook of Fluorescent Probes and Research Chemicals.* Molecular Probes, Eugene, OR.

Haüy, R.-J. (1807). *An Elementary Treatise on Natural Philosophy.* Printed for George Kearsley, London.

Hawley, M., I. D. Raistrick, J. G. Berg, and R. J. Houlton. (1991). Growth mechanism of sputtered films of $YBa_2Cu_3O_7$ studied by scanning tunneling microscopy. *Science* **251**: 1587–1589.

Hawryshyn, C. W. (1992). Polarization vision in fish. *American Scientist* **80**: 164–175.

Hayashi, T. (1957). Some dynamic properties of the protoplasmic streaming in *Chara. Protoplasma* **102**: 1–9.

Heald, M. A. and J. B. Marion. (1989). *Classical Electromagnetic Radiation,* third edition. Brooks/Cole, Australia.

Heaviside, O. (1892). *Electrical Papers.* Macmillan, New York.

Heaviside, O. [1898]. (1922). Appendix G. Note on the motion of a charged body at a speed equal to or greater than that of light. In: *Electromagnetic Theory.* Volume II. Electromagnetic Theory. Benn Brothers, London.

Heaviside, O. (1922). *Electromagnetic Theory.* Benn Brothers, London.

Hecht, E. and A. Zajac. (1974). *Optics.* Addison-Wesley, Reading MA.

Heisenberg, W. (1930). *The Physical Principles of the Quantum Theory.* Dover, Chicago. IL.

Heisenberg, W. (1958). *Physics and Philosophy. The Revolution in Modern Science.* Harper and Row, New York.

Heisenberg, W. (1971). *Physics and Beyond. Encounters and Conversations.* Harper and Row, New York.

Heisenberg, W. (1979). *Philosophical Problems of Quantum Physics*. Ox Bow Press, Woodbridge, CT.

Heitler, W. (1944). *The Quantum Theory of Radiation*, second edition. Oxford University Press, Oxford.

Hell, S. W. and J. Wichmann. (1994). Breaking the diffraction resolution limit by stimulated emission: Stimulated-emission-depletion fluorescence microscopy. *Opt. Lett.* **19**: 780–782.

Helmholtz, H. von. (1954). *On the Sensations of Tone*. Dover, New York.

Helmholtz, H. von. (2005). *Treatise on Physiological Optics*. Volumes I–III. Dover, Mineola, NY.

Herapath, W. B. (1852). On the optical properties of a newly-discovered salt of quinine, which crystalline substance possess the power of polarizing a ray of light, like tourmaline, and at certain angles of rotation of depolarizing it, like selenite. *Phil. Mag.* **3**(4th ser.): 161–173. and one plate.

Herschel, J. F. W. (1845). Light. *Encyclopedia Metropolitana.* **4**: 341–586. Written and circulated in 1827. [1827].

Herschel, J. F. W. (1845a). On a case of superficial colour presented by a homogeneous liquid internally colourless. *Phil. Trans. Roy. Soc. Lond.* **135**: 143–145.

Herschel, J. F. W. (1845b). On the epipolic dispersion of light, being a supplement to a paper entitled On a case of superficial colour presented by a homogeneous liquid internally colourless. *Phil. Trans. Roy. Soc. Lond.* **135**: 147–153.

Herschel, J. F. W. (1876). *Popular Lectures on Scientific Subjects*. Daldy, Isbister & Co, London.

Herschel, J. F. W. (1840). On the chemical action of the rays of the solar spectrum on preparations of silver and other substances, both metallic and non-metallic, and on some photographic processes. *Phil. Trans. Roy. Soc. Lond.* **130**: 1–59.

Herschel, J. F. W. (1966). *A Preliminary Discourse on the Study of Natural Philosophy*. Johnson Reprint Corp, New York.

Herschel, W. (1800a). Investigation of the powers of the prismatic colours to heat and illuminate objects. *Phil. Trans.* **90**: 255–283.

Herschel, W. (1800b). Experiments on the refrangibility of the invisible rays of the sun. *Phil Trans.* **90**: 284–292.

Hertz, H. (1893). *Electric Waves*. Macmillan and Co, London.

Herzberg, G. (1944). *Atomic Spectra and Atomic Structure*. Dover, New York.

Hewlett, P. S. (1983). Three-colour dark-ground images of transparent colourless objects. *Microscopy* **34**: 522–528.

Hildebrand, J. A., D. Rugar, R. N. Johnston, and C. F. Quate. (1981). Acoustic microscopy of living cells. *Proc. Natl. Acad. Sci. U.S.A.* **78**: 1656–1660.

Hindle, B. and H. M. Hindle. (1959). David Rittenhouse and the illusion of reversible relief. *Isis* **50**: 135–140.

Hiramoto, Y. (1967). Observations and measurement of sea urchin eggs with a centrifuge microscope. *J. Cell Physiol.* **69**: 219–230.

Hiramoto, Y., Y. Hamaguchi, Y. Shoji, T. E. Schroeder, S. Shimoda, and S. Nakamura. (1981). Quantitative studies on the polarization optical properties of living cells. II. The role of microtubules in birefringence of the spindle of the sea urchin egg. *J. Cell Biol.* **89**: 121–130.

Hiramoto, Y. and I. Kaneda. (1988). Diffusion of substances in the cytoplasm and across the nuclear envelope in egg cells. *Protoplasma* **2**(Suppl.): 88–94.

Hirose, A. and K. E. Lonngren. (1991). *Introduction to Wave Phenomena*. Krieger, Malabar, FL.

Hoffman, R. and L. Gross. (1970). Reflected-light differential-interference microscopy: Principles, use and image interpretation. *J. Microscopy* **91**: 149–172.

Hoffman, R. and L. Gross. (1975a). The modulation contrast microscope. *Nature* **254**: 586–588.

Hoffman, R. and L. Gross. (1975b). Modulation contrast microscope. *Applied Optics* **14**: 1169–1176.

Hoffman, R. (1977). The modulation contrast microscope: Principles and Performance. *J. Microscopy* **110**: 205–222.

Hogg, J. (1898). *The Microscope. Its History, Construction, and Application*. George Routledge & Sons, London.

Hooke, R. (1665). *Micrographia*. Printed by Jo. Martyn and Ja. Allestry. London.

Hooke, R. (1678). *Microscopium*. Cutlerian lecture, London.

Hooke, R. (1705). *Posthumous Works,* London.

Hornberger, B., M. Feser, and C. Jacobsen. (2007). Quantitative amplitude and phase contrast imaging in a scanning transmission X-ray microscope. *Ultramicroscopy* **107**: 644–655.

Horowitz, P. and W. Hill. (1989). *The Art of of Electronics*, second edition. Cambridge University Press, Cambridge.

Horváth, G. and D. Varjú. (2004). *Polarized Light in Animal Vision. Polarization Patterns in Nature*. Springer, Berlin.

Hovestadt, H. (1902). *Jena Glass and Its Scientific and Industrial Applications*. Macmillan, London.

Howarth, M., K. Takao, Y. Hayashi, and A. Y. Ting. (2005). Targeting quantum dots to surface proteins in living cells with biotin ligase. *Proc. Natl. Acad. Sci. USA,* **102**: 7583–7588.

Howells, M. R., J. Kirz, and D. Sayre. (1991). X-ray microscopes. *Scientific American.* **265**: 88–94. February.

Houstoun, R. A. (1930). *A Treatise on Light*. Longmans, Green, London.

Hubbell, J. H. (2006). Electron-positron pair production by photons: A historical overview. *Radiation Physics and Chemistry* **75**: 614–623.

Hughes, A. F. and M. M. Swann. (1948). Anaphase movement in the living cell. *J. Exp. Biol.* **25**: 45–70.

Hughes, D. (1987). Immunogold-silver staining with epipolarisation microscopy. *Medical Technology and Scientist*: 38–39.

Hughes, D. A., P. J. Morris, S. Fowler, and A. J. Chaplin. (1991). Immunogold-silver staining of leucocyte populations in lung sections containing carbon particles requires cautious interpretation. *Histochemical J.* **23**: 196–199.

Hund, F. (1974). *The History of the Quantum Theory*. Harrap, London.

Hunt, B. J. (1991). *The Maxwellians*. Cornell University Press, Ithaca, NY.

Huxley, A. (1943). *The Art of Seeing*. Chatto and Wl\indus, London.

Huxley, A. F. (1952). Applications of an interference microscope. *J. Physiol.* **117**: 52–53P.

Huxley, A. F. (1954). A high-power interference microscope. *J. Physiol.* **125**: 11–13P.

Huxley, A. F. (1974). Muscular contraction. *J Physiol.* **243**: 1–43.

Huxley, A. F. and R. Niedergerke. (1954). Interference microscopy of living muscle fibres. *Nature* **173**: 971–973.

Huxley, A. F. and R. Niedergerke. (1958). Measurement of the striations of isolated muscle fibres with the interference microscope. *J. Physiol.* **144**: 403–425.

Huxley, H. E. and J. Hanson. (1954). Changes in the cross-striations of muscle during contraction and stretch and their structural interpretation. *Nature* **173**: 973–976.

Huxley, H. E. and J. Hanson. (1957). Quantitative studies on the structure of cross-striated myofibrils. I. Investigations by interference microscopy. *Biochem. Biophys. Acta* **23**: 229–249.

Huygens, C. [1690]. (1945). *Treatise on Light*. University of Chicago Press, Chicago, IL.

Ingelstam, E. (1957). Some considerations concerning the merits of interference microscopes. *Exp. Cell Res.* **4**(Suppl.): 150–157.

Inoué, S. (1951). A method for measuring small retardations of structure in living cells. *Exp. Cell Res.* **2**: 513–517.

Inoué, S. (1952a). The effect of colchicine on the microscopic and submicroscopic structure of the mitotic spindle. *Exp. Cell Res.* **2**(Suppl.): 305–318.

Inoué, S. (1952b). Studies on depolarization of light at microscope lens surfaces. *Exp. Cell Res.* **2**: 513–517.

Inoué, S. (1953). Polarization optical studies of the mitotic spindle. I. The demonstration of spindle fibers in living cells. *Chromosoma* **5**: 487–500.

Inoué, S. (1959). Motility of cilia and the mechanism of mitosis. 402–408. In: *Biophysical Science—A Study Program*. J. L. Oncley, F. O. Schmitt, R. C. Williams, M. D. Rosenberg, and R. H. Holt, eds. John Wiley & Sons, New York.

Inoué, S. (1964). Organization and function of the mitotic spindle. 549–598. In: *Primative Motile Systems in Cell Biology*. R. D. Allen and N. Kamiya, eds. Academic Press, New York.

Inoué, S. (1981). Cell division and the mitotic spindle. *J. Cell Biol.* **91**: 131s–147s.

Inoué, S. and A. Bajer. (1961). Birefringence in endosperm mitosis. *Chromosoma* **12**: 48–63.

Inoué, S. and K. Dan. (1951). Birefringence of the dividing cell. *J. Morphol.* **89**: 423–456.

Inoué, S. and E. D. Salmon. (1995). Force generation by microtubule assembly/disassembly in mitosis and related movements. *Mol. Biol. Cell,* **6**: 1619–1640.

Inoué, S. (1961). Polarizing microscope. Design for maximum sensitivity. 480–485. In: *The Encyclopedia of Microscopy*. G. L. Clarke, ed. Reinhold, New York.

Inoué, S. (1986). *Video Microscopy*. Plenum, New York. NY.

Inoué, S. and W. L. Hyde. (1957). The simultaneous realization of high resolution and high sensitivity with the polarizing microscope. *J. Biophys. Biochem. Cytol.* **3**: 831–838.

Inoué, S., M. Goda, and R. A. Knudson. (2001a). Centrifuge polarizing microscope. II. Sample biological applications. *J. of Microscopy* **201**: 357–367.

Inoué, S., R. A. Knudson, M. Goda, K. Suzuki, C. Nagano, N. Okada, H. Takahashi, K. Ichie, M. Iida, and K. Yamanaka. (2001b). Centrifuge polarizing microscope. I. Rationale, design and instrument performance. *J. of Microscopy* **201**: 341–356.

Inoué, S. and H. Sato. (1967). Cell Motility by labile association of molecules. *J. Gen. Physiol.* **50**(Suppl.): 259–292.

Inoué, S. and K. R. Spring. (1997). *Video Microscopy. The Fundamentals*, second edition. Plenum Press, New York.

Israel, H. W., R. G. Wilson, J. R. Aist, and H. Kunoh. (1980). Cell wall appositions and plant disease resistance: Acoustic microscopy of papillae that block fungal ingress. *Proc. Natl. Acad. Sci. USA* **77**: 2046–2049.

Izzard, C. S. and L. R. Lochner. (1976). Cell-to-substrate contacts in living fibroblasts: An interference reflection study with an evaluation of the technique. *J. Cell Sci.* **21**: 129–159.

Jackson, J. D. (1962). *Classical Electrodynamics*. John Wiley & Sons, New York.

Jackson, S. A., M. L. Wang, H. M. Goodman, and J. Jiang. (1998). Application of fiber-FISH in physical mapping of Arabidopsis thaliana. *Genome* **41**: 566–572.

Jaffe, B. (1960). *Michelson and the Speed of Light*. Anchor Books, Garden City, NJ.

James, D. I. (1980). The effect of reciprocity failure on exposure and contrast in macrophotography. *Proc. Roy. Microscope Soc.* **15**: 50–53.

James, J. (1976). *Light Microscopic Techniques in Biology and Medicine*. Martinus Nijhoff Medical Division, The Netherlands. 180–192.

James, J. and H. Dessens. (1962). Immersion-refractometric observations on the solid concentration of erythrocytes. *J. Cell Comp. Physiol.* **60**: 235–241.

James, T. H. (1952). Photographic development. *Sci. Amer.* **187**(11): 30–33.

Jammer, M. (1966). *The Conceptual Development of Quantum Mechanics*. McGraw-Hill, New York.

Jammer, M. (1974). *The Philosophy of Quantum Mechanics*. John Wiley & Sons, New York.

Jardine, L. (2005). *The Curious Life of Robert Hooke: The Man Who Measured London*. Harper Perennial, New York.

Jeans, J. H. (1905a). On the partition of energy between matter and aether. *Phil. Mag,* **10**: 91–98.

Jeans, J. H. (1905b). A comparison between the two theories of radiation. *Nature* **72**: 293–294.

Jeans, J. H. (1905c). On the laws of radiation. *Proc. Roy. Soc. Lond.* **76**: 526–545.

Jeans, J. H. (1924). *Report on Radiation and the Quantum-Theory*, second edition. Fleetway, London.

Jeans, J. H. (1927). *The Mathematical Theory of Electricity and Magnetism*. Cambridge University Press, Cambridge.

Jeans, J. (1968). *Science & Music*. Dover, New York.

Jefimenko, O. D. (1966). *Electricity and Magnetism*. Appleton-Century-Crofts, New York.

Jelinski, L. W., R. W. Behling, H. K. Tubbs, and M. D. Cockman. (1989). NMR Imaging: From whole bodies to single cells. *American Biotechnology Laboratory* April: 34–41.

Jenkins, F. A. and H. E. White. (1937). *Fundamentals of Optics*. McGraw-Hill, New York.

Jenkins, F. A. and H. E. White. (1950). *Fundamentals of Optics*, second edition. McGraw-Hill, New York.

Jenkins, F. A. and H. E. White. (1957). *Fundamentals of Optics*, third edition. McGraw-Hill, New York.

Jenkins, F. A. and H. E. White. (1976). *Fundamentals of Optics*, fourth edition. McGraw-Hill, New York.

Jenner, C. F., Y. Xia, C. D. Eccles, and P. T. Callaghan. (1988). Circulation of water within wheat grain revealed by nuclear magnetic resonance micro-imaging. *Nature* **366**: 399–402.

Jensen, W. A. (1962). *Botanical Histochemistry*. W. H. Freeman and Co, San Francisco, CA.

Johannsen, A. (1918). *Manual of Petrographic Methods*. McGraw-Hill, New York.

Johansen, D. A. (1940). *Plant Microtechnique*. McGraw-Hill, New York.

Johnson, D. S., L. Bai, B. Y. Smith, S. S. Patel, and M. D. Wang. (2007). Single-Molecule Studies Reveal Dynamics of DNA Unwinding by the Ring-Shaped T7 Helicase. *Cell* **129**: 1299–1309.

Johnson, L. V., M. L. Walsh, and L. B. Chen. (1980). Localization of mitochondria in living cells with rhodamine 123. *Proc. Natl. Acd. Sci. USA,* **77**: 990–994.

Johnson, L. V., M. L. Walsh, B. J. Bockus, and L. B. Chen. (1981). Monitoring of relative mitochondrial membrane potential in living cells by fluorescence microscopy. *J. Cell Biol.* **88**: 526–535.

Johnston, R. N., A. Atalar, J. Heiserman, V. Jipson, and C. F. Quate. (1979). Acoustic microscopy: Resolution of subcellular detail. *Proc. Natl. Acad. Sci. U.S.A.* **76**: 3325–3329.

Johnston, S. F. (2001). *A History of Light and Colour Measurement. Science in the Shadows.* Institute of Physics, Bristol.

Jones, H. C. (1913). *A New Era in Chemistry.* D. Van Nostrand Co., New York.

Jordan, P. (1928). Die Lichtquanten hypothese. *Erg. Der Exakt. Nature.* **17**: 158.

Jordan, P. (1935). Zur Neutrinotheorie des Lichtes. *Zeits. f. Phys.* **93**: 464–472.

Jordan, P. (1936a). Über die Wechselwirkung von Spinorteilchen. *Zeits. f. Phys.* **98**: 709–713.

Jordan, P. (1936b). Lichtquant und Neutrino. *Zeits. f. Phys.* **98**: 759–767.

Jordan, P. (1936c). Zur Herleitung der Vertauschungsregeln in der Neutrinotheorie des Lichtes. *Zeits. f. Phys.* **99**: 109–113.

Jordan, P., R. de, and L. Kronig. (1936). Lichtquant und Neutrino. II. Dreidimensionales Strahlungsfeld. *Zeits. f. Phys.* **100**: 569–583.

Jovin, T. M. and D. J. Arndt-Jovin. (1989). Luminescence digital imaging microscopy. *Ann. Rev. Biophys. Biophys. Chem.* **18**: 271–308.

Jowett, B. (1908). *The Republic of Plato.* Clarendon Press, Oxford.

Joyce, K. (1995). *Astounding Optical Illusions.* Sterling, New York.

Juster, N. (1963). *The Dot and the Line: A Romnace in Lower Mathematics.* Random House, New York.

Kacher, B. (1985). Asymmetric illumination contrast: A method of image formation for video light microscopy. *Science* **227**: 766–768.

Kalinin, S. and A. Gruverman, eds. (2007). *Scanning Probe Microscopy.* Volumes 1 and 2. Springer Science + Business Media, New York.

Kallenbach, E. (1986). *The Light Microscope. Principles and Practice for Biologists.* Charles C. Thomas, Publisher, Springfield, IL.

Kamitsubo, E., M. Kikuyama, and I. Kaneda. (1988). Apparent viscosity of the endoplasm of characean internodal cells measured by the centrifuge method. *Protoplasma* **1**(Suppl.): 10–14.

Kamitsubo, E., Y. Ohashi, and M. Kikuyama. (1989). Cytoplasmic streaming in internodal cells of *Nitella* under centrifugal acceleration: a study done with a newly constructed centrifuge microscope. *Protoplasma* **152**: 148–155.

Kamiya, N. and K. Kuroda. (1957). Cell operation in *Nitella*. III. Specific gravity of the cell sap and endoplasm. *Proc. Jpn. Acad.* **33**: 403–406.

Kamiya, N. and K. Kuroda. (1958). Cell operation in *Nitella*. IV. Tension at the surface of the effused endoplasmic drops. *Proc. Jpn. Acad.* **34**: 435–438.

Kaneda, I., E. Kamitsubo, and Y. Hiramoto. (1987). Mechanical properties of the cytoplasm in echinoderm eggs determined from the movement of inserted gold particles caused by centrifugal acceleration. *J. Muscle Res. Cell Motil.* **8**: 285.

Kaneda, I., E. Kamitsubo, and Y. Hiramoto. (1990). The mechanical structure of the cytoplasm of the Echinoderm egg determined by "gold particle method" using a centrifuge microscope. *Develop. Growth and Differ.* **32**: 15–22.

Kangro, H. (1976). *Early History of Planck's Radiation Law.* Taylor Francis, London.

Kargon, R. H. (1982). *The Rise of Robert Millikan. Portrait of a Life in American Science.* Cornell University Press, Ithaca.

Kaufman, L. and I. Rock. (1962). The moon illusion. *Science* **136**: 953–961.

Kelvin, L. (1904). *Baltimore Lectures on Molecular Dynamics and the Wave Theory of Light.* C. J. Clay and Sons, London.

Kepler, J. [1611]. (1966). *A New Year's Gift; or, On the Six-Cornered Snowflake.* Clarendon Press, Oxford.

Kingsbury, B. F. (1944). 1944. Simon Henry Gage. *Science* **100**: 420–421.

Kingsbury, B. F. and O. A. Johannsen. (1927). *Histological Technique.* John Wiley & Sons, New York.

Kirchhoff, G. (1860). On the simultaneous emission and absorption of rays of the same refrangibility. *Phil. Mag. Ser. 4* **19**: 193–197.

Kirchhoff, G. (1861). On a new proposition in the theory of heat. *Phil Mag. Ser. 4* **21**: 240–247.

Kirchhoff, G. and R. Bunsen. [1860]. (1901). Chemical analysis by spectral observations. 99–126. In: *The Laws of Radiation and Absorption; Memoirs by Prévost, Stewart, Kirchhoff, and Kirchhoff and Bunsen.* D. B. Brace, ed. Amer. Book Co., New York.

Klar, T. A., S. Jakobs, M. Dyba, A. Egner, and S. W. Hell. (2000). Fluorescence microscopy with diffraction resolution limit broken by stimulated emission. *Proc. Natl. Acad. Sci. USA* **97**: 8206–8210.

Klar, T. A., E. Engel, and S. W. Hell. (2001). Breaking Abbe's diffraction resolution limit in fluorescence microscopy with stimulated emission depletion beams of various shapes. *Phys. Rev. E* **64**: 066613.

Klein, H. A. (1970). *Holography with an Introduction to the Optics of Diffraction, Interference, and Phase Differences.* J.B. Lippincott, Philadelphia, PA.

Klein, M. J. (1964). Einstein and the Wave-Particle Duality. *The Natural Philosopher* **3**: 1–49.

Kleppner, D. (2005). Rereading Einstein on radiation. *Physics Today* February: 30–33.

Knight, C. (1867). *Arts and Sciences.* Vol. VI. Bradbury, Evans, and Co., London.

Knowles, D. (1994). *The Secrets of the Camera Obscura.* Chronicle Books, San Francisco, CA.

Koch, R. (1880). *Investigations into the Etiology of Traumatic Infective Diseases.* The New Sydenham Society, London, UK.

Kock, W. E. (1965). *Sound Waves and Light Waves. The Fundamentals of Wave Motion.* Doubleday, Garden City, NJ.

Koester, C. J. (1961). Interference microscopy: Theory and techniques. 420–434. In: *Encyclopedia of Microscopy.* G. L. Clark, ed. Reinhold Publishing, New York.

Köhler, A. [1893]. (1993). A new system of illumination for photomicrographic purposes. *Proc. RMS* **28**: 181–186.

Können, G. P. (1985). *Polarized Light in Nature.* Cambridge University Press, Cambridge.

Koonce, M. P., R. A. Cloney, and M. W. Berns. (1984). Laser irradiation of centrosomes in newt eosinophils: Evidence of centriole role in motility. *J. Cell Biol.* **99**: 1999–2010.

Koshland, D. E., T. J. Mitchison, and M. W. Kirschner. (1988). Polewards chromosome movement driven by microtubule depolymerization *in vitro*. *Nature* **331**: 499–504.

Kramers, H. A. and H. Holst. (1923). *The Atom and the Bohr Theory of Its Structure*. Alfred A. Knopf, New York.

Krug, W., J. Rienitz, and G. Schulz. (1964). *Contributions to Interference Microscopy*. Hilger & Watts, London.

Kruyt, H. R. and H. S. van Klooster. (1927). *Colloids. A Textbook.* J. Wiley & Sons, New York.

Kuhn, T. S. (1978). *Black-Body Theory and the Quantum Discontinuity*. University of Chicago Press, Chicago, IL. 1894–1912.

Kuroda, K. and K. Kamiya. (1989). Propulsive force of Paramecium as revealed by the video centrifuge microscope. *Exptl. Cell Res.* **184**: 268–272.

LaChapelle, E. R. (1969). *Field Guide to Snow Crystals*. University of Washington Press, Seattle, WA.

Lambert, W. E. and M. H. Sussman. (1965). The infinity-corrected microscope. *The Microscope* **14**: 482.

Land, E. (1951). Some aspects of the development of sheet polarizers. *J. Opt. Soc. Am.* **41**: 957–963.

Lang, W. (1968). Nomarski Differential Interference-Contrast Microscopy. Zeiss Info #70. I. Fundamentals and Experimental Designs and Zeiss Info #71. II. Formation of the Interference Image.

Langley, K. H., R. W. Piddington, D. Ross, and D. B. Sattelle. (1976). Photon correlation analysis of cytoplasmic streaming. *Biochem. Biophys. Acta.* **444**: 893–898.

Langley, S. P. (1881). On a thermal balance. *The Chemical News and Journal of Physical Science* **43**: 6.

La Porta, A. and M. D. Wang. (2004). Optical torque wrench: angular trapping, rotation, and torque detection of quartz microparticles. *Phys. Rev. Lett.* **92**: 190801.

Leach, J., M. J. Padgett, S. M. Barnett, S. Franke-Arnold, and J. Courtial. (2002). Measuring the orbital angular momentum of a single photon. *Phys. Rev. Lett.* **88**: 257901.

Lebedew, P. (1901). Untersuchungen über die Druckkräfte des Lichtes. *Ann. d. Phys. Ser. 4* **6**: 433–458.

Lee, A. B. (1921). *The Microtomist's Vede-Mecum, eighth edition*. P. Blakiston's Son & Co., Philadelphia, PA.

Lee, D. (2007). *Nature's Palette. The Science of Plant Color*. University of Chicago, Chicago.

Leeman, F. (1977). *Hidden Images. Games of Perception, Anamorphic Art, Illusion, From the Renaissance to the Present*. Harry N. Abrams, New York.

Leeuwenhoek, Antoni van, (1673). The figures of some of Mr. Leeuwenhoek's microscopical observations. London.

Leeuwenhoek, Antoni van. (1798). The Select Works of Antony van Leeuwenhoek Containing his Miscroscopical [sic] Discoveries in Many of the Works of Nature. Printed by Henry Fry, London.

Leeuenhoek, A. and A. van. (1674–1716). In: *Antony van Leeuwenhoek and His 'Little Animals'*. C. Dobell, ed. Russell & Russell, New York. 1958.

Lemons, R. A. and C. F. Quate. (1974). Acoustic microscope—scanning version. *Appl. Phys. Lett.* **24**: 163–165.

Lemmons, R. A. and C. F. Quate. (1975). Acoustic microscopy: Biomedical applications. *Science* **188**: 905–911.

Leslie, J. (1804). *Experimental Inquiry into the Nature and Propagation of Heat*. Printed for J. Mawman, London.

Leslie, R. J. and J. D. PickettHeaps. (1983). Ultraviolet microbeam irradiations of mitotic diatoms: Investigation of spindle elongation. *J. Cell Biol.* **96**: 548–561.

Lewis, A., M. Isaacson, A. Harootunian, and A. Murray. (1984). Development of a 500 Å spatial resolution light microscope. I. Light is efficiently transmitted through λ/16 diameter apertures. *Ultramicroscopy* **13**: 227.

Lewis, G. N. (1926a). The conservation of photons. *Nature* **118**: 874–875.

Lewis, G. N. (1926b). The nature of light. *Proc. Natl. Acad. Sci. USA* **12**: 22–29.

Libbrecht, K. (2007). *The Art of the Snowflake: A Photographic Album*. Voyageur Press, St. Paul, MN.

Lieberman, K., S. Harush, A. Lewis, and R. Kopelman. (1990). A light source smaller than an optical wavelength. *Science* **247**: 59–61.

Lindberg, D. C. (1968). The theory of pinhole images from antiquity to the thirteenth century. *Archive for History of Exact Sciences* **5**: 154–176.

Lindberg, D. C. (1976). *Theories of Vision from al-Kindi to Kepler*. University of Chicago Press, Chicago.

Lindberg, D. C. (1983). The theory of pinhole images in the fourteenth century. *Studies in the History of Medieval Optics*. Variorum Reprints, London.

Lindberg, D. C. and G. Cantor. (1985). *The Discourse of Light from the Middle Ages to the Enlightenment*. Castle Press, Pasadena, CA.

Lindley, D. (2007). *Uncertainty. Einstein, Heisenberg, Bohr, and the Struggle for the Soul of Physics*. Doubleday, New York.

Lindsay, R. B. (1960). *Mechanical Radiation*. McGraw-Hill, New York.

Lindsay, R. B. (1966). The story of acoustics. *J. Acous. Soc. Amer.* **39**: 629–644.

Lindsay, R. B. (1969). *Concepts and Methods of Theoretical Physics*. Dover, New York.

Lindsay, R. B., ed. (1973). *Acoustics. Benchmark Papers in Acoustics: Historical and Philosophical Development*. Dowden, Hutchinson & Ross, Stroudsburg, PA.

Lintilhac, P. M. and T. B. Vesecky. (1984). Stress-induced alignment of division plane in plant tissues grown *in vitro*. *Nature* **307**: 363–364.

Lister, J. J. (1830). On some properties in achromatic object-glasses applicable to the improvement of the microscope. *Phil. Trans. Roy. Soc. Lon.* **120**: 187–200.

Liu, Z., W. R. Bushnell, and R. Brambl. (1987). Potentiometric cyanine dyes are sensitive probes for mitochondria in intact plant cells. *Plant Physiol.* **84**: 1385–1390.

Livingston and D. Michelson. (1973). *The Master of Light*. University of Chicago Press, Chicago, IL.

Lloyd, C. W. (1987). The Plant Cytoskeleton: The impact of fluorescence microscopy. *Ann. Rev. Plant Physiol.* **38**: 119–159.

Lloyd, F. E. (1924). The fluorescent colors of plants. *Science* **59**: 241–248.

Lloyd, H. (1873). *Elementary Treatise on the Wave-Theory of Light*. Longmans, Green, London.

Locke, J. (1690). *Essay Concerning Human Understanding*. Printed by Eliz. Holt for Thomas Basset. London.

Lodge, O. (1907). *Modern Views of Electricity*, third edition. Macmillan, London. Revised.

Lodge, O. (1909). *The Ether of Space*. Harper & Brothers, London.

Lodge, O. (1925). *Ether and Reality*. Hodder and Stoughton, London.

Lommel, E. (1888). *The Nature of Light, with a General Account of Physical Optics*. Kegan, Paul Trench, & Co., London.

Lorentz, H. A. (1923). *Clerk Maxwell's Electromagnetic Theory*. Cambridge University Press, Cambridge.

Lorentz, H. A. (1924). The radiation of light. *Nature* **113**: 608–611.

Lorentz, H. A. (1927). *Lectures on Theoretical Physics*. Volume I. Macmillan, London.

Lorentz, H. A. (1952). *The Theory of Electrons and its Application to the Phenomena of Light and Radiant Heat*. Dover, New York.

Loudon, R. (1983). *The Quantum Theory of Light*, second edition. Clarenden Press, Oxford.

Loudon, R. (2003). What is a photon? *Optics & Photonics News*. October: S6–S11.

Luby-Phelps, K., F. Lanni, and D. L. Taylor. (1986). Probing the structure of cytoplasm. *J. Cell Biol.* **102**: 2015–2022.

Luby-Phelps, K. (1989). Preparation of fluorescently labeled dextrans and ficolls. *Methods in Cell Biology* **29**: 59–73.

Luby-Phelps, K., F. Lanni, and D. L. Taylor. (1988). The submicroscopic properties of cytoplasm as a determinant of cellular function. *Ann. Rev. Biophys. Biophys. Chem.* **17**: 369–396.

Luckiesh, M. (1965). *Visual Illusions*. Dover, New York.

Lüers, H., K. Hillmann, J. Litniewski, and J. Bereiter-Hahn. (1991). Acoustic microscopy of cultured cells. Distribution of forces and cytoskeletal elements. *Cell Biophys.* **18**: 279–293.

Lummer, O. and E. Pringsheim. (1899a). Die Vertheilung der Energie im Spectrum des schwarzen Körpers. *Verh. d. D. Phys. Ges.* **1**: 23–41.

Lummer, O. and E. Pringsheim. (1899b). Die Vertheilung der Energie im Spectrum des schwarzen Körpers und des blanken Platins. *Verh. d. D. Phys. Ges.* **1**: 215–235.

Lummer, O. and E. Pringsheim. (1900). Über die Strahlung des schwarzen Körpers für lange Wellen. *Verh. d. D. Phys. Ges.* **2**: 163–180.

MacCullagh, J. (1880). *The Collected Works of James MacCullagh*. Hodges, Figgis, & Co., Dublin.

Mach, E. (1926). *The Principles of Physical Optics*. Dover, New York.

Mack, J. E. and M. J. Martin. (1939). *The Photographic Process*. McGraw-Hill, New York.

MacKenzie, J. M., M. G. Burke, T. Carvalho, and A. Eades. (2006). Ethics and Digital Imaging. *Microscopy Today* **12**: 40–41.

Miao, J., P. Charalambous, J. Kirz, and D. Sayre. (1999). An extension of the methods of x-ray crystallography to allow imaging of micron-size non-crystalline specimens. *Nature* **400**: 342–344.

Malpighi and Marcello. (1675–1679). *Anatome Plantarum*. Johannis Martyn, Londoni.

Malpighi, M. (1686). *Opera Omnia*. Thomam Sawbridge, Londoni.

Marti, O. and M. Amrein. (1993). *STM and SFM in Biology*. Academic Press, San Diego.

Martin, B. (1742). *Micrographia Nova: Or, A New Treatise on the Microscope, and Microscopic Objects*. J. Newbery and C. Micklewright, Reading.

Martin, B. (1761). *The Description and Use of an Universal Microscope, Answering All the Purposes of Single, Compound, Opake, and Aquatic Microscopes*. Printed for, and sold by the author, London.

Martin, B. (1774). *The Description and Use of an Opake Solar Microscope*. Printed for, and sold by the author, London.

Martin, L. C. (1966). *The Theory of the Microscope*. Blackie & Son, London.

Martin, L. V. (1988). Early history of dark ground illumination with the microscope. *Microscopy* **36**: 124–138.

Martin, L. V. (1993a). Brownian motion and Robert Brown's Priority. *Proc. RMS.* **28**: 143–144.

Martin, L. V. (1993b). The Abbe condenser and illuminating apparatus. *Quekett Journal of Microscopy* **37**: 7–12.

Mason, P. (1981). *The Light Fantastic*. Penguin Books, Middlesex, England.

Mattern, J. (2005). *George Eastman and the Story of Photographic Film*. Mitchell Lane Publishers, Hockessin, DE.

Mathews, W. W. (1953). The use of hollow-cone illumination for increasing image contrast in microscopy. *Trans. Amer. Microscop. Soc.* **72**: 190–195.

Matzke, M. A. and A. J. M. Matzke. (1986). Visualization of mitochondria and nuclei in living plant cells by the use of a potentialsensitive fluorescent dye. *Plant, Cell and Environment* **9**: 73–77.

Maurolico, F. [1611]. (1940). *The Photismi de Lumine of Maurolycus. A Chapter in Late Medieval Optics*. Translated by H. Crew. Macmillan, New York.

Maxwell, J. C. (1865). A dynamical theory of the electromagnetic field. *Phil. Trans. Roy. Soc. Lond.* **155**: 459–512.

Maxwell, J. C. [1891]. (1954). *A Treatise on Electricity and Magnetism*. Dover, New York.

Maxwell, J. C. [1891]. (1897). *Theory of Heat*. Longmans, Green, New York. New Edition

Maxwell, J. C. (1931). *A Commemorative Volume*. 1831–1931, with contributions by J. J. Thomson, M. Planck, A. Einstein, J. Larmor, J. Jeans, W. Garnett, A. Fleming, O. Lodge, R. T. Glazebrook, and H. Lamb. Macmillan, New York.

McCauley, M. M. and P. K. Hepler. (1990). Visualization of the endoplasmic reticulum in living buds and branches of the moss *Funaria hygrometrica* by confocal laser scanning microscopy. *Development* **109**: 753–764.

McClung, C. E., ed. (1937). *Handbook of Microscopical Technique*, second edition. Paul B. Hoeber, New York.

McCormick, D. L. and L. D. McCormick. (1990). Applications of scanning tunneling microscopy to biological materials. *Am. Bio. Lab.* May: 55–58.

McCormick, J. B. (1987). *18th Century Microscopes. A Synopsis of History and Workbook*. Science Heritage, Lincolnwood, IL.

McCrone, W. C., L. B. McCrone, and J. Delly. (1979). *Polarized Light Microscopy*. Ann Arbor Science Publishers Inc.

McCrone, W. C., L. B. McCrone, and J. Delly. (1984). *Polarized Light Microscopy*. McCrone Research Institute, Chicago, IL.

McCrone, W. C. (1990). 1500 forgeries. *Microscope* **38**: 289–298.

McCrone, W. C. (1991). Calibration of EC slide hotstage. *The Microscope* **39**: 43–52.

McCrone, W. C. (1992). Why use the polarized light microscope? *Amer. Laboratory* April: 17–21.

McElheny, V. K. (1999). *Insisting On the Impossible: The Life of Edwin Land*. Basic Books, New York.

McKenna, N. M. and Y.-L. Wang. (1989). Culturing cells on the microscope stage. *Methods in Cell Biology* **29**: 195–205.

McLaughlin, R. B. (1977). *Special Methods in Light Microscopy*. Microscope Publications, London.

McNeil, P. L. (1989). Incorporation of macromolecules into living cells. *Methods in Cell Biology* **29**: 153–173.

Mellors, R. C. and R. Silver. (1951). A micro-fluorometric scanner for the differential detection of cells; application of exfoliative cytology. *Science* **114**: 356–360.

Mellors, R. C. and J. Hlinka. (1955). Quantitative cytology and cytopathology. IV. Interferometric measurement of the anhydrous organic mass (dry weight) of genetic material in sperm nuclei of the mouse, the rat and the guinea pig. *Exp. Cell Res.* **9**: 128–134.

Menzel, D. H. ed. (1955). *Fundamental Formulas of Physics*. Prentice-Hall, New York.

Menzel, D. H. ed. (1960). *Fundamental Formulas of Physics*. Dover, New York.

Mermin, N. D. (2004). Could Feynman have said this? *Physics Today* May 57(5): 10–11.

Merrium, R. W. and W. Koch. (1960). The relative concentration of solids in the nucleolus, nucleus, and cytoplasm of the developing nerve cell of the chick. *J. Biophys. Biochem. Cytol.* **7**: 151–160.

Meyer, C. F. (1949). *The Diffraction of Light, X-rays, and Material Particles*. J. W. Edwards, Ann Arbor, MI.

Michelson, A. A. (1907). *Light Waves and Their Uses*. University of Chicago Press, Chicago, IL.

Michelson, A. A. (1962). *Studies in Optics*. University of Chicago Press, Chicago, IL.

Michette, A., G. Morrison, and C. Buckley. (1992). *X-Ray Microscopy III*. Springer-Verlag, Berlin.

Miller, A. I. (1994). *Early Quantum Electrodynamics. A Source Book*. Cambridge University Press, Cambridge.

Miller, D. C. (1916). *The Science of Musical Sounds*. Macmillan, New York.

Miller, D. C. (1935). *Anecdotal History of the Science of Sound to the Beginning of the 20th Century*. Macmillan, New York.

Miller, D. C. (1937). *Sound Waves. Their Shape and Speed*. Macmillan, New York.

Millikan, R. A. (1917). *The Electron. Its Isolation and Measurement and the Determination of Some of Its Properties*. University of Chicago Press, Chicago, IL.

Millikan, R. A. (1924). The electron and the light quant from the experimental point of view. *Nobel Lecture*. May 23, 1924.

Millikan, R. A. (1935). *Electrons (+ and −), Protons, Photons, Neutrons, and Cosmic Rays*. University of Chicago Press, Chicago, IL.

Millikan, R. A. (1950). *The Autobiography of Robert Milliakn*. Prentice Hall, New York, NY.

Millikan, R. A. and H. G. Gale. (1906). *A First Course in Physics*. Ginn & Co., Boston, MA.

Millikan, R. A. and J. Mills. (1908). *A Short University Course in Electricity, Sound, and Light*. Ginn & Co., Boston, MA.

Millikan, R. A., H. G. Gale, and W. R. Pyle. (1920). *Practical Physics*. Ginn & Co., Boston, MA.

Millikan, R. A., H. G. Gale, and J. P. Coyle. (1944). *New Elementary Physics*. Ginn & Co., Boston, MA.

Millikan, R. A., D. Roller, and E. C. Watson. (1937). *Mechanics Molecular Physics Heat, and Sound*. Ginn & Co., Boston, MA.

Minnaert, M. (1954). *The Nature of Light and Color in the Open Air*. Dover, New York.

Minsky, M. (1988). Memoir on inventing the confocal scanning microscope. *Scanning* **10**: 128–138.

Mitchison, J. M. (1953). A polarized light analysis of the human red cell ghost. *J. Exp. Biol.* **30**: 397–432.

Mitchison, J. M. (1957). The growth of single cells. Schizosaccharomyces. *Exp. Cell Res.* **13**: 244–262.

Mitchison, J. M. and M. M. Swann. (1953). Measurements on sea-urchin eggs with an interference microscope. *Quart. J. Microscop. Sci.* **94**: 381–389.

Mitchison, T. J. (1989). Polewards microtubule flux in the mitotic spindle: Evidence from photoactivation of fluorescence. *J. Cell Biol.* **109**: 637–652.

Moffitt, J. R., Y. R. Chemla, D. Izhaky, and C. Bustamante. (2006). Differential detection of dual traps improves the spatial resolution of optical tweezers. *Proc. Natl. Acad. Sci. USA* **103**: 9006–9011.

Moiseev, L. M. S. Ünlü, A. K. Swan, B. B. Goldberg, and C. R. Cantor. (2006). DNA conformation on surfaces measured by fluorescence self-interference. *Proc. Nat. Acad. Sci. USA* **103**: 2623–2628.

Molyneux, W. (1687). Concerning the apparent magnitude of the Sun and Moon, or the apparent distance of two stars, when nigh the horizon, and when higher elevated. *Phil. Trans.* **15**(187): 314–323.

Molyneux, W. (1692). *Dioptrica Nova. A Treatise of Dioptricks, in Two Parts. Wherein the Various Effects and Appearances of Spherick Glasses, both Convex and Concave, Single and Combined, in Telescopes and Microscopes, Together with Their Usefulness in Many Concerns of Humane Life, are Explained*. Printed for Benj, Tooke, London.

Molyneux, W. (1709). *Dioptrica Nova. A Treatise of Dioptricks, in Two Parts. Wherein the Various Effects and Appearances of Spherick Glasses, both Convex and Concave, Single and Combined, in Telescopes and Microscopes, Together with Their Usefulness in Many Concerns of Humane Life, are Explained*. Printed for Benj, Tooke, London.

Moore, R. (1966). *Niels Bohr*. Alfred A. Knopf, New York.

Moore, W. (1989). *Schrödinger. Life and Thought*. Cambridge University Press, Cambridge.

Moran, B. R. and J. F. Moran. (1987). An inexpensive digital temperature monitoring device for the poor-microscopists hotstage. *The Microscope* **36**: 43–52.

Morikawa, K. and M. Yanagida. (1981). Visualization of individual DNA molecules in solution by light microscopy: DAPI staining method. *J. Biochem.* **89**. 693–396.

Morise, H., O. Shimomura, F. Johnson, and J. Winant. (1974). Intermolecular energy transfer in the bioluminescent system of *Aequorea*. *Biochemistry* **13**: 2656–2662.

Morita, S. (2007). *Roadmap of Scanning Probe Microscopy*. Springer-Verlag, Berlin.

Morris, V. J., A. R. Kirby, and A. P. Gunning. (1999). *Atomic Force Microscopy for Biologists*. Imperial College Press, London.

Moss, V. A. (1988). Image processing and image analysis. *Proc. Roy. Microsc. Soc.* **23**: 83–88.

Murphy, D. B. (2001). *Fundamentals of Light Microscopy and Electronic Imaging*. John Wiley & Sons, New York.

Musikant, S. (1985). *Optical Materials: An Introduction to Selection and Application. Engineering Series*. Volume 6. CRC Press.

Mustacich, R. V. and B. R. Ware. (1974). Observation of protoplasmic streaming by laserlight scattering. *Phys. Rev. Lett.* **33**: 617–620.

Mustacich, R. V. and B. R. Ware. (1976). A study of protoplasmic streaming in *Nitella* by laser Doppler spectroscopy. *Biophys. J.* **16**: 373–388.

Mustacich, R. V. and B. R. Ware. (1977). Velocity distributions of the streaming protoplasm in *Nitella flexilis. Biophys. J.* **17**: 229–241.

Muybridge, E. (1887). *Animal Locomotion: An Electro-Photographic Investigation of Consecutive Phases of Animal Movements 1872–1885.* University of Pennsylvania. Printed by the Photogravure Company of New York.

Muybridge, E. (1955). *The Human Figure in Motion.* Dover, New York.

Naegeli, C. and S. Schwendener. (1892). *The Microscope in Theory and Practice.* Swan Sonnenschein & Co., London.

Nakaya, U. (1954). *Snow Crystals. Natural and Artificial.* Harvard University Press, Cambridge, MA.

Nakazono, M., F. Qiu, L. A. Borsuk, and P. S. Schnable. (2003). Laser-capture microdissection, a tool for the global analysis of gene expression in specific plant cell types: Identification of genes expressed differentially in epidermal cells or vascular tissues of maize. *Plant Cell* **15**: 583–596.

Nagendra Nath, N. S. (1936). Neutrinos and light quanta. *Proc. Indian Acad. Sci.* **3**: 448.

Neblette, C. B. (1952). *Photography. Material and Processes,* fifth edition. D. Van Nostrand Co., Princeton, NJ.

Needham, G. H. (1958). *The Practical Use of the Microscope. Including Photomicrography.* Charles C. Thomas, Springfield, IL.

Needham, J. (1962). *Science and Civilization in China.* Volume 4. *Physics and Physical Technology.* Physics. Cambridge University Press, Cambridge.

Nelson, E. M. (1891). The substage condenser: Its history, construction, and management: And its effect theoretically considered. *Journal of the Royal Microscopical Society:* 90–108.

Nelson, T., S. L. Tausta, N. Gandotra, and T. Liu. (2006). Laser microdissection of plant tissue: What you see is what you get. *Annu. Rev. Plant Biol.* **57**: 181–201.

Nernst, W. (1923). *Theoretical Chemistry from the Standpoint of Avogadro's Rule of Thermodynamics.* Macmillan, London.

Neter, J., W. Wasserman, and M. H. Kutner. (1990). *App-lied Linear Statistical Models,* Third Edition. Irwin, Homewood, IL.

Neuenschwander, D. E. (2001). The book of lumen. *Radiations Magazine* **7**(2): 7–10.

Neuman, K. C. and S. M. Block. (2004). Optical trapping. *Rev. Sci. Instr.* **75**: 2787–2809.

Newhall, B. (1937). *Photography 1839–1937.* The Museum of Modern Art, New York.

Newhall, B. (1949). *The History of Photography from 1839 to the Present Day.* The Museum of Modern Art, New York.

Newton, I. (1687). *Mathematical Principles of Natural Philosophy.* Translated into English by Andrew Mott in 1729 and revise by F. Cajori. Univeristy of Califormia Press, Berkeley, CA.

Newton, I. [1730]. (1979). *Opticks.* Dover, New York.

Nichols, E. F. and G. F. Hull. (1903a). The pressure due to radiation. *Phys. Rev.* **17**: 26–50.

Nichols, E. F. and G. F. Hull. (1903b). The pressure due to radiation. *Phys. Rev.* **17**: 91–104.

Nichols, E. L. and D. T. Wilber. (1921a). Flame excitation of luminescence. *Phys. Rev.* **17**: 453–468.

Nichols, E. L. and D. T. Wilber. (1921b). The luminescennce of certain oxides sublimed in the electric arc. *Phys. Rev.* **17**: 707–717.

Nicol, W. (1828). On a method of so far increasing the divergency of the two rays in calcareous spar that only one image may be seen at a time. *Edinburgh Philosophical J.* **6**: 83–84.

Nicol, W. (1834). On a method of so far increasing the divergency of the two rays in calcareous spar, that only one image may be seen at a time. *Edinburgh Philosophical J.* **16**: 372–376.

Nicol, W. (1839). Notice concerning an improvement in the construction of the single vision prism of calcareous spar. *Edinburgh Philosophical J.* **27**: 332–333.

Niedrig, H. (1999). *Optics.* Walter de Gruyter, Berlin, Germany.

Niven, W. D., ed. (2003). *The Scientific Papers of James Clerk Maxwell.* Volumes I and II. Dover, New York.

Noll, D. and H. H. Weber. (1934). Polarisationsoptik und molekularer Feinbau der Z-Abschnitte des Froschmuskels. *Pflügers Archiv Gesamte Physiol. Menschen Tiere* **235**: 234–246.

Nossal, R. and S. H. Chen. (1973). Effects of chemoattractants on the motility of *Escherichia coli. Nature New Biol.* **244**: 253–254.

Oersted, H. C. (1820). Experiments on the effect of a current of electricity on the magnetic needle. *Ann. Phil.* **16**: 273.

Oiwa, K., S. Chaen, E. Kamitsubo, T. Shimmen, and H. Suzie. (1990). Steady-state force-velocity relation in the ATP-dependent sliding movement of mysosin-coated beads on actin cables *in vitro* studied with a centrifuge microscope. *Proc. Natl. Acad. Sci. U.S.A.* **87**: 7893–7897.

O'Kane, D. J., R. E. Kobres, and B. A. Palevitz. (1980). Analog subtraction of video images for determining differences in fluorescence distributions. *J. Cell Biol.* **87**: 229a.

Oldfield, R. (1994). *Light Microscopy. An Illustrated Guide.* Wolfe, London.

Omar, S. B. (1977). *Ibn al-Haytham's Optics.* Bibliotheca Islamica, Minneapolis, MN.

Opas, M. (1978). Interference reflection microscopy of adhesion of *Amoeba proteus. J. Microscopy* **112**: 215–221.

Osborn, M. and K. Weber. (1982). Immunofluorescence and immunocytochemical procedures with affinity purified antibodies: Tubulincontaining structures. *Methods in Cell Biology* **24A**: 97–132.

Oster, G. (1955). Birefringence and dichroism. 439–460. In: *Physical Techniques in Biological Research.* Volume 1 Optical techniques. G. Oster. and A. W. Pollister, eds. Academic Press, New York.

Osterberg, H. (1955). Phase and interference microscopy. 377–437. In: *Physical Techniques in Biological Research.* Volume 1 Optical techniques. G. Oster. and A. W. Pollister, eds. Academic Press, New York.

Pagano, R. E. and R. G. Sleight. (1985). Defining lipid transport pathways in animal cells. *Science* **229**: 1051–1057.

Pagano, R. E. (1988). What is the fate of diacylglycerol produced at the Golgi apparatus? *Trends in Biochem. Sci.* **13**: 202–205.

Pagano, R. E. (1989). A fluorescent derivative of ceramide: Physical properties and use in studying the golgi apparatus of animal cells. *Methods in Cell Biology* **29**: 75–102.

Pais, A. (1982). *Subtle is the Lord.... The Science and the Life of Albert Einstein.* Clarendon Press, Oxford.

Pais, A. (1986). *Inward Bound. Of Matter anf Forces in the Physical World.* Clarendon Press, Oxford.

Pais, A. (1991). *Niels Bohr's Times, in Physics, Philosophy, and Polity.* Clarendon Press, Oxford.

Palevitz, B. A. and P. K. Hepler. (1976). Cellulose microfibril orientation and cell shaping in developing guard cells of *Allium*: The role of microtubules and ion accumulation. *Planta* **132**: 71–93.

Paley, W. (1803). *Natural Theology*. Daniel and Samuel Whiting, Albany, NY.

Palmer, A. E. and R. Y. Tsien. (2006). Measuring calcium signaling using genetically targetable fluorescent indicators. *Nature Protocols* **1**: 1057–1065.

Pankove, J. I. (1971). *Optical Processes in Semiconductors*. Dover, New York.

Panofsky, W. K. H. and M. Phillips. (1955). *Classical Electricity and Magnetism*. Addison-Wesley, Reading MA.

Park, D. (1997). *The Fire within the Eye*. Princeton University Press, Princeton, NJ.

Parthasarathy, M. V. (1985). F-actin architecture in coleoptile epidermal cells. *Eur. J. Cell Biol.* **39**: 1–12.

Parthasarathy, M. V., T. D. Perdue, A. Witztum, and J. Alvernaz. (1985). Actin network as a normal component of the cytoskeleton in many vascular plant cells. *Amer. J. Bot.* **72**: 1318–1323.

Pasteur, L. [1860]. (1915). *Researches on the Molecular Asymmetry of Natural Organic Products*. Alembic Club, Edinburgh.

Pasteur, L. (1878). La théorie des germes et ses applications à la chirurgie. *Bull. Acad. Natl. Méd. ser. 2.* **7**: 166–167.

Pasteur, L. (1879). *Studies on Fermentation: The Diseases of Beer, Their Causes, and the Means of Preventing Them.* Macmillan, London.

Patzelt, W. J. (1985). *Polarized Light Microscopy. Principles, Instruments, Applications*, third edition. Ernst Leitz. Wetzlar, West Germany.

Pauli, W. (1973). *Pauli Lectures on Physics*. Volume 1. Electrodynamics. Volume 2. Optics and the Theory of Electrons. Dover, New York.

Pauling, L. and S. Goudsmit. (1930). *The Structure of Line Spectra*. McGraw-Hill, New York.

Pawley, J. B. (1990). *Handbook of Biological Confocal Microscopy*. Plenum, London.

Payne, B. O. (1954). *Microscope Design and Construction*. Cooke, Troughton & Simms, Ltd, York, England.

Peacock, G. (1855). *Miscellaneous Works of the Late Thomas Young*. Volumes I and II. John Murray, London.

Pearson, H. (2005). CSI: Cell Biology. *Nature* **434**: 952–953.

Peat, F. D. (1997). *Infinite Potential. The Life and Times of David Bohm*. Addison Wesley, Reading, MA.

Pereira, J. (1854). *Lectures on Polarized Light*, second edition. Longman, Brown, Green, and Longmans, London.

Peres, M. R. and C. D. Meitchik. (1988). Zero perspective imaging. *Industrial Photography* March: 30–31.

Periasamy, A. and R. N. Day. (2005). *Molecular Imaging. FRET Microscopy and Spectroscopy*. Oxford University Press, Oxford.

Perrin, J. B. (1909). *Réalité moléculaire* (Molecular reality). http://web.lemoyne.edu/~giunta/perrin.html.

Perrin, J. B. (1923). *Atoms*, second English edition, revised. Van Nostrand, New York.

Phillips, W. D. (1997). Laser cooling and trapping of neutral atoms. *Nobel Lecture,* (December 8): 1997.

Piekos, W. B. (1999). Diffracted-light contrast enhancement: A re-examination of oblique illumination. *Microscopy Research and Techniques* **46**: 334–337.

Pieribone, V. and D. Gruber. (2006). *Aglow in the Dark: The Revolutionary Science of Biofluorescence*. Belknap Press, Cambridge, MA.

Planck, M. (1920). The genesis and present state of development of the quantum theory. *Nobel Lecture*, (June 2): 1920.

Planck, M. (1932). *Introduction to Theoretical Physics*. Volume III. *Theory of Electricity and Magnetism*. Volume IV. Theory of Light; Macmillan, London.

Planck, M. (1936). *The Philosophy of Physics*. W. W. Norton & Co., New York.

Planck, M. (1949a). *Scientific Autobiography and Other Papers*. Philosophical Library, New York.

Planck, M. (1949b). *Theory of Heat. Introduction to Theoretical Physics,* Volume V. Macmillan, London.

Plato, A. (1965). *Timaeus and Critias*. Penguin Books, New York.

Ploem, J. S. (1967). The use of a vertical illuminator with interchangeable dichroic mirrors for fluorescence microscopy with incident light. *Zeitschr. f. wiss. Mikroskopie* **68**: 129–142.

Ploem, J. S. and H. J. Tanke. (1987). *Introduction to Fluorescence Microscopy*. Oxford University Press, Oxford.

Pluta, M. (1988). *Advanced Light Microscopy*. Volume 1. Principles and Basic Properties. Elsevier, Amsterdam.

Pluta, M. (1989). *Advanced Light Microscopy*. Volume 2. Specialized Methods. Elsevier, Amsterdam.

Pluta, M. (1993). *Advanced Light Microscopy*. Volume 3. Measuring Techniques. Elsevier, Amsterdam.

Pohl, D. W., W. Denk, and M. Lanz. (1984). Optical stethoscopy: Image recording with resolution $\lambda/20$. *Appl. Phys. Lett.* **44**: 651.

Polyak, S. (1957). *The Vertebrate Visual System*. University of Chicago Press, Chicago, IL.

Pool, R. (1990). Making light work of cell surgery. *Science* **248**: 29–31.

Pope, A. (1745). *An Essay on Man*. Printed for John and Paul Knapton, London.

Popular Science Staff. (1934). *Wonders through the Microscope*. Popular Science Publishing Co., New York.

Porta and G. B. della. [1589]. (1957). *Natural Magick*. Basic Books, New York.

Porter, A. B. (1906). On the diffraction theory of microscopic vision. *Phil. Mag. Ser 6* **11**: 154–166.

Poynting, J. H. (1904). Radiation in the solar system: Its effect on temperature and its pressure on small bodies. *Phil. Trans. R. Soc. Lon. Ser. A* **202**: 525–552.

Poynting, J. H. (1910). *The Pressure of Light*. Soc. For Promoting Christian Knowledge, London.

Poynting, J. H. (1920). *Collected Scientific Papers*. Cambridge University Press, Cambridge.

Poynting, J. H. and J. J. Thomson. (1922). *A Text-Book of Physics. Sound*, eighth edition. Charles Griffin, London.

Pramanik, A. (2006). *Electromagnetism. Problems with Solutions*. Prentice-Hall of India, New Delhi, India.

Prasher, D., V. Eckenrode, W. Ward, F. Prendergast, and M. Cormier. (1992). Primary structure of the Aequorea victoria green-fluorescent protein. *Gene* **111**: 229–233.

Prendergast, F. and K. Mann. (1978). Chemical and physical properties of aequorin and the green fluorescent protein isolated from *Aequorea forskålea*. *Biochemistry* **17**: 3448–3453.

Priestley, J. (1772). *The History and Present State of Discoveries Relating to Vision, Light, and Colours*. Printed for J. Johnson, London.

Preston, R. D. (1952). *The Molecular Architecture of Plant Cell Walls*. John Wiley & Sons, New York.

Preston, T. (1895). *The Theory of Light*, second edition. Macmillan, London.

Pringle, J. R., R. A. Preston, A. E. M. Adams, T. Stearns, D. G. Drubin, B. K. Haarer, and E. W. Jones. (1989). Fluorescence microscopy methods for yeast. *Methods in Cell Biology* **31A**: 357–435.

Ptolemy, A. (1936). *Tetrabiblos or Quadripartite Being Four Books of the Influence of the Stars*. The Aires Press, Chicago.

Purcell, E. M. (1985). *Electricity and Magnetism*. McGraw-Hill, Boston, MA.

Pye, D. (2001). *Polarized Light in Science and Nature*. Institute of Physics, Bristol.

Quader, H., A. Hofmann, and E. Schnepf. (1989). Reorganization of the endoplasmic reticulum in epidermal cells of onion bulb scales after cold stress: Involvement of cytoskeletal elements. *Planta* **177**: 273–280.

Quekett, J. (1848). *A Practical Treatise on the Use of the Microscope, Including the Different Methods of Preparing and Examining Animal, Vegetable and Mineral Structures*. Bailliere, London.

Quekett, J. (1852). *A Practical Treatise on the Use of the Microscope, Including the Different Methods of Preparing and Examining Animal, Vegetable and Mineral Structures*, second edition. Bailliere, London.

Querra, J. M. (1990). Photon tunneling microscopy. *Appl. Opt.* **29**: 3741–3752.

Rawlins, D. J. (1992). *Light Microscopy*. Bios Scientific Publishers, Oxford.

Rayleigh, L. (1870). On the light from the sky, its polarization and colour. Copy of typed manuscript in Cornell University Library. *Phil. Mag.* (1871) **16**: 107–120, 274–279.

Rayleigh, L. (1871). On the scattering of light by small particles. *Phil. Mag.* **41**: 447–454.

Rayleigh, L. and J. Strutt. (1872). On the diffraction of object glasses. *Monthly Notices of the Royal Astronomical Society* **33**: 59–63.

Rayleigh, L. (1891). On pin-hole photography. *Phil. Mag. Ser 5* **31**: 87–99.

Rayleigh, L. [1894]. (1945). *The Theory of Sound*. Dover, New York.

Rayleigh, L. [1900]. (1964). *Optics*. Encyclopedia Britannica, Vol. 17. In: Scientific Papers by John William Strutt, Baron Rayleigh Vol. II 1881–1887 Cambridge Univ. Press.

Rayleigh, L. (1905a). The dynamical theory of gases. *Nature* **71**: 559.

Rayleigh, L. (1905b). The dynamical theory of gases and radiation. *Nature* **72**: 54–55.

Rayleigh, L. The constant of radiation as calculated from molecular data. *Nature* 72: 243–244.

Reade, J. B. (1837). On a new method of illuminating microscopic objects. *Goring and Pritchard's Micrographia*, containing practical essays on reflecting, solar, oxy-hydrogen gas microscopes, micrometers, eye-pieces, etc. 227–231. Whittaker and Co., London.

Reade, J. B. (1854). Letter from the Rev. J. B. Reade to H. Fox Talbot Esq. *The Photographic Journal* **2** (July 21): 9–10, 1854.

Reed, M. A. (1993). Quantum dots. *Scientific American* January: 118–123.

Reichert, E. T. (1913). *The Differentiation and Specificity of Starches in Relation to Genera, Species, Etc.* Carnegie Institute of Washington, Washington, DC.

Rennie, J. (1991). A small disturbance. Did experimental obstacles leave Brown motionless? *Scientific American* (August) **265**: 20.

Revel, J.-P. (1993). Evolution and revolution in microscopy II. *Microscopy Today* **93**: 1–2.

Rheinberg, J. (1896). On an addition to the methods of microscopical research, by a new way of optically producing color. *J. Roy. Microscop. Soc.* **373**.

Richards, B. M. and A. Bajer. (1961). Mitosis in endosperm: Changes in nuclear and chromosome mass during mitosis. *Exp. Cell Res.* **22**: 503.

Richards, O. W. (1950). Phase microscopy. 687–695. In: *McClung's Handbook of Microscopical Technique*, third edition. R. McClung Jones, ed. Paul B. Hoeber, New York.

Richards, O. W. (1963). *AO-Baker Interference Microscope Reference Manual*, second edition. Am. Opt. Co., Buffalo, NY.

Richards, O. W. (1964). Measurement with the AO-Baker interference microscope. *Appl. Phys.* **3**: 1027–1030.

Richards, O. W. (1966). An introduction to the theory of interference microscopy. 43–61. In: *Introduction to Quantitative Cytochemistry*. G. L. Wied, ed. Academic Press, New York.

Richardson, J. H. (1991). *Handbook for the Light Microscope. A User's Guide*. Noyes Publications, Park Ridge, NJ.

Richtmyer, F. K. (1928). *Introduction to Modern Physics*, first edition. McGraw-Hill, New York.

Richtmyer, F. K. and E. H. Kennard. (1942). *Introduction to Modern Physics*, third edition. McGraw-Hill, New York.

Richtmyer, F. K. and E. H. Kennard. (1947). *Introduction to Modern Physics*, fourth edition. McGraw-Hill, New York.

Richtmyer, F. K., E. H. Kennard, and T. Lauritsen. (1955). *Introduction to Modern Physics*, fifth edition. McGraw-Hill, New York.

Richtmyer, F. K., E. H. Kennard, and J. N. Cooper. (1969). *Introduction to Modern Physics*, sixth edition. McGraw-Hill, New York.

Riordan, M. and L. Hoddeson. (1997). *Crystal Fire: The Invention of the Transistor and the Birth of the Information Age*. Norton, New York.

Rittenhouse, D. (1786). Explanation of an optical deception. *Trans. Am. Philos. Soc.* **2**: 37–42.

Ritter, J. W. [1801]. (1968). Die Entdeckung der ultravioletten Strahlen. In: Ostwald's Klassiker der Exakten Wissenschaften. Die Begründung der Elektrochemie. *Academische Verlagsgesellschaft Frankfurt am Main*.

Robertson, J. K. (1941). *Introduction to Physical Optics*, third edition. D. van Nostrand Co., New York.

Robinson, A. (2006). *The Last Man Who Knew Everything*. Pi Press, New York.

Robinson, H. M. (1935). *Science versus Crime*. Bobbs-Merrill Co., Indianapolis. This book was also released the same year with the title: *Science Catches the Criminal*. Blue Ribbon Books, New York.

Robinson, G. M., R. D. M. Perryand, and W. Peterson. (1991). Optical interferometry of surfaces. *Scientific American* (July) **265**: 66–71.

Rochow, T. G. and E. G. Rochow. (1978). *An Introduction to Microscopy by Means of Light, Electrons, X-Rays, or Ultrasound*. Plenum Press, New York.

Röentgen, W. (1899). On a new kind of rays. In: *Roentgen Rays: Memoirs by Roentgen, Stokes and J. J. Thomson.* Harper's Scientific Memoirs. Vol III. Harper & Brothers, New York.

Ronchi, V. (1991). *Optics. The Science of Vision.* Dover, New York.

Rosengren, B. H. O. (1959). Determination of cell mass by direct x-ray absorption. *Acta Radiological* **178**(Suppl.): 1–62.

Rossner, M. and K. M. Yamada. (2004). What's in a picture? The temptation of image manipulation. *J. Cell Biol.* **166**: 11–15.

Ross, K. F. A. (1954). Measurement of the refractive index of cytoplasmic inclusions in living cells by the interference microscope. *Nature* **174**: 836–837.

Ross, K. F. A. and G. Galavazi. (1964). The size of bacteria, as measured by interference microscopy. *J. Roy. Microscop. Soc.* **13–25**.

Ross, K. F. A. (1964). Nucleolar changes in differentiating myoblasts. Quart. *J. Microscop. Sci.* **105**: 423–447.

Ross, K. F. A. (1967). *Phase Contrast and Interference Microscopy for Cell Biologists.* Edward Arnold, London.

Ross, K. F. A. (1988). Phase contrast and interference microscopy. *Microscopy* **36**: 97–123.

Rost, F. W. D. (1995). *Fluorescence Microscopy.* Cambridge University Press, Cambridge.

Rossotti, H. (1983). *Colour. Why the World Isn't Grey.* Princeton University Press, Princeton, NJ.

Rowe, W. F. and James. E. Starrs. (2001). Microscopical examination of Jesse W. James' hair. *The Microscope* 29.

Roukes, N. (1974). *Plastics for Kinetic Art.* Watson-Guptill Publications, New York.

Ruch, F. (1966). Birefringence and dichroism of cells and tissues. 57–86. In: *Physical Techniques in Biological Research*, second edition. Volume III, Part A. Cells and Tissues. A. W. Pollister, ed. Academic Press, New York.

Ruppersberg, J. P., J. K. H. Horber, C. Gerber, and G. Binnig. (1989). Imaging of cell membranous and cytoskeletal structures with a scanning tunneling microscope. *FEBS Lett.* **257**: 460–464.

Russ, J. (2004). Seeing the Scientific Image (parts 1,2,3). *Proc. Roy. Microscopy Soc.* **39**(2): 1–18. **39**(3): 1–15; **39**(4):1–15.

Russell, J. C. (1848). On certain effects produced on sound by the rapid motion of the observer. Notices and Abstracts of Communications to the British Association for the Advancement of Science at the Swansea Meeting, August 1848. *British Assoc. Reports* **18**: 37–38.

Ruthmann, A. (1970). *Methods in Cell Research.* Cornell University Press, Ithaca, NY.

Sabra, A. I. (1989). *The Optics of Ibn Al-Haytham.* The Warburg Institute, University of London, London.

Sachs, M. (1988). *Einstein versus Bohr.* Open Court, La Salle, IL.

Sack, F. D. and A. C. Leopold. (1985). Cytoplasmic streaming affects gravity-induced amyloplast sedimentation in maize coleoptiles. *Planta* **164**: 56–62.

Sack, F. D., M. M. Suyemoto, and A. C. Leopold. (1985). Amyloplast sedimentation kinetics in gravity stimulate maize roots. *Planta* **165**: 295–300.

Sacks, O. (1984). *A Leg To Stand On.* Summit Books, New York.

Salmon, E. D. (1975a). Pressure-induced depolymerization of spindle microtubules. I. Spindle birefringence and length changes. *J. Cell Biol.* **65**: 603–614.

Salmon, E. D. (1975b). Spindle microtubules: Thermodynamics of in vivo assembly and role in chromosome movement. *Ann. N. Y. Acad. Sci.* **253**: 383–406.

Salmon, E. D. and G. W. Ellis. (1976). Compensator transducer increases ease, accuracy, and rapidity of measuring changes in specimen birefringence with polarization microscopy. *J. Microscopy* **106**: 63–69.

Sanderson, J. B. (1992). A modern Abbe test plate. *Proc. R.M.S.* **27/3**: 193–194.

Sato, H. G. W. Ellis, and S. Inoué. (1975). Microtubular origin of mitotic spindle form birefringence. *J. Cell. Biol.* **67**: 501–517.

Sattelle, D. B. and P. B. Buchan. (1976). Cytoplasmic streaming in *Chara corallina* studied by laser light scattering. *J. Cell Sci.* **22**: 633–643.

Saunders, M. J. and P. K. Hepler. (1981). Localization of membraneassociated calcium following cytokinin treatment in *Funaria* using chlorotetracycline. *Planta* **152**: 272–281.

Schacht, H. (1853). *The Microscope in its Special Application to Vegetable Anatomy and Physiology.* Samuel Highley, London.

Schaeffer, HF. (1953). *Microscopy for Chemists.* Dover, New York.

Schaffner, K. F. (1972). *Nineteenth-Century Aether Theories.* Pergamon Press, Oxford.

Scheele, C. W. (1780). *Chemical observations and experiments on air and fire.* Printed for J. Johnson, London.

Schellen, H. (1885). *Spectrun Analysis.* Longmans, Green & Co., London.

Schellen, H. et al. (1872). *Spectrum Analysis Explained.* Estes and Lauriat, Boston, MA.

Scheuerlein, R., R. Wayne, and S. J. Roux. (1988). Early quantitative method for measuring germination in non-green spores of *Dryopteris paleacea* using an epifluorescence-microscope technique. *Physiol Plant.* **73**: 505–511.

Schiering, D. W., E. F. Young, and T. F. Byron. (1990). An FTIR microscope. *American Laboratory* (November): 26–39.

Schleiden, M. (1849). *Principles of Scientific Botany; or, Botany as an inductive science.* Longman, Brown, Green, and Longmans, London.

Schmahl, G. and D. Rudolph, eds. (1984). *X-ray microscopy II.* Springer-Verlag, Berlin.

Schmidt, J. W. (1924). *Die Bausteine des Tierkörpers in polarisiertem Lichte.* Fr. Cohen, Bonn, Germany.

Schmitt, F. O. (1990). *The Never-Ceasing Search.* American Philosophical Society, Philadelphia, PA.

Schöppe, G., P. Fischer, and D. Schau. (1987). Jenalval-interphako and jenavert-interference microscopes of the series 250-cf Jena microscopes. *Jena Review* **32**: 11.

Schrödinger, E. (1922). Dopplerprinzip und Bohrsche Frequenzbedingung. *Physik. Zeit.* **11**: 170–176.

Schrödinger, E. (1928). *Four Lectures on Wave Mechanics.* Blackie & Son Ltd., London.

Schuster, A. (1898). Potential matter. *Nature* **58**: 367–618.

Schuster, A. (1904). *An Introduction to the Theory of Optics.* Edward Arnold, London.

Schuster, A. (1909). *An Introduction to the Theory of Optics*, second edition. Edward Arnold, London.

Schuster, A. (1932). *Biographical Fragments.* Macmillan, London.

Schuster, A. and J. W. Nicholson. (1924). *An Introduction to the Theory of Optics*, third edition. Edward Arnold & Co., London.

Schütz, W. (1966). Ernst Abbe. University teacher and industrial physicist. *Jena Review* **11**: 13–23.

Schwann, T. (1847). *Microscopical Researches into the Accordance in the Structure and Growth of Animals and Plants.* The Sydenham Society, London.

Schwartz, E. I. (2002). *The Last Lone Inventor: A Tale of Genius, Deceit & the Birth of Television.* HarperCollins, New York.

Schwartz, E. R. (1934). *Textiles and the Microscope.* McGraw-Hill, New York.

Scientific Imaging with Kodak films and plates. (1987). Eastman Kodak Co., Rochester, NY.

Seckel, A. (2000). *The Art of Optical Illusions.* Carlton Books, London.

Seckel, A. (2001). *More Optical Illusions.* Carlton Books, London.

Seckel, A. (2002). *The Great Book of Optical Illusions.* Firefly Books, Toronto, CA.

Seckel, A. (2004a). *Armchair Puzzlers: Optical Teasers: Sink Back and Solve Away!* (Armchair Puzzlers). University Games.

Seckel, A. (2004b). *Masters of Deception: Escher, Dali & the Artists of Optical Illusion.* Sterling, New York.

Seebeck, T. J. (1821). *Ueber den magnetismus der galvenische kette, Abh. K. Akad.* Wiss, Berlin.

Segré, G. (2007). *Faust in Copenhagen. A struggle for the Soul of Physics.* Viking, New York.

Serway, R. A., C. J. Moses, and C. A. Moyer. (2005). *Modern Physics,* third edition. Thomson, Brooks/Cole, Belmont, CA.

Shack, R., R. Baker, R. Buchroeder, D. Hillman, R. Shoemaker, and P. H. Bartels. (1979). Ultrafast laser scanning microscope. *J. Histochem. Cytochem.* **27**: 153–159.

Shadbolt, G. (1850). The annular condenser for the microscope. *Trans. Micro. Soc. Lond.* **2**: 132–133.

Shadbolt, G. (1851). Observations upon oblique illumination; with a description of the author's sphaero-annular condenser. *Trans. Micro. Soc. Lond.* **3**: 154–159.

Shadowitz, A. (1975). *The Electromagnetic Field.* Dover, New York.

Shaner, N., P. Steinbach, and R. Tsien. (2005). A guide to choosing fluorescent proteins. *Nature Methods* **2**: 905–909.

Shapiro, D., P. Thibault, T. Beetz, V. Elser, M. Howellsand, C. Jacobsen, J. Kirz, E. Lima, H. Miao, A. M. Neiman, and D. Sayre. (2005). Biological imaging by soft x-ray diffraction microscopy. *Proc. Natl. Acad. Sci. USA* **102**: 15343–15346.

Shaw, P. J. (1993). Computer reconstruction in three-dimensional fluorescence microscopy. 211–230. In: *Electronic Light Microscopy.* D. Shotton, ed. Wiley-Liss, New York.

Shedd, J. C. (1906). The index of refraction (Snell's Law). *School Science and Mathematics* **6**: 678–680.

Shockley, W. (1956). Transistor technology evokes new physics. *Nobel Lecture* (December 11): 1956.

Shotton, D. (1993). An introduction to digital image processing and image display in electronic light microscopy. 39–70. In: *Electronic Light Microscopy.* D. Shotton, ed. Wiley-Liss, New York.

Shreve, A. P., J. K. Trautman, T. G. Owens, and A. C. Albrecht. (1990). Two-photon excitation spectroscopy of thylakoid membranes from Phaeidactylum tricornutun: Evidence for an in vivo two-photon-allowed carotenoid state. *Chemical Physics Letters* **170**: 51–56.

Shropshire, W. Jr. (1959). Growth responses of *Phycomyces* to polarized light stimuli. *Science* **130**: 336.

Shropshire, W. Jr. (1962). The lens effect and phototropism of *Phycomyces. J. Gen. Physiol.* **45**: 949–958.

Shropshire, W. Jr. (1963). Photoresponses of the fungus *Phycomyces. Physiol. Reviews* **43**: 38–67.

Schurcliff, W. A. and S. S. Ballard. (1964). *Polarized Light.* D. Van Nostrand Co., Princeton, NJ.

Shurkin, J. N. (2006). *Broken Genius: The Rise and Fall of William Shockley, Creator of the Electronic Age.* Palgrave Macmillan, New York.

Siedentopf, H. (1903). On the rendering visible of ultra-microscopic particles and of ultra-microscopic bacteria. *J. Roy. Microscopical Soc.* 573–578.

Siedentopf, H. and R. Zsigmondy. (1903). über Sichtbarmachung und Grössenbestimmung ultramikroskopischer Teilchen, mit besonderer Anwendung auf Goldrubingläser. *Annalen d Physik* **10**: 1–29.

Skilling, H. H. (1942). *Fundamentals of Electric Waves.* John Wiley & Sons, New York.

Skirius, S. A. (1984). A poor-microscopists hot stage. *The Microscope* **29**: 117–120.

Slater, C. and E. J. Spitta. (1898). *An Atlas of Bacteriology Containing One Hundred and Eleven Original Photomicrographs with Explanatory Text.* Scientific Press, London.

Slayter, B. N. (1957). The star test, its interpretation and value. *Jour. Quekett Mic. Club* **4**: 415–422.

Slayter, E. M. (1970). *Optical Methods in Biology.* WileyInterscience, New York.

Smith, A. M. (1996). *Ptolemy's Theory of Perception: An English Translation of the Optics.* The American Philosophical Society, Philadelphia, PA.

Smith, D. E. (1959). *A Source Book in Mathematics.* Dover, New York.

Smith, F. G. (1856). *This Microscope and its Revelations.* Blanchard and Lea, Philadelphia, PA.

Smith, F. H. (1955). Microscopic interferometry. *Research* **8**: 385–395.

Smith, H. G. (1956). *Minerals and the Microscope.* Thomas Murby and Co., London.

Smith, R. (1738). *A Compleat System of Opticks in Four Books, viz. a Popular, a Mathematical, a Mechanical, and a Philosophical Treatise. To which are added remarks upon the whole.* Printed for the author and sold there by Cornelius Crownfield, Cambridge.

Smith, R. F. (1987). A Tribute: 'The Four Horsemen of Microscopy'. *Functional Photography* September/October.

Smith, S. and J. Glaister Jr. (1931). *Recent Advances in Forensic Medicine.* P. Balkiston's Son & Co., Philadelphia, PA.

Sommerfeld, A. (1923). *Atomic Structure and Spectral Lines.* Methuen, London.

Sommerfeld, A. (1964). *Lectures on Theoretical Physics.* Volume III. Electrodynamics. Academic Press, New York.

Sommerfeld, A. (2004). *Mathematical Theory of Diffraction.* Birkhäuser, Boston.

Spencer, M. (1982). *Fundamentals of Light Microscopy.* Cambridge University Press, Cambridge.

Spencer, M. W., S. A. Casson, and K. Lindsey. (2007). Transcriptional profiling of the Arabidopsis embryo. *Plant Physiol* **143**: 924–940.

Spitta, E. J. (1907). *Microscopy. The Construction, Theory and Use of the Microscope.* E. P. Dutton and Co., New York.

Spring, K. R. and R. J. Lowy. (1989). Characteristics of low light level television cameras. *Methods in Cell Biology* **29**: 269–289.

Starr, M. A. (1896). *Atlas of Nerve Cells*. Macmillan, New York.

Staves, M. P., R. Wayne, and A. C. Leopold. (1995). Detection of gravity-induced polarity of cytoplasmic streaming. *Protoplasma* **188**: 38–48.

Stern, M. D. (1975). *In vivo* evaluation of microcirculation by coherent light scattering. *Nature* **254**: 56–58.

Stephenson, L. G., T. W. Bunker, W. E. Dubbs, and H. D. Grimes. (1998). Specific soybean lipoxygenases localize to discrete subcellular compartments and their mRNAs are differentially regulated by source-sink status. *Plant Physiol.* **116**: 923–933.

Stokes, G. G. (1852). On the Change of Refrangibility of Light. I. *Phil. Trans. Roy. Soc. Lond.* **142**: 463–562.

Stokes, G. G. (1853). On the Change of Refrangibility of Light. II. *Phil. Trans. Roy. Soc. Lond.* **143**: 385–396.

Stokes, G. G. (1884). *On Light. First Course: On the Nature of Light*. Macmillan, London.

Stokes, G. G. (1885). *On Light. Second Course: On Light as a Means of Investigation*. Macmillan, London.

Stoney, G. J. (1896). Microscopic vision. *Phil. Mag. Ser. 5* **42**: 332–349. 423–442; 499–528.

Strange, A. (1989). Rheinberg stop contrast illumination. *Microscopy* **36**: 337–340.

Strasburger, E. (1875). *Uerer [sic] Zellbildung und Zelltheilung*. H. Dabis, Jena, Germany.

Stratton, J. A. (1941). *Electromagnetic Theory*. McGraw-Hill, New York.

Strong, C. L. (1968). Two methods of microscope lighting that produces color. *Scientific American* **218**(4): 124–130.

Strong, J. (1958). *Concepts of Classical Optics*. W.H. Freeman and Co., San Francisco, CA.

Summers, D. (2007). *Vision, Reflection, & Desire in Western Painting*. University of North Carolina Press, Chapel Hill, NC.

Sutton, T. (1858). *A Dictionary of Photography*. Sampson, Low, Son & Co., London.

Svensson, G. (1957). Scanning interference microphotometry. *Exp. Cell Res.* **4**(Suppl.): 165–171.

Svoboda, K. and S. M. Block. (1994). Biological application of optical forces. *Annual Reviews of Biophysics and Biomolecular Structure* **23**: 247–285.

Swammerdam, J. (1758). *The Book of Nature or, The history of Insects Reduced to Distinct Classes, Confirmed by Particular Instances, Displayed in the Anatomical Analysis of Many Species*. Printed for C. G. Seyffert, London.

Swann, M. M. (1951a). Protoplasmic structure and mitosis. I. *J. Exp. Biol* **28**: 417–433.

Swann, M. M. (1951b). Protoplasmic structure and mitosis. II. *J. Exp. Biol.* **28**: 434–444.

Swann, M. M. (1952). The spindle. 119–133. In: *The Mitotic Cycle*. A. Hughes, ed. Academic Press, New York.

Swenson, L. S. Jr. (1972). *The Ethereal Aether*. University of Texas Press, Austin, TX.

Synge, E. H. (1928). A suggested method for extending microscopic resolution into the ultra-microscopic region. *Phil. Mag. Ser. 7* **6**: 356–362.

Synge, E. H. (1932). An application of piezoelectricity to microscopy. *Phil. Mag.* **13**: 297.

Takagi, S., E. Kamitsubo, and R. Nagai. (1991). Light-induced changes in the behavior of chloroplasts under centrifugation in *Vallisneria* epidermal cells. *J. Plant. Physiol.* **138**: 257–262.

Takagi, S., E. Kamitsubo, and R. Nagai. (1992). Visualization of a rapid, red/far-red light-dependent reaction by centrifuge microscopy. *Protoplasma* **168**: 153–158.

Talbot, W. F. (1834a). Experiments on light. *Phil. Mag. 3rd Series 5* **29**: 321–332.

Talbot, W. F. (1834b). On Mr. Nicol's polarizing eye-piece. *Phil. Mag.* April 1834: 289–290.

Talbot, W. F. (1834c). Facts relating to optical science. I. *Phil. Mag.* **4**(20): 112–114.

Talbot, W. F. (1836a). Facts relating to optical science. III. *Phil. Mag.* **9**(54): 1–4.

Talbot, W. F. (1836b). On the optical phaenomena of certain crystals. *Phil. Mag.* **9**(54): 288–291.

Talbot, W. F. (1837). On the optical phenomena of certain crystals. *Phil Trans. Part 1*: 25–27. 29–35.

Talbot, W. F. (1839a). On analytic crystals. *Phil. Mag.* **14**: 19–21.

Talbot, W. F. (1839b). Some account of the art of photogenic drawing. *Phil. Mag.* **14**: 196–208.

Talbot, W. H. F. (1839c). An account of the processes employed in photogenic drawing, in a letter to Samuel H. Christie, Esq. Sec. R.S. *Phil Mag.* **14**: 209–211.

Talbot, W. H. F. (1844–1846). The Pencil of Nature. Reprinted in: *Henry Fox Talbot. Selected Texts and Bibliography*. Edited by M. Weaver, G. K. Hall & Co., Boston, MA.

Tanaka, T., C. Riva, and I. BenSira. (1974). Blood velocity measurements in human retinal vessels. *Science* **186**: 830–831.

Taylor, C. A. and H. Lipson. (1964). *Optical Transforms, Their Preparation and Application to X-Ray Diffraction Problems*. Cornell University Press, Ithaca, NY.

Taylor, D. L. (1975). Birefringence changes in vertebrate skeletal muscle. *J. Supramol. Struc.* **3**: 181–191.

Taylor, D. L. (1976). Quantitative studies on the polarization optical properties of striated muscle. I. Birefringence changes of rabbit psoas muscle in the transition from rigor ro relaxed state. *J. Cell Biol.* **68**: 497–511.

Taylor, D. L., M. Nederlof, F. Lanni, and A. S. Waggoner. (1992). The new vision of light microscopy. *American Scientist* **80**: 322–335.

Taylor, R. B. (1984). Rheinberg Updated. *Proc. Roy. Micro. Soc.* **19**: 253–256.

Tazawa, M. (1957). Neue Nethode zur Messung des osmotischen Wertes winer Zelle. *Protoplasma* **48**: 342–359.

Temple, P. A. (1981). Total internal reflection microscopy: A surface inspection technique. *Appl. Optics.* **20**: 2656–2660.

Ter Haar, D. (1967). *The Old Quantum Theory*. Pergamon Press, Oxford.

Terasaki, M., J. Song, J. R. Wong, M. J. Weiss, and L. B. Chen. (1984). Localization of endoplasmic reticulum in living and glutaraldefixed cells with fluorescent dyes. *Cell* **38**: 101–108.

Terasaki, M., L. B. Chen, and K. Fujiwara. (1986). Microtubules and the endoplasmic reticulum are highly interdependent structures. *J. Cell Biology* **103**: 1557–1568.

Terasaki, M. (1989). Fluorescent labeling of endoplasmic reticulum. *Methods in Cell Biology* **29**: 125–135.

Thibault, P. V. Elser, C. Jacobsen, D. Shapiro, and D. Sayre. (2006). Reconstruction of a yeast cell from X-ray diffraction data. *Acta Cryst.* **A62**: 248–261.

Thompson, B. (1794). A method of measuring the comparative intensities of the light emitted by luminous bodies. *Phil. Trans. Roy. Soc.* **84**: 67–82.

Thomson, J. J. (1897). Cathode rays. *Phil. Mag.* **44**: 293.

Thompson, S. P. (1901). *Michael Faraday. His Life and Work.* Cassell and Co., London.

Thompson, S. P. (1904). *Elementary Lessons in Electricity & Magnetism.* Macmillan, New York.

Thompson, S. P. (1905). On the Nicol prism and its modern varieties. *Proceedings of the Optical Convention, 1905.* Norgate & Williams, London.

Thomson, G. P. (1930). *The Wave Mechanics of Free Electrons.* McGraw-Hill, New York.

Thomson, J. J. (1895). *Elements of the Mathematical Theory of Electricity and Magnetism.* Cambridge University Press, Cambridge.

Thomson, J. J. (1909). *Elements of the Mathematical Theory of Electricity and Magnetism*, fourth edition. Cambridge University Press, Cambridge.

Thurgood, G. (1995). Landscapes in microscopy as an art form. *Quekett Journal of Microscopy* **37**: 455–466.

Time-Life. (1970). *The Camera. Life Library of Photography.* Time-Life Books, New York.

Tokunaga, M. and T. Yanagida. (1997). Single molecule imaging of fluorophores and enzymatic reactions achieved by objective-type total internal reflection fluorescence microscopy. *Biochemical and Biophysical Research Communications* **235**: 47–53.

Tolansky, S. (1968). *Interference Microscopy for the Biologist.* Charles C. Thomas, Publisher, Springfield, IL.

Towne, D. H. (1988). *Wave Phenomena.* Dover, New York.

Turner, G. L. E. (1982). Frits Zernike, 1888–1966. *Proc. Roy. Micr. Soc.* **17**: 100–101.

Tsien, R. Y. (2003). Imagining imaging's future. *Nature Reviews Molecular Cell Biology* **4**: SS16–SS21.

Tsien, R. Y. (2005). Breeding molecules to spy on cells. *The Harvey Lectures: Series 99* 2003–2004. Harvey Society.

Tsien, R. Y., L. A. Ernst, and A. S. Waggoner. (2006). Fluorophores for confocal microscopy: Photophysics and photochemistry. 351–365. In: *Handbook of Confocal Microscopy.* J. Pawley, ed. Plenum Press, New York.

Tyndall, J. (1873). Lectures on Light. Delivered in the United States in 1872–'73. D. Appleton and Co., New York.

Tyndall, J. (1876). Professor Tyndall on Germs. *American Journal of Science* **64** (April): 1876.

Tyndall, J. (1887). *Notes of a Course of Nine Lectures on Light*, twelfth edition. Longmans, Green, London.

Ueda, T., Y. Mori, T. Nakagaki, and Y. Kobatake. (1988). Changes in cAMP and cGMP concentration, birefringent fibrils and contractile activity accompanying UV and blue light photoavoidance in plasmodia of an albino strain of *Physarum polycephalum. Photochem. Photobiol.* **47**: 271–275.

Upsky, N. G. and P. E. Pagano. (1985). A vital stain for the Golgi apparatus. *Science* **228**: 745–747.

Valaskovic, G. (1991). A simple control unit for the EC slide hotstage. *The Microscope* **39**: 43–52.

Valentin, G. (1861). *Die Untersuchung der Pflanzen- und der Thiergewebe in Polarisirtem Lichte.* Wilhelm Engelmann, Leipzig.

Van't Hoff, J. H. (1874). A suggestion looking to the extension into space of the structural formulas at present used in chemistry, and a note upon the relation between the optical activity and the chemical constitution of organic compounds. *Archives neerlandaises des sciences exactes et naturelles* **9**: 445–454.

van't Hoff, J. H. (1967). *Imagination in Science.* Springer, Berlin.

van Spronsen, E. A., G. V. Sarafis, J. Brakenhoff, H. T. M. van der Voort, and N. Nanninga. (1989). Three-dimensional structure of living chloroplasts as visualized by confocal scanning laser microscopy. *Protoplasma* **148**: 8–14.

van der Voort, H. T. M., J. A. C. Valkenburg, E. A. van Spronsen, C. L. Woldringh, and G. J. Brakenhoff. (1987). Confocal microscopy in comparison with electron and conventional light microscopy. In: *Correlative Microscopy in Biology: Instrumentation and Methods.* M. A. Hayat, ed. Academic Press, New York.

Varlamov, V. V. (2002). About algebraic foundations of Majorana-Oppenheimer quantum electrodynamics and de Broglie-Jordan neutrino theory of light. *Annales de la Fondation Louis de Broglie* **27**: 273.

Varley, E., M. Lenz, S. J. Lee, J. S. Brown, D. A. Ramirez, A. Stintz, S. Krishna, A. Reisinger, and M. Sundaram. (2007). Single bump, two-color quantum dot camera. *Appl. Phys. Lett.* **91**: 081120–1—081120–3.

Vaughan, J. G. (1979). *Food Microscopy.* Academic Press, New York, NY.

Veldkamp, W. B. and T. J. McHugh. (1992). Binary optics. *Scientific American* **266**: 92–97.

Vogel, H. (1889). The Chemistry of Light and Photography. D. Appleton and Co., New York, NY.

Vogelmann, T. C., J. F. Bornman, and D. J. Yates. (1996). Focusing of light by leaf epidermal cells. *Physiol. Plant.* **98**: 43–56.

Vogelmann, T. C. (1993). Plant tissue optics. *Ann. Rev. Plant Physiol. & Mol. Biol.* **44**: 231–251.

von Frisch, K. (1950). Die Sonne als Kompaß im Leben der Bienen. *Experientia* **6**: 210–221.

von Rohr, M. ed. (1920). Geometrical investigation of the formation of images in optical instruments. His Majesty's Stationary Office, London.

von Rohr, M. (1936). *Abbe's Apochromats.* Carl Zeiss, Jena.

Waggoner, A., R. DeBiasio, P. Conrad, G. R. Bright, L. Ernst, K. Ryan, M. Nederlof, and D. L. Taylor. (1989). Multiple spectral parameter imaging. *Methods in Cell Biology* **30**: 449–478.

Walker, C. E. (1905). *The Essentials of Cytology.* Archibald Constable & Co., Ltd., London.

Walker, M. I. (1971). *Amateur Photomicrography.* Focal Press, New York.

Walter, R. J. Jr. and M. W. Berns. (1986). Digital image processing and analysis. Chapter 10. In: *Video Microscopy.* S. Inoué, ed. Plenum Press, New York.

Wang, Y. L., J. M. Heiple, and D. L. Taylor. (1982). Fluorescent analog cytochemistry of contractile proteins. *Methods in Cell Biology* **25B**: 1–11.

Wang, Y.-L. (1985). Exchange of actin subunits at the leading edge of living fibroblasts: Possible role of treadmilling. *J. Cell Biol.* **101**: 597–602.

Wang, Y.-L. (1989). Fluorescent analog cytochemistry: Tracing functional protein components in living cells. *Methods in Cell Biology* **29**: 1–12.

Wayne, R. and M. P. Staves. (1991). The density of the cell sap and endoplasm of *Nitellopsis* and *Chara. Plant Cell Physiol.* **32**: 1137–1144.

Wayne, R. and M. Staves. (1996). A down to earth model of gravisensing or Newton's Law of Gravitation from the apple's perspective. *Physiol. Plant.* **98**: 817–921.

Wayne, R., M. P. Staves, and A. C. Leopold. (1990). Gravity-dependent polarity of cytoplasmic streaming in *Nitellopsis*. *Protoplasma* **155**: 43–57.

Webb, W. W. (1986). Light microscopy–A modern renaissance. *New York Acad. Sci.* **483**: 387–391.

Weber, W. and R. H. Kohlrausch. (1856). Ueber die Elektricitätsmenge, welche bei galvanischen Strömen durch den Querschnitt der Kette fließt. *Annalen der Physik* **99**: 10–25.

Wedgewood, T. (1792). Experiments and observations on the production of light from different bodies, by heat and attrition. *Phil. Trans. Roy. Soc. Lond.* **82**: 28–47.

Weiss, D. G., W. Maile, and R. A. Wick. (1989). Video Microscopy. In: *Light Microscopy in Biology. A Practical Approach.* A. J. Lacey, ed. ILR Press, Oxford.

Welland, M. E., M. J. Miles, N. Labbert, V. J. Morris, J. H. Coombs, and J. B. Pethica. (1989). Structure of the globular protein vicilin revealed by scanning tunneling microscopy. *Int. J. Biol. Macromol.* **11**: 29–32.

Wensberg, P. C. (1987). *Land's Polaroid: A Company and the Man Who Invented It.* Houghton Miflin, New York.

Wheatley, D. (1992). Brown motionless. (Letter) *Biologists* **39**: 124.

Wedgwood, T. and H. Davy. (1802). An account of a method of copying paintings upon glass, and of making profiles, by the agency of light upon nitrate of silver. *J. Royal Institution of Great Britain* **1**: 170–174.

Weiser, H. B. (1939). *Colloid Chemistry. A Textbook.* J. Wiley & Sons, New York.

Wenham, F. H. (1850). On the illumination of transparent microscopic objects on a new principle. *Trans. Micr. Soc. Lond.* **3**: 83–90.

Wenham, F. H. (1854). On the theory of the illumination of objects under the microscope with relation to the aperture of the object-glass, and the properties of light; with practical methods for special differences of texture and colour. *Quart. Jour. Micr. Sci.* **2**: 145–158.

Wenham, F. H. (1856). On a method of illuminating opaque objects under the highest powers of the microscope. *Trans. Micr. Soc. Lond./ Quart. Jour. Micr. Sci.* **4**: 55–60.

Whittaker, E. (1951). *A History of the Theories of Aether and Electricity.* Volume I. The Classical Theories. Thomas Nelson and Sons, London.

Whittaker, E. (1953). *A History of the Theories of Aether and Electricity.* Volume II. The Modern Theories, 1900–1926. Thomas Nelson and Sons, London.

Wick, S. M., R. W. Seagull, M. Osborn, K. Weber, and B. E. S. Gunning. (1981). Immunofluorescence microscopy of organized microtubule arrays in structurally stabilized meristematic plant cells. *J. Cell Biol.* **89**: 685–690.

Wickramasinghe, H. K. (1989). Acoustic microscopy. *Advances in Optical Electron Microscopy* **11**: 153–182.

Wien, W. (1893). Eine neue Beziehung der Strahlung schwarzer Körper zum zweiten Hauptsatz der Wärmetheorie. *Berl. Ber.*: 55–62.

Wien, W. (1896). Über die Energievertheilung im Emissionspectrum eines schwarzen Körpers. *Ann. d. Phys.* **58**: 662–669.

Wien, W. (1911). On the laws of thermal radiation. *Nobel Lecture,* December 11: 1911.

Wiesendanger, R. ed. (1998). *Scanning Probe Microscopy. Analytical Methods.* Springer-Verlag, Berlin.

Williams, L. P. (1987). *Michael Faraday.* Da Capo Press, Inc., New York.

Williamson, R. E. (1991). Orientation of cortical microtubules in interphase plant cells. *Int. Rev. Cytol.* **129**: 135–206.

Williamson, S. J. and H. Z. Cummins. (1983). *Light and Color in Nature and Art.* John Wiley and Sons, New York.

Wilson, E. B. (1895). *An Atlas of the Fertilization and Karyokinesis of the Ovum.* Macmillan, New York.

Wilson, L. and W. W. McGee. (1988). Construction and calibration of a demountable 'Poor-Microscopist's Hotstage'. *The Microscope* **36**: 125–131.

Wilson, T. (1990). *Confocal Microscopy.* Academic Press, New York.

Winton, A. L. (1916). *The Microscopy of Vegetable Foods, with Special Reference to the Detection of Adulteration and the Diagnosis of Mixtures.* John Wiley & Sons, New York.

Woan, G. (2000). *The Cambridge Handbook of Physics Formulas.* Cambridge University Press, Cambridge.

Woll, K., L. A. Borsuk, H. Stransky, D. Nettleton, P. S. Schnable et al. (2005). Isolation, characterization, and pericycle-specific transcriptome analyses of the novel maize lateral and seminal root initiation mutant rum1. *Plant Physiol* **139**: 1255–1267.

Wollaston, W. H. (1824). On the apparent direction of the eyes in a portrait. *Phil. Trans. Roy. Soc. Lond.* **114**: 247–256.

Wolniak, S. M., P. K. Hepler, and W. T. Jackson. (1980). Detection of membrane-calcium distribution during mitosis in *Haemanthus* endosperm with chlorotetracycline. *J. Cell Biol.* **87**: 23–32.

Wolniak, S. M., P. K. Hepler, and W. T. Jackson. (1981). The coincident distribution of calciumrich membranes and kinetochore fibers at metaphase in living endosperm cells of *Haemanthus. Eur. J. Cell Biol.* **25**: 171–174.

Wood, A. and F. Oldham. (1954). *Thomas Young. Natural Philosopher 1773–1829.* Cambridge University Press, Cambridge.

Wood, R. D. (1971a). J. B. Reade, F.R.S., and the Early History of Photography, Part I: a re-assessment on the discovery of contemporary evidence. *Annals of Science* **27**: 13–45.

Wood, R. D. (1971b). J. B. Reade, F.R.S., and the Early History of Photography, Part 2: Gallic acid and Talbot's calotype patent. *Ann. Sci.* **27**: 47–83.

Wood, R. W. (1905). *Physical Optics.* Macmillan, New York.

Wood, R. W. (1914). *Physical Optics,* New and Revised Edition. Macmillan, New York.

Wood, R. W. (1961). *Physical Optics,* third revised edition. Dover, New York.

Woods, R. T., M. J. Hennessey, E. Kwok, and B. E. Hammer. (1989). NMR Microscopy-A new biological tool. *Biotechniques* **7**: 616–622.

Wright, A. E. (1907). *Principles of Microscopy.* Macmillan, New York.

Wu, F. S. (1987). Localization of mitochondria in plant cells by vital staining with rhodamine 123. *Planta* **171**: 346–357.

Xu, C. and W. W. Webb. (1996). Measurement of two-photon excitation cross-sections of molecular fluorophores with data from 690 nm to 1050 nm. *J. Opt. Soc. Am. B.* **13**: 481–491.

Yang, X., M. A. Miller, R. Yange, D. F. Evans, and R. D. Edstrom. (1990). Scanning tunneling microscopic images show a laminated structure for glycogen molecules. *FASEB J.* **4**: 3140–3143.

Yavetz, I. (1995). *From Obscurity to Enigma: The Work of Oliver Heaviside*. Birkhäuser, Boston, MA.

Young, H. D. and R. A. Freedman. (2000). *Sears and Zemansky's University Physics*. Volume One. Addison Wesley, San Francisco, CA.

Young, I. T. (1989). Image fidelity: Characterizing the image transfer function. *Methods in Cell Biology* **30**: 1–45.

Young, T. (1794). Description of a new species of *Opercularia*. *Trans. Linn. Soc. Lond.* **3**: 30–32. and one plate.

Young, T. (1800). Outlines of experiments and inquiries respecting sound and light. *Phil. Trans. Roy. Soc. Lond.* **90**: 106–150.

Young, T. (1801). A letter respecting sound and light in reply to some observations of Prof. Robison. *J. Nut. Phil. Chem. Arts* **5**: 161–167.

Young, T. (1802). The Bakerian Lecture. On the theory of light and colours. *Phil. Trans. R. Soc. Lond.* **92**: 12–48.

Young, T. (1804a). Experiments and calculations relative to physical optics. *Phil. Trans. R. Soc. Lond.* **94**: 1–16.

Young, T. (1804b). Reply to the animadversions of the Edinburgh reviewers on some papers published in the Philosophical Transactions. J. Johnson, London.

Young, T. (1807). A Course of Lectures on Natural Philosophy and the Mechaical Art. Printed for Joseph Johnson, London.

Zajonc, A. (1993). *Catching the Light. The Entwined History of Light and Mind*. Bantam Books, New York.

Zajonc, A. (2003). Light reconsidered. *Optics & Photonics News* October: S2–S6.

Zankel, K. L., P. V. Burke, and M. Delbruck. (1967). Absorption and screening in Phycomyces. *J. Gen. Physiol.* **50**: 1893–1906.

Zernike, F. (1942a). Phase-contrast, a new method for microscopic observation of transparent objects. I. *Physica* **9**: 686–698.

Zernike, F. (1942a). Phase-contrast, a new method for microscopic observation of transparent objects. II. *Physica* **9**: 974–986.

Zerike, F. (1946). Phase contrast, a new method for the microscope observation of transparent objects. 116–135. In: *Achievements in Optics*. A. Bouwers, ed. Elseveier, New York.

Zernike, F. (1948). Observing the phases of light waves. *Science* **107**: 463.

Zernike, F. (1955). How I discovered phase contrast. *Science* **121**: 345–349.

Zernike, F. (1958). The wave theory of microscopic image formation. 525–536. In: *Concepts of Classical Optics*. J. Strong, ed. W.H. Freeman and Co., San Francisco, CA.

Zhang, D., P. Wadsworth, and P. K. Helper. (1990). Microtubule dynamics in living dividing plant cells: Confocal imaging of microinjected fluorescent brain tubulin. *Proc. Natl. Acad. Sci. USA* **87**: 8820–8824.

Zhang, X., S. Madi, L. Borsuk, D. Nettleton, R. J. Elshire, B. Buckner, D. Janick-Buckner, J. Beck, M. Timmermans, P. S. Schnable, and M. J. Scanlon. (2007). Laser microdissection of narrow sheath mutant maize uncovers novel gene expression in the shoot apical meristem. *PLoS Genet* **3**(6): e101.

Zhong, X., P. F. Fransz, E. J. Van Wennekes, P. Zabel, A. Van Kammen et al. (1996). High-resolution mapping on pachytene chromosomes and extended DNA fibers by fluorescence *in-situ* hybridisation. *Plant Mol. Biol. Rep.* **14**: 232–242.

Zhong, X., P. F. Fransz, J. Wennekes-Van Eden, M. S. Ramanna, A. Van Kammen et al. (1998). FISH studies reveal the molecular and chromosomal organization of individual telomere domains in tomato. *Plant J.* **13**: 507–517.

Zieler, H. W. (1972). *The Optical Performance of the Light Microscope*. Microscope Publications Ltd., London.

Zimmer, M. (2005). *Glowing Genes: A Revolution in Biotechnology*. Prometheus Books, Buffalo, NY.

Zumbahlen, H., ed. (2008). *Linear Circuit Design Handbook*. Newnes.

A Final Exam

The following exam is similar in style to the midterm exam I give to my students. I no longer give a traditional final exam, but ask my students to do a creative writing project. They must gear their project to someone who knows nothing about light microscopy and initially does not want to know anything about it, but after they read the story, they will learn something about light microscopy, find it interesting enough, and want to know more. I have received children's stories, spy stories, adventure stories, science fiction stories, and poetry. One of my students went on to become a science writer!

This is an open book exam. There are 20 questions; each one is worth 20 points. Good luck!

1a. Assume we have a concave parabolic mirror. Draw and describe the image of an object placed at the center of curvature. Will the image be magnified or minified; erect or inverted; real or virtual? Give an example of a place in a bright-field microscope where you would find this optical arrangement.

1b. Assume we have a double convex lens. Draw and describe the image of an object placed in front of the front focal plane of the lens. Will the image be magnified or minified; erect or inverted; real or virtual? Give an example of a place in a bright-field microscope where you would find this optical arrangement.

1c. Assume we have a double convex lens. Draw and describe the image of an object placed at the front focal plane of the lens. Will the image be magnified or minified; erect or inverted; real or virtual? Give an example of a place in a bright-field microscope where you would find this optical arrangement.

1d. Assume we have a double convex lens. Draw and describe the image of an object placed between the front focal plane and the lens. Will the image be magnified or minified; erect or inverted; real or virtual? Give an example of a place in a bright-field microscope where you would find this optical arrangement.

2. René Descartes wrote, "...*there are no images that must represent in every respect the objects they represent*...." More than 200 years later, Ernst Abbe concluded that the image formed by a microscope is related to a nonexistent specimen whose complete diffraction pattern precisely matches the one collected by the objective lens and not the specimen itself. Using Abbe's experimental evidence to support your arguments, briefly discuss the importance of Descartes' statement when determining the structure of a specimen with a microscope.

3. In 1611, Maurolico wrote the following: "*If you view the light of a candle, placed not too far through a white feather from a dove or from some other bird, when placed opposite the eye you will see, between the lines on the feather and those branches, a certain distinct cross with a wonderful variety of colors, such as are seen in the rainbow. This can happen only through the light being received between the small grooves of the feather tufts, and there multiplied, continually incident and by turns reflected.*" Maurolico died before the development of the wave theory of light. Using the wave theory of light, draw diagrams to show Maurolico why the feather caused the formation of the wonderful variety of colors.

4. What is the modulation transfer function? How is it determined and what does it tell you?

5. Under ideal conditions, the human eye can resolve two high-contrast black dots separated by 0.07 mm on a white background. What would be the total magnification necessary for the eye to resolve two points in a specimen that are 250 nm apart? What must the minimum numerical aperture of the objective lens be in order to resolve these two points when viewed with 500 nm oblique illumination? Use either the Abbe criterion or the Rayleigh criterion.

6a. How will a 6000 nm thick specimen with a refractive index of 1.35383 look in a positive phase-contrast microscope when the mounting medium is water (n = 1.3330) and we observe the specimen with 500 nm light?

6b. How will the same specimen look if we observe it with a negative phase-contrast microscope?

6c. How will the same specimen look if we observe it with a positive phase-contrast microscope with 400 nm light?

6d. How will the same specimen look if we observe it with a positive phase-contrast microscope with 500 nm light, but change the refractive index of the medium by adding 1% bovine serum albumin to it?

7. There is a dishonest member of the chamber of commerce in a city in Denmark. This city makes all its money from tourists coming to a spa. People going to the spa are getting sick. You believe that there are poisonous bacteria living in the waters of the spa, and the spa should be shut down. The dishonest member does not want to give up the money coming in from the tourists, so he gets a phase-contrast microscope, but he finds that there are poisonous bacteria in the water. He has an idea, and determines that the refractive index of these 1000 nm in diameter spherical bacteria is 1.463 and the refractive index of the mineral-rich spa water is 1.338. He then sets up the phase-contrast microscope with 500 nm light in order to show the other members of the chamber that the water is pure and that no bacteria are living in it. He claims that since he has shown that there are no bacteria, the spa should remain open. What do the rest of the members see when they look in the microscope? If you were on the committee, what would you suggest doing with the phase-contrast microscope in order to be sure that there are no bacteria?

8. In your own words, what are the important messages of Plato's *Allegory of the Cave* for microscopists?

9a. Using diagrams, describe the physics of fluorescence.

9b. Briefly discuss how our knowledge of the physics of fluorescence helps us to select the excitation filter, the chromatic beam splitter, and the barrier filter.

9c. What are the advantages of the epi-fluorescence microscope over a bright-field fluorescence microscope and a dark-field fluorescence microscope?

10a. Imagine you are observing a 10,000 nm thick specimen that has a n_e of 1.34925 and a n_o of 1.3355 with a polarized light microscope using 550 nm light. Is the specimen positively or negatively birefringent?

10b. How would the specimen appear when its optical axis is oriented +45 degrees relative to the azimuth of maximal transmission of the polarizer?

10c. How would the specimen appear when its optical axis is oriented parallel to the azimuth of maximal transmission of the polarizer?

10d. How would you describe the polarized light that emerges from the specimen when its optical axis is oriented +45 degrees relative to the azimuth of maximal transmission of the polarizer?

10e. How would you describe the polarized light that emerges from the specimen when its optical axis is oriented −45 degrees relative to the azimuth of maximal transmission of the polarizer?

10f. How many degrees would you have to rotate a de Sénarmont compensator counterclockwise in order to bring the specimen oriented at +45 degrees relative to the azimuth of maximal transmission of the polarizer to extinction?

10g. Could you use a $\lambda/20$ Brace-Köhler compensator to measure the retardation of this specimen? Why or why not?

11a. Draw the light rays from the specimen to the objective lens when the specimen is illuminated with dark-field illumination. Briefly describe how a microscope with dark-field illumination generates contrast.

11b. Draw the light rays from the specimen to the objective lens when the specimen is illuminated with Rheinberg illumination. Briefly describe how a microscope with Rheinberg illumination generates contrast.

11c. Draw the light rays from the specimen to the objective lens when the specimen is illuminated with oblique illumination. Briefly describe how a microscope with oblique illumination generates contrast.

12. Imagine that a freshman is coming to the laboratory tonight to use a bright-field microscope. Write a flow chart of the things that he or she must do to set up Köhler illumination.

13a. Describe how a pseudo-relief image is generated in a differential interference microscope.

13b. Describe how a pseudo-relief image is generated in a Hoffman modulation contrast microscope.

13c. Describe how a pseudo-relief image is generated with an analog image processor.

13d. Describe how a pseudo-relief image is generated in a digital image processor.

14a. Describe how sound waves are produced and focused in an acoustic microscope.

14b. Which kinds of information about a specimen can you get with an acoustic microscope?

15. How does a near-field scanning optical microscope differ from a bright-field microscope?

16. Describe the difference between a two-photon confocal microscope and a standard confocal microscope. What are the advantages of a two-photon confocal microscope over a standard confocal microscope?

17. Describe how a digital CCD camera generates a numerical array of intensities.

18a. Describe how the reference wave is created in an image duplication microscope that is not based on polarized light.

18b. Describe how the reference wave is generated in an image duplication microscope based on polarized light.

19. Light is the tool we use to probe the specimens in a light microscope. In your own words, describe the nature of light.

20.   Identify the following people:

Thomas Young   Lord Rayleigh

Fritz Zernike   Shinya Inoué

Ernst Abbe   Christiaan Huygens

Max Planck   Isaac Newton

Albert Einstein   James Clerk Maxwell

A Microscopist's Model of the Photon

At the end of the semester, after teaching all that was known about physics, Arthur Schuster (1898, 1932) would advise his students to think and dream of things that are not known and not in the textbook. In fact, one of the things he asked his students to consider in 1898 was antimatter. In this spirit, I will present ideas about light that are not in any textbook. They may be right and they may be wrong, but they are definitely thought-provoking.

After centuries of struggling to understand the wave and corpuscular aspects of light (Cajori, 1916; Gillispie, 1960; Azimov, 1966; Herschel, 1966; Schaffner, 1972; Swensen, 1972; Bitter and Medicus, 1973; Crew, 1981; Mason, 1981; Buchwald, 1989; Chen and Barker, 1992; Zajonc, 1993; Hakfoort, 1995; Chen, 1997; Neuenschwander, 2001), the two aspects were combined in the quantum theory of radiation by Albert Einstein (Einstein, 1905a, 1909a, 1909b, 1917; Schrödinger, 1928; Thomson, 1930; Fermi, 1932; Heitler, 1944; Klein, 1964; Jammer, 1966, 1974; Moore, 1966; Pais, 1982, 1986, 1991; Loudon, 1983; Bialynicki-Birula, 1992). According to Einstein's quantum hypothesis, the corpuscular properties of light, including energy ($h\nu$) and momentum (h/λ), were not only mechanical properties but also wave-like properties since they depended on the frequency and wavelength of light, respectively. Einstein treated the photon as a mathematical point that enshrouded the wave-particle controversy. Einstein's hypothesis was supported by Robert Millikan's (1950) experiments on the photoelectric effect that showed that the energy of a photon was related to its frequency, and Arthur Compton's (1923) scattering experiments that indicated that energy and momentum were conserved when photons scattered off electrons as if the photons were little billiard balls.

Treating the photon as a mathematical point bothered many of Einstein's contemporaries, including Max Planck (1920), Niels Bohr (1922), Bohr et al. (1924), Hendrik Lorentz (1924), and Robert Millikan (1924b), who believed that a theory of light should also be able to explain interference and diffraction effects (Miller, 1994). Realizing the inadequacy of his theory, Einstein (in Campos, 2004) wrote to Lorentz in 1909 stating that, "I am not at all of the opinion that one should think of light as being composed of mutually independent quanta localized in relatively small spaces." As an alternative to Einstein's corpuscular model, Bohr developed a model of the photon as part of the Copenhagen interpretation of quantum mechanics. The Copenhagen interpretation has become the orthodox version of quantum mechanics (Heisenberg, 1930, 1958, 1971, 1979; Bohr, 1934, 1958; Feynman, 1965; Fine, 1986; Sachs, 1988; Beller, 1999; Griffiths, 2005). The Copenhagen interpretation states that the photon does not really exist in any location at all—it is the act of measurement that gives it existence (Segré, 2007). The photon will appear as one or the other of its complementary properties depending on the measurement device. The photon will appear as a particle if one uses a measuring device like a photomultiplier tube to measure the photon and it will appear as a wave if a measuring device like a diffraction grating is used to measure the photon. According to Pascual Jordan, "Observations not only disturb what is to be measured, they produce it…. We compel (the particle) to assume a definite position" (in Griffiths, 2005).

Wave mechanics provides a method for describing a photon as a wave packet, which is something between a point-like particle and an infinite plane wave. According to wave mechanics, a photon is formed by the integration of a continuous and wide distribution of infinite plane waves (de Broglie, no date). The more wavelengths that are included in the integration, the more point-like the photon will be. The linear momentum (p) of a wave is related to the wavelength (λ) through Planck's constant (h) according to the following equation:

$$p = h/\lambda$$

Consequently, a point-like photon is composed of an infinite number of waves with an infinite number of wavelengths, each with a unique linear momentum, whereas an infinite plane wave has a single wavelength and definite linear momentum, but no localization. With this interpretation of a photon, the interaction of light with matter depends on the probability that a photon with a given wavelength will be created where the matter is. If we accept the uncertainty principle as the foundation of physics, this is as visual and intuitive a model of the photon as possible with wave and quantum mechanics. Although the fundamental nature of the uncertainty principle and the probabilistic interpretation of physical entities along with the anthem "shut up and calculate" are accepted by the majority of physicists, this Copenhagen interpretation was never accepted by Planck, Einstein, Schrödinger, de Broglie, and David Bohm (Moore, 1989; Peat, 1997; Bolles, 2004; Mermin, 2004; Born, 2005; Camilleri, 2006; Lindley, 2007). Perhaps these physicists were stick-in-the-muds, or perhaps there is room for alternative heuristic hypotheses that give rise to a model of the photon that is not based on the fundamental nature of probability and uncertainty. Microscopists like to make things visible, and by taking into consideration a number of the properties of light that are of interest to microscopists, Ben-Zion Ginzburg and I have been working on a visual and intuitive model of the particle-like and wave-like properties of the photon. I will present some salient aspects of this model here.

According to Richard Feynman (1988), "Every particle in nature has an amplitude to move backwards in time, and therefore has an anti-particle… Photons look exactly the same in all respects when they travel backwards in time… so they are their own anti-particles." We start with the assumption that the photon may not necessarily be an elementary particle, but a binary structure composed of matter and antimatter. In this way, the photon is reminiscent of molecular hydrogen (Cannizzaro, 1858) or a meson (Fermi and Yang, 1949), which were both originally thought to be elementary particles. We posit that the two particles that make up a photon are conjugate particles in terms of mass and sense of rotation (parity). According to our hypothesis, the photon is not an elementary particle and its own antiparticle as posited by Feynman's quantum electrodynamics, but a complex structure composed of a particle and its conjugate antiparticle.

Using, in an unconventional manner, an equation equivalent to that which de Broglie (1925, 1946, 1953, 1955, 1962) created to kick off wave mechanics, we claim that the mass (m) of an isolated particle is related to its angular frequency (ω) or spin in the following way:

$$m^{\pm} = \hbar\omega^{\pm}/c^2$$

where $\hbar$ is Planck's constant divided by 2π and c is the speed of light. We arbitrarily assign a positive sense of rotation (counterclockwise as it is moving toward you) to a material particle and a negative sense of rotation (clockwise as it is moving toward you) to an antimaterial particle. We define matter to have positive mass and antimatter to have negative mass. Although negative mass is a legitimate concept in physics (Bondi, 1957), it has been an unwelcome concept (Dirac, 1930, 1933). The traditional use of the de Broglie equation is for relating the wavelength (λ), and not angular quantities of a particle to its mass:

$$m = hc/(\lambda c^2) = h/(\lambda c)$$

We posit that the reason that a photon is massless is because the sum of the masses of the two particles that make up a photon vanishes. That is, the masses of the two

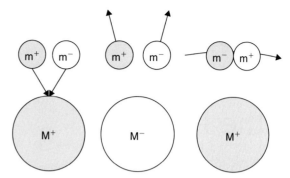

FIGURE A-1 Direction of acceleration between a large body mass and test masses.

particles that make up a photon are equal in magnitude and opposite in sign ($m^+ = -m^-$) and it follows that the sum of the angular frequencies of the two particles that make up a photon must also vanish. For this to happen, the angular frequencies of the two particles must be equal in magnitude and opposite in sign ($\omega^+ = -\omega^-$).

According to Newton's second law, mass (m) is the ratio of force (F) to acceleration (a):

$$m = F/a$$

where both the force and the acceleration are vector quantities. For positive masses, the vector of acceleration is parallel to the force vector, and for negative masses, the two vectors are antiparallel. Because the vectors are parallel for matter and antiparallel for antimatter, we could claim that antimatter reacts to a given force like matter going "backwards in time".

How do particles of negative and positive mass interact with themselves and with each other (Figure A-1)? If we initially consider the particles to have mass but not charge, we can use Newton's law of gravitation in an unconventional manner to describe the causal force and Newton's second law (Newton, 1687) to determine how any two particles, with masses of arbitrary sign, respond to the causal force and accelerate relative to each other. By equating the gravitational force to the inertial force, using Newton's law of gravitation and Newton's second law, we get:

$$F = (G/r^2)m_1 m_2 = m_2 g$$

where F is the gravitational force (in N), r is the distance between the two particles (in m), G is the gravitational constant ($6.67300 \times 10^{11}\,m^3kg^{-1}s^{-2}$), m_1 is the mass of a large body like the earth or the sun (in kg), m_2 is the mass of a test particle (in kg), and g is the acceleration due to gravity of the test particle relative to the large body (in m/s^2). The test body accelerates toward the large body when g > 0, and the test body accelerates away from the large body when g < 0.

When the masses of a large body and a test particle are positive, there will be an attractive force (F > 0), and the positive mass test particle will accelerate toward the large positive mass body (g > 0). When the mass of a large body is positive and the mass of the test particle is negative, there will be a repulsive force (F < 0), and the negative mass test particle will accelerate toward the large positive mass body (g > 0). When the masses of a large body and a test particle are negative, there will be an attractive force (F > 0), and the negative mass test particle will accelerate away from the large negative mass body (g < 0). When the mass of a large body is negative and the mass of the test particle is positive, there will be a repulsive force (F < 0), and the positive mass test particle will accelerate away from the large negative mass body (g < 0).

If the magnitudes of the masses of a negative mass particle and a positive mass particle are the same, the positive mass particle will accelerate away from the negative mass particle (g < 0) and the negative mass particle will accelerate toward the positive mass particle (g > 0). Consequently, the negative mass particle will chase the positive mass particle (Bondi, 1957). We posit that the gravitational force between the two conjugate particles that make up the photon provides the motive force that causes a photon to move. Moreover, as shown in Figure A-1, a massless photon has weight and will be bent by the sun as Dyson et al. (1920) and Eddington (1920) observed.

However, if the particles only had the properties of mass and angular frequency, a photon made of two conjugate particles would accelerate to infinite velocity. As I will discuss later, the particles must also have charge to constrain the velocity. The gravitational force-induced movement of the charged particles causes an electromagnetic force that is responsible for reducing the velocity of the photon to the speed of light.

Because of the intrinsic mass and angular velocity of the particles, each of the particles that compose the photon has an angular momentum or spin. Since the experimentally measured angular momentum of a photon is $\hbar$ (Beth, 1936; Leach et al., 2002), we posit that the angular momentum of each of the particles is $\hbar/2$. That is, the photon is a spin-1 boson, whereas the particles that make up the photon are spin-½ fermions (de Broglie, 1962). Classically, the angular momentum of a particle is equal to $mv\Gamma r$, where m is the mass of the particle, v is its angular velocity, r is its radius, and Γ is a dimensionless geometric factor that equals one for a point mass at the end of a mass-less string of radius r (Young and Freedman, 2000). Assuming $\Gamma = 1$, and since $v = \omega r$, the angular momentum (L) of a photon is given by the following equation:

$$\hbar = \hbar/2 + \hbar/2 = (m^+/2)\omega^+r^2 + (m^-/2)\omega^-r^2$$

For a given photon, the products of (m/2) and ω have the same sign for each particle since both the mass and the sense of rotation have opposite signs for each particle (Figure A-2). The angular frequency of the entire photon is equal to the absolute value of the angular frequency of each particle and ω is also the angular frequency of the photon itself.

$$L = \hbar = (m^+/2)\omega^+r^2 + (m^-/2)\omega^-r^2 = m\omega r^2$$

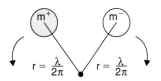

FIGURE A-2   The radius of a rotating mass with angular momentum of $\hbar/2$.

The angular frequency of the photon is related to its optically measured properties of frequency and wavelength by the following equation:

$$\omega = 2\pi\nu = 2\pi c/\lambda$$

The sign of the angular momentum of a photon propagating in free space is undetermined and can be determined only through its interaction with matter or antimatter. We posit that the angular momentum of a photon measured by its interaction with matter will be $+\hbar$ and its angular momentum measured by its interaction with antimatter will be $-\hbar$. Likewise, the effective mass, linear momentum, and energy of a photon propagating in free space can be determined only after its interaction with matter or antimatter. Again, we posit that these quantities are positive when a photon interacts with matter that has positive kinetic energy ($\frac{1}{2}mv^2$) and a positive specific heat and negative when a photon interacts with antimatter that has negative kinetic energy ($\frac{1}{2}mv^2$) and a negative specific heat.

Assuming that a photon interacts with matter, its energy is given by $\hbar\omega > 0$ and its apparent mass is given by $\hbar\omega/c^2 > 0$. Substituting this mass for the mass term in the equation that describes the angular momentum of a photon, we can calculate the radius of a photon using the following equation:

$$\hbar = m\omega r^2 = \hbar\omega^2r^2/c^2$$

Solving for the radius of the photon, we get:

$$r = c/\omega = 1/k = \lambda/2\pi$$

and since the angular wave number $\omega/c = k$, the radius of the photon is equal to the reciprocal of the wave number or $\lambda/2\pi$, giving the photon extension in space. The circumference of a cylindrical photon is equal to its wavelength since the circumference is given by $2\pi r = 2\pi\lambda/(2\pi) = \lambda$. The transverse extension of a photon in space, with a cross-section of $\pi r^2 = \lambda^2/(4\pi)$ helps us visualize why there is a limit to how close two point-like objects can be before we cannot resolve them as two separate points with a microscope. It also allows us to visualize how two nearby photons can interfere with one another and how a single photon can go through two slits, separated by a distance equal to $\lambda/2\pi$, at the same time. The fact that the radius of a photon is related to its wavelength by the constant 2π makes it understandable why, heretofore, we have thought about resolution, diffraction, and interference solely in terms of wavelength instead of wavelength and wave width.

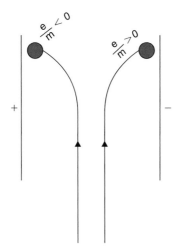

FIGURE A-3 A particle with a charge-to-mass ratio greater than zero bends toward the negative pole of an electric field and a particle with a charge-to-mass ratio less than zero bends toward the positive pole of an electric field.

Photons in free space, in the idealized case where there is neither emitter nor absorber, are considered electrically neutral, and consequently the elementary charge is not included in Maxwell's equations for radiation in free space. However, this should give us pause since photons carry electromagnetic energy. Einstein (1909a) wrote,

> We should remember that the elementary quantum ε of electricity is an outsider in Maxwell-Lorentz electrodynamics…. The fundamental equation of optics …will have to be replaced by an equation in which the universal constant ε (probably its square) also appears in a coefficient…. I have not yet succeeded in finding a system of equations…suitable for the construction of the elementary electrical quantum and the light quanta. The variety of possibilities does not seem so great, for one to have to shrink from this task.

The definition of charge is problematic in that its sign actually depends on the sign of the mass. J. J. Thomson (1897) characterized the charge-to-mass ratio of the electron from the way it bent in an electric field. Carl Anderson (1932, 1933, 1999) discovered the anti-electron or positron because he found a particle that bent in a direction opposite to that of an electron in the electric field (Figure A-3). It was as if its charge-to-mass ratio had the same magnitude but opposite sign as the electron. Indeed, it is only a matter of bookkeeping whether we interpret the anti-electron (=positron) as a particle with positive mass and positive charge or as a particle with negative mass and negative charge. If we interpret the anti-electron as having a positive charge, it must have a positive mass to accelerate toward the negative pole of the field, parallel to the attractive force and antiparallel to the repulsive force. If we interpret the anti-electron as having a negative mass, then it must have a negative charge to accelerate toward the negative pole, parallel to the repulsive force, and antiparallel to the attractive force.

Once we realize that the definition of charge depends on the assumptions about mass, we have to rewrite Coulomb's Law to make the assumptions about mass explicit:

$$F = -(1/(4\pi\varepsilon_0 r^2))\mathcal{A}q\mathcal{A}_{test}q_{test}$$

where ε_0 represents the electrical permittivity of the vacuum; q represents the charge; and $\mathcal{A}$ which is dimensionless, represents the sign of the mass of the charge in question. According to this modified version of Coulomb's Law, like charges of positive mass and like charges of negative mass exert repulsive forces (F < 0) on each other, and opposite charges of positive mass and opposite charges of negative mass exert attractive forces (F > 0) on each other. To model how the charges accelerate in response to the electrostatic forces, we set the modified Coulombic force equal to the inertial force using Newton's second law.

$$m_{test}a = -(1/(4\pi\varepsilon_0 r^2))\mathcal{A}q\mathcal{A}_{test}q_{test}$$

where a is the acceleration of the test particle with mass (m_{test}) and mass-independent charge ($\mathcal{A}_{test}q_{test}$) relative to a stationary mass-independent charge ($\mathcal{A}q$). The test particle accelerates toward the stationary charge when a > 0 and away from the stationary charge when a < 0.

Solving the previous equation for a homogeneous system made of positive mass, a test mass will accelerate away from a stationary mass with a like charge, and a test mass will accelerate toward a stationary mass with an opposite charge. By contrast, in a homogeneous system made of negative mass, a test mass will accelerate toward a stationary mass with a like charge, and a test mass will accelerate away from a stationary mass with an opposite charge.

In a heterogeneous system like the photon, which we posit to be made of positive and negative mass, a negative mass (of either charge) will chase the positive mass, as long as the particles have opposite charges. The positive mass photon will chase the negative mass photon when the two particles have masses with opposite sign but like charge. Binary photons with either longitudinal polarization will remain in motion indefinitely in the absence of external forces.

Hans Christian Oersted (1820) serendipitously discovered that linearly moving charges generate circular magnetic fields. We posit that the particles that make up the photon also generate circular magnetic fields, whose magnitudes depend on the velocity of the particles. Let us assume that the two particles that make up a photon initially are traveling with opposite velocities and separated by a distance of $\lambda/2$ (Figure A-4). Let us also assume that the midpoint of the two particles always travels along the propagation vector at 2.99792458×10^8 m/s. The midpoint of the two particles must travel at the speed of light, however the individual velocities of the two particles are not constrained by the speed of light—one travels faster than the speed of light, and the other travels slower such that the average velocity of the two particles is invariant and equal to 2.99792458×10^8 m/s.

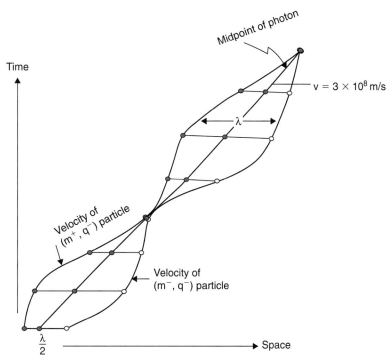

FIGURE A-4   The movement through time and space of the two conjugate particles that make up a photon.

As the positive mass particle accelerates away from the midpoint, it generates a magnetic field, whose magnitude is proportional to its velocity. Ignoring the accelerations (v^2/r) for the time being, the following equation, based on the work of André-Marie Ampere, Felix Savart, and Jean-Baptiste Biot, gives the magnitude of the magnetic field perpendicular to the velocity (v) of a charged particle.

$$\mathbf{B} = (\mu_o/4\pi r^2)(\mathscr{A}q\mathbf{v})$$

As a particle with positive mass and negative charge accelerates away from the midpoint, it generates a magnetic field with a counterclockwise circulation from the perspective of the midpoint. By contrast, as a particle with negative mass and negative charge accelerates away from the midpoint it produces a magnetic field with a clockwise circulation from the perspective of the midpoint. The circulating magnetic field of each particle will generate an electromotive force (emf) in a cross-sectional area (A) according to Faraday's Law:

$$emf = -(A\mu_o/(4\pi r^2))d(\mathscr{A}q\mathbf{v})/dt$$

As the particle with positive mass and negative charge travels away from the midpoint, the magnetic field it generates will generate an electromotive force that is oriented with the positive pole at the midpoint (Figure A-5). As the particle with negative mass and negative charge travels away from the midpoint, the magnetic field it generates will generate an electromotive force that is oriented with the negative pole at the midpoint. Since a particle with positive mass and negative charge accelerates toward the positive pole of an electric field, the induced electromotive force will retard

the movement of the particle with positive mass and negative charge that is moving away from the midpoint. Since a particle with negative mass and negative charge accelerates toward the negative pole of an electric field, the induced electromotive will retard the motion of the particle with negative mass and negative charge that is moving away from the midpoint.

Let us assume that the velocity of the positive mass, negatively charged particle, and the velocity of the negative mass, negatively charged particle decelerate to 0 m/s relative to the midpoint and the velocities vanish when each particle is $\lambda/2$ away from the midpoint. At this point, the distance between the two particles is equal to λ. If the energy of an isolated photon is a conserved quantity, the kinetic energy of each particle must be converted into potential energy. The kinetic energy of a charged particle is inseparable from the magnetic energy and the potential energy of the charged particle is inseparable from the electrical energy. If the two particles form a harmonic oscillator, the potential energy of the electric field will accelerate each particle toward the midpoint. As the particles approach the midpoint, they will generate magnetic fields, which will induce an electromotive force that will slow them down to 0 m/s as they reach the midpoint. At the midpoint, where the kinetic energy is completely converted into the potential energy and the magnetic energy is completely converted into electrical energy, the particles begin again to move apart because of the electromotive force that the moving particles created. Thus, the photon is like a longitudinal standing wave relative to the midpoint, and a longitudinal traveling wave from the perspective of any

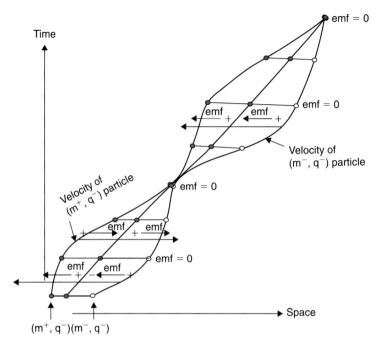

FIGURE A-5   The position-dependent electromotive forces established by the particles that make up a photon.

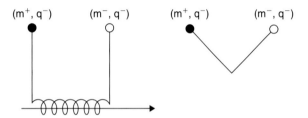

FIGURE A-6   A side view (A) and a front view (B) of a binary photon.

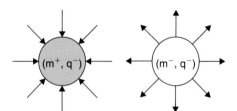

FIGURE A-7   The electric fields produced by the particles that make up a photon.

observer. From a mechanical perspective, the kinetic energy is reversibly converted into potential energy and from an electromagnetic perspective; the magnetic energy is reversibly converted into electrical energy. Consistent with the principle of energy conservation, the two forms of energy that make up a couple are out-of-phase with each other by $\lambda/4$.

The midpoint propagates at velocity c. Its propagation is promoted with infinite potential by the inertial properties of the particles and the gravitational force. However, the infinite velocity is not realized by the photon and its propagation is constrained in terms of direction and speed, by the electrical properties of the particles and the electromagnetic force. This constraint is determined exclusively by the elementary charge (e), the magnetic permeability of the vacuum (μ_o), and the electrical permittivity of the vacuum (ε_o). Indeed Maxwell deduced that light was an electromagnetic wave because the speed of light could be obtained from electrical parameters as given by the following equation (Maxwell, 1864):

$$c = 1/\sqrt{(\mu_o \varepsilon_o)}$$

I have just discussed the longitudinal polarization of the electric field. If this microscopist's model of the photon is

representative of reality, the wave-particle duality can be clarified by picturing the photon as a binary entity with the distance between the two particles oscillating between 0 and λ (Figure A-6). The transverse polarization, however, is of particular interest to microscopists doing polarized light microscopy. I will show how the electrical properties of the particles combined with their respective angular frequencies form an electric dipole that varies in time and space to produce a photon with a transverse, linearly polarized electric field. The magnitude and direction of an electric field (E) surrounding a particle is given by the following equation:

$$\mathbf{E} = -\mathbf{F}/\mathscr{A}q = (1/4\pi\varepsilon_o r^3)\mathbf{r}\mathscr{A}q$$

where $\mathbf{r}$ is a unit vector pointing away from the semi-photon. When $\mathbf{E}$ is positive, the electric field points away from the particle and when $\mathbf{E}$ is negative, it points toward the particle. When the sign of the mass ($\mathscr{A}$) is positive and the charge (q) is negative, the electric field points toward the particle and is antiparallel to the unit vector. When the sign of the mass ($\mathscr{A}$) and the charge (q) are both negative, the electric field points away from the particle (Figure A-7).

We posit that the mass of the particle causes its electric field to decrease with distance in a manner analogous to

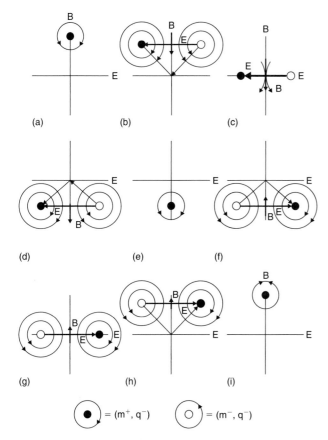

(a)   (b)   (c)

(d)   (e)   (f)

(g)   (h)   (i)

$\bullet = (m^+, q^-)$ $\circ = (m^-, q^-)$

FIGURE A-8 The rotation of the particles that make up photon produce a linearly polarized dipolar field that translates back and forth across the photon.

the Yukawa potential (de Broglie, 1962). For a spin ½ particle, where the angular momentum of the particle is equal to $\hbar/2$, the Yukawa potential falls off as $\exp[-2mcr/\hbar]$. If we assume that the mass of each particle is given by $m = \hbar(\omega/2)/c^2$, then the electric field falls off according to $\exp[-2\hbar(\omega/2)cr/c^2\hbar]$. This simplifies to $\exp[-\omega r/c]$. Since $\omega/c = k$, the electric field falls off as $\exp[-kr]$, and equals $1/e$ at a distance equal to $\lambda/(2\pi)$. The mass of the particle provides the force, known as the "Poincare force" that balances the repulsive force within an individual charged particle of finite dimension and keeps the particle from "exploding." Again, we see that the transverse dimension of the photon does not have to either vanish or be infinite, but can be closely related to its wavelength.

We will now look at the sum of the electric fields generated by the two particles within the cross-section of the photon, perpendicular to the axis of propagation. Because a photon is composed of conjugate particles with oppositely directed electric fields and opposite angular frequencies, in the plane of the cross-section, inside the circumference of the photon, the sum of the electric field vectors of the

two particles will generate an electric field, that is transverse to the axis of propagation, and that is linearly polarized (Figure A-8). It as if Faraday's "lines of force" move from the edge of the photon's cross-section to the midline and on to the other edge. At the other edge, the lines of force reverse polarity move back toward the midpoint and back to the original edge. Each back and forth transition of the electric field comprises one period. The electric field sweeps out a wave width as it propagates along a wave length. Looking at the photon head on, the amplitude of the linear electric field is greatest along the midline of the photon and falls off laterally (Figure A-9). The drift velocity of the two particles also generates a magnetic field within the circumference of the photon. The magnetic fields generated by the two particles reinforce each other along an azimuth perpendicular to the azimuth of the electric field. The transverse magnetic field that results from the drift velocity is in-phase with the transverse electric field.

When light passes through a static electric field generated by positive masses, the $(+m/-q)$ particle will accelerate toward the positive pole and the $(-m/-q)$ particle will accelerate toward the negative pole, and it is conceivable that an external electric field could separate the binary photons into its components. The fact that photons in gamma rays can be induced to form particle-antiparticle pairs by the electric field of a heavy atom (Blackett and Occhialini, 1933; Blackett, 1948; Hubbell, 2006) and that the annihilation of matter and antimatter results in the formation of massless particles provides additional reasons to support the binary nature of the photon, and to question its elementary nature.

In conclusion, a photon may not be an elementary particle but may be composed of two particles, one made of matter and the other made of antimatter. The gravitational force between the two particles causes them to propagate. Because the particles are charged and moving, they generate electromagnetic forces and the particles themselves are subject to these self-induced forces. The electromagnetic forces constrain the speed of the photon to 2.99792458×10^8 m/s and determine the longitudinal polarity of propagation. The two charged particles act like a harmonic oscillator, in which the magnetic field is reversibly transformed into an electric field.

The scattering of photons is highly wavelength-dependent, being inversely proportional to the fourth power of the wavelength (Rayleigh, 1870, 1871; Tyndall, 1876). Perhaps this high wavelength dependence indicates that, for scattering, the area or volume of the photon is more important than the wavelength alone. The average volume of a cylindrical oscillating photon is $(\lambda/2)\lambda^2/(4\pi) = \lambda^3/(8\pi)$. Consequently, the average energy density of an oscillating photon of a given wavelength is $8\pi hc/\lambda^4$, and the average linear momentum density of an oscillating photon of a given wavelength is $8\pi h/\lambda^4$. Scattering can be modeled by assuming the oscillating photon is corpuscular when the energy density and linear momentum density are high, and by assuming that the oscillating photons make up a plane wave when the energy density and linear momentum density are low. As shown in Chapter 11, the volume of the photons can also be used to derive the pre-factor for Planck's radiation law. If photons are emitted from atoms when the volume of the photon is at its smallest and it only reaches its maximum wave width $\lambda/4$ away from the emitting atom, it is easy to visualize why there is no wavelength-limited resolution in the near field.

According to the relativistic version of Doppler's Principle, the wavelength of a photon is dependent on the relative velocity of the observer. A particle moving toward photons will experience those photons as being blue-shifted and it will experience the photons it is moving away from as being red-shifted. A Doppler-shifted change in the linear momentum density of photons will cause the moving particle to experience the photons in front as having increased linear momentum density and will experience the photons behind it as having decreased linear momentum density. As the velocity of a particle approaches the speed of light, the photons in front of it acquire enough linear momentum density to counter any accelerating force. Einstein (1905a) posited that it was space (x) and time (t) that were velocity-dependent, but we suggest that time and space may be absolute quantities, and given that photons are described by the argument (kx − ωt), it is the complementary properties of angular wave number (k) and angular frequency (ω) that are velocity-dependent. In this way of thinking, light itself prevents particles from moving faster than the speed of light.

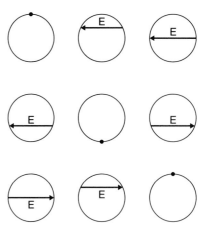

FIGURE A-9 Visualization of the linearly polarized electric field of a photon as it translates back and forth across the photon.

Particles with mass have an angular frequency or spin, which de Broglie (1962) considers to be "the most essential property of particles." This implies that $\hbar$ and not h is the more descriptive constant. All photons have a spin of $\hbar$ and each of its constituents have a spin of $\hbar/2$. The rotation of the particles around a circumference equal to λ sweeps out a particle with a cross-sectional area equal to $\lambda^2/(4\pi)$. The cross-sectional area, as opposed to the wavelength, is what leads to diffraction and interference effects as well as the limit of lateral resolution in the light microscope. The opposing movements of the charged particles around the circumference lead to a photon whose electric field is linearly polarized and suitable for use as a tool in a polarizing light microscope.

Index

Note: cp indicates color plate

Color Plates

Legends for Colored Plates. (All photomicrographs, except the interference duplication micrographs were taken by students on an Olympus BH-2. The interference photomicrographs were taken on a Zeiss Jena microscope.) The students typically use relatively low magnification objectives (e.g., 10x, 20x, and 40x) in order to be able to visualize microscopic objects in context.

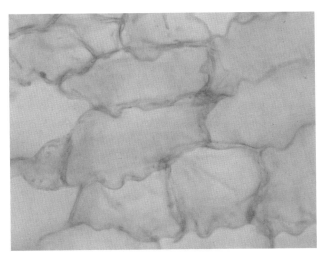

PLATE 1 Bright-field image of cork.

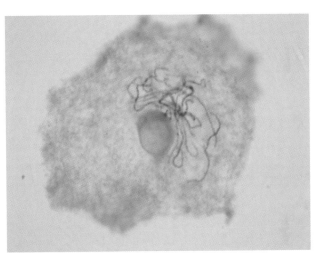

PLATE 2 Bright-field image of chromosomes of *Zea mays* in prophase. Slide prepared by Lester Sharp.

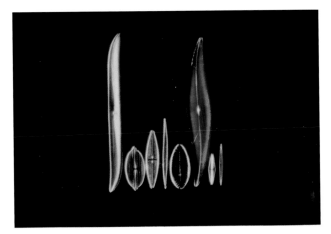

PLATE 3 Dark-field image of diatoms. From left to right: *Gyrosigma balticum* (280, 15), *Navicula lyra* (160, 8), *Stauroneis phoenocenteron* (150, 14), *Nitzchia sigma* (200, 23), *Surirella gemma* (100, 20) *Pleurosigma angulatum* (150, 19), *Frustulia rhomboids* (50, 34), and *Amphipleura pellucida* (80–140, 37). The numbers in parentheses indicate the approximate length in micrometers and the approximate number of striae per ten micrometers, respectively. The colors are due to diffraction.

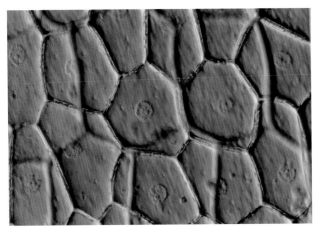

PLATE 4 Photomicrograph of nuclei in the upper epidermal cells of vanilla orchid taken with oblique illumination.

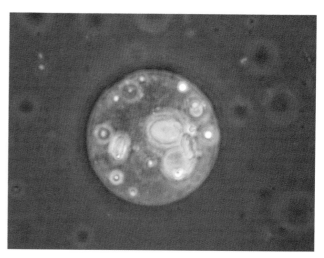

PLATE 5 Photomicrograph of an endoplasmic drop of *Chara* with rotating chloroplasts taken with positive phase-contrast microscopy.

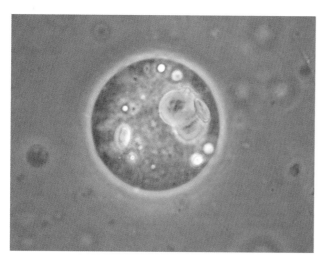

PLATE 6 Photomicrograph of an endoplasmic drop of *Chara* with rotating chloroplasts taken with negative phase-contrast microscopy.

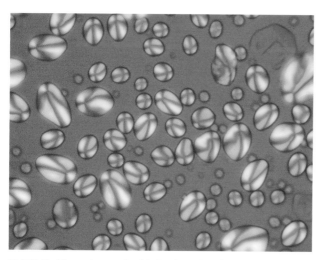

PLATE 7 Photomicrograph of isolated starch grains from potato taken with a polarized light microscope with a first-order wave plate. The slow axis of the compensator is oriented NE to SW, indicating that the positively birefringent starch is radially arranged.

PLATE 8 Photomicrograph of bordered pits in conifer wood taken with a polarized light microscope with a first-order wave plate. The slow axis of the compensator is oriented NE to SW, indicating that the positively birefringent cellulose is tangentially arranged around the bordered pits.

PLATE 9 Photomicrograph of a stomatal complex from the lower epidermis of the vanilla orchid taken with a polarized light microscope.

PLATE 10 Photomicrograph of a stomatal complex from the lower epidermis of the vanilla orchid taken with a polarized light microscope with a first-order wave plate. The slow axis of the compensator is oriented NE to SW, indicating that the positively birefringent cellulose is radially arranged in the guard cells of the stomatal complex and tangentially arranged in the subsidiary cells.

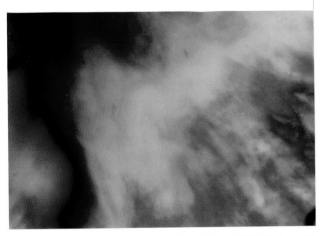

PLATE 11 Photomicrograph of deoxyribonucleic acid (DNA) taken with a polarized light microscope with a first-order wave plate.

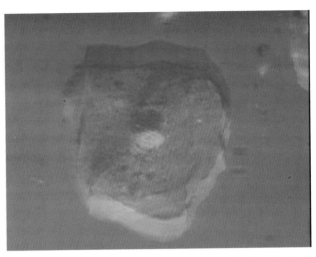

PLATE 12 Image duplication interference micrograph of a cheek cell before compensation.

PLATE 13 Image duplication interference micrograph of a cheek cell after compensation.

PLATE 14 Differential interference contrast micrographs of an optically sectioned stomatal complex of *Rhoeo discolor*.

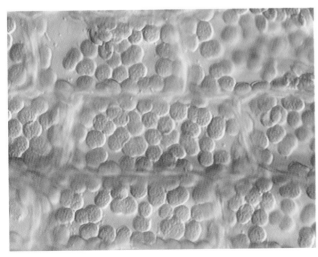

PLATE 15 Differential interference contrast micrograph of the cells of *Elodea* showing chloroplasts with grana.

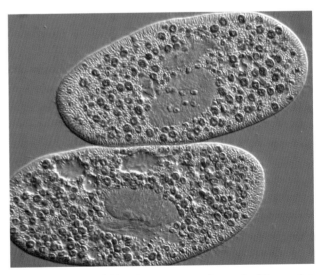

PLATE 16 Differential interference contrast micrograph of *Paramecium bursaria*.

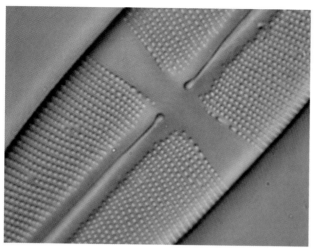

PLATE 17 Differential interference contrast micrograph of *Stauroneis phoenocenteron* oriented perpendicular to the direction of shear.

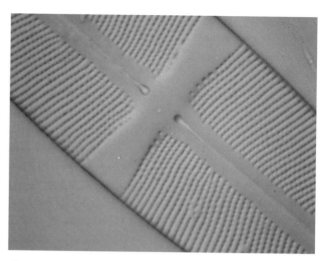

PLATE 18 Differential interference contrast micrograph of *Stauroneis phoenocenteron* oriented parallel to the direction of shear.

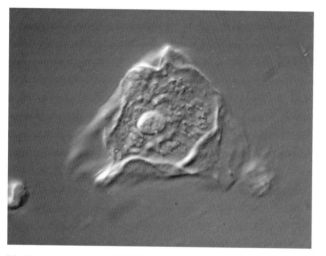

PLATE 19 Hoffman modulation contrast micrograph of a cheek cell.

PLATE 20 The endoplasmic reticulum in an onion epidermal cell stained with $DiOC_6$ (3) and viewed with a fluorescence microscope.

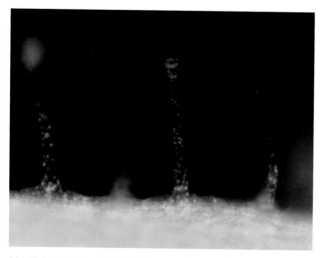

PLATE 21 Epidermal hairs of tobacco cells expressing green fluorescent protein targeted to the mitochondria viewed with fluorescence microscopy. The red fluorescence is due to the autofluorescence of chlorophyll in the chloroplasts. Maureen Hanson produced the transformed tobacco plants.

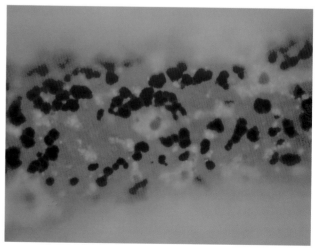

PLATE 22 The endoplasmic reticulum in an onion epidermal cell stained with $DiOC_6$ (3) and viewed with a fluorescence microscope.